DATA FOR SCIENCE AND TECHNOLOGY

DATA FOR SCIENCE AND TECHNOLOGY

Proceedings of the Eighth International CODATA Conference
Jachranka, Poland, 4-7 October 1982

At the invitation of the Polish Academy of Sciences

edited by

PHYLLIS S. GLAESER

Executive Secretary, Committee on Data for Science and Technology (CODATA), Paris

1983

NORTH-HOLLAND PUBLISHING COMPANY – AMSTERDAM · NEW YORK · OXFORD

ISBN: 0 444 86668 x

Published by:

North-Holland Publishing Company
P.O. Box 1991, 1000 BZ Amsterdam, The Netherlands

for:

Committee on Data for Science and Technology
51, Boulevard de Montmorency, 75016 Paris, France

Sole distributors for the U.S.A. and Canada:

Elsevier Science Publishing Company, Inc.
52 Vanderbilt Avenue
New York, N.Y. 10017, U.S.A.

Library of Congress Cataloging in Publication Data

International CODATA Conference (8th : 1982 : Jachranka, Poland)
Data for science and technology.

Bibliography: p.
Includes indexes.
1. Science--Information services--Congresses.
2. Technology--Information services--Congresses. 3. Information storage and retrieval systems--Science--Congresses. 4. Information storage and retrieval systems--Technology--Congresses. I. Glaeser, Phyllis S. II. Polska Akademia Nauk. III. International Council of Scientific Unions. Committee on Data for Science and Technology. IV. Title.
Q224.I56 1982 025'.065 83-7993
ISBN 0-444-86668-X (Elsevier)

PRINTED IN THE NETHERLANDS

CONTENTS

ACKNOWLEDGEMENTS ix

MEMBERS OF THE ORGANIZING COMMITTEE x

EDITOR'S NOTE xiii

FOREWORD xv

SELECTED PROCEEDINGS

PRESIDENTIAL ADDRESS - MASAO KOTANI, President CODATA, Science University of Tokyo, Japan 1

KEYNOTE ADDRESSES

Implementation of CODATA Mission and Objectives in Poland, TOMASZ PLEBAŃSKI, Director, Polish Committee of Standardization and Measures, Poland 3

Role of a Government Geological Survey in Providing Data to Guide Mineral Resource Development, J.M. FRANKLIN and W.W. HUTCHISON, Canada 5

GEOSCIENCES AND NATURAL RESOURCES

Resource Assessment and Mineral Development, WILLIAM A. VOGELY, Pennsylvania State University, U.S.A. 16

Resource Data Systems for Developing Nations, ALLEN L. CLARK, International Institute for Resource Development, Austria 20

Development of a Mineral Resources Information System (MRIS) and Its Application in the Optimal Long-Term Utilization of Mineral Resources, LÁSZLÓ SZIRTES, Central Institute for the Development of Mining, Hungary 28

The SUEZ- An Information System for Storing and Retrieval of Data on Polish Minerals, RYSZARD CICHY, Central Board of Geology, Poland 35

Mineral Resource Information Systems, ALLEN L. CLARK, International Institute for Resource Development, Austria 40

New Ways of Estimating Mineral Reserves and Resources, VÁCLAV NĚMEC, Geoindustria, Czechoslovakia 49

Data Banks of the Polish Geological Survey - Their Principles and Utilization, MACIEJ K. RAJECKI, Geological Institute, Poland 53

Data Management and Manipulation in Mineral Exploitation, MICHAEL J. SHULMAN, Freie Universität Berlin, Federal Republic of Germany 59

A Method for Shadow Removal from Multispectral Scanner Data and Its Use in Mineral Exploration, R. SINDING-LARSEN, Norwegian Institute of Technology, Norway 66

The Problems of Thermodynamics in Geochemistry and Cosmochemistry, I.L. KHODAKOVSKY, Vernadsky Institute of Geochemistry and Analytical Chemistry, U.S.S.R. 68

JAFOV: Data Base on the Japanese Fossil Vertebrates, N. NISHIWAKI, K. YAMAMOTO and T. KAMEI, Kyoto University, Japan 75

MATERIALS SCIENCE AND METALLURGY

Data for Materials Selection and Substitution, N.A. WATERMAN, Michael Neale and Associates, United Kingdom 81

Cooperation in Developing Computerized Materials Data Bases, J.H. WESTBROOK, General Electric, U.S.A. 91

Data for Alloy Design, S. IWATA, University of Tokyo, Japan 99

Estimation of Some Materials Data by Artificial Intelligence, Z. HIPPE, I. Łukasiewicz Technical University, Poland 107

A Semiconductor Database and Its Application to Heterostructure Design, M. YAMAZAKI, S. YOSHIDA, H. IHARA and S. GONDA, Electrotechnical Laboratory, Japan 111

Materials Data Base for Energy Applications - II, S. IWATA, A. NOGAMI, S. ISHINO and Y. MISHIMA, University of Tokyo, Japan 119

Mechanical Properties: A Review of Some Recent Developments, G.C. CARTER, National Academy of Sciences, U.S.A. 123

A Data Base System for Properties of Iron and Steel, HANS ROHLOFF, VDEH-Institut für Angewandte Forschung GmbH, Federal Republic of Germany 127

The ASM/NBS Binary Phase Diagram Evaluation Program, T.B. MASSALSKI, Carnegie-Mellon University H. BAKER, ASM, and J. RUMBLE, NBS, U.S.A. 132

The National System on Numerical Data Concerning the Properties of Materials in Czechoslovakia, M. ŠULCOVÁ, Czechoslovak Institute of Standardization and Quality, Czechoslovakia 135

Structural Materials Corrosion Data Bank, M. DOBRUCKI, J.M. KASPRZYK, E. KROP, J. LISKOWACKI, J. RAKOWSKI and K. WRÓBLEWSKI, Poland 138

The Computer Aided System of Materials Selection, T. FLORCZAK, "Tekoma", Poland 143

Generic Representation of Materials in a Polymer Database (PCMRDB), Y. FUJIWARA, T. NAKAYAMA, A. AMADA, K. IIDA, K. HATADA, Y. HIROSE, A. NISHIOKA, N. OHOBO, I. SUZUKI, T. YASUGAHIRA and S. FUJIWARA, Japan 150

COMPUTER SCIENCE AND DATA BANKS

Conceptual Description of System Transformations and Reactions, J.E. DUBOIS, ITODYS, France 155

Recent Trends in Computer Systems Architecture and General Language, A. JANICKI, Institute of Mathematical Machines, Poland 167

Queuing Network Models: A Tool for Relational Data Base Design, D. MAIO and C. SARTORI, CIOC - C.N.R. and Istituto di Elettronica, Italy 178

The Relational Interface to the CDS ISIS Data Base, J. BAŃKOWSKI, K. BIESAGA, J. DOBOSZ, A. FIGURA, S. ROMAŃSKI, J. RYBNIK, M. SULEJ, A. WOJEWODA and E. ZABŻA-TARKA, Institute for Scientific, Technical and Economic Information, Poland 183

Status and Trends of Numeric Data Banks, J. RUMBLE, NBS, U.S.A. 188

Organization of Data Banks in the U.S.S.R., V.V. SYTCHEV, Y.S. VISHNYAKOV, L.V. GURVICH, A. D. KOZLOV and N.G. RAMBIDI, U.S.S.R. 193

Microcomputer Spatial Data Bases for Human Settlements Planning, J. COINER, I. ARMILLAS and V. ROBINSON, U.N. Centre for Human Settlements, Kenya, and Hunter College, U.S.A. 198

Technological Evolution of Environmental Data Acquisition Systems, G. BUCCI, G. NERI and F. BALDASSARRI, Italy 205

The Chesapeake Bay Information Systems (BIF) for Environmental Data Storage and Access, A. BAILEY and R. COLWELL, University of Maryland, U.S.A. 210

Data Bank ASTRA for the Investigation of Chemical Equilibrium, B.G. TRUSOV, S.A. BADRAK, I.M. BARISHEVSKAYA, V.F. PERCHIK and V.P. TUROV, U.S.S.R. 219

Data Bank on Thermophysical Properties of Fluids, V.V. SYTCHEV, A.D. KOZLOV and G. SPIRIDONOV, U.S.S.R. 224

Evaluation of the Efficiency of Computer-Aided Spectra Search Systems, K. SCHAARSCHMIDT, German Democratic Republic 226

DARC System: Transformational Relationships and Hyperstructures in Chemistry, R. PICCHIOTTINO, G. SICOURI, J.E. DUBOIS, ITODYS, France 229

Coding of a Chemical Graph, M. UCHINO, Tokyo Institute of Technology, Japan 235

GHS - A Computer Program for the Evaluation of Thermodynamic Parameters of Complex Equilibria, P.V. KRISHNA RAO, R. SAMBASIVA RAO and A. SATYANARAYANA, Andhra University, India 237

DATA FOR THE CHEMICAL AND POWER INDUSTRIES

Gasification and Liquefaction of Coal and Data Needed for Plant Design and Operation, H. JÜNTGEN, Bergbau-Forschung GmbH, Federal Republic of Germany 241

Nuclear Data for Nuclear Science and Technology, J.J. SCHMIDT, International Atomic Energy Agency, Austria 252

Data Needs, Source and Methodologies for Water Quality Planning and Control, J.F. SIMONSEN and P. SCHJØDTZ HANSEN, Water Quality Institute, Denmark 263

Thermophysical Property Data for Organic Compounds of Industrial Importance - A New Data Project, M.J. HIZA, B. LE NEINDRE, L. SOBEL, C.F. SPENCER and A.M. SZAFRAŃSKI 272

Sequential Parameter Estimation for Constructing Parameter Tables of Group Contribution Models, S. KEMÉNY and G. CHIKÁNY, Technical University of Budapest, Hungary 280

Computer Aided Data Approximation and Prediction for G^E and H^E, K. SÜHNEL, Karl Marx University, German Democratic Republic 286

The Use of Characteristic Data for Predicting Phase Equilibrium Behaviour, H. SCHUBERTH, Martin Luther Universität Halle-Wittenberg, German Democratic Republic 291

Calculation of Vapor-Liquid Equilibrium by Equation of State in the Alcohol and Semi-Inert Compound Systems, H. WENZEL and A. SKRZECZ, Federal Republic of Germany and Poland 297

Estimation of Critically Evaluated Model Parameters for the Calculation of Multicomponent Liquid-Liquid Equilibria, D. LEMPE, M. GRASSMANN, H. KRÜGER and C. CARL, Technical University "Carl Schorlemmer" Leuna-Merseburg, German Democratic Republic 301

Correlation of Binary Liquid-Liquid Equilibrium and/or Solubility Data, R. STRYJEK and M. ŁUSZCZYK, Institute of Physical Chemistry, Poland 307

Model Testing of N/Ch/-I Phase Equilibrium, Z.R. SZCZEPANIK and P. MILLER, School of Planning and Statistics, Poland 311

Thermodynamic Analysis and Optimization of Titanium Carbide Synthesis, S.A. BADRAK, I.M. BARISHEVSKAYA, V.L. BRISKIN, T.Y. KOSOLAPOVA, V.F. PERCHIK, E.V. PRILUTSKY, A.I. ROSENFELD and V.P. TUROV, U.S.S.R. 314

System of Experimentally Based Equations for the Calculation of Thermodynamic Properties of Fluids, V.V. SYTCHEV, A.A. VASSERMAN, A.D. KOZLOV, G.A. SPIRIDONOV and V.A. TSYMARNIY, U.S.S.R. 317

LIFE SCIENCES

The Accessibility of Nutritional Data - Necessity and Realization, H. HAENDLER, Documentation Centre of Hohenheim University, Federal Republic of Germany 321

Advancement of a National Information System of Laboratory Organisms (NISLO), H. SUGAWARA and Y. TATENO, Institute of Physical and Chemical Research, Japan 325

PHYSICAL SCIENCES

International Collaboration in Data Compilation for Elementary Particle Physics, F.D. GAULT, University of Durham, United Kingdom 333

Data File for Disordered Crystal Structures, K.O. BACKHAUS, H. SCHRAUBER and H. GRELL, Central Institute of Physical Chemistry, German Democratic Republic 337

AUTHOR INDEX 341

SUBJECT INDEX 342

LIST OF PARTICIPANTS 344

ACKNOWLEDGEMENTS

CODATA gratefully acknowledges support for the Conference from the following:

CONFERENCE ORGANIZERS

Polish National Committee for CODATA
Chairman: Professor A. Bylicki
Scientific Secretary: J. Šach

Institute of Physical Chemistry of the Polish Academy of Sciences
Director: Professor W. Zielenkiewicz

CO-ORGANIZERS

Institute of Geology
Director: Professor Waclaw Ryka

Institute of Industrial Chemistry
Director: Dr. Jerzy Kopytowski

Centre for Scientific, Technical and Economical Information
Deputy Director: Dr. Adam Wysocki

Research and Development Centre for Standard Reference Material WZORMA
Director: Professor Tomasz Plebański

SUPPORTING INSTITUTIONS

Central Geological Board
Chairman: Dr. Zdzisław Dembowski

Central Statistical Office
Chairman: Professor Wiesław Sadowski

Ministry of Chemical Industry
Minister: Professor Edward Grzywa

Polish Academy of Sciences
Scientific Secretary: Professor Zdzislaw Kaczmarek
Secretary of Department of Earth Sciences: Professor Roman Teisseyre
Secretary of the Department of Technical Sciences: Professor Bohdan Ciszewski

MEMBERS OF THE ORGANIZING COMMITTEES

INTERNATIONAL SCIENTIFIC PROGRAM COMMITTEE

Professor A. Bylicki, Chairman (Poland)
Professor R. Sinding-Larsen, Vice-Chairman (Norway)
Professor V.V. Sytchev, Vice-Chairman (USSR)
Dr. J.H. Westbrook, Vice-Chairman (USA)
Professor Z. Dembowski (Poland)
Professor J.E. Dubois (France)
Professor A.S. Kertes (Israel)
Dr. D.R. Lide, Jr. (USA)
Professor Y. Mashiko (Japan)
Professor C.N.R. Rao (India)
Professor W. Schirmer (GDR)
Professor M. Szulczewski (Poland)

ORGANIZING COMMITTEE

Co-chairman:	Professor Z. Hippe	I. Łukasiewicz Technical University, Rzeszów
Co-chairman:	Professor A. Bylicki	Institute of Physical Chemistry, Polish Academy of Sciences, Warsaw
Secretary:	Dr. J. Małczyński	Research and Development Centre for Standard Reference Materials, Warsaw
Secretary:	Dr. J. Šach	Scientific Information Centre, Polish Academy of Sciences, Warsaw
Members:	Dr. A. Mączyński	Institute of Physical Chemistry, Polish Academy of Sciences, Warsaw
	Professor T. Plebański	Research and Development Centre for Standard Reference Materials, Warsaw
	M. Rajecki, M.Sc.	Geological Institute, Warsaw
	Dr. A. Szafrański	Institute for Industrial Chemistry, Warsaw
	Dr. W. Trąbczyński	Research and Development Centre for Standard Reference Materials, Warsaw
	Dr. Z. Werner	Geological Institute, Warsaw
	Dr. A. Wysocki	Centre for Scientific, Technical and Economical Information, Warsaw
	Professor K. Zięborak	Institute for Industrial Chemistry, Warsaw
	J.M. Żytowiecki, M.Sc	Polish Academy of Sciences, Warsaw

CODATA EXECUTIVE COMMITTEE 1982

President:	Professor Masao Kotani (Japan)
Vice-President:	Professor J.E. Dubois (France)
Vice-President:	Professor V.V. Sytchev (U.S.S.R.)
Secretary General:	Professor E.F. Westrum, Jr.
Treasurer:	Dr. D. G. Watson
Members:	Professor A. Bussard (IUIS)
	Professor Andrzej Bylicki (Poland)
	Dr. Dorothy L. Duncan (IUNS)
	Dr. David R. Lide, Jr. (U.S.A)
	Dr. M. Menaché (IUGG)
	Professor C.N.R. Rao (India)
	Professor W. Schirmer (GDR)

EDITOR'S NOTE

The papers included in this volume were prepared by the authors on camera-ready copy. However, for the sake of clarity, 65% of these were retyped at the Secretariat in the short period allotted prior to publication. While many corrections were made, many more might have been desirable. I hope the readers will be tolerant and contact the authors for further information.

These proceedings are the result of the diligent work of the Conference Program Committee, presided by Professor Andrzej Bylicki and the Organizing Committee who, under very difficult circumstances brought to fruition the 8th International CODATA Conference in Jachranka, Poland.

The 59 papers presented herein were selected from the 124 presented at the Conference by the following section editors:

Geosciences and Natural Resources	Professor R. Sinding-Larsen
Materials Science and Metallurgy	Dr. J. H. Westbrook
Computer Science and Data Banks	Professor J.E. Dubois
Data for the Chemical and Power Industries	Professor E. F. Westrum, Jr. and Dr. D. R. Lide, Jr.
Life Sciences and Physical Sciences	Dr. D. G. Watson

To them, I offer my sincere thanks and gratitude for the time and effort they contributed.

I would like to take this opportunity to deeply acknowledge the encouragement and guidance I have received from Professor Edgar F. Westrum, Jr. who, in 1982, terminated a 9-year term as Secretary-General of CODATA.

I also wish to express my appreciation to the authors, the majority of whom do not have English as their native tongue, for their contributions.

And finally, Ms. Sarah Penwarden has earned very special thanks for her careful retyping of many papers, her cheerful assistance in the preparation of this volume, and her patience.

Phyllis Glaeser

Paris, February 1983

FOREWORD

The Committee on Data for Science and Technology (CODATA) was established by the International Council of Scientific Unions in 1966 to provide a focus for world-wide efforts on the compilation, evaluation, and dissemination of technical data. The goal of CODATA is to promote more effective handling of data and to foster collaboration on both an international and interdisciplinary basis. CODATA maintains an interface with the scientific and technical community through the scientific unions of ICSU and through various national organizations. One of its key activities is the organization of a biennial conference to address a wide range of data-related topics. This volume contains selected papers from the 8th International CODATA Conference.

The 8th CODATA Conference, held at Jachranka/Zegrzynek, Poland on 4-7 October 1982 was attended by 176 participants from 22 countries. The scientific program was organized by a committee headed by Professor Andrzej Bylicki of the Institute of Physical Chemistry, Polish Academy of Sciences, and contained representatives from France, Israel, United States, Japan, India, German Democratic Republic, and Poland. A total of 124 papers was presented during the four-day conference.

The primary theme of the Conference was Data on Natural Resources -- Their Use for the Development of Society. Emphasis was given to data needed for the estimation of mineral resources and for the efficient utilization of raw materials to produce the goods and services required by society. Many of the papers dealt with the use of modern computer and telecommunication technology for the storage, retrieval, and dissemination of diverse types of technical data. Papers were also presented on several topics concerning data in the physical and life sciences.

This book, which contains 59 papers selected from those presented at the conference, is intended to give a sample of the topics discussed. Representative papers describe newly-developed data banks, dealing especially with mineral resources and other geological information. At the resource utilization level, a number of papers discuss data bases of chemical and material properties pertinent to the metallurgical and chemical process industries. Another group of papers addresses advances in computer methodology, including data base design and management.

While we hope this selection gives the flavor of the subjects covered, readers are invited to contact the Editor for further details of the 8th CODATA Conference.

David R. Lide, Jr.
Secretary General, CODATA

Washington, February 1983

Data for Science and Technology, P.S. Glaeser (ed.)
North-Holland Publishing Company

PRESIDENTIAL ADDRESS

Professor Masao Kotani
Science University of Tokyo, Tokyo, Japan

It is a great pleasure for me to attend the Opening Ceremony of the 8th International CODATA Conference with our colleagues in Poland and from other countries to the East, West, North and South, united by a common interest in CODATA and a common recognition of the important role of quantitative data for the development and application of science.

My first duty here is to express, on behalf of CODATA and representative participants, our hearty appreciation to the Polish National Authorities, Polish Organizing Committee and the Program Committee. I would like to express grateful thanks to Professor Zdzisław Kaczmarek, Scientific Secretary of the Academy of Sciences, Professor Maciej Nałecz, Deputy Scientific Secretary of the Academy of Sciences, Professor Roman Teisseyre, Scientific Secretary of the Division of Earth Sciences and Professor Bohdan Ciszewski, Scientific Secretary of the Technical Science Division of the Academy and to all those who have made every effort to overcome the various difficulties encountered and have taken great trouble to realize a well-organized conference, in such a beautiful environment. As far as I know, there have been no denials of Polish visas and thus the ICSU principle of free circulation of scientists has been fully respected, for which I appreciate the kind consideration of the government agency concerned. I am sure that this 4-day conference will proceed very smoothly and achieve a fruitful success. Warm thanks to all those Polish members, in particular Professor Andrzej Bylicki, who have been most enthusiastically and actively engaged in the necessary arrangements for the successful organization of the conference.

And now I should like to report to you on some major events in CODATA which have occurred since the last Conference in Kyoto two years ago.

Firstly, I must inform you of the tragic death of Professor Boris Vodar in February this year. As all of us know, he was one of the six founders of CODATA and was deeply involved in its administration. In particular, he was the second President of CODATA, succeeding Professor Rossini, and during his presidency he organized the Conference at Tsakhcadzor and also the Conference at Le Creusot. His perspectives were not confined to Western ideologies alone for he was deeply interested in Russian and Oriental cultures and played an important role in the promotion of mutual understanding between East and West within CODATA. He performed the difficult task of moving the CODATA Secretariat to Paris. His warm and friendly personality endeared him to many and as a scientist he was a major figure in high pressures physics and chemistry. He was highly active in research and administration in this field, working in close cooperation with CODATA.

To our profound regret, I have to report further that Professor Touloukian, pioneer of the Task Group on Transport Properties, passed away on June 2nd.

Professor Hans Jancke, distinguished physicist and a very important member of the G.D.R.'s National Committee, left the earthly world on May 29th.

Dr. G.T. Armstrong, Chairman of the Task Group on Biothermodynamic Data, died suddenly on March 9th.

On behalf of the participants in this Conference I should like to pay tribute to the memory of the late Professor Boris Vodar and three other members.

My only regret concerning this Conference is that many U.S. scientists, including Dr. Edward Brady and Dr. David Lide, were not permitted to be here with us by a decision of the Department of State to prohibit government scientists visiting Poland to participate in our Conference and General Assembly. This decision was made in spite of eager wishes and efforts of our U.S. colleagues to avoid this unfortunate situation. I believe that progress in science has been made, and will continue to be made also in the future, by the cooperation of scientists throughout the whole world, irrespective of the political conditions of the countries in which they reside, and our important duty is to esteem and defend the freedom of fellow scientists to participate in multilateral global scientific meetings to the utmost of our capability. Traditionally ICSU has been deeply concerned with this matter and established the principle of free circulation of scientists. It was for this reason that we appealed strongly to the Polish Academy of Sciences not to deny entry visas. Now, faced by the unfortunate event in the U.S., I thought it my duty to take due action, since the event is in conflict with the ICSU principle. On behalf of our Officers I expressed formally our deep regret at the decision to the U.S. Academy of Sciences and the CODATA National Committee, and further informed the ICSU Committee on the Free Circulation of Scientists of this issue.

As science penetrates more and more into various aspects of life, it is increasingly probable that freedom of scientists' participation in international meetings is made subordinate to political diplomacy: in such a political climate we must work even harder to avoid unfortunate events.

It goes without saying that CODATA is an international organization. CODATA works in close collaboration with scientists from many different countries and in cooperation with national academic and governmental organizations. CODATA shares its international nature with many scientific unions.

I should like to emphasize another characteristic feature of CODATA: it is an interdisciplinary organization. We know that astronomical data obtained by Tycho Brahe on the motion of planets were analysed by Copernicus, Kepler and Newton, which led to the birth of mechanics and exact science. In our era, which is characterized by progress made in science and technology, data which originate in one particular field or discipline are very frequently required in other disciplines where they have an essential role as condensed elements of knowledge; e.g. decay constants of some radioactive nuclei are of basic importance in geology and archeology, and data on thermodynamic and transport properties are vitally important in the chemical industry. Besides these examples, the recent emergence of what is known as "mission oriented research" has greatly enhanced systematic demand for high quality data, free of any potential disciplinary barriers. The main theme of the present conference:

Data on Natural Resources

Their Use for the Development of Society

reminds us of the importance of the interdisciplinary nature of CODATA. Furthermore, methodology of handling data developed in one discipline can be transferred, with necessary adaptation, to other disciplines. Particularly with the rapid development of computer technology and telecommunication systems, information science is providing a common technical link connecting many disciplines.

Thus CODATA is both international and interdisciplinary and I believe that these dual qualities are essentially important for CODATA. For me personally, it has always been a special pleasure to attend CODATA meetings; to learn problems and ways of thinking of scientists in other disciplines through data activities, and to renew friendships with good friends from many disciplines. In my opinion, CODATA will, in its long history, contribute to the strengthening of the consciousness of a single integrated science. If I may borrow the phrase of ONLY ONE SCIENCE from a recent publication of the National Science Foundation of the U.S., which originated from Louis Pasteur, I should like to express internationality and interdisciplinarity as aims of CODATA in the following words: ONLY ONE WORLD AND ONLY ONE SCIENCE.

Data for Science and Technology, P.S. Glaeser (ed.)
North-Holland Publishing Company

IMPLEMENTATION OF CODATA MISSION AND OBJECTIVES IN POLAND

Tomasz Plebański

Director, Polish Committee of Standardization and Measures
ul. Elektoralna 2, Warsaw, Poland

The Polish National Committee on CODATA has given me the floor to characterize briefly the sources and framework of the cooperation of Poland within the CODATA worldwide data program.

The roots of our scientific interest in reliable data grew up from the IUPAC Commission on Physicochemical Data which was established in 1934 on the recommendation of - and presided over by - Professor Wojciech Świetoslawski, a prominent Polish physical- and thermochemist. Later on, his two students, Professor Wojciech Zielenkiewicz and Professor Andrzej Bylicki, were elected consecutive chairmen of our National Committee on CODATA, two more are actively engaged in the Task and Working Groups of CODATA, if not to mention the others involved in data research, and myself.

At the origin of our data system, besides physicochemists, metrologists also were clearly seeing the absolute necessity of critically evaluated consistent values of fundamental constants and reference data to define measuring scales and standards, and to maintain a uniform system of measurements.

Our first approach to the CODATA mission resulted in 1966 at the initiative of Dr. Guy Waddington, Director of the CODATA Washington Office from 1966 through 68, whose name is printed in golden type in our country, because of his unusual role at the early stage of the consolidation of CODATA, and due to the impact of the *CODATA International Compendium of Numerical Data Projects*, the authorship of which we attribute to him.

On the agreed upon common application from the Polish Academy of Sciences and the National Board for Quality Control and Measures, on whose behalf Professor J. Nowacki, Professor M. Śmiałowski and Mr. Z. Ostrowski were acting, Poland became a Member Country of CODATA in 1969, thus placing itself among the ten nations engaged in the CODATA scientific program. This was preceded by three visits to our country by Professor Frederick D. Rossini, the first President, also called the "father" of CODATA, assisted by Dr. Guy Waddington.

In the same year the first National Committee on CODATA was formed under the chairmanship of Professor Maciej Nałęcz, actually the Deputy Secretary General of the Polish Academy of Sciences. Thanks to his authority and specific ability to attract scientists from various disciplines, the initial physicochemical and metrological grounds of our data activity were extended over computer sciences, solid state physics, crystallography and others. Also in 1969, a small Office of Standard Reference Data was established within the National Board for Quality Control and Measures. This Office is now attached to the Research and Development Centre for Standard Reference Materials which acts under the supervision of the Polish Committee for Standardization, Measures and Quality Control.

Our young data program had at its beginning several godparents from abroad, who came to Poland, lectured and helped us in various ways. By not repeating those named before, these were: Academician M.A. Styrikovich, Academician G.B. Bokii and Professor W.A. Medvedev from the U.S.S.R., Professor R. Norman Jones from Canada, Professor B.J. Zwolinski from the U.S.A. and Professor Y. Mashiko from Japan. Later on advice and scientific support came from other distinguished CODATA scientists: Professor Masao Kotani, the actual President of CODATA, from Japan, Professor V.V. Sytchev, Professor L.V. Gurvich and Dr. A.D. Kozlov from the U.S.S.R., the much regretted President of CODATA, Professor B. Vodar and Professor J.E. Dubois from France, Dr. J.D. Cox from the U.K., Dr. E.L. Brady and Professor E.F. Westrum, Jr. from the U.S.A., Dr. W.W. Hutchison from Canada, and the Paris Secretariat of CODATA headed by the late lamented Mr. B. Dreyfus and the actual Executive Secretary, the pride of CODATA, Mme. Phyllis Glaeser. Excuse me for being unable to list them all.

It is our impression, however, that to some extent Poland has contributed, on the other hand, to the development of the CODATA program and goals. Permit me to note the following events:

- the organization in 1969 in Warsaw of the International Symposium on Numerical Reference Data for Science and Technology with about 100 scientists from 15 countries in attendance, among them Members of the Bureau and Delegates to CODATA were present; proceedings were published;

- co-organization in 1977 at Baranowo near Poznań of the CODATA - UNISIST International Training Course on Treatment of Experimental Data in the Biological Sciences, Engineering, Physics and Chemistry, with 93 participants in attendance,

- organization in 1980 by Professor R. Hippe at the Technical University, Rzeszów, of a Summer

School on Data Processing in Chemistry with a broad attendance from various countries. Finally, we were privileged to play the role of hosts to this Conference.

The National Committee on CODATA has also organized local Courses and Conferences which reflected the CODATA objectives and program. Systematically announced are Symposia on Physicochemical Data for Chemical Engineering, on Data Banks and on Data for Solid State Physics. Proceedings of all CODATA Conferences are rediscussed and reviewed.

Among the many CODATA publications and recommendations the broadest scope of attention was given to the following:

- as mentioned before, the *International Compendium of Numerical Data Projects*,

- The Recommended Consistent Values of the Fundamental Physical Constants which were announced as a compulsory standard,

- The Key Values for Thermodynamics,

- The Guide for the Presentation in the Primary Literature of Numerical Data Derived from Experiments,

- Selected papers relevant to energy problems,

- Man-Machine Communication in Scientific Data Handling,

- Geological Data Files,

- *The CODATA Directory of Data Sources for Science and Technology*,

- Data Bases in Molecular Spectroscopy,

- *Data Handling for Science and Technology*

and the recently published *Inventory of Data Sources in Science and Technology* . It is our feeling, however, that CODATA has so far put too small an emphasis on data determination, generation and quality assessment.

Let me now describe briefly some areas of data activity in which we are engaged at the present moment. These are:

- computerized evaluation of property data with inbuilt consistency testing for maverick substances of industrial interest, and subsequently a maverick chemicals data book,

- thermodynamic data banks for technology comprising vapour-liquid equilibrium data and saturated vapour pressures of organic compounds. The results are also being published.

- investigations intended for determining phase equilibrium data,

- computer research conducted towards various purposes,

- creation of data banks in several application fields including geosciences, materials science, metallurgy, chemistry and others,

- multipurpose investigations on geoscientific data,

- investigations aimed at data for industry with special emphasis placed on petro- and carbochemicals,

- development of modern information systems,

- reference data bases for metrology and measuring techniques,

- particular data problems concerning crystal structures.

The new fields of future activity will include data concerning energy and life sciences. Following through on the ideas of Dr. Adam Wysocki, I would also like to stress the growing necessity for linking data activities with universal information systems under the auspices of Unesco.

From what has been said before, I would risk drawing the conclusion that CODATA's mission and objectives have found a sensitive ground in Poland and have been to some extent promoted with benefits to physics, chemistry, industry, natural-, life-, and geosciences. The very differentiated functions of CODATA, listed in its constitution, were implemented with varying degrees of importance with the following list of priorities:

1. Application of new methods for data handling, storage and retrieval to the evaluation, preparation, organized production and dissemination of data; increasing personal contacts among workers and the awareness of the importance of data.

2. Promotion of quality control of data, achieving coordination among data projects, stimulating wider distribution of compilations of high quality, and training evaluators and compilers.

Other functions of CODATA still require much more effort to become effective in our country.

What we expect from CODATA is that it continue to respond to the arising new tendencies and situations in the development of science and technology. This is of particular importance in view of the growing significance of data prediction and processing methods with the ever-increasing role of computer sciences and data banks.

Data for Science and Technology, P.S. Glaeser (ed.)
North-Holland Publishing Company

ROLE OF A GOVERNMENT GEOLOGICAL SURVEY IN PROVIDING DATA TO GUIDE MINERAL RESOURCE DEVELOPMENT

J.M. Franklin[1] and W.W. Hutchison[2]

[1] *Geological Survey of Canada, 601 Booth Street, Ottawa,*
[2] *Earth Sciences Sector, Energy, Mines and Resources, 508 Booth Street, Ottawa, Canada*

During the past 10 to 15 years there has been a significant shift in the way in which data are handled due to new geological theories and the use of computers. These advances have led to the formulation of more accurate models for the origin of deposits and so aided exploration or resource evaluation. This presentation focuses on a number of Canadian approaches ranging from a review of how data were collected historically as well as more recently (such as the CANMINDEX mineral deposit file). Three models are examined: volcanogenic massive sulphide deposits (Cu-Zn-Ag), gold deposits and unconformity related uranium deposits.

INTRODUCTION

The role of national geoscience agencies has changed during the past fifteen years. Previously effort was focused on regional mapping to describe the distribution of rock types and piece together the local geological history. This, combined with research in universities, industry and government, led to new concepts and processes - many of which could not be tested because of the lack of unifying models.

Formulation of the concept of plate tectonics in the 1960's allowed rationalization of many phenomena. It became the unifying global model which enticed geoscientists from many different disciplines to study problems of common interest. This was healthy and refreshing for the science resulting in what many perceived to be a revolution in our understanding of geological processes. Whether or not this was a valid conclusion, the concept has provided an immense stimulus for the science in general and has had a continuing influence on the programs of national geological agencies.

In the 1970's these national geological agencies received a different stimulus with the challenge presented by the various oil crises. Many suddenly found themselves being asked to assess not only how much proven reserves (if any) of oil, gas, coal or uranium their country contained but also the potential resources of these commodities.* Concerns soon arose not only about energy but also about resource adequacy of other commodities including various metallic minerals, fertilizer minerals and water. This was a particularly difficult problem because many scientists considered that it was not possible to build sound scientific models as a basis for prediction.

*The difference between reserves and undiscovered resources is fundamental. Reserves could be likened to the assets of a 50 year old person while potential resources is the estimate of what a 6 year old child might have as assets by age 50.

The purpose of this paper is to sketch some of the special aspects of data and information in the geological sciences, to stress the way in which data are collected within the context of a proposed model, to describe some actions that were taken in Canada, and then use three models which are considered to be useful in mineral exploration as well as resource evaluation.

NATURE OF GEOLOGICAL DATA AND INFORMATION

Stratigraphy

Most national geological agencies use a stratigraphic approach to recording and displaying information and data. This simply means that for any region the data and information for different rock units are ordered according to a historical chronology using 'key values' for natural, globally identifiable events marking the break between two systems. Random sampling on the ground or photography or other types of remote sensing from the air, could allow construction of a map showing the distribution of different rock units. But for geological purposes such a distribution map, although useful in reconnaissance mapping, is meaningless because it lacks the vital information on absolute or relative ages of the rock or map units. Geological data must be referenced in space and time (x,y,z and t).

To transpose a distribution map into a geological map requires the intervention of a geologist to discriminate among the data sets and study the relationships between them. When a geologist working in the field realizes that he has moved from one data set (rock unit A) to another data set (rock unit B), he or she will backtrack to determine the nature of the relationship or contact. It could be for example a normal relationship (Fig.1), an unconformity, a thrust fault or an intrusive contact. Once this is established it is normally possible to deduce which of the adjacent rock units is older.

Provided the contact is a first order discontinuity (sharp) and can be traced on the ground (is not covered by soil or growth) it can be recorded with precision and accuracy and so is subject to evaluation.

Rock Types

Each rock unit may be equated with a specific event or period on the geological time scale and may comprise one or more rock types classified according to a particular scheme. Some schemes are chemical while others depend on the proportion of different minerals present (mineralogical) and incorporate such features as texture of the rock. Nowadays a chemical approach is quicker, cheaper and provides greater precision but it is still not sufficient by itself for many classification purposes. For example a volcanic andesite extruded as a lava flow on the earth's surface may have the same chemical composition as a coarse granitic rock such as tonalite emplaced at a depth of 15 km or so.

Mineralogy

Critical for deducing the conditions under which certain rocks formed is the presence or absence of certain diagnostic minerals for example in metamorphic rocks. An example shown in Figure 2 illustrates the significance of the presence of kyanite and staurolite in one rock type H1 and sillimanite without muscovite in another rock type H2 in part of the Prince Rupert area of coastal British Columbia (1). This information indicates that some of these relatively young metamorphic rocks were formed at depths in the earth's crust as great as 25 to 30 km below the surface (at pressures of 7×10^5kPa to 9×10^5kPa).

Minerals

Variations within component minerals of rocks reveal even more information. Many silicate minerals are part of complex solid solution series. The amount of a particular cation in a mineral may be temperature and/or pressure dependent. Furthermore, the sharing (partition coefficient) of a common cation between two co-existing minerals may also have significance. At a smaller scale yet still the internal zoning of crystals such as plagioclase, staurolite and garnet can be determined using an electron microprobe. This information too may provide vital information for deducing the conditions under which these rocks formed.

Inclusions in Minerals

Fluid and solid inclusions are common within some silicate minerals. Studies of fluid inclusions (using heating or cooling microscope stages) can be used to determine the temperature and/or pressure of formation of the rock. Some solid inclusions such as zircon, sphene and apatite are used to determine the age of the rock by measuring the amount of a particular daughter radioisotope present in the inclusion. In geology it is a curious fact that the smaller the object being studied then the more likely it is to be quantified precisely.

There are two points in this discussion. One is that the most fundamental units in geology, the map units or domains, can be defined in space quantitatively but the nature of the domain boundaries is usually not quantifiable (and is therefore described). Although the nature of the domain boundary may be uncertain (if it is obscured by soil or vegetation) yet it may have overriding significance for the precise data and information on rocks and minerals within domains. (A biological analogy would be for a geologist to find that the phenomenon being studied is the trunk instead of the leg of an elephant).

REVIEWING DATA AND INFORMATION NEEDS

The normal flow of data in geological sciences has been for working notes or files of data to be compiled for a particular project. Publication occurs when significant progress can be reported or on completion of the project. At the close of a project the working notes and in certain instances the data may be archived or discarded. Only rarely are they incorporated in part of any sort of active master file for multi-user access.

In the mid-1960's a number of needs were anticipated and initiatives were made to commence building files of 'hard' data for institutional purposes as opposed to the personal project files. By the early 1970's the Survey had built a number of files. For example it had 16 different files on mineral deposits. A number of the files were for specific commodities but the common characteristic of the files was that they could not be related to one another. Accordingly the Survey had to re-evaluate its data needs for petroleum and mineral resources. Those of us who became involved in a project called CANMINDEX (2) took care to tune the system design to meet not only individual scientists' needs but to also allow inter-file analysis. A number of pilot projects were established and at the end of one year an evaluation took place which led to the concept of a "shallow" mineral deposit file that acted both as an index to mineral deposits as well as a medium to allow "deeper files" (on such aspects as chemistry of deposits, ore grade or on reserves) to be attached and interrelated through the index file itself.

The following contribution by R. Laramée summarizes the philosophy and the current form of the file. Most computer-based files on mineral deposits and occurrences, assembled in the late sixties and the early seventies, can be described as tabular: each column represented a data item and each row represented a mineral deposit. Simple file management software was available to handle this sort of data organization. The lack of standards in building those files made them unsuitable for use in any project other than the one for which they were initially conceived. Yet essentially the same data made up the core of all these independent files.

CANMINDEX (Canadian Index of Mineral Occurrences) was designed to be the common denominator of all other mineral deposits and occurrences files, whether computer- or paper-based. It can be described as a tabular file of those data items found in most other files: deposit name, location, main commodities, short description of geology, references. Since this type of information is available for almost all deposits or occurrences, all of them can be accommodated in the file. Also, once a deposit or occurrence is recorded in CANMINDEX, using all available techniques of data validation and quality control, there is no need for recording it again. The type of information recorded in CANMINDEX is of a non-volatile nature; for example, if the location of a mineral deposit is carefully recorded, it never needs to be brought up to date, as opposed to annual production figures.

In CANMINDEX itself, there is no provision for recording mineralogy, reserve or production data, analyses, etc. However other files linked to CANMINDEX can be built to store these various kinds of data. The only constraint put on these other files is that they must contain the CANMINDEX identification number. These files can contain extremely detailed information about few deposits, or they can be designed for very specific projects, long term or short term. Even if the application for which these specialized files were made terminates, the basic information about mineral deposits (the CANMINDEX data items) remains available for use by others. In summary CANMINDEX acts both as an index file to mineral deposits or as a "medium" which links or interrelates deeper files on mineral deposits. Figure 3(a) shows several mineral deposits file in the classical tabular format while Figure 3(b) shows how CANMINDEX can help in resolving problems of file integration.

CANMINDEX (2) has been operational for six years and now holds records on approximately 19 000 mines, deposits and occurrences in Canada. Although there may be as many as 100 000 separate occurrences in Canada, it is believed that CANMINDEX currently contains most of the more significant occurrences. The project has been highly dependent on input from provincial agencies and the dedication of a small core group in Ottawa. The file is now sufficiently comprehensive that most geologists will consult it before starting a research project or work on a specific application.

THREE MINERAL DEPOSIT MODELS

As new concepts evolve (such as plate tectonics) and new models are proposed, so too, new data are required for interpretation or for testing the models.

In the following section three models are described each of which has been improved by recent scientific discoveries. For example the spectacular submarine hot springs depositing high grade sulphides along fractures in the East Pacific Rise of the eastern Pacific Ocean, is confirmation of a process postulated a few years ago in studying mineral deposits over a billion (10^9) years old. The nature of the process is now under active study in a variety of environments and should provide valuable data to further refine the models. Similarly studies of ancient land surfaces have been invaluable in providing vital data for studying possible uranium concentrations at a major Precambrian unconformity.

The following geological models for three important deposit types illustrate some aspects of the current research efforts in Canada, and also the application of this research to exploration and resource assessment.

1. Volcanic-associated massive sulphide deposits are important sources of copper, zinc, lead, silver and gold in Canada. These stratabound sulphide bodies occur in a wide variety of submarine volcanic and volcaniclastic sequences. Canadian districts are in Precambrian greenstones* belts, the Paleozoic strata of New Brunswick and Newfoundland, and the Jurassic of British Columbia. Other well known major districts occur in the Green Tuff belt of Miocene volcanic rocks in Japan, the Iberian pyrite belt, the Tasman geosyncline, and the Troodos ophiolite district in Cyprus. This deposit type has been extensively studied (refer to comprehensive reviews in 3, 4, 5).

Precambrian volcanic-associated massive sulphide deposits occur in those parts of greenstone belts that contain a small but almost ubiquitous amount of felsic volcanic rocks, situated within a dominantly basaltic terrain. These deposits typically contain very low contents of lead, and vary considerably in Zn/Cu ratio.

Phanerozoic (less than 600 million years old) deposits can be divided into two groups. Those in sequences dominated by felsic volcanic rocks, such as the Kuroko district of Japan and the Tasman geosyncline, or those in areas of both felsic volcanic rock and greywacke, such as the Bathurst, New Brunswick district contain recoverable lead, and belong to the Zn-Pb-Cu type. Those in ophiolite terrains, such as Cyprus are copper-rich, with minor zinc and very high Zn/Pb ratios, and belong to the same Cu-Zn group as most of the Precambrian deposits.

Both the Cu-Zn and Ph-Zn-Cu deposits are layered accumulations of sulphide minerals (primarily pyrite) which formed on or near the seafloor by precipitation close to the discharge site of hydrothermal fluids. Precipitation appears to have occurred rapidly because, although some deposits occur at distinct breaks in the stratigraphic record commonly marked by depositions distant from the source, most occur near vents within volcanic sequences that formed a continuum during sulphide deposition. The discharge sites are typically along volcanic fault zones which accompanied volcanism (some formed during cauldron-subsidence events).

The discharge sites are marked by intense alteration, including disseminated and "stringer-type" mineralization. The rock under the deposits may contain a large, laterally extensive zone of only slightly altered rock now interpreted to be hydrothermal alteration of rocks which possessed at least moderate primary permeability. The same strata also contain large subvolcanic intrusions, now recognized as an integral part of the process.

*Greenstones belts are extensive areas of old volcanic rock which are commonly weakly to moderately metamorphosed and contain green coloured minerals, such as chlorite, epidote and amphibole.

The principal elements of the genetic model for volcanic-associated massive sulphide deposits are illustrated in Figure 4, and include the following:

1) metals were leached from permeable strata under the deposits ("reservoir" rocks) by heated seawater; the leaching reaction caused widespread alteration;

2) the heat was supplied by high-level subvolcanic intrusions;

3) in domains of highly permeable underlying strata, a "cap" to the reservoir was needed in order to allow the solutions to be trapped and reach an appropriate leaching temperature;

4) metalliferous brines were rapidly removed from their reservoirs, by "seismic pumping" (6) along synvolcanic faults.

Precise U/Pb age determinations of zircons combined with geochemical studies, including the rare-earth elements, have been used to identify the subvolcanic intrusions. These are typically elongate granite bodies (trondhjemite-quartz diorite), which may locally contain disseminated copper mineralization. These bodies occupy the base of the "productive" volcanic cycles. They usually have only a very small contact metamorphic aureole, indicative of rapid, efficient heat transfer upward from the intrusion. Such transfer occurs within volcanic and sedimentary sequences that have high porosity and are near surface. Interpore water heated to 300°C or more by the volcanic intrusions probably reacted with the rock constituents, leaching both metals and sulphur to form a metal-rich brine. Lead isotope studies indicate a correlation of isotopic composition with local source rocks. The important guidelines for exploration or mineral resource evaluation are the presence of:

1) submarine volcanic rocks;

2) subvolcanic intrusions and accompanying felsic volcanic rocks;

3) altered volcanic strata as an indication of reservoir zones;

4) synvolcanic faults, commonly accompanied by fault-scarp talus, located in reactivated volcanic caldera complexes; near ore deposits, such fault zones may be intensively altered.

2. Gold deposits occur throughout many areas of the Superior and Slave structural provinces of the Canadian Shield. They are commonly present in ancient Precambrian (Archean) volcanic and sedimentary sequences, and are relatively rare in younger Precambrian (Proterozoic) volcanic rocks of the Churchill, Southern, Grenville and Bear provinces. Gold is present in veins and shear zones in volcanic, sedimentary and intrusive rocks. Although disseminated sulphide is commonly associated with the gold, the deposits are not amenable to discovery by most conventional geophysical prospecting techniques. Few comprehensive descriptions are available for this deposit type. (7) describes some general characteristics, and (8) provides a review of Superior Province deposits.

The major gold-bearing districts within Superior Province, Canada (Figure 5) are elongate to linear zones that conform closely to major transcurrent faults, or secondary faults related to these major features. These faults commonly separate volcanic and sedimentary domains. Those that occur within volcanic belts (such as the Destor-Porcupine fault in the Timmins district) may have had a long-lived history; their earliest movement was vertical and had a surface expression that resulted in formation of sedimentary basins surrounded by volcanic rocks and bounded by conglomerate wedges. Later lateral movement of the faults caused widespread deformation, including brittle fracturing of the more competent lithologies. Granitic intrusions were injected along these fault zones, and commonly contain gold. Gold was emplaced into brittle fracture zones and plastically deformed rocks in shear zones; the latter are usually adjacent to the main faults.

Gold in some areas is more concentrated where the shear zones transect rocks containing magnetite, a mineral that reacted with the gold-bearing solution causing gold precipitation. In most districts, gold enrichment is accompanied by widespread alteration of surrounding rocks. Those deposits in shear zones are usually associated with increased micaceous minerals.

A few important gold deposits are evidently layered and may be products of hot springs which emanated onto the seafloor or land surface. Modern hot springs in New Zealand contain anomalous gold contents, and precipitates from such springs may produce gold deposits.

The principal elements of the genetic model for gold deposits, as shown in Figure 6, are much less rigorously tested than those for many other deposit types, but include the following:

1) gold-bearing fluid generation: the ore fluids were generally deficient in base metals, and from limited fluid inclusion data, contained very little chlorine but a higher content of CO_2. Thus these fluids were not derived from seawater. Typically, deposition occurred close to 400°C. Oxygen isotope data indicate that the fluids formed by a release of water from the rocks during progressive burial (metamorphism). This metamorphic-release mechanism would provide a small amount of fluid in a large volume of rock - conditions well suited to formation of gold-rich, base metal-poor interpore solutions.

2) focusing mechanism: the metamorphically generated fluid may have collected in, and moved upwards through, the deep penetrating fault systems that typify most of the vein-gold districts. As many of these faults are long-lived, the time of fluid collection and movement may vary considerably. Some early fluids may have been expelled

on surface, forming bedded gold deposits. Most formed probably from 50 to 200 million years after volcanism and sedimentation.

3) precipitation mechanism: the destruction of magnetite, with accompanying formation of pyrite and/or hematite that occurs adjacent to many gold zones, indicates that interaction between solutions and surrounding rocks may be one mechanism for gold precipitation. Simple cooling (or mixing with surface water), or adiabatic expansion may also control precipitation.

Oxygen isotope data from gold-related veins and altered rocks indicate a distinctive increase in $\delta^{18}O$, possibly due to metamorphogenic fluids as the principal gold-transporting medium. Lead isotope data indicate that most deposits formed 50 to 200 million years after their enclosing rocks: sulphur isotope data indicate that sulphides directly associated with gold mineralization have distinctive, usually positive, $\delta^{34}S$ compositions. Generally the various types of isotopic data indicate that the gold mineralizing process was quite different from the process of formation of other types, in that gold was emplaced considerably later than the immediately surrounding rocks.

3. Unconformity-related uranium deposits: This class of deposits has become well known in the past fifteen years as a result of major new discoveries in the Athabaska region of Canada and the Alligator River district in Australia. These deposits typically occur at or near the unconformity (see Figure 9) between relatively undeformed continental to shallow water marine middle Proterozoic sandstone (1300-1500 million years) and more deformed early Proterozoic basement rocks (1700-2300 million years). The latter are usually metamorphosed shallow marine mudstone, sandstone and graphitic carbonate-rich rocks. (Excellent descriptions are provided by 9, 10, 11).

The deposits are composed of massive to disseminated pitchblende (UO_2), accompanied by quartz, chlorite, adularia, hematite, coffinite (hydrated uranium silicate) and minor amounts of sulphide minerals. The Key Lake and Midwest Lake deposit of the Athabaska district are rich in nickel; other deposits are enriched in Te, Se, and Bi. In general, most are enriched in a broad suite of transition metals. The uranium deposits are typically elongate, and some are confined to major fault zones within the early Proterozoic "basement" rocks. Others are at, or immediately above, the intersection of the unconformity with specific strata within the basement, typically carbon-rich mudstone. Within the entire Athabaska district, uranium mineralization occurs in sub-equal amounts in both the basement rocks and overlying sandstone.

Rocks surrounding the pitchblende accumulations are intensely altered. The earliest alteration formed due to oxidative weathering of the unconformity surface, and consists of hematized remnants of basement rocks. This ancient soil and rubble ("paleosol") is quite variable in vertical extent, but is ubiquitous in the Athabaska area. In the Alligator River, Australia area, intense leaching of the early Proterozoic rocks occurred prior to deposition of the middle Proterozoic sandstone. The alteration produced by middle Proterozoic weathering occurred under tropical (equatorial) conditions typified by a semi-arid climate. Superimposed locally on this paleosol-related alteration are later alterations in part related to ore deposition.

The principal elements of the genetic model for an unconformity-related uranium deposit shown in Figure 9, are:

1) sediments in the "basement" rocks and locally derived paleosol and granular sedimentary material were the source of uranium (and other anomalously-enriched elements) for the deposits. Specific rock types such as metamorphosed mudstones, phosphate- and halite-rich rocks were probable sources.

2) equatorial weathering provided near-surface, low temperature mobilization into groundwater allowing release of uranium for later remobilization into veins; (an alternative possibility is that some supergene enrichment, forming vein-type deposits within the basement, may have taken place).

3) oxidized water within the middle-Proterozoic sandstone leached uranium from the basement (and paleosol) rocks. Reduction, localized near carbonaceous metamorphosed mudstones and in fault zones, caused uranium precipitation at or near the unconformity surface.

4) post-depositional circulation of meteoric (rain) water within the sandstone caused local migration of uranium mineralization upwards into the sandstone, forming small pitchblende accumulations throughout the rocks above the major deposits.

In addition to the various types of geological research, much effort is being placed in developing new remote-sensing tools as an aid to exploration. Four principal tools have been tested:

1) Lake sediment geochemical studies have proven effective in delineating uranium-rich districts, even where the deposits are deeply buried beneath sandstone and glacial debris.

2) Airborne gamma-ray spectrometer surveys have similarly outlined uraniferous districts.

3) Airborne magnetic gradient surveys (using the gradiometer designed by the Geological Survey of Canada) have proven very effective in tracing the basement strata and structures beneath more than 200 m of sandstone and glacial debris. These surveys outline lithological units with exceptional precision.

4) Deep-penetrating airborne electromagnetic surveys have been developed by the private sector, and are currently being tested. These surveys, if

successful, will aid in tracing the carbonaceous mudstones units beneath the sandstone.

The important guidelines for exploration for this deposit type are:

1) the presence of uranium-enriched source rocks

2) evidence of middle Proterozoic paleo-equatorial weathering, followed by deposition of continental to shallow-water marine sandstones

3) evidence of major fault zones which transect the uranium-rich source rocks, and which may also have been reactivated to displace the sandstone-cover sequence.

CONCLUSION

This paper has focused on three models related to the origin of specific types of deposits of copper, lead and zinc, uranium and gold. The continuing hunt to replace these metals where reserves are exhausted must be accompanied by relentless research efforts in the hope of finding significant clues leading to new exploration insight. We must gain a better understanding of the mineralization processes and also those geological events that have affected the rock surrounding the mineral deposits. Thus we must continue the regional geological studies which provide vital information on how parts of continents formed by accretion of mini-continents which became welded to one another (like large icebergs blown together), on the conditions of metamorphism deep in the earth in the recent past, and on current rates of uplift in the Himalayas (of 30-50 cm/100 years) which are significantly higher than previously recorded; these are all important aspects of understanding the processes of continental accretion. Concepts derived from these studies have important implications in understanding the mineralizing processes.

The impact of these major geological discoveries can be immense and may seriously perturb the conceptual framework within which certain data are being collected and used. For example, the suites of fossils of identical age in rocks now only a few miles apart may have been thousands of miles apart 100 million years ago. Similarly, areas of crust rich in mineral deposits may have been broken into blocks, some of which moved a comparable distance; recognition of this movement may guide exploration efforts to new areas of the world.

All these data and concepts underscore the almost regular restlessness of our planet Earth which continually challenges geologists and frustrates those trying to measure its motions now and in ancient times. What is vital is that those responsible for directing work on data collection be keenly aware of the evolution of the science; failure to do so might stultify further inquiry and lead inadvertently to acquisition of "meaningless" data.

Consequently the role and responsibility of a national geological agency is to ensure that vital reference data are systematically recorded and evaluated. Needs for data are anticipated to ensure that human and financial resources are strategically used and that the basic systems are tuned to serve a variety of applications.

SELECTED BIBLIOGRAPHY

(1) Hutchison, W.W., 1982: Geology of the Prince Rupert-Skeena map area, British Columbia, Geological Survey of Canada, Memoir 394, 116p.

(2) Picklyk, D.D., Rose, D.G., Laramée, R., 1978: Canadian Mineral Occurrence Index (CANMINDEX) of the Geological Survey of Canada. Geological Survey of Canada, Paper 78-8, 27p.

(3) Franklin, J.M., Lydon, J.W., and Sangster, D.F., 1981: Volcanic associated massive sulphide deposits; Economic Geology, 75th Anniversary vol., B.J. Skinner, editor, p.485-627.

(4) Klau, W. and Large, D.E., 1980: Submarine exhalative Cu-Pb-Zn deposits, a discussion of their classification and metallogenesis: Geol. Sahrb., sec. D., no. 40, p. 13-58.

(5) Solomon, M., 1976: "Volcanic" massive sulphide deposits and their host rocks - a review and explanation, in Wolf, K.A., ed. Handbook of strata-bound and stratiform ore deposits, II, Regional studies and specific deposits: Amsterdam, Elsevier, p.21-50.

(6) Hodgson, C.J., and Lydon, J.W., 1977: The geological setting of volcanogenic massive sulphide deposits and active hydrothermal systems: some implications for exploration: Canadian Institute of Mining and Metallurgy Bulletin, vol. 70, p.95-106.

(7) Franklin, J.M. and Thorpe, R.I., 1982: Comparative metallogeny of the Superior, Slave and Churchill provinces; Geological Assoc. of Canada Special Paper 25, R.W. Hutchinson, C.D. Spence and J.M. Franklin, editors, p.1-35.

(8) Hodgson, C.J., and MacGeehan, P.J., 1982: A review of the geological characteristics of "gold-only" deposits in the Superior Province of the Canadian Shield: Canadian Institute of Mining and Metallurgy, Spec. Vol. No. 24, R.W. Hodder and W. Petruk, editors, Geology of Canadian Gold Deposits, p.211-229.

(9) Hoeve, J. and Sibbald, T.I.I., 1978: On the genesis of Rabbit Lake and other unconformity-type uranium deposits in northern Saskatchewan, Canada: Economic Geology vol. 73, no. 8, p.1450-1473.

(10) McMillan, R.H., 1978: Genetic Aspects and classification of important Canadian uranium deposits; Mineralogical Association of Canada, Short Course in Uranium deposits, M. Kimberly, editor, p.187-204.

(11) Clark, L.A., and Burrill, G.H.R., 1981: Unconformity-related uranium deposits, Athabaska area, Saskatchewan, and East Alligator River area, Northern Territory, Australia: Canadian Institute of Mining and Metallurgy, vol. 74, no. 831, p.63-72.

(12) Hollister, L.S., Lappin, A., and Hampson, J., 1975: Physical conditions and settings for the generation of tonalite plutons in the Central Coast Ranges of British Columbia (abstr.); American Geophysical Union, Transactions, vol. 56, p.1080.

TYPES OF CONTACTS

IN EACH FIGURE ROCK TYPE B IS YOUNGER THAN A

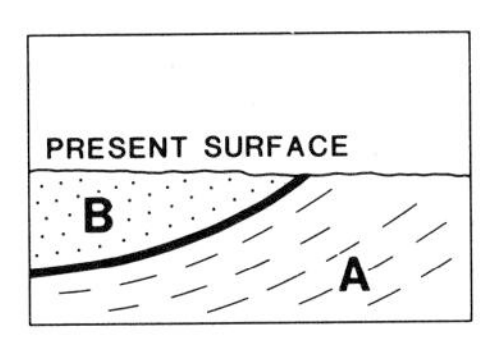

a NORMAL SEQUENCE

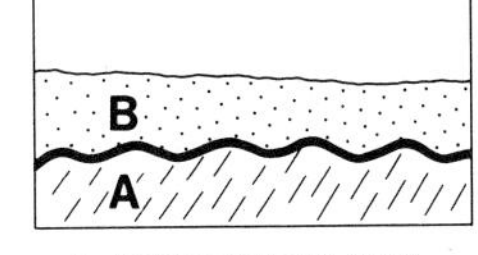

b UNCONFORMITY

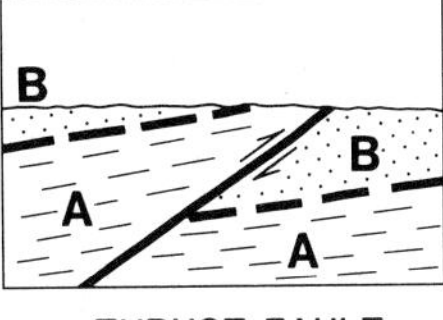

c THRUST FAULT

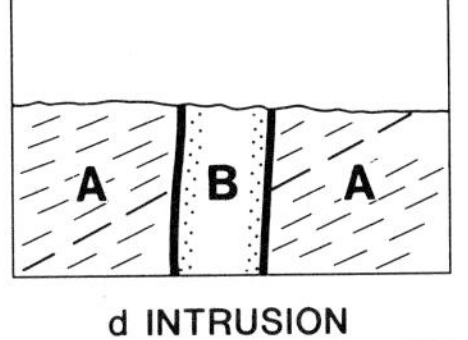

d INTRUSION

GSC

Figure 1. Types of contacts between rock units A and B defined by thick line. In all cases rock unit A is older than rock unit B. a) Normal sequence. b) Unconformity. c) Thrust fault (dashed line is normal sequence; solid line is fault). d) Intrusion - in this case B may have been molten at time of emplacement.

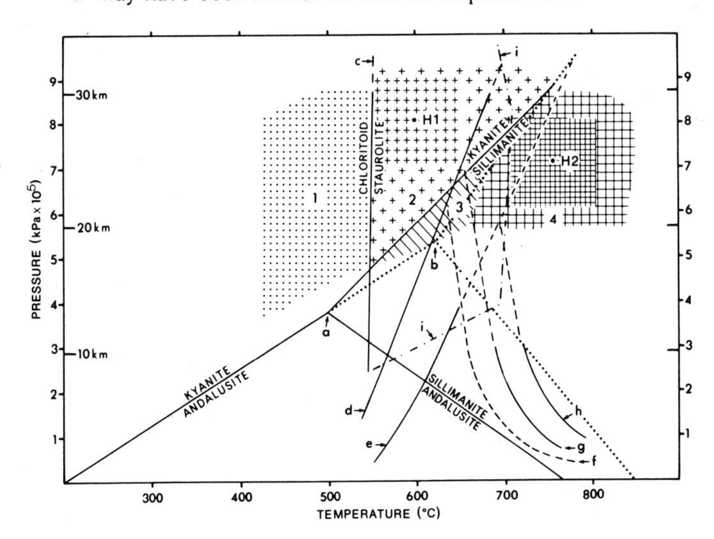

Figure 2. Inferred P-T conditions of higher grade metamorphic rocks on mainland of Prince Rupert - Skeena, British Columbia. H1 and H2 are points 1 and 2 of (12). Reference stability data are contained in (1).

FIGURE 3(a)

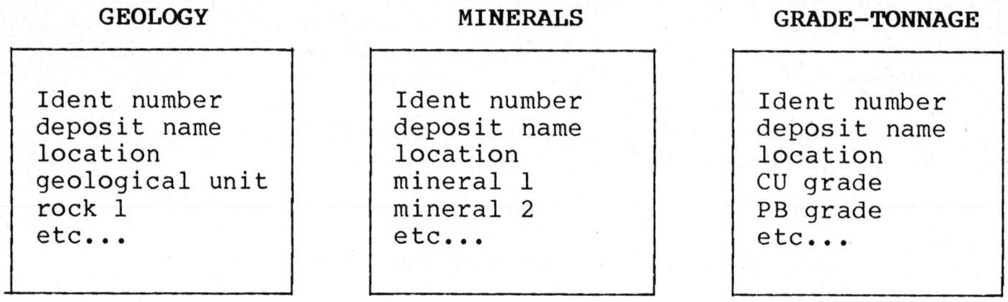

GEOLOGY	MINERALS	GRADE-TONNAGE
Ident number deposit name location geological unit rock 1 etc...	Ident number deposit name location mineral 1 mineral 2 etc...	Ident number deposit name location CU grade PB grade etc...

FIGURE 3(b)

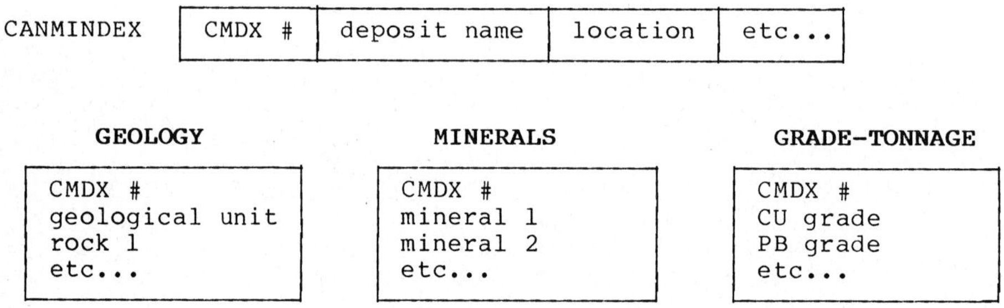

CANMINDEX

CMDX #	deposit name	location	etc...

GEOLOGY	MINERALS	GRADE-TONNAGE
CMDX # geological unit rock 1 etc...	CMDX # mineral 1 mineral 2 etc...	CMDX # CU grade PB grade etc...

Figure 3. Two approaches to computerized data files. a) multiple files, each with a "front end" of similar basic data. b) CANMINDEX file; the basic file contains fundamental name and locational data, plus references; other files may be attached by using CANMINDEX link number only, thus avoiding duplication of information.

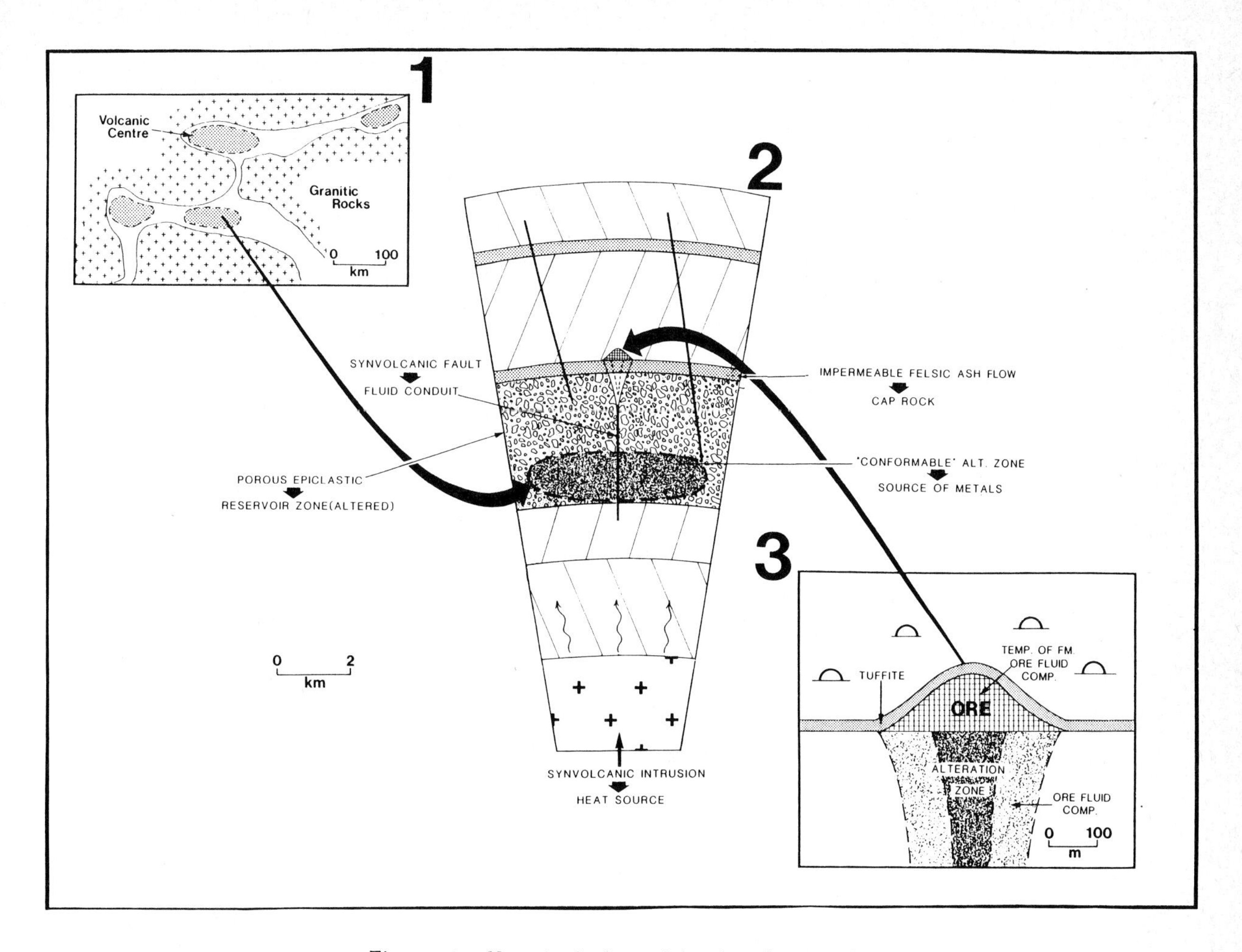

Figure 4. Hypothetical model of volcanic-related massive sulphide mineralization. 1) Volcanic centres within submarine volcanic and sedimentary sequences have the highest potential. 2) The subvolcanic intrusion provides heat to the water saturated reservoir zone; metals are leached from the rock, and moved to surface along a synvolcanic fault, passing through the impermeable cap rock. 3) The massive sulphide precipitates as a mound on the seafloor, immediately above the intensely altered conduit (fault).

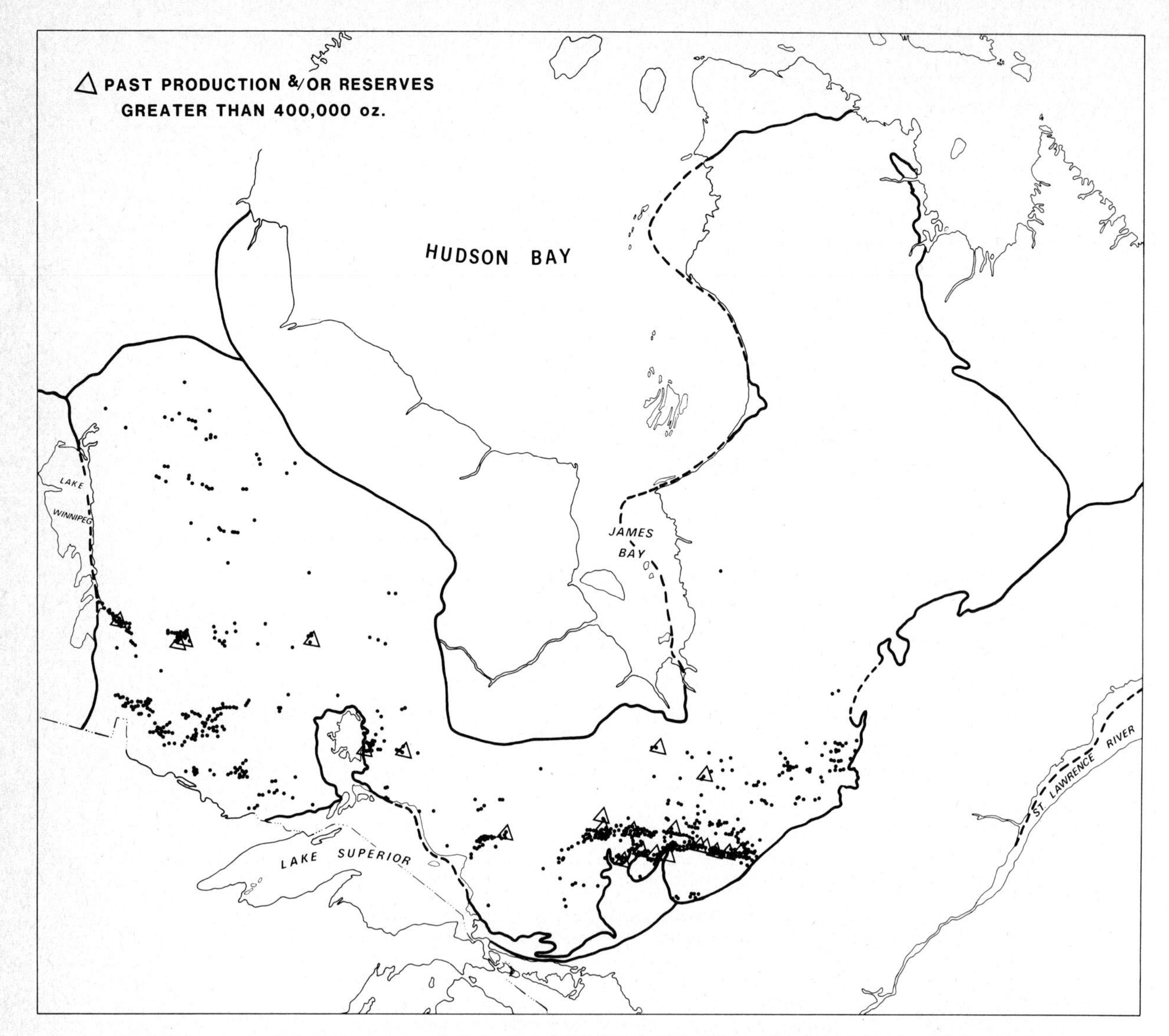

GOLD DEPOSITS AND OCCURRENCES IN THE SUPERIOR PROVINCE

Figure 5. Distribution of 1600 gold deposits and occurrences in the Superior Province of the Canadian Shield. Note that major producers are shown with a triangular symbol. Data were retrieved from the CANMINDEX computerized data storage system, and computer plotted.

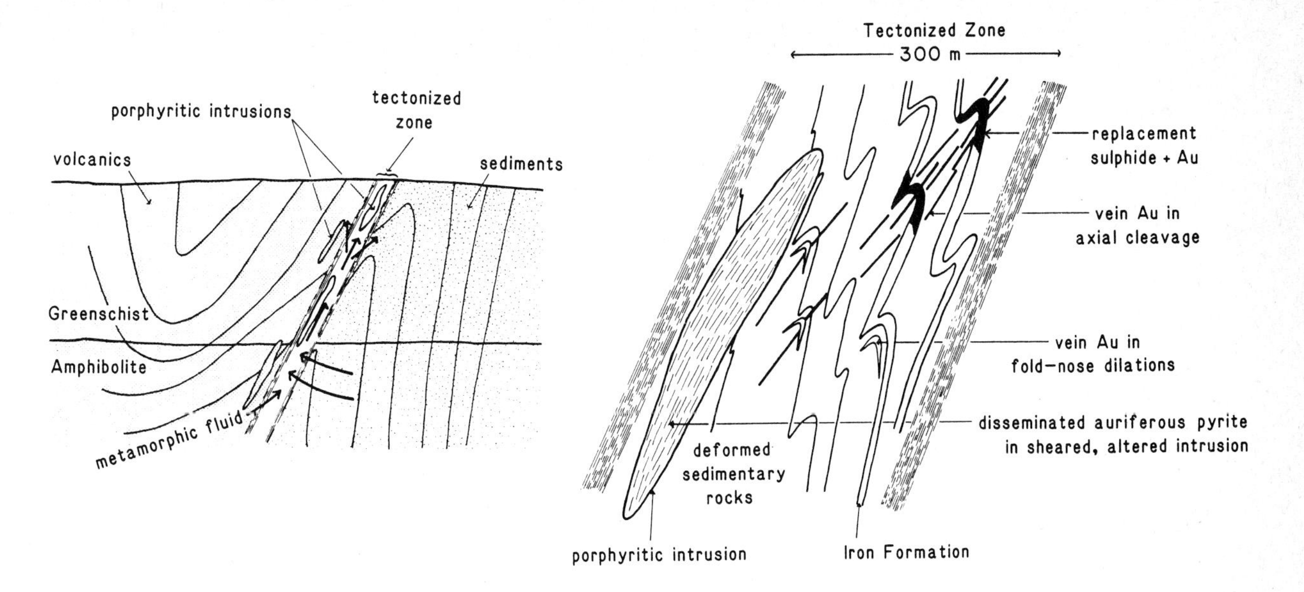

Figure 6. Model of gold mineralization in typical vein and shear zone-related deposits of the Canadian Shield. a) Gold-rich fluid is generated at the greenschist-amphibolite metamorphic boundary; fluids are focused in major fault zones, and move up to the zone of gold precipitation.

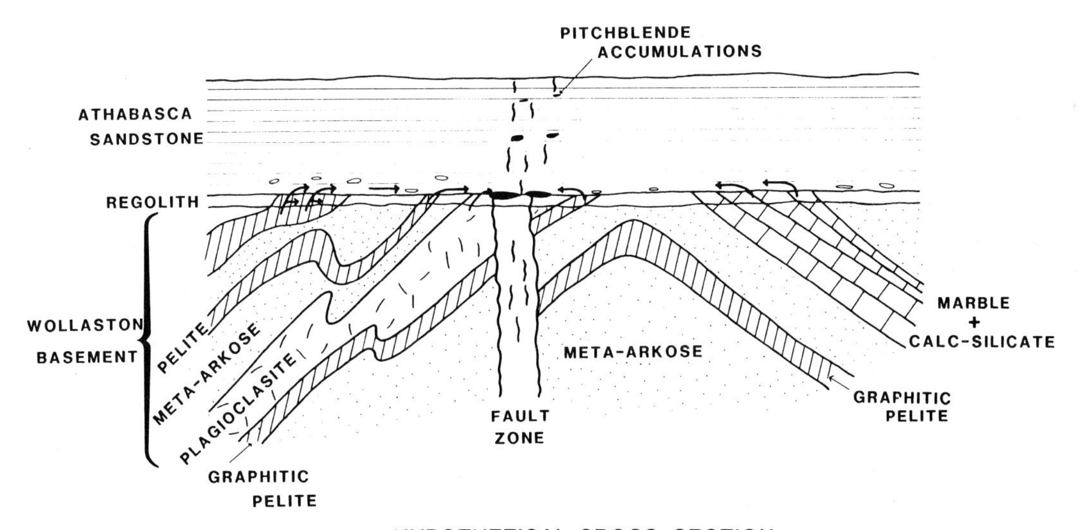

HYPOTHETICAL CROSS-SECTION

ATHABASCA URANIUM DISTRICT

Figure 7. Hypothetical setting for unconformity-type uranium deposits. Uranium is primarily enriched in pelitic and some plagioclasite layers in the metamorphosed basement rock. It is made available for later mobilization in percolating rainwater by oxidative weathering which forms the paleosol. Ground water flow through the basal units of the cover rock allowed uranium to move to local zones of reduction, such as graphitic mudstones and fault zones. Eventually, minor amounts of uranium are remobilized up through the sandstone.

Data for Science and Technology, P.S. Glaeser (ed.)
North-Holland Publishing Company

RESOURCE ASSESSMENT AND MINERAL DEVELOPMENT

William A. Vogely

Pennsylvania State University
University Park, Pennsylvania, U.S.A.

Mineral supply functions are the framework within which data needs are developed. The mineral supply function can be broken into six stages, each requiring a different kind of data. Geologic science cannot now provide the information needed for resource endowment description, so that stage and the stage of exploration cannot be specified. Thus, resource assessments must use indirect methodologies that fall far short of adequacy. This situation means that policies must be undertaken in ignorance of their outcomes, leading to difficult strategy issues.

Introduction

The nature of resource data depends upon the stage of the supply function under consideration. This paper reviews the characteristics of mineral supply functions; identifies the nature of the data associated with each stage of that function; examines the methodologies of making resource assessments; relates information needs for policy formulation to the level of industrial development; and discusses the role of resource assessment in host government - foreign investor negotiations.

Mineral Supply Functions

Mineral supply functions differ from those usually described in economic theory in several important respects. These differences flow from the fact that minerals must be produced from deposits that occur in the earth's crust or waters. Thus, it is not solely a matter of capital investment to create productive capacity; the existence of a specific deposit in nature and the technological ability to find that deposit is a prerequisite to additional capacity. Further, each deposit is a unique one so that investment in research specifically related to that deposit's chemical and physical characteristics is usually required before investment in capacity can be undertaken.

Mineral supply functions can be described as a six stage process. Such a categorization is, of course, arbitrary in nature, and the description used here is designed to highlight the data requirements at each stage - other breakdowns might be better suited for other purposes.

1) Resource endowment: Production must be related to a nature provided ore body. The distribution of ore bodies in the crust in terms of size, grade, chemical composition, physical parameters, and location in three dimensions make up the resource endowment.

2) Exploration: To find a commercial deposit existing in the resource endowment requires exploration. Exploration is a costly activity, and the success of exploration in terms of value of findings relative to costs is the discovery function. The nature of the discovery function depends on the resource endowment, knowledge of that endowment, exploration technology and constraints placed upon exploration by land use and other societal goals.

3) Development: After a deposit is found, investments must be made to design the mine and the processes to win the mineral from the host material. As each deposit is unique, such expenditures may be large, and have a major bearing on the value of the deposit.

4) Ore production: This is the first step of the standard production function where investment and production rates are determined by economic considerations.

5) Smelting and refining: The final step to produce the finished raw material. This too is controlled by economics.

6) Scrap recovery: Materials obtained from obsolete durable goods returned to the supply stream. Other than the fact that capacity for scrap is limited by stock, that is scrap is not produced as an activity, this stage is also controlled by economic factors.

Data Associated with Each Stage

1) Resource endowment: For known mineral deposits the complex of geological, geophysical, geochemical, geographical, spatial and derived information such as tonnage, grade, ore type, and reserves is required to describe the deposit. These data are currently collected in many computerized data banks throughout the world, and are also contained in compendiums on ore-bodies of the world, journal articles, company records and other assorted sources.

2) Exploration: Data required to understand the

level and effectiveness of exploration activity is generally not available. Exploration activity could be measured by kind of activity and level of activity. For example, magnetic surveys could be described in terms of miles flown, density of flight lines and area covered. The results of the survey could be stored for retrieval under key characteristics of the results, and interpreted data could also be similarly stored. Unfortunately, not even the most simple or gross data are available. For petroleum, some information on expenditures is collected, and some well data are available for some countries. For other minerals only very scattered information is available.

There is major work to be done to design a useful set of data definitions and standards around which an information system could be specified. Work in this area is only beginning, although as pointed out above, understanding of exploration activity and the associated discovery function is critical to the formulation of a supply function.

3) Development: There are two distinct kinds of activity involved in development. The first is the design of the mining system; the second the design of the method of separation of the desired materials from the waste rock. Both of these activities involve engineering design, with all of the data requirements implied by that activity. In addition, the second requires data on the physical and chemical characteristics of the ore so that a method of separation can be determined to design the mill circuit. Costs of alternative mine and mill designs, with the trade offs of recovery efficiency are needed. Thus, this stage requires both a large amount of physical data and economic parameters.

4) Ore production: It is at this stage that the information needs become almost entirely economic in nature. Once the scale of the mine and mill has been established in the development stage, and the facility built, the level of production within the capacity constraints is a matter of policy, guided by the objectives of the management of the mine. Costs, revenues, employment, transportation, markets, stocks and other such matters are the data required.

5) Smelting and refining: Since this stage is akin to a manufacturing operation, usually using standard technology, the data needs are economic in nature.

6) Scrap recovery: The collection of obsolete scrap is a simple business driven by economics. The special data needs arise only from the delineation of the reservoir of materials contained in the durable goods that are the source of the scrap. Information on the material content of durable consumer and capital goods, the historical production levels of such goods and the average useful life of each major class of goods is required to measure the stock of obsolete scrap. The rate of recovery from that stock is driven by economic criteria.

<u>Resource Assessment Methodologies</u>

The usual way that resources are classified is by geologic knowledge of existence and economic feasibility of production. Reserves are those resources that are fully demonstrated to exist and are economic to mine under current price/cost conditions. Other groups are marginal and submarginal known resources, inferred economic resources, and undiscovered resources of all economic classes. For this paper "resource assessments" refer to estimates of the undiscovered economic class.

By definition the task of resource assessment is to make a set of estimates about undiscovered deposits. Following the nature of the mineral supply function, the best approach would be to start from resource endowment data and through a discovery function make an estimate of the production capacity the unknown resources can sustain over time. This is not possible because the state of the art in the geological sciences does not permit a description of the resource endowment that could be the basis of a discovery function. Thus resource assessments, if they are to be made, must use indirect methodologies.

The major methods used for resource assessments are analogies, both geographic and geologic; time rate analysis; crustal abundance models; and subjective probability analysis.

1) Geographic analogy: There are two major applications of this technique. One uses the known resources of a well explored area for a mineral that occurs in several geologic environments, such as uranium, and applies the volume of discovered resources per unit of area to unexplored regions. The second sums the real value of mineral production per unit in a fully developed region, and by comparison with the value in a less developed area estimates the potential value to be developed in that area. This second method is not mineral specific.

2) Geologic analogy: Here the mineral production and reserves of a geologic environment that has been fully explored and developed are applied to a similar environment in undeveloped regions. The most common examples are those which use the oil-in-place and recoverable oil of a fully developed sedimentary basin on a cubic volume of sediments basis to estimate the oil potential of undeveloped basins. More complex analogies, using the deposit size distributions and discovery rates of the developed areas are being used for unexplored areas in the United States for oil, natural gas, and the major metals.

3) Time rate analysis: This is a method that en-

joyed much popularity about ten years ago. The method is based upon the observation that a mineral deposit usually shows a normal bell shape function for production over the life of the deposit. If this is taken as the model for all deposits taken together, then a time series of discoveries and production will determine the cross-over point, and predict the inflection point of the production curve, which in turn allows the calculation of life time production. King Hubbert of the U.S. Geological Survey, using this method, predicted that oil production in the United States would peak in 1970; the prediction was made in the mid 1950's and statistically was correct in the early 1970's. The method was considered proven, and the results of its application were widely used for policy purposes.

However, flaws in the logic of the argument were pointed out, and now the method is not widely used for resource assessments. These flaws involved generalization to all deposits of the behaviour of a single deposit, and the great sensitivity of the estimate to the precise determination of the inflection point.

4) Crustal abundance models: This class of techniques relates the crustal abundance - clarke - of a mineral to the ultimate recovery of that mineral. The assumed relationship is linear in logs. Using the cumulative production plus reserves for a mineral that has been widely sought for centuries and the abundance of that mineral as the basis for the relationship, ultimate recovery as a function of abundance for each mineral is computed. It is assumed that this assessment applies to any large land area in the world.

5) Subjective probability analysis: This is now the most widely used method for resource assessment. Stated simply, the judgements of geologists familiar with the mineral occurrence in a geologic region are sought, interaction between the experts is assured, and an estimate of potential economic resources for a mineral by region is prepared. A probablility range around the estimate based, it must be assumed, upon the variability of the expert opinion, is determined. Perhaps the most fully developed application of the methodology is the recurrent assessments of the remaining petroleum to be discovered in the United States done by the U.S. Geological Survey.

It is obvious that all of the indirect methodologies have shortcomings. These are related to the problem that a fact about which society is truly ignorant - not just uncertain - is being estimated. Each method applies some kind of historical data to an assumed fundamental relationship. So long as that relationship is not based in the science of ore deposition, the method is no more than a guess, with an unknown, but very large, variance.

A troublesome conclusion follows from the evaluation of resource assessement methodologies. The results are so imprecise that a policy based upon them may be worse than a policy that takes complete ignorance as a basis. For example, the United States mandated consumption shifts away from natural gas and continued price controls because the resource assessments indicated only a small amount of potential production. Had this policy not been changed, it would have been self-fulfilling in that the search for natural gas would have stopped. A policy that preserved the process of search on the grounds that the amount to be discovered is truly unknown would have been wiser.

But all is not lost! True ignorance is a function of geologic knowledge. For resource estimates in areas where there is information such as geologic maps, geophysical and geochemical surveys and other data, analogies become more applicable. For many purposes resource potentials by quality ranges are quite sufficient for policy decisions. The point is that it is a mistake to believe that society knows more than it really does, and to act on that belief.

Development and Data Needs

Data needs for policy are different at various stages of development. A less developed nation will seek to use its resource endowment as a multiplier of the productivity of investment. An industrialized nation is likely to need data on the trade offs between domestic resource production and imports; land use conflicts between mining and alternative uses; and industrial policy issues.

The differing needs can be related to the mineral supply functions described above. A less developed nation will need on a priority basis information on the resource endowment of its land and the required levels of exploration investments associated with better discovery outcomes. After discovery, the information for development of the ore and for ore production must be available. Such a nation will not have a priority need for data on the later stages of the supply function.

On the other hand, an industrialized nation will face policy issues about the trade offs between imports and domestic production involving environmental impacts, national security considerations, trade policies, regional economic impacts and many other factors. The size of the information needs and the complexity of the data system increases as development proceeds.

Government Investor Negotiations

The relationships between host nations and foreign investors are a question of strategy. At one extreme, the government can offer exploration rights to large areas without any preliminary geological study. At the other extreme, the government can undertake its own exploration effort and offer only the right to develop a discovered, evaluated deposit. At issue is the information cost to the host nation relative to the benefit to be received.

Under the first extreme, the host nation can receive a payment for the right to explore, and pre-established returns from the economic rents of any discoveries. Under the second extreme, the host nation will bear the expense of the exploration program, and can obtain a larger share of the economic rents, if any. The difference lies in the party taking the risks of exploration. Following the first course, the host nation takes no risks, but forgoes some downstream rents. The second course involves the host nation accepting the risk of exploration, and getting the bulk of the economic rents.

Most nations follow a mixed strategy. Basic geologic and other data will be acquired on government account and used to narrow the area for exploration. This will allow a higher return for exploration grants, and will reduce the amount at risk by the nation in case of failure to find commercial deposits. A resource assessment using a flawed but inexpensive methodology could be used as a guide for adopting the mixed or the first strategy. However, the state of the art of resource assessment does not permit the clear choice of the second strategy.

References

[1] Brobst, D.A., Fundamental Concepts for the Analysis of Resource Availability, in Smith, V.K., (ed.), Scarcity and Growth Reconsidered (John Hopkins, Baltimore, 1979).

[2] Vogely, W.A., Issues in Mineral Supply Modeling, in Amit, R. and Avriel, M., (eds.) Prospectives on Resource Policy Modeling (Ballinger, Cambridge 1982).

Data for Science and Technology, P.S. Glaeser (ed.)
North-Holland Publishing Company

RESOURCE DATA SYSTEMS FOR DEVELOPING NATIONS

Allen L. Clark

IIRD, International Institute for Resource Development
Vienna, Austria

The resource data systems required by most developing nations differ significantly in objectives from those in many developed nations. Specifically resource data systems for developing nations can and should be (a) smaller in scope, although expandable through time, (b) multi-purpose in use, (c) easily programmable and portable (d) inexpensive and (e) tailored to the nations' needs.

Resource data systems for most developing countries would be used initially to undertake a national inventory of mineral resources. Secondly such a system would support broad regional exploration programs and finally would be used in national resource assessment programs to estimate both the known and unknown recoverable resources. Implementation of such a program is a sequential program consisting of 8 basic steps with several concurrent activities.

Examples of resource data systems in developing countries include a mineral inventory of Mexico primarily for geologic and metallogenic purposes; a non-metallic resource assessment of Kenya; and an inventory of small mines in Bolivia which ultimately lead to a complete information system. Small, efficient, and economic resource data systems are becoming increasingly available and represent a major technological breakthrough which is easily transferred to developing nations.

INTRODUCTION

The importance to developing nations of mineral and energy resources management, based on adequate resource data and data systems, can be vividly demonstrated by the fact that the approximately 110 developing nations of the world include over 2800 million people, living on 51 percent of the earth's surface, who have an annual per capita income equivalent to US$ 280. The economic disparity between the industrialized and developing nations is further demonstrated by the fact that the developing nations account for only 20 percent of the total gross world product, 7 percent of world industry and only 1 percent of research and services. Specific to the role of resources is that approximately 75 percent of foreign exchange earnings of developing nations are from raw materials which are processed and consumed elsewhere and over half of the developing nations depend on a single mineral or energy commodity for over half of their export earning.

However, while the developing nations are serving as major suppliers of raw materials (iron, bauxite, copper, tin manganese, cobalt) these nations are, in general, not receiving the full benefits which such developments should supply. This loss of mineral wealth, coupled with problems of internal distribution of revenue to other sectors of the economy, has only served to widen the gap between the developing and industrialized countries. In part many of these problems can be related directly to the general lack in most developing countries of resource data upon which realistic national planning for resource development can be based. Recognition of this problem has led to a widespread and increasing need, within both developing and industrialized nations, to develop resource data files and data systems relative to non-renewable energy and mineral resources.

CREATION OF RESOURCE DATA FILES

A major problem in the development of a resource data system within any nation is to plan and implement a national program for the creation of the basic resource data files. Although normally conceived to be an easy and rapid task; experience has shown that the development of resource data files is fraught with a wide spectrum of problems. Indeed it is not uncommon that many nations and organizations develop data files that fail to meet their purpose and are either abandoned or have to be recreated. In large part this occurs for the following reasons: resource data files are a relatively new field of endeavor for most organizations, they are improperly designed to meet either their intended use or future uses, hardware, software and methodologies are rapidly evolving, the potential uses and the user community were not properly defined and too often the beginning of a resource data file was the development of an input format to capture the maximum amount of data: an almost absolute guarantee of failure. To avoid these problems and accommodate a rapidly evolving technology in the field of resource management, the initial design and implementation of a resource data file is absolutely critical to its successful use.

MAJOR PROGRAM ELEMENTS

Step I. Definition of uses, products and user community - before any activity is undertaken relative to the development of a data file a detailed analysis of proposed uses, products to be produced and the potential user community must be undertaken and completed. Such a study should include both the present and future, to the extent possible, uses of the data file as the methodologies to be employed and the desired estimates will change with time and the data file must be constructed to meet these changing needs. This initial step is essential not only for defining the uses, products and users but will also form the basis for assessing the capability of existing, or required, hardware, software, peripheral equipment and orgware necessary to support the resource assessment program.

Step II. Data Definition - Having defined the anticipated uses and products to be derived from use of the data file a detailed list of data items, required to meet the need of the proposed uses, should be compiled. The following two examples, one for a resource map and the other for the estimation of reserves and resources are illustrative.

Date Items for Resource Map

- Location
- Type of Deposit
- Major Product/s
- Host rock
- Tectonic Setting
- Age

Data Items Reserve/Resource

- Location
- Major Products
- Minor Products
- Production (a) past
  - (b) present
- Reserve estimate
- Resource estimate

Completion of this exercise for all potential uses will create a comprehensive list of data items required to support the uses of the file. Although not all uses or data items are of immediate concern detailed loans must be made if they are to be accommodated later without restructuring the initial data file.

Step III. Data Availability Analysis- It must always be remembered that the uses and content of a data file are totally dependent on the availability of data for file construction. Therefore, having completed a prioritized list of uses and required data elements, it is necessary to ascertain what data are available to develop the resource data files.

Data availability analysis will, at least in most cases, considerably limit the scope of applications for the resource data files because a large proportion of the required data will normally not be available. This will necessitate a major decision with respect to the data file program, i.e. the data file program can proceed utilizing only existing data and thereby with limited uses or a program can be initiated to acquire new data, through specific data generating programs, and thereby expanding the data available to meet the potential uses. Normally this decision should be made prior to proceeding to Step IV.

Step IV. Data Format Development- Having completed the analyses required in Steps I, II and III it is possible to begin the development of a data capture format. The format is developed by combining all of the data items, which are available from existing data sources or that are to be generated by specific data generation programs, required to support the major uses. The list of data items should then be grouped into major subdivisions such as:

General Identifiers	Production
Location	Reserves/Resources
Ownership	Recovery Method
Geologic factors	Infrastructure
Deposit type	Etc.
Host rock	

Following the grouping of data items they can be ordered and a generalized data capture format developed; patterned after numerous existing formats.

Step V. Standards and Definitions - Having developed the data capture format it is essential that it not be utilized until specific standards and definitions have been compiled for each of the individual data items. This is an absolutely essential step and unless implemented may result in tremendous inconsistencies in the data file and seriously limit the file's utility. Two examples serve to illustrate this point:

Location - must be given in standard units such as U.T.M., longitude and latitude, or national grid. In addition the accuracy of location (nearest tenth of unit or nearest .05 seconds for decimal degrees) must be specified to ensure that data within the file can be plotted consistently on maps.

Reserves - Each estimate of reserves must be accompanied by a cut-off figure in the case of metals or, as in the case of oil, a definition of whether it is primary, secondary or tertiary recovery. In addition standard units such as short or long tons, barrels of oil or barrels of oil equivalent must be specified for each reserve figure.

Step VI. Data Capture - Having completed the capture format, standards, and definitions the actual process of filling in the data forms can be initiated. This job is best done either by technical personnel, who are geologists or petroleum engineers, or by clerical personnel who work under the direct supervision of the geologists or petroleum engineers. It must be recognized that the compilation of data requires a great deal of subjective evaluation and interpretation and therefore cannot be left entirely to clerical personnel. The data file will only be as good as the information that goes into it and therefore the initial data entry is a singularly critical step.

CONCURRENT ACTIVITIES

The preceeding six steps are essential and represent the primary activities in the development of a resource data file. However, there are a series of additional activities that must be undertaken concurrently the most important of which are:

Data input methodology - The method by which data is taken from the input formats and put into a machine processable file must be determined; normally concurrently with the development of the data input format. Numerous techniques are available ranging from standard key punching and the development of card decks to highly sophisticated optical scanners which input data directly onto tape or disc. The construction of the format will vary depending on the techniques used.

In general it makes very little difference how the data are input; normally it is advisable to use whatever capacity is available. Given an option however it is recommended that a data be input through the use of a programmable terminal, whereby data are typed onto a format which is displayed on a Cathode Ray Tube (CRT) screen. This system allows for data input directly onto cassettes or tape and also provides for simultaneous editing on input.

Hardware, software and peripherals-Based on the analysis performed in Step I, and the results of Step III, a program can be developed for the acquisition of hardware, software and peripherals that meet immediate needs but that can also be expanded to meet future uses. This philosophy of sequential implementation provides for continuity in the program, allows for long-term planning and budgeting, reduces the need for duplication and provides a stable system for the training of operating personnel and users.

Field and Laboratory Data Collection-Upon the completion of Step III attention should be given to the development and implementation of standardized procedures, to be used in field and laboratory studies, which will insure that required data are developed. It cannot be emphasized too strongly that the most effective data file program is one which initially provides for the capture of present and future data, so that the data files can be continously updated with minimal effort, and only secondarily begins to recover past data.

MAJOR PROGRAM ELEMENTS (cont.)

Step VII.Pilot Program - Initially focus should be placed on the development of a pilot program once step IV is proceeding, and sufficient data are available for supporting one or two of the proposed uses. This step is particularly critical as it (a) serves as a check on the data and the system to ascertain both support the proposed uses and (b) provides a visible product which can be used to show the utility of the system and practical applications; both essential to insuring continued support to the activity by the professionals and the administrative personnel.

Step VIII. Operational System-Following modifications to the system based on experience with the pilot programs, it is appropriate to make the entire operation essentially "turn key" with respect to procedures and uses. Once this has been done for the pilot program the system can be expanded to other uses as time, money, personnel and data permit. It is essential that procedures must be implemented which insure that the data file and system are regularly updated and maintained through time.

Just as the development of resource data files is a sequential program of activities it should be emphasized that the designing and implementing of a resource assessment program is similarly sequential. Virtually all resource assessment programs proceed from aggregated to disaggregated estimates and from the subjective to the quantitative. Rarely is it possible in a resource assessment to have sufficient data available to launch a large integrated synthesis resource assessment. Therefore as the data file grows, so does the resource assessment methodology: careful planning will ensure that the former provides the latter with the required inputs.

INITIAL USES OF RESOURCE DATA FILES

In the development of a resource data file the most often asked question is "what will I gain from developing this data file?". Then follows a chicken and egg development where funds and personnel are only available when a product can be produced to prove the file's value yet no product is available unless there is support to create the file. This is an ongoing problem which virtually every geoscience organization faces in the initial development of its resource data files. In the case of data files to support resource assessments the problem is compounded in many cases by the need to initially acquire large amounts of data, either commercially or through institutional effort, which is expensive, time consuming and may result in delays in producing the resource assessment. Therefore to resolve the "chicken and egg" problem and provide for the continued support of a resource data file considerable thought should be given to immediate products that can be derived while proceeding to complete the file for its use in an overall resource assessment program.

In general a large number of useful products can be derived from a resource data file depending on whether one undertakes a single data element analysis or compound data element analyses.

Single Element Analyses: An often overlooked aspect of resource data files is the number of unique and useful products which can be produced from the portrayal or analysis of a single data item or field within a data file. Obvious examples of graphic products are location maps, production distribution and other graphic representations of individual data elements. Equally important however are the products derived from an analysis of a single data field where the best known examples are (a) historical discovery rate analysis in a petroleum well data file using the data field "year of discovery", or (b) the size distribution of deposits using "total reserves field" in an oil and gas pool file or a deposit file of coal resources. Similarly an analysis of the number of producing, non-producing and development prospects can be determined from an analysis of the "status of deposit" field in a mineral resource data file. Clearly these represent rather simple application of a resource data file but they are products which can be rapidly and cheaply produced to show the value of files early in their development history and normally long before their use in a resource assessment.

Compound Data Analysis: Similar to the single element analyses a large number of graphic and analytical products are available from the use of two or more data elements within a resource data file. As common examples long term supply analysis for a deposit utilizing the fields "total reserves" and "productive capacity" or a reserve accumulating analysis using the fields of "reserves discovered" and "yearly production". These two examples combined provide an enormous amount of information relative to the long term supply of any energy or mineral resource, yet require analyses of only 4 data fields within a data file which would normally be created to support a resource assessment.

In the development of data files for resource assessment considerable effort should be spent in defining the short and intermediate term uses and products of the data file. This will partially resolve the "chicken and egg" problem, provide a basis for prioritizing data capture and in many cases provide a useful check on the data and the information system as the file is created.

Having reviewed the procedures for developing resource data files, which are the basic building blocks of a resource data system it is appropriate to look at the present status of resource data systems, ongoing developing programs in resource data systems development and finally to look at a major application of resource data systems i.e. national resource assessments.

PRESENT STATUS OF RESOURCE DATA SYSTEMS

It is difficult, if not impossible, to specify the level of data system development within developing or developed nations for in both cases systems range from simple data files to complex computer systems. There are however a number of general parameters which typify resource data systems in many developing nations; excluding those nations at the extremes of simplicity or complexity. Among the most significant parameters are:

(a) Dedicated computer facilities specifically for resource data are uncommon and the majority of computing is done in a large main-frame computer in a national computer center or major ministry.

(b) Resource data files are poorly developed; with few files available which cover a specific area of information in any detail.

(c) Access to computer facilities is administratively difficult and technical assistence in computer science is either lacking or difficult to obtain.

(d) Little or no software is available which is specific to the handling of resource data and acquisition and installation is difficult.

(e) Resource data analysis priorities are almost exclusively on an individual project basis with little or no interaction with other co-workers nationally or internationally.

(f) Knowledge of other data systems in developing or developed nations is poor to non-existent and cooperative projects are rare.

Although there are numerous exceptions to the above generalizations they are in the author's experience sufficiently common to provide a basis for discussion of the need for and impact of new technological developments on resource data systems in developing countries.

FUTURE TRENDS IN RESOURCE DATA SYSTEMS

A detailed discussion of the advancements being made in resource data systems, which are applicable to developing nations, is beyond the scope of this paper. However some of the more important are worthy of summary comment and are discussed in the following. There is however one factor which overshadows the technological advances being made in resource data systems, even though it is a direct result, the advances are at ever decreasing costs. A prevailing trend in ADP activities is increasing capabilites at decreasing costs and it is this single factor which brings resource data systems within the financial reach of geoscience organizations in developing nations.

DECREASING COST

Overshadowing the technological advances, although a direct result of them, is the present and future trend toward increased capacity at lower costs within resource data systems hardware and to a lesser extent software. However this single factor is perhaps the major determinant in bringing resource data systems within the financial and technical reach of geoscience organizations within developing nations. Access to dedicated computer capacity will reduce one of the major bottlenecks in computer access within the developing nations as workers will no longer be dependent on remote computer centers which were administratively difficult to access. Increased access, because of lower cost, has the additional benefit of developing a critical mass of users which is a major step in resolving the isolation that many developing country scientists experience when involved in resource data programs.

HARDWARE TRENDS

The rapid and continuing evolution of the mini- and microcomputer industry coupled with the introduction of small computers by the major computer firms has led to systems which have the computing and storage capacity of previous main-frame computers yet require only limited space and maintenance. In the context of developing nations the advances in reduced maintenance are equally as important as the reduced size with increased capacity in that present hardware can be used in much less controlled environments relative to heat, dust, moisture and other physical factors: the exact conditions found in many developing nations where hardware service and repair is different if not impossible to acquire within a reasonable time frame.

A major trend in hardware at the present is toward that of the personal computer and clearly a large effort is being made within the computer market to bring more computer capability to the individual. At the present, for economic and access reasons, this is not a major or immediate concern within many developing nations although the author has observed personal computers being used for mineral resource inventories in Chile and Mexico and for geochemical and geophysical data interpretation in Jordan.

SOFTWARE TRENDS

The rapid evolution of hardware and of analytical programs within the developed countries has resulted in increased access to software for developing nations. Correspondingly the software is more easily adapted because of a trend toward the development of software that is modular, transportable, relocatable and most important maintainable. This trend has resulted in the development of libraries and software where programs are accessible, shared and/or sold, and well documented.

Perhaps of most importance to developing nations is the availability of more and more pre-programmed software which not only provides the user with the analytical power of the software program but has the capacity to lead the user through the program step by step. Although not in common use for resource data systems it represents a major vehicle for the transfer of capability and knowledge to developing nations.

TELECOMMUNICATION TRENDS

Telecommunications represent one of the potentially most exciting opportunities for developing nations to expand their resource data systems by linkage with existing systems elsewhere in the world. However unlike hardware and software advances which have come with reducing costs, more so in hardware than software, telecommunications are still extremely costly in developing nations if accessible at all. There can be little doubt that for some developing nations access to telecommunications will provide a major vehicle for advancement, however in most nations the infrastructure costs of telecommunications are prohibitive for the near to intermediate term.

Where accessible telecommunications have provided easy access to additional hardware, software and most importantly basic data for use in specialized analyses, future developments, particularly in terms of distributive data bases, will significantly reduce basic data collection costs and provide a more consistent base for international resource assessments and analyses.

FUTURE RESOURCE DATA SYSTEMS

Technology, economics and politics are all working, albeit at highly variable rates, to facilitate the development of resource data systems within developing nations. The high level of effort however must be coupled with an orderly development and implementation which insures that the nations resource data system meets its present and future objectives. To accomplish this considerable thought must be given to:

1. Applicable data systems - Increased capability at lower cost is all too often translated into larger and larger program objectives for a nation's resource data system. Nothing could be more detrimental to orderly development than such a program which fails to meet its objectives because of insufficient data, unacceptable costs or excessive time requirements for development. Therefore present data systems would be designed to be expandable through time but of reasonable scope initially to meet realistic objectives.

2. Multi-Purpose in Use - Virtually all resource data systems which are successful begin with achievable objectives which require data from a limited subset of information within a discipline, i.e., mineral resource data for resource assessments. However through time a natural evolution will require multi-disciplinary data, i.e., mineral resource, economic, infrastructure and environmental for regional development planning. Therefore present systems should be created with the capacity to handle multi-disciplinary data through time.

3. Tailored to nation's needs - Any resource data system should be tailored to meet specific national needs but all too often the needs are near term and inadequate thought is given to future needs as the nation develops and diversifies. Critical to this problem is to insure that national standards for data systems are adhered to which will allow for the development of integrated data systems with time.

RESOURCE DATA PROGRAMS IN DEVELOPING NATIONS

Several developing nations have embarked upon the development of resource data files and data systems and have been very successful in their efforts. In most cases the initial activity has centered around the development of a nation's inventory and subsequently the effort has been expanded to undertake national resource assessment programs. Among the most successful which represent a diverse spectrum of uses are the following:

Bolivia - An inventory of approximately 8000 small mines (Clark, A. and Cook, J., 1979) was undertaken and completed in Bolivia from 1976 to 1978. The existing inventory is being updated to include data on the approximately 250 medium- and large-scale mines in Bolivia. When completed in 1980 the inventory will represent coverage of all the major mineral occurrences in Bolivia. As a direct outgrowth of this program the Fondo National de Exploration Minera (FONDO) was founded and Bolivia has committed US$ 25 million to the small mining sector.

In addition to data on the mineral deposits, the inventory includes data on transportation, power availability, regional demography and development potential. At present, the Bolivian inventory is the most complete and comprehensive study that has been done on a national basis.

Mexico - The mineral resource inventory of Mexico represents one of the largest inventory efforts presently underway anywhere in the world. At present data have been compiled on approximately 30 000 mineral occurrences (Lee-Moreno, Consejo de Recursos Naturales de Mexico, written communications, 1978) and the data file is expected to exceed 100 000 entries when completed. The mineral inventory of Mexico closely follows the format for data collection utilized in the Bolivia inventory and by the U.S. Geological Survey CRIB System. This adoption allows the resultant file to be used both for geologic analyses and for mineral resource exploration, and development. The mineral resource inventories of Bolivia and Mexico are perhaps the best examples of broad based mineral resource inventory programs and serve as models for other national programs. However, both programs require large amounts of capital and manpower and a comprehensive information base to build upon, factors that limit their use in many developed and developing countries. There are, nevertheless, several other national inventory programs that are underway to support a wide range of national needs. Representative of these programs are:

Colombia - A national mineral inventory program (Rosas, H. et al., 1980) has been developed and implemented for Colombia that attempts to integrate a wide range of geoscientific data with the mineral inventory. Specifically, the program is being developed in conjunction with the development of a digital geographic file for the plotting and analysis of mineral data, and data files on geochemistry, geophysics and basic field geology. The primary purpose of the Colombia mineral inventory activity is to support exploration for and development of mineral resources. At present, data are compiled on two main areas and the program is expanding to acquire a total national coverage by 1985.

Additional national mineral resource inventory programs are presently underway in Japan, Korea, Peru, Turkey, Kenya, Cyprus, Venezuela, Egypt, Indonesia, Malaysia, Israel, and Newfoundland, Canada, all of which are similar to, or represent combinations of the programs described above.

All of these programs, and many soon to be initiated, are providing a wealth of valuable data on the world's present and potential sources of mineral resources.

NATIONAL RESOURCE ASSESSMENTS

A rapid evolving field of endeavor, in both developing and developed nations, utilizing resource data files and systems as basic tools, is in national resource assessments. These assessments attempt to estimate the resource base of a nation in terms of the location, quality and quantity of resources known or to be discovered. A review of the assessment methodologies and their data file requirements has been discussed by the author (Clark, A.L., 1980, In Press). Resource assessments do however represent a major advance in providing valuable analyses to developing nations because each estimate of resources can and does serve as a direct input into a nation's overall minerals program. Unfortunately the value of such inputs has not been clearly articulated or demonstrated because of a lack of adequate and widespread applications. The benefits of a national resource assessment program extend beyond those provided by having an inventory (evaluation of known deposits) and an estimation of undiscovered recoverable mineral and energy resources. These additional benefits are almost immediately accruable to a nation because of their usefulness as a basis for defining and implementing specific resource development programs and policies. Specific examples include:

1. A comprehensive national resource assessment defines commodities for both internal consumption and those for export as foreign exchange earners. Additionally such data allow for the balanced development of both an internal minerals industry and an export industry. In many instances such balanced development will provide a basis for the export of processed metals and energy sources rather than the export of raw materials.

2. The thorough knowledge of the development potential of specific commodities or projects places the viable programs in a

more competitive position to attract development funds.

3. Knowledge of resource and development options, shown by a resource assessment, provide a more solid basis for the formulation of national minerals and energy policy and legislation which will further development in line with the nation's overall development priorities.

4. The ability to develop several commodities or to follow alternate development scenarios, provides a logical basis for the development of a multi-commodity rather than a single commodity mineral economy. This diversification of the national development program materially reduces the vulnerability of the nation relative to fluctuations in export prices and earnings of a single commodity and reduces the variability in export earnings.

5. The resource assessment will provide a refined basis for the logical expansion and development of the geoscience community required to support the program; including the need for imported technology, training and technical assistance.

6. A well formulated initial resource assessment provides a basis for additional assessments thereby materially reducing their costs and providing the nation with continous policy and planning inputs.

7. The Basic data and regional assessments within a national program provide useful and necessary sector inputs into national planning and economic development programs.

The above examples are but a limited number of the potential benefits to be derived from a national resource assessment but demonstrate both the diversity and impact of possible applications. Although the benefits from a resource assessment will depend in large part on the actual resource endowment of the individual nation it is clear that a resource assessment is a necessary and critical element of resource policy formulation required by policy and management personnel. An operational and responsive resource data system is a major pre-requisite in reaching these objectives.

REFERENCES:

(1) Clark, A.L. and Cook, J.L., 1979, Global Resource Data Systems, U.S. Geological Survey Project Report (IR) NC-68 p. 44

(2) Clark, A.L., 1982 (In Press) The importance of Data Banks for Resource Assessment, Proceedings of the Third International Symposium on "New Paths to Mineral Exploration" Hannover, F.R.G.

Data for Science and Technology, P.S. Glaeser (ed.)
North-Holland Publishing Company

DEVELOPMENT OF A MINERAL RESOURCES INFORMATION SYSTEM (MRIS) AND ITS APPLICATION IN THE OPTIMAL LONG-TERM UTILIZATION OF MINERAL RESOURCES

László Szirtes

Central Institute for the Development of Mining
1525 Budapest, P.O. Box 83, Hungary

The paper deals with concepts of the Mineral Resources Information System, reviews the problems encountered in the development of existing subsystems and summarizes the results obtained using the system. Three examples are given. On the mine level, the use of the subsystems on rock mechanics and geological analysis; on the company level, the subsystem on economic analysis; on the mining and energy sector level, the model developed for the optimal long-term planning of mineral resources utilization is illustrated. The criterion of optimality for this model is the minimization of the total long-term demand for capital expenditures for a given supply.

1. WHAT IS "MINERAL RESOURCES INFORMATION SYSTEM"

In 1979, under the supervision of the Ministry of Heavy Industries (now the Ministry of Industry) responsible in Hungary for the utilization of mineral resources, the development of a computerized information system meeting the major information requirements of solid minerals exploitation was started.

The aim of the development has been to gather such information on the exploration-mining-processing-consumption process of mineral resources that is necessary to monitor, supervise and control the procedure of minerals utilization. To reach this goal, information is necessary from mines, mining companies (trusts) and from the sectoral (mining and energy sector) level.

Concerning the development of the information system of mineral resources utilization, two types of interpretation of the Mineral Resources Information System (the Hungarian abbreviation is NYIR) were used rather than to accept the alluring concept of heirarchic, integrated system development. *In a wider sense*, the Mineral Resources Information System (MRIS) describes the whole process and all levels of mineral resources utilization.

MRIS, in a wider sense, is a loosely coupled system of partly computerized, partly manual information systems of mine plants, mining companies and ministries where transfer of information between the different levels takes place mainly on documents. Within the wider MRIS program, methodological assistance is given for the automation of the highly data- or computation-intensive activities on all levels and for all utilization phases. Our aim here, is that the systems to be implemented be data and process compatible with each other.

MRIS, *in a narrower sense*, realizes a mapping function for a part of the exploitation process, the mining and the processing of mineral resources on the sector level.

2. THE DEVELOPMENT OF MRIS IN A WIDER SENSE

The aim of MRIS development is to form an information base which, in the whole verticality of mineral resources utilization, provides the most important information for *planning, supervision* and *management* regularly and with a relatively low requirement for manual intervention. The description of the relations within mineral resources utilization, its system and function approach is a directional aid of the development work (1).

The supervisory function of MRIS, in the verticality of mineral resources utilization, is based on data acquisition and primary data processing. The computer system should, as far as possible, fulfil this function automatically. The managerial function can be fulfilled with a series of decisions. MRIS contributes to decision-making by providing information necessary for optimal decisions.

MRIS is in multiple relation with other control and management systems. The multiple relation primarily means exchange of information and methodological standardization. As the automation and maturity levels of the "related" systems are different, in the development of MRIS, dynamic forms of the information relation have to be dealt with.

The information system marked with I in Fig. 1 indicates the Geological Information System (GIS) to be developed under the direction of the Central Office for Geology. To underline the function of MRIS in connection with mineral exploration, the interface between the two systems is called, in both systems, a common system of MRIS-GIS. System II marks the information systems of mining companies (in some cases trusts), III stands for the information system of mineral resources consumption (in the Central Office for Statistics).

For lack of regular operation of the independent

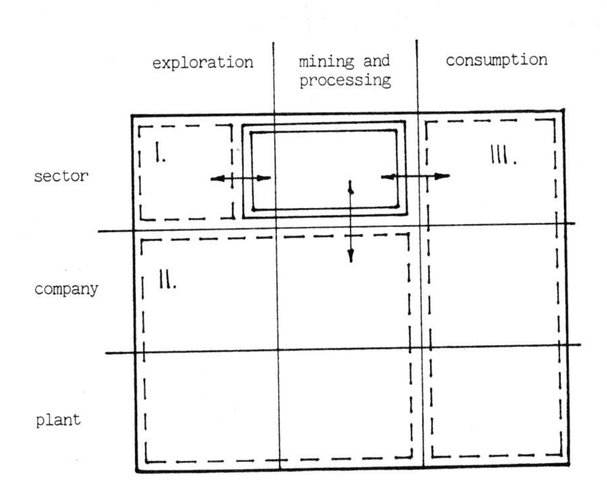

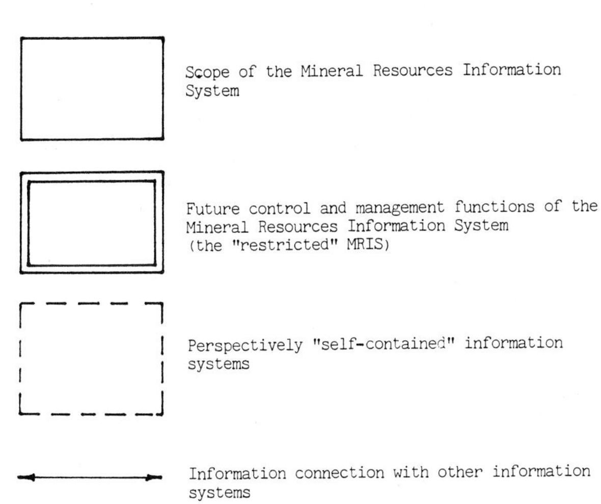

Figure 1: Scope of MRIS with the "independent" information systems to be developed in the long run.

computerized information systems, in the development of MRIS, some information flows have to be *simulated*. This work, in the majority of cases, is inconceivable without temporary provision for the function of independent systems. The use of this *system development strategy* and the relatively little domestic experience in the field of computer-aided decision making have made it necessary to realize MRIS with the definition and establishment of so-called *prototype systems* rather than direct implementation of the functions of MRIS (2). The prototype systems can fulfil major functions of MRIS even without the operation of the "independent" information systems (I, II, and III in Fig. 1) but they operate in so-called model areas so their services do not cover all domestic mineral deposits. Their data files (the so-called cadasters) form the basis of the MRIS data base and, during its development their maintenance is also provided.

3. MAIN POINTS OF MRIS DEVELOPMENT

The typical functions of the information systems - operational control, management control, and strategic planning (3) - can be found on all levels of MRIS but their importance - which is decisive from development priority assignment - differs from level to level. On the mine level, for the close connection to the mineral deposit and mining, the importance is placed on operational computations; on the company level, it is placed on technical-economic planning; on the sectoral level, the importance of long-term planning is determinant. The examples illustrating the results obtained to date in the MRIS development have been compiled so that the main points of development are clarified. Part of the prototype systems shown in the examples contain methods that can be coupled to several data files of MRIS; others are a combination of data files and procedures fulfilling certain functions of MRIS (Fig. 2).

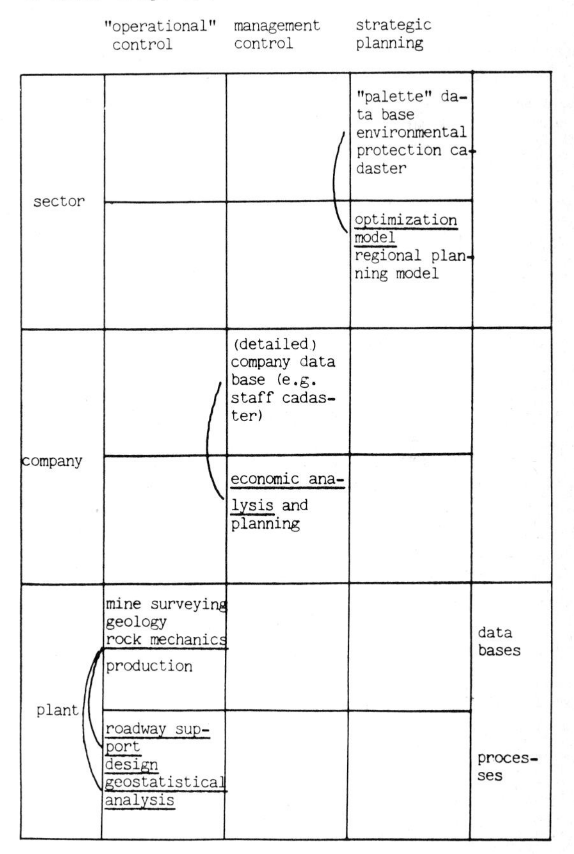

Figure 2: Essential elements of MRIS (the underlined data bases and processes are illustrated by examples).

4. THE MINE LEVEL OF MRIS

On the mine level, the MRIS contains procedures, e.g. the geostatistical analysis system and complete information subsystem, e.g. the rock mechanical cadaster with the connected roadway support design system.

a) Geostatistical Analysis in the Mineral Resources Information System

The geostatistical analysis system of MRIS allows primary processing of the data of geosciences (basic statistical evaluation) and performs some, rather frequent, geostatistical analyses.

b) Rock Mechanical Cadaster of MRIS

The rock mechanical cadaster of the MRIS stores a maximum of 83 parameters of borehole samples. The parameters are grouped into six classes:
- data to identify the location of the sample (10 data);
- geometric and phase parameters (16 data);
- direct material parameters (9 data);
- rock mechanical (elasticity and rheology) material parameters (15 data);
- strength (plasticity, breaking and soil mechanical) parameters (21 data);
- stress parameters (12 data);

Of the 83 parameters, 10 are computed ones. The majority of the information requested from the rock mechanical cadaster can be ordered *to given space ranges*.

5. COMPANY (TRUST) LEVEL OF MRIS

On the company (trust) management level of MRIS from the technical-economic planning function some means of economic planning will be discussed below.

The improvement of the economic control system of Hungary started in 1968, which, due to the ever accelerating changes in the world economy, has involved frequent modification of the regulators in recent years. Within the frame of the MRIS, *simulation aids* are provided for the *mining companies,* with which the effect of the control system can be analysed or plan alternatives can be simply generated. The simulation models on the company level are *exact mapping* of the economic control system.

We are simulating
- formation and utilization of the development, divisible and reserve funds;
- formation of participation funds;
- sources and utilization of technical development fund;
- formation of incomes above basic wages of senior officials;
- wage control vs. company performance;
- formation of producer's prices for industrial products.

A logistic diagram of the formation and utilization of development-, divisible- and reserve funds can be seen in Fig. 3. Given the value of input parameters, the specialist performing economic planning can evaluate plan varieties in an interactive mode and can automatically generate plan varieties respectively.

6. SECTOR LEVEL OF MRIS

On the mining sector level of mineral resources utilization, improvement of *long-term* planning methods - development of optimal mineral resources utilization plans - is a priority objective. The sudden (and, for the importers, painful) changes of the last decade on the energy sources market have underlined the importance of long-term supply planning of mineral resources.

With regard to the optimal utilization of mineral resources and energy sources, *four decision points* can be marked in the process of economic management (Fig. 4):
- control of utilization of domestic resources (I);
- distribution of end-products and control of their transformation structure (II);
- control of their implementation on the basis of realistic export and import conditions (III);
- determination of the target system of mineral resources utilization strategy (IV).

To optimize the process of mineral resources utilization, the process itself must first be described. The different units (phases) of the process can be described with so-called *technological blocks*.

The technological blocks of mineral resources utilization transform the material and energy flows. These transformations are described by mathematical functions. The functions describe the connections between the input and output product flows of technological blocks. A group of functions represent *physical relations* by describing the amount of input and output products as well as other byproducts transferred to the environment. The other group of functions provides *economic* transformations (Fig. 5) since, for instance to create a given block, social expenditures are necessary, the two basic groups of which are manpower and capital. During its operation, a given technological block can have different effects on its surroundings. The aim is obviously to establish a system which meets the social requirements described by demand functions with a minimal expenditure and which does not affect adversely the environmental conditions constraining mineral resources utilization, e.g. lowering of water level, limiting sulphur emission, etc.

Possible effects of the blocks on the environment are described by limiting functions. In order to determine the optimal system of mineral resources utilization, the *domestic resources* should be mapped first. Connected to this, are geological prospecting and the creation of alternative designs of the exploitation units to be located on the proven mineral reserves. These designs can, of course, include several opening, development and mining plans for the same mine. The *verticalities of processing* should also be described by alternative projects, e.g. electric-

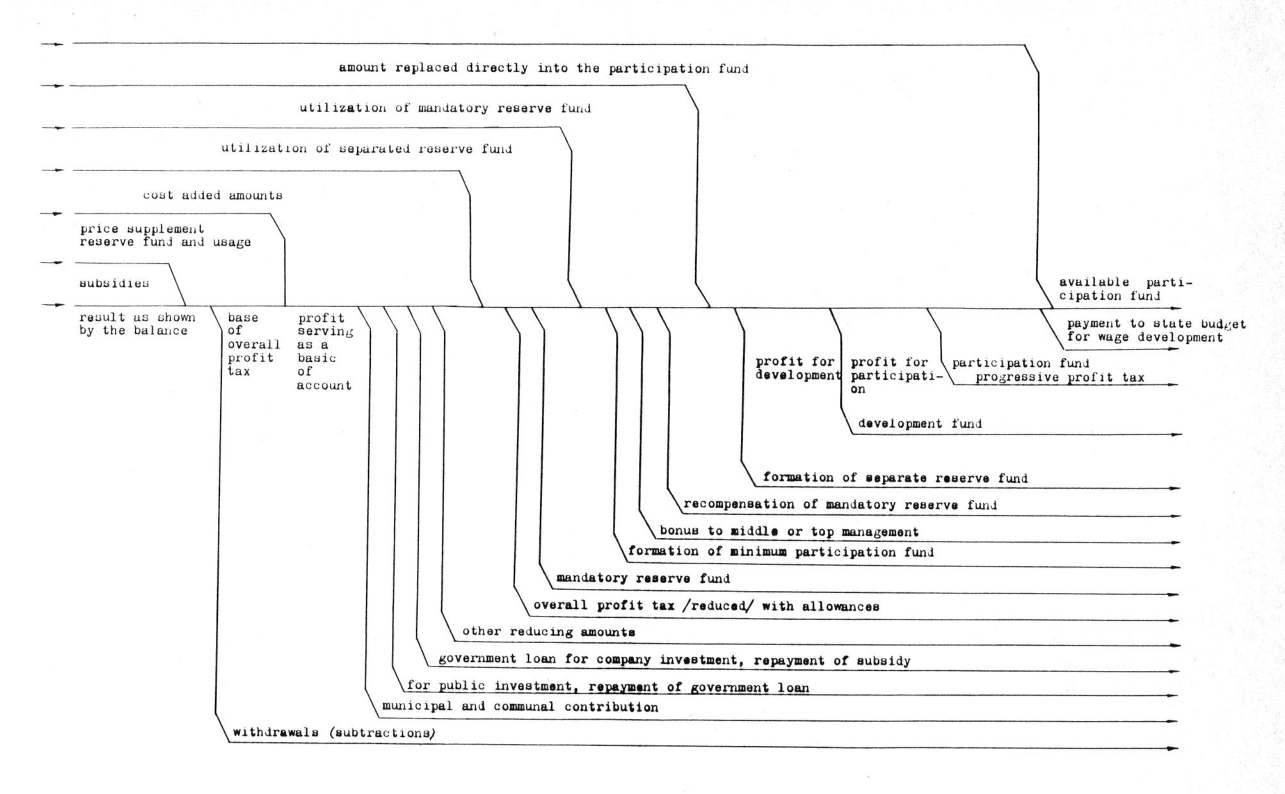

Figure 3: Model of reserve-, development- and participation fund formation

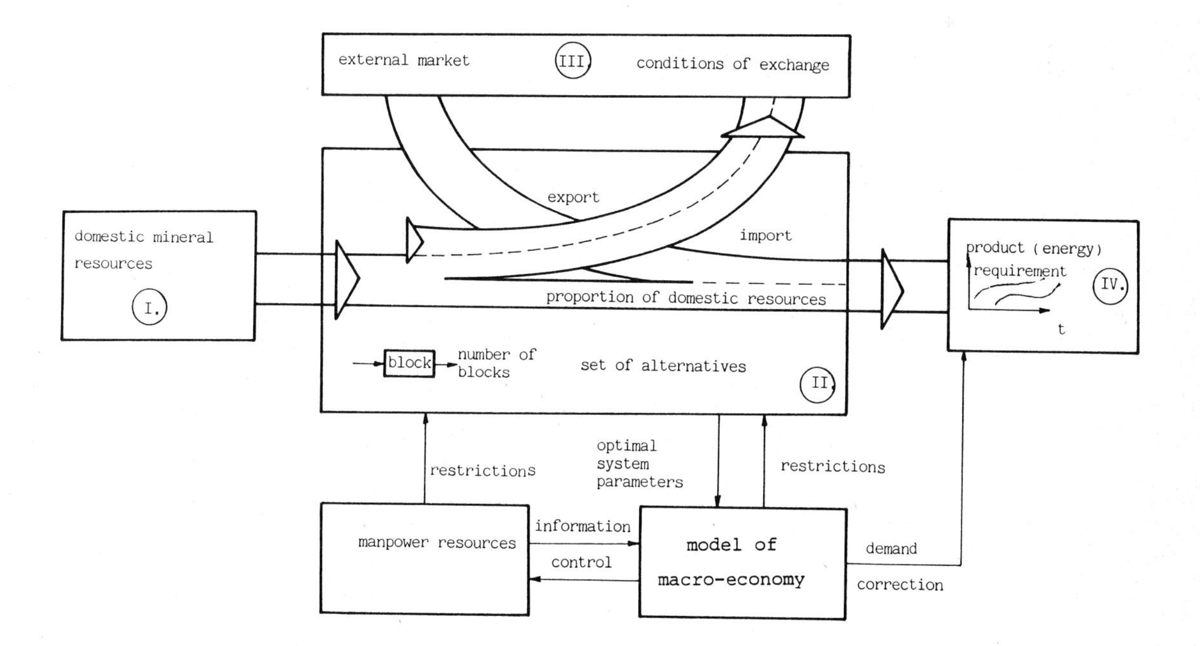

Figure 4: Main decisions in the optimization of mineral resources utilization

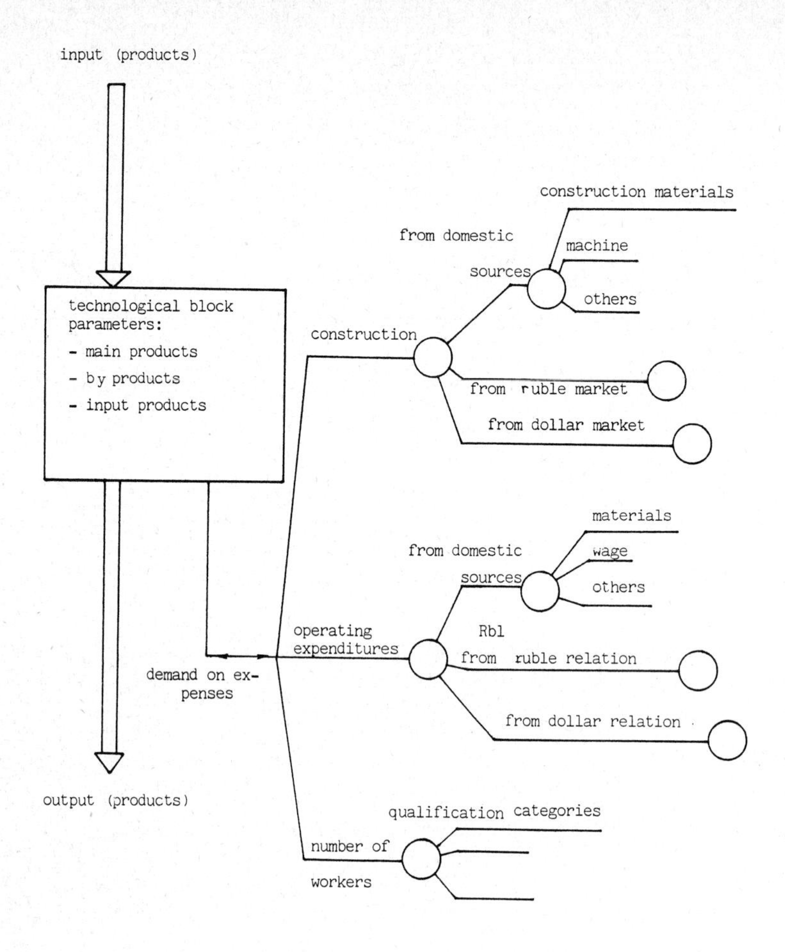

Figure 5: Parameters of the technological block description

ity generation is possible from pulverized brown coal, coal-oil mixture, crude oil, etc.

This starting - or most extensive technological block set should contain the export block, the import compensating block which concentrates information on export-import conditions, and terms of trade. The requirement and demand function in Fig. 4 are generated in the model of the macroeconomy, and they are the input data of the optimization.

The objective functions and constraints of the optimization model can be chosen from the following list (a detailed mathematical description of the model can be found in [4]).

Possible objective functions:

- Minimization of all capital and operating costs of discounted foreign type.
- Minimization of necessary dollar-ratio of investments.
- Minimization of the dollar content of operating costs.
- Minimization of the number of workers. It is possible to minimize the average staff number, the maximum staff number during a given period or create a well-balanced manpower market.

Possible constraints

- limits of investment according to different foreign exchange types (annual limits, cumulative limits for several years);
- staff number limits;
- annual ruble and dollar balances;
- export-import restrictions, determining the market type for each product;
- limits for exportable and importable products for both dollar and ruble market.

In Fig. 6 technological blocks of a coal utilization system of the mining sector are illustrated with a definition of the products and coal demands.

7. CONCLUSIONS

The MRIS, as an aggregate of data bases and data processing functions, spans the whole range of mineral resources utilization. In the implemented subsystems on the plant level, problems of operational control are emphasized; on the company level, those of managerial control; and on the sectoral level, strategic planning tasks are em-

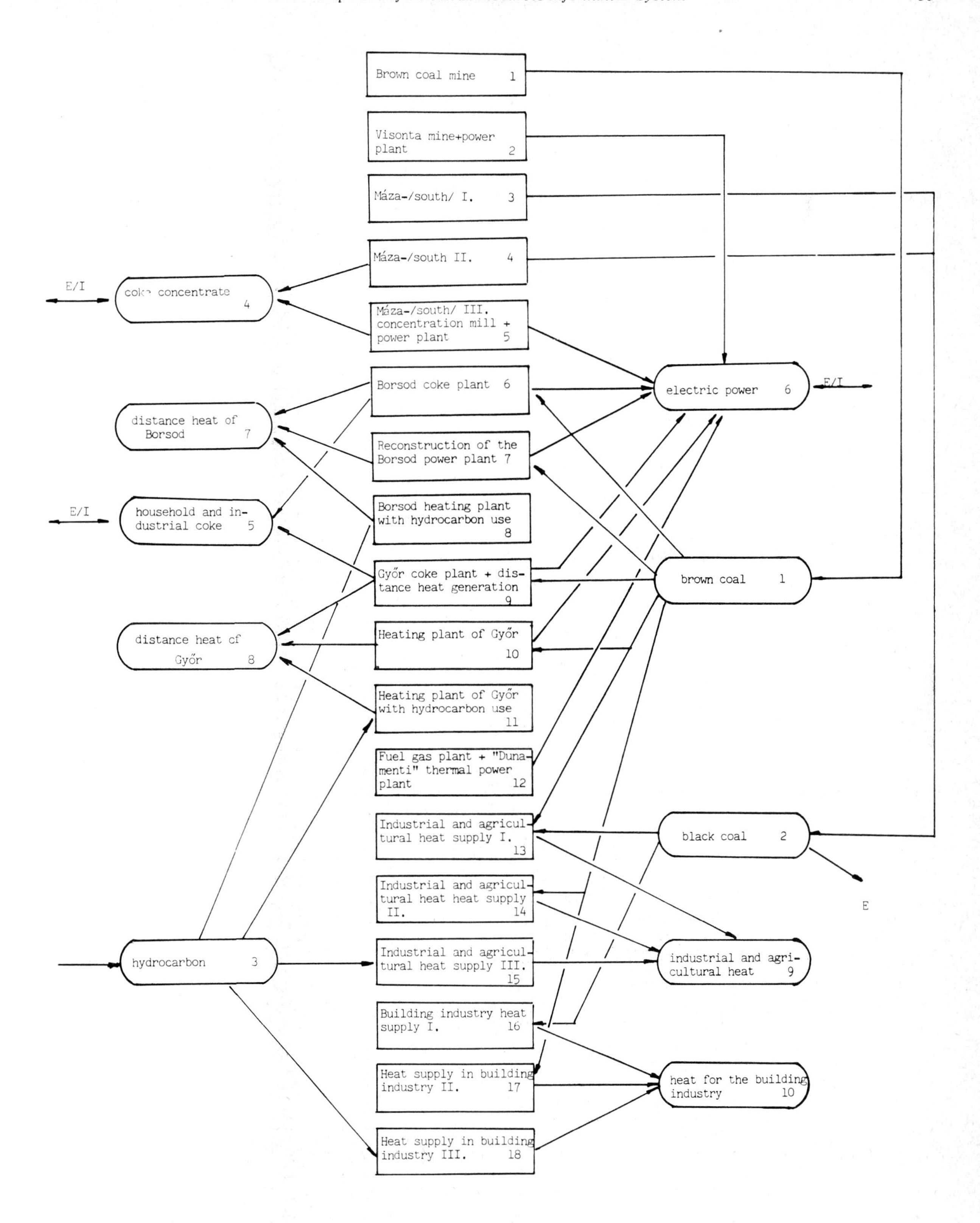

Figure 6: Model of a coal utilization system.

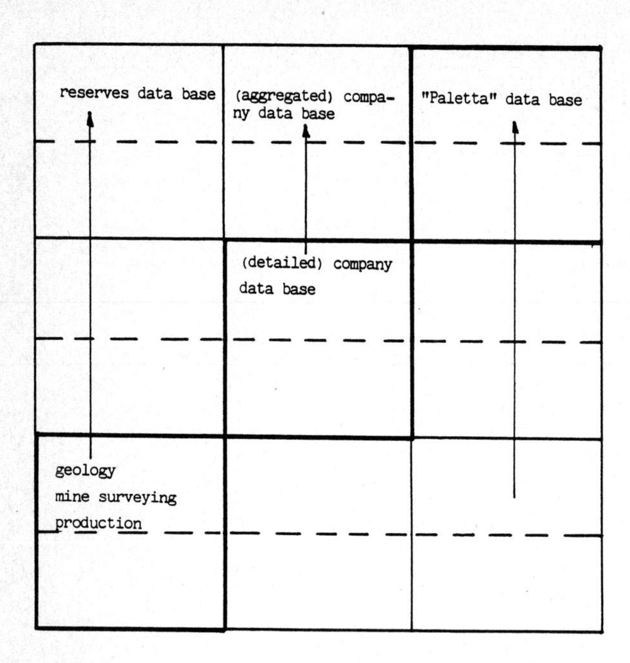

Figure 7: Processes of data acquisition and aggregation of information between data bases of MRIS.

phasized. This does not exclude usage of non-stressed levels for solving problems. In Fig. 7, according to the structure of Fig. 2, vertical arrows are used to denote the process of data-acquisition and information aggregation. The data base of the long-term planning of mineral resources utilization (the so-called "palette" data base) for instance, is constructed from mine plant and company data (data acquisition) or the national balance of mineral reserves is built upon balances of mining companies (aggregation).

In the development of the information system which has begun on different levels of the MRIS, small computers (mine plants, mining company levels) and medium-scale computers (sector level) are used, although some functions, e.g. geostatistical analysis) are realized on both categories of computers.

REFERENCES:

[1] Kapolyi L., System and function approach of mineral resources, (Academic Press, Budapest, 1981). In Hungarian.

[2] Szirtes L. and Gál, I., Requirements and objectives of the development of a mineral resources information system. 17th International Symposium on the Application of Computers and Mathematics in the Mineral Industries, Moscow, Oct. 20-25, 1980 (Lecture No. B. 31., 451-460). In Russian.

[3] McCosh, A.M., Rahman, M. and Earl, M.J., Developing managerial information systems, (MacMillan Press Ltd., London, 1981).

[4] A system-model suggested for the utilization of the fuel and nonfuel resources. (Central Institute for the Development of Mining, Budapest, March, 1980).

Data for Science and Technology, P.S. Glaeser (ed.)
North-Holland Publishing Company

THE SUEZ – AN INFORMATION SYSTEM FOR STORING AND RETRIEVAL OF DATA ON POLISH MINERALS

Ryszard Cichy

Central Board of Geology, Warsaw, Poland

The SUEZ is an information retrieval system designed to collect, process, update and publish mineral statistics data in Poland. The system has been successfully applied by the Central Board of Geology as an instrument of government policy in raw-materials in a centrally-planned economy. Its main function is to draw up plans for geological exploration, mineral supply, and development of mining industries exploiting the domestic deposits.

For a centrally planned economy to be run smoothly, it is vitally important to have quick, easy access to reliable data on the actual quantity and availability of mineral reserves and the degree of identification of reserves.

In this country, the Central Board of Geology (CBG) is a state agency in charge of exploration and conservation of domestic reserves.

By virtue of the Geological Law of Poland, passed in 1964, the CBG is responsible for collecting and making available information on the actual inventory of the country's reserves and for making estimates of how these reserves are capable of meeting domestic demands and export requirements.

The above mentioned law also makes all state mining companies which exploit minerals accountable to the CBG for submitting relevant data on the inventory of reserves indespensable to reaching its objectives.

Thus, the balancing constitutes part of a vital statistical survey necessary for the country's administration, and the CBG, its main agency in this regard, has a virtual monopoly on recording data and producing a full range of mineral statistics.

Since data on mineral reserves had been increasing in volume in recent years, side by side with a growing interest in availability of raw materials, a need finally arose to use computers for more efficient storing and processing of that data. In 1974, the initial assumptions were made for the SUEZ (a Polish acronym of a System for Storing and Retrieval of Data on Minerals).

DATA COLLECTION

The overall assumptions cover a number of subjects in the fields of general geology, geology of deposits, technology and mining. Therefore, it was essential, at the very outset, to compile seven lists of unambiguous items to enable communication between the data sources and to make it possible for the data bank to operate efficiently.

The lists were helpful in converting the descriptive features of minerals into compact numeric and alphanumeric codes with a limited number of characters. Because of a high degree of geological assurance with regard to the data and the significant role minerals play in the nation's economy, the system has been designed to cover all identified deposits.

As of now, data on 3700 deposits of over 50 kinds of minerals are within the operational capacity of the system's processor. However, the data is collected by means of the only locally feasible, traditional way, which is filling in statistical forms since the technological base of the country is rather modest by modern standards. Users of the deposits are required to fill out these forms separately for each deposit carefully following the detailed instructions.

There are seven kinds of forms used, which vary with the subject matter they cover, its scope, and function.

However, there are three kinds of forms for the data on deposits which are to enter the data bank, which differ in their way of describing the deposit parameters such as hard rock minerals or bitumens. The forms contain the so called "fixed data", which are only subject to updating whenever a given deposit is documented anew which, in practice, occurs once every five years. The information in the forms concerns data on address and organization - 7 points, data on geological and mining parameters of the deposit - 18 points, data on quality of the minerals - 4 points, and a set of data on the estimate of reserves which have been officially confirmed by the CBG - 30 points. It should be added that the deposit parameters are of the mean value, determined by the calculations based on the geological documentation of the deposit. The proper documentation requires, according to the regulations, special card files on the deposit. Such cards are identical with the deposit balance forms entered in the data bank.

There is still another kind of form, the so called "update sheets", whose data on the quantity and changes of the reserves is updated by the user of the deposit at the end of each year. The data describe the initial quantity of reserves in different categories, changes in volume such as additions or loss due to the cause specified on the card e.g. a more detailed identification, extensions of the known deposit, new safety pillar sizing as a result of mining operations, exploitation of the reserves etc., and the qauntity of reserves at the end of the annual surveying period. Since the computer technology in this country is still in the budding state, the majority of users who fill the forms out do not use data carriers yet, and, hence, the forms are sent by mail to the computing center within three months following the end of each yearly surveying period.

DATA RELIABILITY ESTIMATES

The data sources, that is, mining companies, are obliged to observe strictly the mining industry surveying regulations. Hence, all data concerning production, additions and losses, changes and actual quantity of reserves are a result of measurements made at the mine exploiting areas and periodic reports are prepared by the mine management (that is daily, monthly, and quarterly output reports, which show if the work is going according to plan).

On the legal side, if the data are false or have been "doctored", the perpetrator is punished with fines since there are severe laws and regulations in the penal code safeguarding the data's not being tampered with.

At the source, the data are checked for reliability by the management overseeing the exploitation of reserves and later, as it moves along, by the appropriate statistical service units. The statistical forms have document status and, as such, are under the protection of the law.

The second stage of data control occurs when the CBG makes periodic comparative data analyses which consist in checking whether the data sent from the users of reserves meets the requirements of the current national production plan and how it compares with the figures of the previous surveying period. In case of discrepancies or glaring mistakes, the report sheet is discussed with the user and undergoes a thorough examination.

Lists of items subject to updating are made during the initialization, and faulty records or false variables are deleted and corrected when the check-out routines are used. This is done when, for instance, the data indicates an imbalance of reserves although the production rates appear to be of the standard rank; or the total of output loss and gain seems to exceed the figures reflecting the corresponding overall depletion of reserves, etc.

From the practical point of view, over the years, though labor intensive, the multistage data control methods applied have proved to be efficient as far as the very system and its area of application are concerned. The reliability of the data is further confirmed by an independent input of data from other statistical sources concerned, for instance, with production or consumption of mine-produced raw materials.

CATEGORIES AND DIVISIONS OF DATA

The basic concern of the data is the documented deposit or the so called "exploitation area". The definition of the term, documented deposit, is derived from the Polish regulations concerning identification of a deposit in connection with its preparation for mine planning and design.

In view of these regulations the degree of resource identification of the deposit is expressed by categories marked with alphanumeric characters C_2, C_1, B, A, which indicate from C_2 onwards a declining degree of the investment risk involved. Moreover, there is a binary economic division of reserves into economic reserves, non-economic reserves and another binary technical division into industrially recoverable and industrially unrecoverable ones.

There is still another quality division within each category and division mentioned above, specific to each mineral.

Such a quality division for hard coal, for example, consists of 14 classes and groups, which vary with the calorific value, caking powder, dilation, etc. With regard to crude oil, there is a binary division into paraffin and base crude oil, wax free crude.

Natural gas is categorized into condensate, wet gas, dry gas, and mine gas.

Another criterion used for making categories of gas is the nitrogen content in gas. There are five grades covering divisions from 15% to over 75% of the nitrogen content.

The classification of zinc and lead ores depends on determining how they are amenable to industrial processing, which yields the oxide and sulphide types. For the same reasons, iron ores are classified into sedimentary and magmatic types to satisfy the genetic criteria. In the case of each ore the reserves cited concern the quantity on the one hand and the metal content on the other.

Resistance to fire is the basis for the classification of fire-clays into six quality categories. There are five quality categories in glass, sand, etc. The majority of the quality parameters are determined by the Polish Standards.

To use all classification possibilities of the system would result in obtaining a considerable number of combinations sufficient to properly estimate the reserve base as well as the scale of utility of minerals in different technological processes.

Before the data are entered for processing, the deposits are classified into five groups on the basis of the actual state of investment in the deposit and its current stage of development. This division criterion is essential for an overall economic analysis to determine the development potential of each branch of the mining industry. All changes such as depletion or additions of reserves are recorded throughout all the above mentioned division groups.

Each record of mineral production is marked whether it comes from the reserves within the safety pillars or from the outside and includes data on its group or class.

The classification applies equally to the base mineral recovered as well as to the by-product.

The measurement units used in the estimates are metric tons and cubic meters. Records of production and reserves quantity are made with the accuracy of up to 10 tons. Besides the numeric data to describe the quantity of reserves and deposit parameters, the system allows for some non-digitized data in the form of additional succinct remarks on the pertinent characteristics of the deposit.

HOW THE SYSTEM OPERATES

The SUEZ has been designed for the ECLIPSE C/300 minicomputer by Data General. The computer with its working store capacity of 60 K words of 16·bits is praised for its high performance.

Besides the central processor, the configuration includes the following peripherals:

- a magnetic disk memory of the Diablo model 44 with the capacity of 10 MB,

- a magnetic tape memory on 9 tracks of the format Nr 22; the recording density of 800 bpi,

- a card reader of the Documentation M-600 L type which reads 600 80-column cards per minute,

- a line printer of the Data Products 2230 type, which can print 300 lines per minute with 132 characters on a line,

- a monitor display with a console of Data General 6052 type.

The software is written in the FORTRAN IV and V languages and consists of 89 programs and 88 subroutines.

Communication with the computer is in the conversational mode. Since the languages are of the FORTRAN family, the computer is provided with the interactive version.

There are nine disk packages for working files and, all in all, 16 reels of magnetic tape are necessary to record entries of the main data base and working files to prepare an annual balance of resources.

On the whole, the SUEZ gathers data on 3700 deposits of 50 kinds of minerals classified into over 300 types, categories and modifications. The bulk is industrial rocks and construction raw materials, which constitute 86% of the total number of deposits; bitumens 5%, solid fuels (coals) 4.7%; metal ores 2.3%; chemical raw materials (sulphur, salts and others) 1.4%. Every year data on 300 new deposits, on the average, are entered into the system. Because of major changes, data on at least 1800 deposits are subject to major updating; and in case of the other 2000, the data require minor updating; be it a change of address, a change of user, a different quality parameter, etc.

ISSUE OF PRINTOUTS

In principle, the SUEZ is destined for inside use in the CBG, and since it contains classified information of economic relevance withheld from general circulation, it is off limits to outside users unless they are officially authorized. In such a case the user sends his application by mail and is later charged an appropriate fee.

Information on deposits is printed on 8 typical tabulograms with a standardized content which varies with the scope of the printed information, the content layout, and the degree of the data integration. The printout is provided by the line printer, printed on telex tape or shown on the monitor display. The user makes his query, specifies the kind of information he is searcing, and chooses the output device and the kind of tabulogram. The printout of the typical tabulograms is issued in a matter of microseconds but some specific information may require a dialogue with the program.

Experience has shown that the content and the layout of the typical tabulograms meet the needs of the majority of users' inquiries sufficiently. Also the speed and efficiency of performance seems to satisfy the requirements of the regular users.

USE OF THE SYSTEM

The data in the bank of the SUEZ are put to a large variety of uses in the field of national economy, that is in scheduling geographical research, regional planning, in preparing programs for development in mineral production and industrial processing of minerals, in management of deposits, in solving geological and environmental conservation as well as other scientific problems based on mineral statistics. The data are also used in finding solutions in other less related fields like plotting the transportation network and other large scale projects. Recently, the SUEZ has been used to determine the amount of money the mining industry is expected to donate to the Geological Research Projects fund.

The so called "deposit information card" is the most frequently retrieved set of data indispensable for work at the CBG at present. It contains an updated set of data on the deposit, its reserves and production rates over a given surveying period. Next in demand are cumulative estimates of the national reserves and production rates of a given commodity. Data on the rate of the reserve's growth in a given year of the system operation are also in great demand to help estimate the profitability of the completed geological studies and to define the aims, objectives, and scope of exploration in the near future.

Every year a detailed analysis is made, using data provided by the system, of the degree of reserves identification to make the necessary adjustments for the documented deposits to meet the requirements of the mining industry investement plan. The analysis concerns two such basic factors as the degree of identification of the reserves (e.g. C_1+B cat.) and quantity of the deposit reserves that guarantees a planned operational mine output at its full production potential and is also worthy of the investment risk. Also taken into account are viable options for a planned mine-site, provided the degree of geological identification of the mineral deposit is worth consideration.

Naturally, the principle of the minimum investment dominates the planning philosophy of mineral production development of the whole mining industry and data from the system are indispensable in the search for the optimum solutions.

To reach this objective the reserves are estimated by classifying the deposits into mines in full operation (or in the process of being modernized), standby fields at the operating mines, and deposits with undeveloped mine sites.

Small firms producing construction materials are quite often interested in deposits abandoned by large contractors when production on a large scale becomes unprofitable. If such is the case, the system helps to find such deposits for potential users, who might now exploit them in small pits at a smaller investment cost (rather than start a capital-intensive development of new deposits).

However, an extensive use of the system is made in overseeing the estimate and changes of reserves of the mines in operation. This is an essential task of the Administration's policy in raw materials and would be unfeasible without decent statistical work in tracing constant changes in quantity of all the classified categories of commodities. A developed classification of reserves taking note of the annual growth and depletion along with an analysis of the origin of these changes are reliable assets when on wants to present a true picture of mineral deposit economy on the national scale. The difference between the overall depletion of reserves because of exploitation and the overall production is a particularly significant indicator. Comparing these two magnitudes over the years gives a clear picture of the state of economy of the deposits and the amount of losses suffered in connection with the exploitation system applied. Another indicator instrumental in estimating the development potential of a given mining branch is numerical estimates of the quantity of reserves in the near future by means of the reported loss limits, the planned production, and the present estimate of reserves. Similar indicators concerning ABC_1 to C_2 categories ratio of identification are very helpful in planning geological exploration to properly map the deposits before new mines are designed; such exploration being especially useful in upgrading deposits initially classified as C_2.

Recently, a full set of data on deposits stored in the SUEZ has been used in determining the rates and control of payments to the Geological Research Projects Fund mentioned above. The money for the fund comes from taxes on production paid by state firms recovering minerals. The amount of payment is strictly dependent on the degree of proper economic management of deposits; which means that the moment losses go beyond the established limit the firm in question is liable to pay to the fund 25% more than the originally assigned quota...So much for some additional practical applications of the system.

Naturally, the above mentioned cases do not fully exhaust the range of applications of the SUEZ in carrying out the economic policy of raw material production and distribution in the country. The software used allows for a wide variety of new applications in the economics and statistics of raw materials in case the data bank is provided with new data.

CONCLUSIONS

- The SUEZ System has been designed to meet the needs of the CBG, a major government agency established to carry out objectives of economic policy of raw materials and made responsible for the proper geological identification of deposits in the country.

- The data bank of the system stores data on characteristics of 3700 documented deposits of 50 kinds of minerals and also data on production rates of 3000 exploitation areas with undocumented deposits.

- The scope of data collection, the developed classification system, and the software make it possible for the SUEZ to exercise constant control over the use of deposits and to make estimates of the state of their economy.

- The system uses terms, estimates, and unified classification principles adopted jointly by the Comecon member countries.

- Any correlation with statistical estimates used elsewhere in the world may only be made with approximate accuracy.

- The application of the SUEZ in a centrally planned economy covers such areas as geological exploration planning, regional planning, programs for development of the mineral production and processing industries deposit conservation planning of a large scale linear investment network, updating information on the quantity of reserves taking into account quality, region of origin, branches of industry, and taxation.

- The SUEZ is part of the national statistical survey, which guarantees the output data reliability and makes provisions that the data bank be systematically supplied with new pertinent data.

- The ECLIPSE C/300 minicomputer, its peripherals, and software satisfy the requirements of the users with regard to efficiency, access, and scope of information retrieved.

- The use of the system throughout the years has not caused any labor force reductions, but has proved to be an invaluable asset regarding the growth of informational capacity and facilitation of estimation procedures.

Data for Science and Technology, P.S. Glaeser (ed.)
North-Holland Publishing Company

MINERAL RESOURCE INFORMATION SYSTEMS

Allen L. Clark

IIRD, International Institute for Resource Development
Vienna, Austria

Mineral resource information systems are being developed throughout the world to meet the ever increasing need for more quantitative and less aggregated data and analyses. The major mineral resource information systems, machine processable or manual, are described.
To be truly useful a mineral resource information system must provide not only data but must be capable of storing and delivering analyses and aggregated syntheses of data for use in resource management and policy decisions.

INTRODUCTION

Since the earliest history of mankind a knowledge of non-renewable energy and mineral resources has been a critical factor in both material and cultural development. Even in the Stone Age a knowledge of the best sources of chert, from which to make tools and weapons, was essential to survival. In this respect little has changed over the last few thousand years; except that a far greater variety of commodities are required to support today's complex societies and consequently the need for mineral resource data has increased proportionally. In particular the 1970's and 1980's were, and are, marked by an increasing awareness by most nations, on both the national and international level, that non-renewable energy and mineral resources were no longer to be regarded as either unlimited in supply or economically available at marginal prices. This awareness was caused by a number of factors including rapidly increasing world demand, the formation of OPEC, the rise of resource self-determination in the developing countries, and a rapidly evolving highly dislocating international economic system. There arose a national and international need to answer several basic resource related questions among the most important of which were:

(a) What are the present reserves and future resources, on both a national and international level, of individual non-renewable energy and mineral resources,

(b) What is the geographic distribution of these reserves and resources,

(c) What is a realistic time frame for the exploration and development of estimated undiscovered recoverable resources, and

(d) How does one acquire an accurate inventory and resource assessment upon which national policy can be formulated and long term planning based.

The need to answer these questions and many more of equal or greater significance to individual nations led to a major effort in the development of resource inventories and resource assessment methodologies; with these efforts came the concurrent emergence of the need for data banks and Mineral Resource Information Systems to support the required analyses.

MINERAL RESOURCE DATA

With the recognition of the need for more and better mineral resource data several data collection programs were initiated generally with high expectations of gathering the required data in a short period of time and thereby rapidly developing the required mineral resource information systems. The task of data collection however proved to be neither easy nor rapid and all too often to be extremely expensive and highly personnel intensive. The reasons for the difficulties in developing mineral resource information systems are too complex to deal with in detail however several factors, summarized in the following, are of major importance in understanding the difficulties of data acquisition.

1. There is a general lack of specific data, within the published literature on many of the major mines of the world. Therefore any compilation necessarily needs to rely heavily upon industry whose participation to date has been minimal.

2. The mineral industry is much more fragmented than in the energy sector therefore fewer summary studies and statistics are available.

3. Virtually no data are available in the public domain pertaining to cost and other economic data related to a specific mine.

4. There has been an overall lack of emphasis on mineral deposit studies because of a general perceived sufficient supply of most commodities. As a result there is both less interest and less money expended with respect to mineral deposit data.

5. Because of the historic lack of need for mineral resource data, and in particular exploration and development data, much of the data was not preserved and hence a large amount of valuable data will never be available for analysis.

The above factors deal primarily with data acquisition problems which are inherent in the structure of the mining industry. There are however an equal or larger number of problems which are inherent to the data which materially impact the ability to develop effective mineral resource data files. Among the most significant are those associated with mineral resource data exchange which is the only available mechanism by which the amount of global data required can be captured.

INTERNATIONAL DATA STANDARDS

The major technical problem in the exchange of existing resource data is the lack of standardized data definitions and standard measurement and reporting procedures. These problems are particularly acute in the energy and mineral sectors of the geosciences because of the variable occurrence modes of the individual resources, the broad geographic distribution of the resources and the all too often subjective or poorly quantified nature of the resource data. For these reasons international data standards have long been discussed but little progress has been made in developing such standards for even the broadest usage. Particularly illustrative of the other extreme of the problem is the definition of reserves and resources which was conceptually presented by McKelvey (1972, p. 32-40). This effort led to a modified classification by the Canadian Geological Survey in 1974 and an even more modified classification by the UN in 1978; Schanz (1980, p. 307-313). As a result there exists not one but a number of "International Standards" for defining reserves and resources - the two most critical elements of most resource data bases and major items in resource data exchange programs.

The UNESCO sponsored IGCP program # 98 "Computer Applications to Resource Studies" evaluated the problem of data standards for resource data files and was forced to conclude that available standards were too diverse to be generally applicable to the broad spectrum of data elements; data compatibility would then have to be provided by subsequent conversion to a local standard upon receipt of the file. Such a solution is by and large unacceptable but is clearly exemplary of the present status of data standards relative to much of present spatial data.

INCOMPATIBLE DATA FORMATS

Equally troublesome as the diversity or lack of international data standards is the even more diverse data formats by which data are collected and stored. In virtually every organization resource data (geologic, geochemical, geophysical etc.) are collected and stored utilizing a unique format which makes data exchange almost impossible in the worst cases and exceedingly costly and time consuming in even the best cases. Clearly there are many geochemical and geophysical applications in which a unique format is called for in the context of data use, requirements of a data base management system or to facilitate direct data reduction. There are however, many more applications for which standardized data formats would not only be possible but highly desirable, as in the case of field data collection, regional geochemical surveys and deposit evaluations. Unfortunately, both the need for standardization and the methodology for accomplishing it are only now being recognized in many geoscience organizations. Although these are fields which should be pursued actively by the geoscience community there is at present no mechanism by which this can be accomplished internationally, therefore, the problem persists and becomes greater each year.

VALIDITY OF EXISTING DATA

Separate and apart from the problem of insufficient data, which may reflect either lack of collection or simply that the data are not available for a variety of other reasons, is the problem of validity of the data that does exist. Few if any problems relative to data exchange are as complex or more difficult to resolve than ascertaining the *validity of the data within an exchange program. Too often data exchanges take place on the basis that "regardless of the validity of the data that is all there is and we must use it as it exists." Somehow that is supposed to resolve the problem of validity. Obviously this approach is not endorsed by many, but only some scientists. It does however present a major problem in the exchange of data in that all too often data are used by someone unaware of the validity problem and this can and does lead to major problems.*

Although the inherent problems within both the mineral industry and the data itself have proven to be major barriers to the rapid compilation of mineral resource data there are a number of on-going and/or proposed mineral resource data files which are being vigorously promoted and should ultimately provide the required basis for a mineral resource information system. In general present mineral resource data file programs are either

(a) Those that are anticipated to be global in coverage both with respect to geography and commodities.

(b) Those that are geographically global in coverage but are restricted to defined subsets of commodities or to individual commodities.

(c) Those that are not now anticipated to extend beyond national boundaries but may ultimately be expanded to types a or b above.

Regardless of the intent of existing programs, at present none is sufficiently complete in terms of coverage or data to be considered useful on global mineral resource assessments. However, several are sufficiently complete to be used in commodity or regional specific studies. Among the most complete are the following:

Computerized Resource Data Base (CRIB) (Calkins and others, 1973) - The mineral resource data base of the U.S. Geological Survey which contains approximately 40 000 domestic and 6000 international records on individual deposits or mineral occurrences.

The CRIB data file deals primarily with detailed geology, reserve, resource, and production statistics and bibliographic references.

Minerals Availability System (MAS) (Kingston, 1977) - The mineral data base program of the U.S. Bureau of Mines which contains 137 425 domestic and several thousand international records on individual producing mines. Although still incomplete, the MAS system is totally functional with respect to four commodities (copper, aluminum, chromium, and tin) and partially complete with respect to 20 other commodities. Ultimately, the program is designed to include 11 commodities.

Individual records provide detailed information on mineral properties and operations, environmental factors, land use data and reserves, resources, and production.

Canadian Mineral Occurrence Index (CANMINDEX) (Picklyk et al, 1978) - The mineral deposit data base program of the Canadian Geological Survey is presently composed of approximately 250 records, which constitute a test file, of mineral occurrences in the Bathurst area of New Brunswick, Canada. Present plans (Picklyk, et al, 1978) are to expand the CANMINDEX file to a total coverage of Canada.

The present form of the CANMINDEX file requires information in six categories, unique identifiers, location, geologic data, bibliography, remarks, and cross references. The ultimate goal of the file is a national inventory of mineral resources.

Circum-Pacific Mineral Deposit File (Guild, P.A., written communication, 1979) - An international program of cooperation to gather both mineral and energy resource data of the Circum-Pacific region. The present file contains approximately 3000 records with the major coverage being provided from the Southeast Asia areas.

Individual records are intended to provide primary input to the development of metallogenic maps of the Circum-Pacific area. Therefore, the data items are primarily geological, reserve, resource and production data which are highly aggregated.

MANIFILE (Laznika, P., 1975) - A thorough compilation on the major nonferrous metal mines and mineral deposit districts of the world. The present file is composed of approximately 4800 deposits and represents perhaps the most complete coverage of mineral resources data yet available in a single file.

MANIFILE provides data on location, major and minor commodities produced, production, and reserves. General geologic data are provided with respect to host rock and age of individual deposits.

Mineral Deposit Data Bank (MDDB) (Gabert, 1978, p. 429) - The mineral resource data base of the Federal Institute for Geoscience and Natural Resources (BGR) of the Federal Republic of Germany. The present system operates on a test file of approximately 2000 deposits.

The MDDB is intended to serve as a global inventory data base with information on location, mining activities, primary and secondary minerals, reserves, production, and general geology.

International Uranium Geology Information System (INTURGEO) of International Atomic Energy Agency contains data on international uranium resources in four areas, production and resources, exploration activity, uranium deposit occurrences and national level summary data. Data base used for national and international uranium resource studies, reserve assessment, exploration potential analysis and for deposit modeling.

In addition to the data files already discussed there are numerous other major data collection efforts underway, on an international basis, within the individual Geological Surveys, Bureaus of Mines, or private industry. In particular, the activities of the Geological Survey of Japan, the South African Bureau of Mines, the Bureau of Geological Research and Mines (BRGM) of France, and the Institute of Geological Sciences of England all have active mineral resource data base activities. The majority of these efforts are directed toward the creation of manual archives. However, in every case programs are either planned or underway to computerize the files.

These individual manual files are of importance in that they not only contain the normal information on mineral deposits but also contain a large amount of associated data on environmental, political and development factors, data elements that are of particular value in global resource modeling efforts. Unfortunately, unlike these manual archives, the majority of machine processable data files do not contain these ancillary data elements. A notable exception to this tendency is the MAS system of the U.S. Bureau of Mines which pays particular attention to environmental and developmental factors.

INTERNATIONAL STRATEGIC MINERALS INVENTORY

In 1979 the Author (Clark and Cook, 1980, p. 32-44) proposed the creation of an international data file based on the 1052 major mines of the world, as defined by Mining Magazine (1978, p.222-258) which would serve as a basis for inputs into national, regional and global resource models. At present a prototype file (International Strategic Minerals Inventory - ISMI) to implement this program (Greenwood, W.R. 1982, personal communication) for the commodities Ni, Cr, Mn and phosphate rock has been initiated under a cooperative agreement between the Geological Surveys of Canada, Federal Republic of Germany, United States and the U.S. Bureau of Mines.

IMSI represents a truly international effort to create a global mineral resource data file and as such also provides a set of accepted terms and definitions for the compilation of the data and thereby a defined base for present and future data collection. Toward this end the initial data collection has begun using the format shown in Appendix I.

SUMMARY AND CONCLUSIONS

Mineral resource data files and information systems are being developed to meet an ever increasing need for more quantitative and less aggregated data and analyses. Unfortunately structural deficiencies in both the mineral industry and within the available data has seriously impeded the development of both the required data files and the information systems. A partial resolution of the major problems resulting from inadequate mineral resource data is being partially resolved by a large number of specialized data files being developed by various nations and/or international organizations. However the final resolution of the need for mineral resource data and information systems can best be accomplished by a concerted and standardized international program of data collection similar to that proposed and being used in the International Strategic Minerals Inventory.

REFERENCES:

(1) Calkins, J.A., Kays, O., and Keefer, E.K., 1973, CRIB - The Mineral resources data bank of the U.S. Geol. Survey: U.S. Geol. Survey Circ. 681, 39 p.

(2) Clark, A.L., and Cook, J.L., 1979, Global Resource Data Systems. U.S. Geological Survey Project Report (IR) NC-68, 44p.

(3) Gabert, G., 1978, The importance of mineral and energy inventories, in standards for computer applications in Resource Studies, Cargill, S.M., and Clark, A.L. eds: Journal of International Association of Mathematical Geology, v. 10, no. 5, p. 425-432.

(4) Kingston, G., 1977, Reserve classification of identified nonfuel mineral resources by the Bureau of Mines Availability System in standards for computer applications in Resource Studies, Cargill, S.M., and Clark, A.L., eds: Journal of International Association of Mathematical Geology, v. 9, no. 3, p. 273-279.

(5) Laznika, P., MANIFILE: The University of Manitoba computer-based file of the worlds on nonferrous metallic deposits: Canada Geological Survey paper 75-20.

(6) McKelvey, V.E., 1972, Mineral Resource Estimates and Public Policy: American Scientist, V. 60, p. 32-40.

(7) Mining Magazine, 1978, Mining activity in the Western World: Mining Journal limited, London, March, p. 222-258.

(8) Picklyk, D.D., Rose, D.G., and Laramee, R.M., 1978, Canadian mineral occurrenceindex (CANMINDEX) of the Geological Survey of Canada: Canada Geological Survey paper 78-8, 29 p.

(9) Schanz, J.J., 1980, The United Nations endeavor to standardize mineral resource classification, Natural Resources Forum, vol. 4, p. 307-313.

ANNEX I

INTERNATIONAL STRATEGIC MINERALS INVENTORY RECORD FORM

RECORD IDENTIFICATION

RECORD NUMBER S1 <________> DEPOSIT NUMBER S2 <________> RECORD TYPE S4 ________>

FILE LINK IDENT. S3 <________________>

SITE NAME S5 <________________________________>

SYNOMYMS S6 <________________________________>

LOCATION

MINING DISTRICT/AREA L1 <________________________________>

COUNTY L2 <________> STATE L3 <________> COUNTRY L4 <________>

GEODETIC ELEVATION L7 <________> DATUM OF ELEVATION L8 <________>

LATITUDE L5 <- -> POINT OF REFERENCE AND PRECISION L9 <________> AREA L10 <________> km^2

LONGITUDE L6 <- -> LOCATION COMMENTS L11 <________________>

GEOLOGY

DEPOSIT TYPE(S) G1 <________________________________>

HOST ROCK TYPE G2A <________> FORMATION NAME G2B <________> FORMATION AGE G2C <________>

HOST ROCK TYPE G3A <________> FORMATION NAME G3B <________> FORMATION AGE G3C <________>

HOST ROCK TYPE G4A <________> FORMATION NAME G4B <________> FORMATION AGE G4C <________>

AGE OF FIRST MINERALIZATION G5 <____> AGE OF SECOND MINERALIZATION G6 <____> AGE OF THIRD MINERALIZATION G7 <____>

TECTONIC SETTING G8 <________________> LOCAL ENVIRONMENT G9 <________________>

MAJOR POST-ORE GEOLOGIC EVENT G10 <________> AGE OF MAJOR POST-ORE GEOLOGIC EVENT G11 ________>

PRINCIPAL MINERAL ASSEMBLAGES G12 <________________________________>

GEOLOGY COMMENTS G13 <________________________________

____

____

____>

COMMODITY INFORMATION

PRESENT/LAST OWNER(S) C1 < ______ >

PRESENT/LAST OPERATOR C2 < ______ >

YEAR OF DISCOVERY C3 < ___ > NATURE OF DISCOVERY C4 < ___ > MINING METHOD C5 < ___ > DEPTH OF COVER C6 < ___ >

DEPTH OF UNCONSOLIDATED COVER C7 < ___ > PRODUCT UPGRADED BEFORE EXPORT C8 < ___ > YEAR OF FIRST PRODUCTION C9 < ___ >

YEAR OF LAST PRODUCTION C10 < ___ >

COMMODITIES C11 < ƀ ; ƀ ; ƀ ; ƀ ; ƀ ; ƀ ; ƀ ; ƀ ; ƀ ; ƀ ; ƀ ; >

ABILITY TO EXPAND PRODUCTION IN <1 YR. C12 < UNKNOWN; NONE OR MINOR; UP TO 30%; 30%-60%; <60%; (SPECIFY) < ___ >

ABILITY TO EXPAND PRODUCTION IN 1-5 YRS. C13 < UNKNOWN; NONE OR MINOR; UP TO 30%; 30%-60%; <60% (SPECIFY) < ___ >

NON-GEOLOGIC CONSTRAINTS ON DEVELOPMENT OR EXPANSION IN 1-5 YRS. C14 ENVIRONMENT; GOVT. REGULATIONS;

TRANSPORT/ACCESS; MINE PLANT; PROCESSING; LABOR; ENERGY; OTHER (SPECIFY) < ______ >

ACTIVE DEVELOPMENT IN MINE C15 < YES; NO > ACTIVE EXPLORATION IN DISTRICT C16 < YES; NO >

COMMODITY COMMENTS C17 < ______ >

PRODUCTION

TYPICAL ANNUAL PRODUCTION

	ITEM	ACCURACY	AMOUNT	UNITS	YEARS	GRADE
PA1	a. < ___	ƀ >	b. < ___ >	c. < ___ >	d. ___ >	e. < ___ >
PA2	a. < ___	ƀ >	b. < ___ >	c. < ___ >	d. ___ >	e. < ___ >
PA3	a. < ___	ƀ >	b. < ___ >	c. < ___ >	d. ___ >	e. < ___ >
PA4	a. < ___	ƀ >	b. < ___ >	c. < ___ >	d. ___ >	e. < ___ >

SOURCE OF INFORMATION PA7 < ______ >

PRODUCTION COMMENTS PA8 < ______ >

CUMULATIVE PRODUCTION

	ITEM	ACCURACY	AMOUNT	UNITS	YEARS	GRADE
PC1	a. < ___	ƀ >	b. < ___ >	c. < ___ >	d. < ___ >	e. < ___ >
PC2	a. < ___	ƀ >	b. < ___ >	c. < ___ >	d. < ___ >	e. < ___ >
PC3	a. < ___	ƀ >	b. < ___ >	c. < ___ >	d. < ___ >	e. < ___ >
PC4	a. < ___	ƀ >	b. < ___ >	c. < ___ >	d. < ___ >	e. < ___ >

SOURCE OF INFORMATION PC7 < ______ >

CUMULATIVE PRODUCTION COMMENTS PC8 < ______ >

RESERVES/RESOURCES

R1E AND/OR r1E

	ITEM	IN SITU/ RECOVERABLE	AMOUNT	UNITS	YEAR OF EST	GRADE
RR1	a. < ______	______ ƀ >	b. < ______ >	c. < ______ >	d. < ______ >	e. < ______ >
RR2	a. < ______	______ ƀ >	b. < ______ >	c. < ______ >	d. < ______ >	e. < ______ >
RR3	a. < ______	______ ƀ >	b. < ______ >	c. < ______ >	d. < ______ >	e. < ______ >
RR4	a. < ______	______ ƀ >	b. < ______ >	c. < ______ >	d. < ______ >	e. < ______ >

RR7 SOURCE OF INFORMATION < ______ >

RR8 COMMENTS < ______ >

R2E AND/OR r2E

RR9	a. < ______	______ ƀ >	b. < ______ >	c. < ______ >	d. < ______ >	e. < ______ >
RR10	a. < ______	______ ƀ >	b. < ______ >	c. < ______ >	d. < ______ >	e. < ______ >
RR11	a. < ______	______ ƀ >	b. < ______ >	c. < ______ >	d. < ______ >	e. < ______ >
RR11	a. < ______	______ ƀ >	b. < ______ >	c. < ______ >	d. < ______ >	e. < ______ >

RR15 SOURCE OF INFORMATION < ______ >

RR16 COMMENTS < ______ >

R1M AND/OR r1M

RR17	a. < ______	______ ƀ >	b. < ______ >	c. < ______ >	d. < ______ >	e. < ______ >
RR18	a. < ______	______ ƀ >	b. < ______ >	c. < ______ >	d. < ______ >	e. < ______ >
RR19	a. < ______	______ ƀ >	b. < ______ >	c. < ______ >	d. < ______ >	e. < ______ >
RR20	a. < ______	______ ƀ >	b. < ______ >	c. < ______ >	d. < ______ >	e. < ______ >

RR23 SOURCE OF INFORMATION < ______ >

RR24 COMMENTS < ______ >

R2M AND/OR r2M

RR25	a. < ______	______ ƀ >	b. < ______ >	c. < ______ >	d. < ______ >	e. < ______ >
RR26	a. < ______	______ ƀ >	b. < ______ >	c. < ______ >	d. < ______ >	e. < ______ >
RR27	a. < ______	______ ƀ >	b. < ______ >	c. < ______ >	d. < ______ >	e. < ______ >
RR28	a. < ______	______ ƀ >	b. < ______ >	c. < ______ >	d. < ______ >	e. < ______ >

RR31 SOURCE OF INFORMATION < ______ >

RR32 COMMENTS < ______

______ >

COMBINATIONS OF ANY TWO OF THE ABOVE CATEGORIES (SPECIFY)

RR33 a. <____b̸> b. <____> c. <____> d. <____> e. <____>
RR34 a. <____b̸> b. <____> c. <____> d. <____> e. <____>
RR35 a. <____b̸> b. <____> c. <____> d. <____> e. <____>
RR36 a. <____b̸> b. <____> c. <____> d. <____> e. <____>

RR39 SOURCE OF INFORMATION <____>

RR40 COMMENTS <____>

ANNOTATED REFERENCES

REFERENCE 1 RF1 <____>

REFERENCE 2 RF2 <____>

REFERENCE 3 RF3 <____>

REFERENCE 4 RF4 <____>

REPORT DATE REP1 < b̸ >
REPORTER REP2 <____>
(last, first middle initial)

REPORTER AFFILIATION REP3 <____>

GENERAL COMMENTS REP4 ____>

UPDATE REPORT DATE UP1 < b̸ >
UPDATE REPORTER UP2 <____>
(last, first middle initial)
UPDATE REPORTER AFFILIATION UP3 <____>

UPDATE COMMENTS UP4 <____>

Data for Science and Technology, P.S. Glaeser (ed.)
North-Holland Publishing Company

NEW WAYS OF ESTIMATING MINERAL RESERVES AND RESOURCES

Václav Němec

Geoindustria
Prague, Czechoslovakia

The most recent possibilities for estimating mineral reserves and resources consist in applying computers to this activity. Transcription of traditional methods does not represent any substantial progress. Mathematical models enable us to create space models of ore deposits which are closer to reality and also directly usable for exploitation (time models). Any sophisticated purely mathematical approach in many situations does not permit us to achieve the ideal solution because of some space phenomena (inserted subsystems) mostly caused by planetary disjunctive systems. Possible ways are demonstrated which may help to decipher algorithms of the geologic evolution and to increase both effectiveness of searching new resources and reliability of ore reserves estimation.

1. INTRODUCTION

Estimation of mineral reserves and resources is a typical interdisciplinary activity. Its liaison with applied geology and its mathematical background are of paramount importance but - when solving individual cases - many special problems of chemistry, technology, mining engineering, economy, etc. must be taken into consideration as well. In many situations the application of computers to this activity enables a good contact between various disciplines and the necessary decision making becomes much easier. On the other hand, it is no easy matter to create, verify and introduce into practical application a computerized system of estimating mineral reserves and resources. Let us briefly discuss some problems encountered in this activity with the aim of showing desirable further orientation of both research and practical work.

2. APPLYING COMPUTERS FOR ESTIMATING MINERAL RESERVES AND RESOURCES

2.1 Two Possible Ways

Computers may be applied for estimating mineral reserves and resources in two possible ways:
a) transcription of traditional methods of ore reserves calculations,
b) use of mathematical models.

2.2 Transcription of Traditional Methods

It is really very easy to prepare programs for triangular or polygonal methods or for geological or exploitation blocks. In the case of numerous data, repeated evaluation and easy accessibility to computers, this approach is probably effective but in general the increased costs of preparing input data, programming and computing usually are not proportional to what is gained by the results which do not differ at all from those which can be achieved by using the traditional approach. Therefore this way does not represent any substantial progress - at least from the methodological point of view.

2.3 Use of Mathematical Models

The use of mathematical models seems to be progressive. The results can be presented in small space units which - after successive unification and combination - permit us to simulate various mining processes and to choose from among them the optimal exploitation system.

There are numerous mathematical methods which are being used for estimating mineral reserves and resources. Simple linear interpolation and trend surface analysis belong to the most primitive methods. A higher level is represented by the inverse distance (ID) or inverse distance square (IDS) methods and similar weighting procedures (some of them respecting the anisotropy of ore bodies). Geostatistical methods (as developed by the French school of G. Matheron) represent nowadays the most sophisticated approach to the problem.

Another possibility consists in an interactive solution with specially prepared data serving for hand made geometrization (contours of grades, etc.) usually complemented by the effort to discover statistical interrelationships of various components.

3. SPACE MODELS

3.1 Scope of Space Models

The results of mathematical or geometrical modelling represent a space model of the deposit. First of all such a model has to reflect all geological phenomena which may be of interest for designing and controlling mining processes on the deposit. In its final shape the space model has to cover the exploitation model as well.

For purposes of computing ore reserves and examining exploitation processes, the space models of mineral deposits should be transferred into a special system of blocks where various types of blocks are used simultaneously: micro-, macro- and megablocks. In some situations miniblocks should be applied as well.

3.2 Fundamental Units: Micro- and Miniblocks

The microblocks are fundamental units. They have a constant area (usually 10 x 10 m, 10 x 20 m or 20 x 20 m), their height is equal to the height of mining floors. Input data for each microblock are estimated at its center. The system of microblocks is an adaptation of a regular grid which was known earlier in traditional computations of ore reserves. A system of coordinates is used for numerical identification of microblocks.

When a separation of geological and mining conceptions seems to be more convenient and useful (e.g. in order to examine various possible methods of the mining process), then other fundamental units are used - the miniblocks. They also have a constant area (identical with that of microblocks) and except for the surface miniblocks, their thickness is also constant but this dimension represents only a fraction of the height of mining floors (e.g. 5 or 8 or 10 meters). Thus basic information on space distribution of useful and harmful components is given and geological features of the deposit are represented in a geological model. This model may then be used for any mining solution. After determining all exploitation levels the construction of microblocks is then fully automated.

3.3 Summarizing Units: Macro- and Megablocks

The macroblocks represent a summary of microblocks in larger blocks. Some geological and chemical homogeneity of them is called for primarily, but mining homogeneity also has to be taken into consideration when macroblocks are to be created. For their numerical identification a system of coordinates and quasicoordinates is used.

The larger units summarizing macroblocks from various aspects - the megablocks - serve mainly as the basis for examining various exploitation models and for final summaries.

4. TIME MODELS

4.1 Scope of Time Models

It is only a space model of the deposit with fairly detailed and reliable data which permits one to create time models, i.e. to simulate possible future exploitation processes and to choose the optimum from among them. Different space units can be used for time models of differing time units. Therefore in any situation it is possible to operate with rather limited sets of input data without deteriorating the reality and reliability of practical results.

An effective time models system is distinctly represented by the approach from the whole to the detail. This is also the only way to assure the required mutual accordance of long and short term models as well as to avoid any misuse of short term time models (which may occur by preferring short term payment criteria opposed to long term utility).

4.2 Basic Kinds of Time Models

Strategical models permit us to solve a long term concept of exploitation covering either all minable reserves or at least a portion of them needed for exploitation in a period of about 20 or more years. This period is usually segmented into equal 5-year time units.

Tactical models - derived from previous ones - focus detailed attention on annual exploitation plans. In practice they are usually represented by a series of 5 successive 1-year plans which corresponds to the first period of the strategical model.

Operative models are oriented on brief time periods (1-3 months). They have to be derived from the first period of the tactical model.

4.3 General Remarks

The time models at any level have to satisfy various requirements concerning geological, technological, technical and economic conditions valid for the whole deposit or for its respective parts. The mining activity is inevitably connected with some risk but a good time model has to minimize this risk especially by avoiding contemporaneous exploitation in those segments of the deposit where an increased risk can be expected. Various criteria for the time models may be optimized but in many situations it is almost impossible to determine a dogmatic hierarchy for different criteria. Therefore a solution is needed which permits the interaction of experienced experts when an optimal time model is to be chosen.

With the use of the games theory suitable strategical and tactical time models already have been developed and used in practice.

Further development of time models is to be oriented to their use for higher and larger economic structures of both national and international character. Such a development will allow us to take into consideration important interrelationships from the point of view of industrial branches, economic needs, financial possibilities, etc.

5. MAIN PROBLEM - INSERTED SUBSYSTEMS

Unfortunately in many situations some purely mathematical approaches to modelling mineral deposits do not permit us to obtain a fairly good and reliable space model. This is due to some geological phenomena which very often have an apparently random character. In the set of exploration data these phenomena are represented only by a relatively small portion of information which is not always in proportion to their real existence and importance in nature.

If we consider the whole deposit as a "system", then such phenomena may be called "inserted subsystems". They are represented mostly by some heterogeneous zones which are connected with tectonic phenomena. According to recent research some regularity is characteristic for them but it is really very difficult to capture this regularity only by means of exploration. The inserted subsystems are probably the most serious obstacle to a complete automation of the evaluation process. They have a decisive part in the discrepancies between estimated and real quality of ore reserves.

It seems to be difficult to give universal advice on how to incorporate the inserted subsystems into estimations of mineral reserves and resources. Nowadays they may be recognized and evaluated practically only by applying some interactive approach. Nevertheless some possibility exists to solve this problem on a solid basis.

6. POSSIBLE SOLUTION

6.1 Planetary Disjunctive Systems

Already in the past century some regularity of structural patterns was registered by a few authors. In the last 20 years many geologists have studied the problems of regular (mostly equidistant) rupture systems, the planetary character of which can be observed also in photographs taken from satellites.

The author of this paper started his own research of the problem in 1970 but recently has changed his original formula for equidistances into the following one:

$$y_x = \frac{2^{-x}}{3} \pi D \cos \psi$$

where y = equidistances for given order
x = given order in the system
D = diameter of the planet
ψ = paleogeographical latitude

The respective values for equidistances of orders 1 - 10 are given in Table 1. The paleogeographical latitude is to be applied only for equidistances of meridional orientation.

Individual planetary disjunctive systems are connected with different positions of poles in geological history. The basic system is represented by circular structures (E-W) coinciding with the equator and with other critical parallels. Another perpendicular system (N-S) is represented by critical meridians. Diagonal systems of lineaments and structural lines (NW-SE, NE-SW - again in the paleogeographical sense) are perhaps of the same importance and intensity as the basic N-S and E-W systems. They also seem to be of very regular character. It should be highly interesting to discover the interrelationship between planetary disjunctive systems of various generations in the case of their local or global superposition. Probably some "remaining resonance" of older systems will be identified.

Further research of planetary disjunctive systems will undoubtedly contribute not only to the solution of some serious theoretical problems (reconstruction of poles migration, conditions for originating new planetary disjunctive systems, inheritance of geological structures, etc.) but also to the prediction of new mineral resources as well as to improved interpretation of exploration and exploitation problems. As far as exploration is concerned, it is always necessary to avoid coincidence of the exploration grid with that of the inserted subsystems.

6.2 Interpretation by Variograms

The planetary disjunctive systems have been identified already locally, when evaluating some deposits. They are identical with the above mentioned inserted subsystems. One of the basic tools of geostatistics - the variogram - can be used very well for their identification and detection. One only has to make a suitable choice of distances intervals for the necessary computations as well as a careful interpretation of experimental variograms.

6.3 Transfer of Patterns

Another practical experience of the author has resulted in a special approach to the problem of detailed extrapolation - the transfer of patterns. In the case of a deposit of limestones with a rather complicated geological structure and with at least two inserted subsystems the mining activity is accompanied by very dense boring and sampling of rocks in the area under exploration. For microblocks of a constant area 10 x 10 meters the grades of the so-called impurities (SiO_2 + Al_2O_3 + Fe_2O_3) as well as of the $MgCO_3$ are known and transformed into the multicomponent classification. Then a transfer of patterns is effected: each microblock has its "alter ego" in the deposit on the same level as well as on the lower or higher levels. Such a microblock can be classified in the same class as the original microblock. This approach cannot guarantee the absolute precision in any microblock but it assures that the proportion of inserted subsystems in the unexploited parts of the deposit will be very close to reality.

6.4 Possible Ways to the Final Solution

In order to increase both effectiveness of searching new resources and reliability of ore reserves estimation it should be recommended:
a) to create special banks of data (with tecton-

Table 1: Equidistances in km

Order	1	2	3	4	5	6	7	8	9	10
Latitude										
0°	6678	3339	1669	834	417	209	104	52.0	26.1	13.0
15°	6450	3225	1613	806	403	202	101	50.4	25.2	12.6
30°	5783	2892	1446	723	361	181	90	45.2	22.6	11.3
45°	4722	2361	1181	590	295	148	74	36.9	18.4	9.2
60°	3339	1669	834	417	209	104	52	26.1	13.0	6.5
75°	1728	864	432	216	108	54	27	13.5	6.8	3.4

ic and other phenomena for the whole planet),

b) to continue research on planetary disjunctive systems,

c) to develop adequate mathematical models for simulating and deciphering proper algorithms of the geological evolution,

d) to discover interrelationships and interdependencies among planetary, continental, regional and local tectonic patterns,

e) to use in addition data of a regional character (outside the deposit area) for estimating mineral reserves and resources (exploration and sometimes exploitation data do not guarantee the right proportion of the inserted subsystems in the whole population and are not sufficient for reliable purely mathematical models).

7. CONCLUSION

The suggested research of planetary disjunctive systems does not require any special expensive work in the field. On the other hand, some very positive results can be expected. The deciphering of planetary disjunctive systems would represent considerable progress in discovering new resources especially in the developing countries. This aspect alone should be sufficient to indicate research. Space and time models with full respect to the inserted subsystems can help to increase the economic level of the mineral industry.

Data for Science and Technology, P.S. Glaeser (ed.)
North-Holland Publishing Company

DATA BANKS OF THE POLISH GEOLOGICAL SURVEY – THEIR PRINCIPLES AND UTILIZATION

Maciej K. Rajecki

Geological Institute, Computer Science Department
Warsaw, Poland

Storage and retrieval systems for deep boreholes (SAGWIG), hydrogeological wells (HYDRO), and world mineral resources (WORLD RESOURCES) data are presented. Principles underlying these systems, ADP solutions, output forms and utilization of these systems are also described. In conclusion, a philosophy of creating factual-documental ADP systems for geology is discussed.

1. INTRODUCTION

The Polish Geological Survey has a long tradition but has existed since 1919 in its contemporary form. The main center of the Polish Geological Survey is the Geological Institute in Warsaw, organized in 1919. Since 1952, activities of the Polish Geological Survey have greatly increased as the Central Board of Geology began to coordinate all forms of geological exploration in Poland at that time.

In the 1960's, pursuant to government plans for geological investigations, the Geological Institute considerably expanded systematic prospecting in areas favorable for the occurrence of ore and non-ore deposits, hydrogeological explorations and geological mapping. The result was not only discovering many well-known Polish deposits (Zn-Pb, Cu, S, etc.), but also rapidly increasing geological information to be handled, filed and stored. At the end of the 1960's and in the early 1970's when the Geological Institute began the widely planned prospecting of deep structures aimed at discovering oil and gas deposits - the introduction of ADP-technology to data storage and retrieval became imperative. The problem of systematizing and standardising the geological data was the first step to the creation of data banks at the Geological Institute.

In collaboration with geologists from the Polish Academy of Science, the Geological Institute prepared a standard system of recording field descriptions of borecore observations and tried to introduce this system, called STANDARD, into geological practice. STANDARD systematizes the procedure of core description, geological classifications, and field measurements which must be made on field observations. Alphanumeric data from these field observations are recorded in this system according to simple mnemonic codes, specially prepared by taking into consideration the habits of the geologists. For example, to describe rock texture, structure and colors in the STANDARD, special STANDARD tables were introduced. Tentatively STANDARD's introduction into geological practice was unsuccessful - only a small group of Polish geologists accepted and used this system. It was necessary to return to the non-standardized forms of geological descriptions of borecore, which are in the official geological documents. This form, which the Geological Institute accepted when creating the ADP borehole data storage and retrieval system, is the profile of a borehole, contained in geological documentation or reports officially ratified by the State Geological Commission. At the Geological Institue, we had to organize a special group for training geologists, which in 1972 began to code data from geological reports of deep boreholes into the specially prepared precomputer format. The list of authorized codes was established by the staff of the Geological Institute, with the possibility of introducing new terms and codes. The ADP system named SAGWIG (accronym of Polish words created from the full name), was introduced on a pilot basis in 1974, and at the same time the Geological Institute began to realize the ADP hydrogeological well data system, named HYDRO, which was completed in 1976.

2. PRINCIPLES UNDERLYING THE SYSTEMS

Both systems (SAGWIG and HYDRO) have many common features because the principles underlying them are similar. The aims in creating SAGWIG and HYDRO were as follows:

- secondary uniformity of the basic data description included in geological documents, which as an end result allowed us to compare these data,

- creation of magnetic tapes with a uniformly coded data base of all boreholes on Polish territory,

- rendering the archives more mobile and useful by automatization of the geological data output in the different variants which geologists need to handle. The following conceptional assumptions for projecting and realizing both systems were accepted:

1. SAGWIG and HYDRO ought to be computer oriented, factual systems of geological data storage and retrieval (SAGWIG) and of hydrogeological, technical and water utilization data (HYDRO);
2. In both systems data processing programs must

be a separate part of the system, meaning that automatic data processing is an independent part of the systems;

3. Both systems ought to have similarly organized data bases. Organization of the data base depends on dividing the borehole set and the hydrogeological data set for the territorial subsets limited by the map sheet at a scale of 1:100 000, in such a way, that in one subset it is possible to collect 100-150 profiles of deep boreholes or 400-500 profiles of hydrogeographical wells, respectively. Data from one subset are recorded on one magnetic tape, thanks to which it is possible to add the next 100-200 boreholes or 400-500 hydrogeological wells from the same area;
4. SAGWIG and HYDRO have a modular construction. In both systems all the information from boreholes and from wells is divided into homogeneous parts which determine one module or one unit in the computer recording;
5. Both systems have so called "open construction" which means that it is possible to increase the amount of modules around new thematic units;
6. In both systems many formal and logical control programs search the data for errors and practically all of them are detected and rectified in the storage phase of the data base;
7. The operating principle of both systems relies on the user's formulating a question to which the system gives him the answer. In the question, the type and the value of the system's selectors must be determined; with this information, the searching and output of data is achieved.
8. Both systems, in practice, accomplish searches and output of any combination of geological data which exists in the system's data base.

3. SAGWIG's DATA STRUCTURE AND DATA BASE ORGANIZATION

The structure of the data from one borehole's profile is presented in Figure 1 by P. Stenzel (4).

The principles of data organization from one borehole profile and in the data base of the SAGWIG system are as follows:

- the profile of the borehole is divided into the stratigraphic complex according to the international stratigraphic classification and, if it exists, according to the local classification,

- each stratigraphic complex is formulated by name, thickness, depth of top and bottom of the stratum,

- in each stratigraphic complex, all rocks which occur there must be determined.

All the geological information from the borehole profile is ordered to this stratigraphic complex according to the depth of its occurrence and is grouped in the monothematic files in the SAGWIG data base.

4. HYDRO's DATA STRUCTURE AND DATA BASE ORGANIZATION

The structure of the data base in the HYDRO system is presented in Figure 2 (5). The organization of the data base in the HYDRO system in relation to the structure of information on hydrogeological wells and problems in the organization of seven regional HYDRO data banks were described by M.K. Rajecki et al.(3).

In HYDRO, the data are grouped in monothematic blocks as follows:

block A - basic data concerning localization of a well in three dimensions, and technical geological data concerning the main exploitation level,
block B - tubing data,
block C - data concerning the well log in screened sectors and its technical parameters,
block D - results of test pumping,
block E - lithostratigraphic profile and saturated levels,
block F and G - physical (F) and chemical properties of water (G),
block H - results of granulometric analyses,
block I - stable localization and identification data concerning the hydrogeological intake (collaborated wells),
block J - variable data concerning the hydrogeological intake.

5. SAGWIG AND HYDRO PROGRAMS

The SAGWIG and HYDRO programs have been written in Fortran IV language.

SAGWIG is programmed on a CYBER '72 (CDC 6600) computer and also a PDP 11/45 (DEC production), however, HYDRO is on a Polish computer, ODRA 1305, which is compatible with the ICL 1900 computer. At present in the SAGWIG data base, the Geological Institute has more than 1600 deep drillings and in the HYDRO data base - in regional banks - there are about 80 000 of the 100 000 hydrogeological wells in Poland.

The time needed to answer a user's question oscillates from 20 seconds to half an hour, depending generally on how complicated the question is and/or whether one works in a noninteractive mode using magnetic tape to retrieve the data.

The basic computer programs of HYDRO allow us to effect three patterns of searching according to the sequence of the blocks: A-E, C, G, D, F, B, H-I or J, I-A, E, C, G, D, F, B, H and the third one has a regional character.

In both of the discussed systems, the possibility to register output retrieval data in the memory was planned, and after a user's decision, these data or part of them, could be processed.

The data processing programs of both systems, which are an independent part of the ADP system, realize:

- all kinds of statistical calculations,
- regression and factor analysis,
- plotting of contour and isoline maps of the basic data (by a system named MAPART-2 prepared by J. Owczarczyk and others).

6. SAGWIG AND HYDRO UTILIZATION

The philosophy behind the construction of both these systems is similar to the American and Canadian Petroleum Data Systems and the Finnish GEOKU System by G. Gaal and V. Saukonautio (2). Information recorded in the SAGWIG and HYDRO have a factual character. SAGWIG has no utilization, except from time to time, when a geologist asks for information from this data bank. One of the reasons for this situation is the fact that the SAGWIG data base does not have complete information on all the deep boreholes in Poland. The actual content of the SAGWIG data base covers the year 1972. In the geologist's opinion, the SAGWIG data base has much useful data but he prefers the original deep borehole reports or geological documentation. In the HYDRO system, we have another situation. It is very interesting that geologists very often ask about the filled in coding sheets of HYDRO. Actually, in the data base of the HYDRO data bank we have more than 40 000 hydrogeological wells on magnetic tapes and data on 40 000 more wells on coding sheets. The process of data coding in the two regional HYDRO banks is practically finished but in the remaining HYDRO banks, coding completion is expected at the end of 1983. The usefulness of the HYDRO data banks for the hydrogeologist's work has been confirmed and extension of the HYDRO data base, by including data from STATE HYDROGEOLOGICAL OBSERVATIONS POINTS, is planned. Once this is done, the HYDRO data base will provide data for numerical modelling for hydrology.

7. WORLD RESOURCES SYSTEM

In the period of 1978 - 1980, the Geological Institute prepared and tentatively introduced into practice the computer system on storage and retrieval of data on the world deposit of stable mineral products. The WORLD RESOURCES system has a mixed factual-documental character and is designed to collect in both standardized and free forms, geological-economical information from published sources and nonpublished reports, concerning all the world mineral resources, (similar system - see 1).

Data collected in this system are very useful for correlation research on the geology of mineral deposits and for geological and government authorities, who decide upon the export strategy of the Polish geological services.

The data base of the system is divided into five monothematic files.

In the first file, we recorded basic geographic (localization) and essential geologic data. The next two files include information about reserves of the host component, and associate mineral products. Mining data and deposit assessments are collected in the fourth file. All the terms in these files can be used as selectors in the searching process. Updating, additions and accessing data from all the files, excluding the first one, are possible. In the fifth file we put in nonformalized bibliographic and additional information in the thematic order. All the data from this file are retrievable "in extenso". Computer programs of the WORLD RESOURCES system were written in FORTRAN IV. The data base is recorded on discs of PDP 11/45 computer. At present, in the data base of WORLD RESOURCES we have information on 800 African deposits and about 300 on South American deposits. A fragment of the output from the WORLD RESOURCES system is shown in Figure 3.

8. CONCLUSIONS

On the basis of experience gathered in the course of preparing three large factual systems for geology, the following conclusions are proposed by the author and his collaborators from the Computer Science Department of the Geological Institute:

1. The data base of the geological factual ADP systems ought to be divided into monothematic files (subsets) of data recorded in matrix form and constructed with reference to the theory in "An information algebra" by R. Bosak et al., (1962, CACM 5, p. 191-204). Using this theory to construct a computer data base for geological data has solved and simplified the problems of logical and physical structures of information, query language, searching and data processing (operations on matrix), and also programming.

2. Preparation of the geological data base for ADP technology demands coding the large quantity of information. The capacity of work required for data coding very often makes the project of computerization non-realizable.

3. The geologist's habit of using traditional, non-formalized forms of data recording, and the capacity to put all the geological information gathered by them into a computer does not exist.

4. Probably, only a factual-documental computer system tape of a card register, and data banks with analytical data represent the optimal level of computerization needed for geology.

Such a system, called the MINERAL PRODUCT CARD REGISTER, is in preparation at the Geological Institute.

9. REFERENCES

[1] CRIB - Computerized Resources Information

Bank 1974 - Instructions for Reporters. Revision 9. Sept. 1974. Prepared by U.S. Geological Survey, Reston, Virginia.

[2] Gaal, G. and Suokonautio, V. (1973), An automatic data processing system for explorational mapping in Precambrian terrain: GEOKU, Bull. 266 Geological Survey of Finland.

[3] Rajecki, M.K., Tomaszewski, A. and Szczepanski, E. (1973) - Regionalne banki danych hydrogeologicznych. Technika poszukiwan nr 45-46, English summary.

[4] Stenzel, P. (1978) - Komputerowe systemy archiwizacji i udostępniania zrodlowych danych geologicznych w resorcie geologii. Geoinformatyka tom 1, 43-58.

[5] Stenzel, P. and Berestka, A. (1979) - System HYDRO. Regionalne banki danych hydrogeologicznych. t. 1-6. Wyd. Geol. Warszawa.

[6] Stenzel, P. and Rajecki, M.K. (1981) - Organizacja baz danych dla faktograficznych systemow archiwizacji i przetwarzania informacji geologicznych. Organization of data bases for factual systems of geological data storage and processing (English summary). Przeglad Geologiczny, No. 3 p. 129-133.

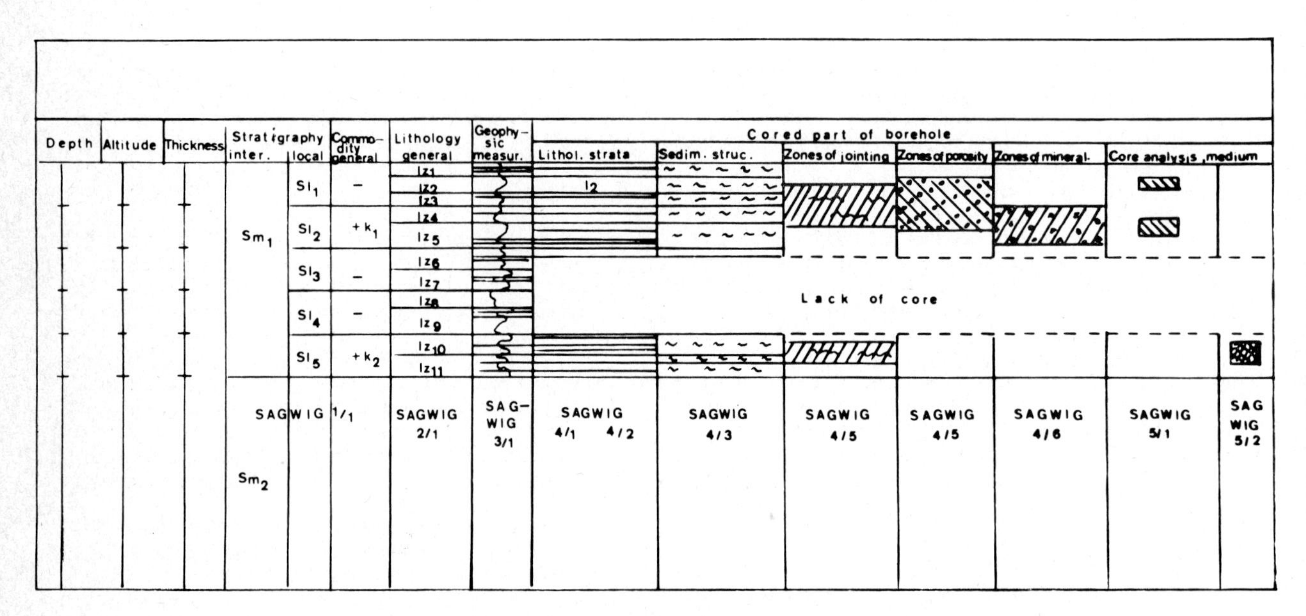

Fig 1. Data structure in borehole
SAGWIG 1/1 – SAGWIG 5/2 – reporting forms and monothematic blocks of data in SAGWIG data base
Sm, Sl – stratigraphic units, L_{1-11} lithological strata

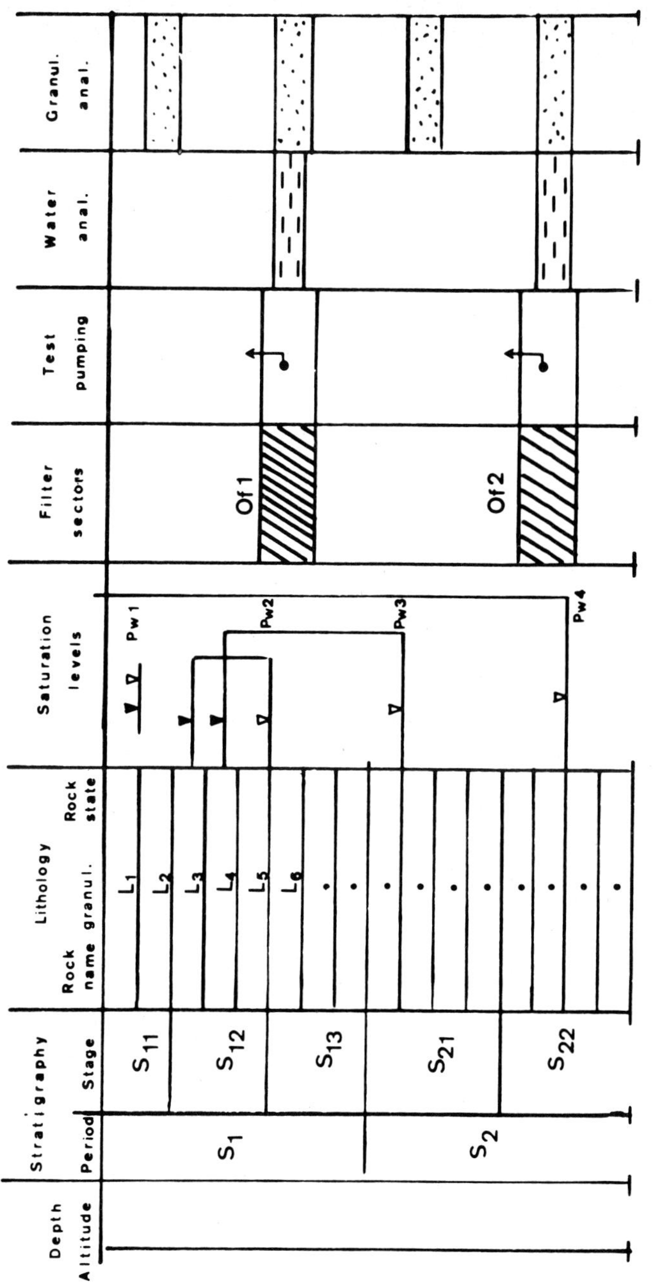

Fig 2. Data structure in hydrogeological well

```
COMPLETE DATA ABOUT DEPOSITS OF BOTSWANA     /FRAGMENT/


DEPOSIT NAME: PIKWE
NAME OF MINERAL PRODUCT: PETLANDITE
DEPOSIT NUMBER IN THE COMPUTER: 1
COUNTRY : BOTSWANA
UNIT NUMBER IN ADMINISTRATIVE DIVISION:  0

ROCK NAME              RELATION TO ORE

AMPHYBOLITE       -      ORE
GNEIS             -      SOURRONDED AMPHYBOLITE

FORMATION:  LIMPOPO BELT
AGE OF FORMATION: ARCHAEAN
DEPOSIT FORM: LINEAR
GEOTECTONIC POSITION: OLD OROGENY
DEPOSIT GENESIS: METAMORFIC DEPOSITS

          R E S E R V E S

STATE AT 1976

TYPE OF RESERVES                       MAGNITUDE OF RESERVES                 AVERAGE CONTENT OF MAIN COMPONENT

PROVED,MEASURED                           22.10  MLN TON                                   14.50  %
OTHERS/PROBABLE,POSSIBLE,SUPPOSED/         9.00  MLN TON                                   11.30  %

          A S S O C I A T E D     M I N E R A L     P R O D U C T S

ASSOCIATED MINERALS          TYPE OF RESERVES          CONTENT OF ASS.MIN.      MAGNITUDE OF RESERVES    RECOVERY

CHALCOPYRITES                  PROVED,MEASURED              11.40 %                22.10  MLN TON          YES
CHALCOPYRITES                  OTHERS                       11.90 %                 9.00  MLN TON          YES

E X P L O I T A T I O N

STATE OF DEPOSIT: IN EXPLOITATION
KIND OF EXPLOITATION: MIXED
ANNUAL PRODUCTION: 2.2 MLN SHORT TON/YEAR

* LOCALIZATION AND GEOGRAPHICAL DATA

ABOUT 50 KM FROM FRANCISTOWN.RAILWAY LINE ABOUT 40 KM FAR FROM.
PROJECT OF CONECTION WITH PIKWE,DITTO ROAD AND ENERGETIC LINE.
SEMIARID AREA WITH AN INSIGNIFICANT POPULATION.

* TECTONIC AND DEPOSIT PARAMETERS

STRONGLY FOLDED REGION PRE-CAMBRIAN TECTONICS,DEPOSIT SITUATED ON INVERTED ANTICLINAL LIMB
STRUCTURE DIPPING TO SW.

* MINERAL CHARACTERISTIC

MASSIVE AND DISSPERSED SULPHIDES.

* MINING

MINING SINCE 1974, OWNER - BAMANGWATO CONCESSIONS LTD,PARTNERS - GOVERNMENT OF BOTSWANA 15%,
OTHERS BOTSWANA RST LTD.
DEPOSIT DEVELOPMENT COST 268 MLN $.

* SURVEY METHODS

FROM 1959 TO 1974 SURVEY ON 26000 SQUARE MILES,SOIL GEOCHEMISTRY,NEXT DRILLING IN ANOMALLY AREA.

* UTILIZATION

AT THE SITE ENRICHMENT AND METALLURGICAL PLANT - ALL PRODUCTS - 46 THOUS.TON/YEAR
OF SULPHIDE ALLOY IS EXPORTED TO USA.

* REFERENCES

BALDOCK,J.W. AND OTHERS 1976,ECON.GEOL.V.71
WORLD MINING.4,1976.
```

Fig. 3

Data for Science and Technology, P.S. Glaeser (ed.)
North-Holland Publishing Company

DATA MANAGEMENT AND MANIPULATION IN MINERAL EXPLOITATION

Michael J. Shulman

Institut für Geologie, Freie Universität Berlin
Berlin, Federal Republic of Germany

A data bank for the processing of geological, mineralogical, geophysical and mining information was established as part of a special research project as an aid in mineral exploration for phosphates. In addition, problems concerning the genesis of phosphates and benefication should be able to be dealt with using the data bank. The data bank not only encompasses information gathered by the institutes involved, but also data furnished by exploration companies, geologic surveys, etc. as well as data previously published in various journals. Using the SIR system, a hierarchical system was deemed superior to a non-hierarchical one and consequently defined. The advantages and disadvantages of the system are shown along with its use in exploration for phosphates.

1. INTRODUCTION

In June 1981, a project, Geologic Problems of Arid Areas, financed by the German Research Association, was initiated in Berlin. As part of this project a data bank was established at the Free University of Berlin to store data concerning the Egyptian and related phosphate deposits. The data bank was not conceived of as a permanent repository, but to serve the individuals involved in the phosphate research during the life of the project. As such, the data bank is problem oriented. The diverse nature of the research interests, varying from mineralogy to genetical modeling of phosphate deposition to developing exploration strategies for phosphates, coupled with the geological complexities of phosphates and with information coming from various external sources, i.e. geological surveys, literature, mining companies, etc., resulted in different attempts being made to develop a data bank that not only contained all the necessary data, but also enabled data to be retrieved without utilizing long or complicated retrieval programs.

Phosphates are of both industrial and scientific interest. The large tonnage of phosphate rock being mined today is primarily used for fertilizer to replenish the phosphorous cycle that has been disrupted by land cultivation. Among the myriad other uses of phosphate or elemental phosphorous are:

1. detergents
2. insecticides
3. oil refining
4. photography
5. water softening
6. steel
7. beverages
8. plasticizers
9. medicines
10. explosives, etc.

The uses of phosphates have been discussed by Jacobs (1). In the United States, approximately 22% of the phosphate rock goes into industrial chemicals with the remainder being used in fertilizers; worldwide, the percentage of rock that is used in fertilizers is higher than in the United States.

In addition to the phosphate content, by-products of the phosphate industry which include fluorine, uranium, vanadium, selenium, etc. can be considered as potential resources due to the large tonnage of phosphate being mined.

Although three types of economic phosphate deposits occur - pegmatite, guano, and phosphorite - phosphate deposits have become synonymous with phosphorite which accounts for approximately 80% of the world production. The world production and reserves of phosphates are shown in Tables 1 and 2. Successful exploration for phosphorite in Turkey and Australia has been based on phosphogenic provinces which are defined by paleography and by oceanic paleocurrents (2) (3). In both cases, economic phosphorite deposits were not known to occur. Training areas, i.e. a well explored area that serves as a model for exploration in an area where data is either lacking or sparse, such as used in oil exploration, are unnecessary due to current knowledge on the origin of phosphorites and the recent success of phosphate exploration. They can, however, be used to isolate critical geological factors associated with economic phosphate deposits that can be applied in formulating exploration stategies. Furthermore, in many cases phosphate beds can be mapped over large distances, Western U.S. and Egypt. Whether searching for unknown phosphate deposits or trying to determine which phosphate outcrops can be developed into an economic deposit, the problem is less where to search than how to search. Exploration strategies that allocate exploration effort optimally must be developed using operations research methods.

Exploration strategies consist of two basic elements: 1) the target distribution and 2) the detection function. The target distribution gives

Land	$x10^3$ T	%
U.S.A.	54406	40.0
U.S.S.R.	25435	18.7
MOROCCO	18770	13.8
P.R. CHINA	6393	4.7
TUNISIA	4761	3.5
JORDAN	3944	2.9
SOUTH AFRICA	3264	2.4
TOGO	2992	2.2
BRAZIL	2856	2.1
ISRAEL	2584	1.9
OTHERS	10609	7.8
	136014	100.0

Table 1: 1980 World Phosphate Production

Land	$x10^6$ T	%
MOROCCO	41985	59.2
U.S.A.	8510	12.0
U.S.S.R	8014	11.3
SOUTH AFRICA	1773	2.5
AUSTRALIA	1489	2.1
TUNISIA	993	1.4
JORDAN	993	1.4
BRAZIL	922	1.3
EGYPT	780	1.1
P.R. CHINA	780	1.1
OTHERS	4681	6.6
	70920	100.0

Table 2: Reserve of Phosphate Rock

the probability that the target is located in a particular cell or area, whereas the detection function gives the probability of detecting the target provided it is in the particular cell or area. The probability of discovery (= target distribution x detection function) is maximized using a Lagrange multiplier with the effort (= cost, time, etc.) available as the auxiliary condition. The extensive data that is incorporated into the data bank is used to calculate the optimal dimensions of discrete exploration cells, target distributions, etc.

2. DATA BANK

2.1. Data Bank Management System

Among the different Data Base Management Systems (DBMS) implemented at the Free University Computing Center, the Scientific Information Retrieval (SIR) System was chosen as most suitable. This system is well documented (4) and particularly well suited for statistical calculations, interfacing with the SPSS and BMDP computer systems. Both statistical systems are widespread and well documented. Although hierarchical as well as non-hierarchical data banks can be formulated, SIR is basically a hierarchical system, i.e. the records at one level of the hierarchy can be said to own records at a lower level, thus resulting in a dendritic pattern. Each record can be uniquely identified by specifying a record key consisting of case identifier (id), record type, and one or more sort-ids. A case is simply a series of records which are grouped together and refer to a single subject, e.g. Egyptian or United States phosphates. A record is a logical grouping of data items such as geochemical, geophysical, stratigraphical, etc. and is identified by a record type number or name. Within a record type, a hierarchy is established by defining additional identifiers called sort-ids. When data is added to the data base, it is stored in sorted order, i.e.

- case identifier
- record type number
- first sort-id value
- second sort-id value
- etc.

Due to the hierarchical relationship, it is not necessary to read all the records during retrieval; for example, if a sort-id value is specified, only those records are read having this sort-id value. Thus retrieval time and cost are held to a minimum. In addition, retrieval programs are relatively simple. Using nested structures the sort-id enables variables from different record types to be retrieved simultaneously.

The disadvantage of the hierarchical system is that once defined it is extremely difficult or time consuming to change. Therefore, the schema definition is of utmost importance and should be approached by taking into consideration not only the hierarchical structure of the data - usually several hierarchies can be envisioned -, but also the complexity of the retrieval programs which must be written to encompass all possible situations. Thus, a simple hierarchy can lead to complicated and time consuming retrieval programs and simple retrieval programs to complicated hierarchies. In addition, it should be mentioned that to take full advantage of the hierarchical structure, extensive bookkeeping is necessary to keep track of the case- and sort-ids. If case- and sort-ids are given without any forethought, much of the advantages gained by the hierarchical structure is lost. An example of the difficulties that can arise was encountered in trying to retrieve data from the Western States phosphate field. Even though locations were numbered according to state, i.e. Montana, Wyoming, etc., and area within the states, retrieving data from selected areas that encompass different numbering blocks cause problems necessitating additional select and reject commands within the retrieval program. Such problems cannot be eliminated, but through careful planning and foresight can be kept to a minimum.

2.2. Phosphate Data Bank

One difficulty encountered in designing a data bank for phosphates was the extreme lateral variation that phosphate beds can exhibit. They not only have a tendency to thin, thicken, or pinch out over a relatively short distance, but also to change facies or facies characteristics. In addition, phosphate-bearing formations contain various ore-bearing horizons which must be correlated at different locations. For example, there are generally 3 to 5 phosphate beds of Maestrichtian age in Syria which increases to 13 in the Khneizir area; the thickness of the phosphatic beds varies from 0.2 to 1.5 m, but may reach 2.5 to 3 m (5). Although particular beds can be followed over large distances, correlation is best accomplished by comparing the vertical sequence of beds (6). To this end, qualitative or descriptive data must be quantified. Chemical data is unreliable in correlating beds since it also exhibits large lateral fluctuations.

A further difficulty in designing the data bank was the data available. Whereas the structure of the data bank could easily be tailored to the information gathered by members of the project, additional valuable and extensive information was received from geologic surveys, mining companies, and from articles, dissertations, and reports. The data deemed necessary or sufficient varies in the reports depending on ther purpose of investigation, methods available, geological aspects of the deposit, etc. Thus, for example, mineralogical and geochemical reports contain not only major element analysis of phosphate beds, but also detailed trace element analysis. Mining reports, on the other hand, generally neglect trace elements or limit themselves to one or more trace elements that can possibly be mined as an accessory constituent; major element analysis is restricted to P205 content and three or four other values. Furthermore, the elements analyzed vary, to a large extent, from deposit to deposit, i.e. one being rich in rare earth elements and another in uranium. The problem is, therefore, how to integrate all the data into the data bank while at the same time maintaining a relatively simple data scheme which does not enlarge the data base with records that are sparsely occupied. Other difficulties such as the terminology used in different reports or sometimes throughout the same report not being consistent - especially in stratigraphic descriptions could be alleviated by converting all deviating terms to current usage.

In designing the hierarchy, a major problem was bookkeeping. In addition to assigning values to variables such as locations or rock names which are relatively straightforward, the quantification of other qualitative or descriptive data was more complicated. It was attempted to keep the classification as simple as possible. Two alternatives were formulated. In one, a simple numerical system was used - with values ranging from 1 to 999 - in conjunction with a detailed list of values and definitions. The second encompassed utilizing a large field in which certain fields are delegated certain functions. For example, the first two fields correspond to a general description and the following fields to attributes according to importance. This simplifies the bookkeeping aspect, however it does not furnish such a detailed description as the first since the number of fields is limited, i.e. differences between descriptions may not come to the fore. This robs the data bank of valuable subjective geological information that may be decisive in formulating exploration strategies. This disadvantage is more than offset by the ease in which individual attributes can be mapped, so that the attribute system was implemented.

The hierarchy that has proven to be most flexible is shown in Figure 1.

The data bank consists of 20 record types of which only 11 have thus far been defined due to lack of data. The choice of the sort-ids was based on both geological and retrieval considerations. Although the significance of the majority of the variables and record types are self evident, several need to be clarified. The case-id, Location, refers to a broad geographical classification, e.g. Egypt, Western United States, etc., whereas Samloc refers to particular sites, e.g. outcrops, wells, mines, etc., within the general area. The sort-id Samloc is, especially for geological data, very important; it enables simple retrieval procedures to be written that, for example, select particular beds or sequences of beds for use in correlation studies, etc. from one or more record types. The hierarchy is site oriented. The variable Cor, correlation number, can only be assigned after correlation studies have been carried out, i.e. all correlated beds receive the same correlation number. Its sole purpose is to be able to retrieve correlated beds for later studies.

In addition to the choice of the sort-ids, the order of the sort-ids is important in simplifying retrieval programs. A logical order has been followed, however, with the sort-id Cor being in the third sort-id position. Otherwise, in order to retrieve correlated beds, all previous sort-ids, i.e. their values, must be given during the retrievals.

The differences in chemical analyses mentioned previously were dealt with by defining two record types each for major and minor elements. The first major element record type encompasses the standard major elements found in the majority of reports. The second major element record type was defined to handle the analysis found in mining reports, etc. where only a limited number of major elements, with possibly one or two minor elements included, was determined. In the second record type the following variables were defined: P_2O_5, Al_2O_3, Fe_2O_3, Acid Insoluble, and Loss on Ignition. In addition, three other variables were defined as code + value. Using this scheme, it was found that relatively few records were sparsely occupied. For example, the reports furnished by the Egyptian Geologic Survey on the drilling results from the Abu Tartur Phosphate Deposit included values for SiO_2, CaO and SO_3, while the extensive investigations of the United States Geological Survey on the Western Phosphate Field as documented in numerous publications contained analyses for C organic, equivalent U and U.

Whereas the inclusion of the major element analyses was rather unproblematic, the integration of the minor elements proved much more difficult. The variation not only in the number of trace elements analyzed, but also their variety is much greater than for major elements. Nevertheless, analogous to the major elements two basic types of chemical analysis were carried out. In the first type, all trace elements were determined in the samples. This is further complicated by a number of trace elements that were sought after, but not detected in any sample. The second type of analyses was carried out for selected trace elements. Within this second group, no common denominator could be found. The first minor elements record type was defined for the 25 most common trace elements found in phosphorites and related rocks. A code + value system was used for the second record element type. The code used is the atomic number of the element. Trace elements that were detected, however not listed on the first record type, were inputted on the second. Since the bed and sample number serve as sort-ids, procedures to retrieve this information were neither complicated nor time consuming. It should be noted that the first record type is, in general, not completely dull due to many trace elements being detected in certain samples and not in others. This data must be included, however, for the various geochemical and geostatistical studies that have to be carried out on data. Utilizing a code system for all trace elements would enlarge storage space and encumber retrieval.

The effectiveness of the data base hierarchy can be judged basically by considering the cost of retrieval, the capability to retrieve all necessary data required by the user, and the complexity of the retrieval programs that must be written. Except for costs, none of the factors can be expressed quantitatively; an estimate of the effectiveness can only be made either by comparing retrieval aspects of different hierarchies formulated for the same data or problem or by comparing its effectiveness with other hierarchical data bases. In both cases, effectiveness can only be judged by constant use and by experience with various data bases.

3. EXAMPLES

In order to demonstrate how the data bank is currently being used to help develop mineral exploration strategies and to illustrate some of the features described, several typical examples along with their retrieval procedures are given. The exploration strategies that are being developed are based on the target probability distribution. Determining this distribution consists of three steps: 1) Isolation of critical geological factors 2) Estimation of each factor's chance to occur 3) Multiplication of the percent chance of all the factors together. The data bank is used primarily for the first step.

3.1 Factor Analysis

A factor analysis of selected trace elements and mapping of the factor scores should be carried out on the phosphate-bearing horizons for the Western United States Phosphate Field. The factor analysis is calculated by the SPSS Computer System and the mapping of the factor scores by an isoline program. The data bank retrieval is:

```
RETRIEVAL
CASE IS              2014
PROCESS REC          1
. MOVE VARS          Samloc
. PROCESS REC        2, from (Samloc, 101, 1,
                     281, 102, 30010000) thru
                     (Samloc, 101, 1, 281, 102,
                     30019999)
. MOVE VARS          Bed
. PROCESS REC        5, with (1, bed)
. MOVE VARS          Ag to Zn
. PERFORM PROCS
. END PROCESS REC
. END PROCESS REC
END PROCESS REC
SPSS SAVE FILE       Filename=fact/
                     Variable=....
END RETRIEVAL
```

The retrieval shows how sort-ids are used and how different record types can be assessed by using a nested structure. It should be noted that the SPSS variable list may include data from any of the different record types if the variaables are converted into summary variables by commands such as compute, move vars, etc. In the above example, the sort-id rock varies from 30010000 to 30019999 which corresponds to phosphoritic rocks, while the others are held constant with the exception of Samloc. Had correlation studies been carried out, the hierarchical retrieval could be replaced by a simple non-hierarchical retrieval, i.e.

```
RETRIEVAL
CASE IS              2014
PROCESS REC          5, with (1)
. MOVE VARS          Ag to Zn
. PERFORM PROCS
END PROCESS REC
SPSS SAVE FILE       Filename= fact/
                     Variable= ....
END RETRIEVAL
```

3.2. Discriminant Function Analysis

A discriminant function analysis should be carried out on a set of new variables. The variables must be defined during the retrieval and encompass data from various record types. It is assumed that all beds have been correlated. Only those beds with a P_2O_5 weight percent greater than 28 are to be assessed.

```
RETRIEVAL
CASE IS              2014
PROCESS REC          3, with (3)
```

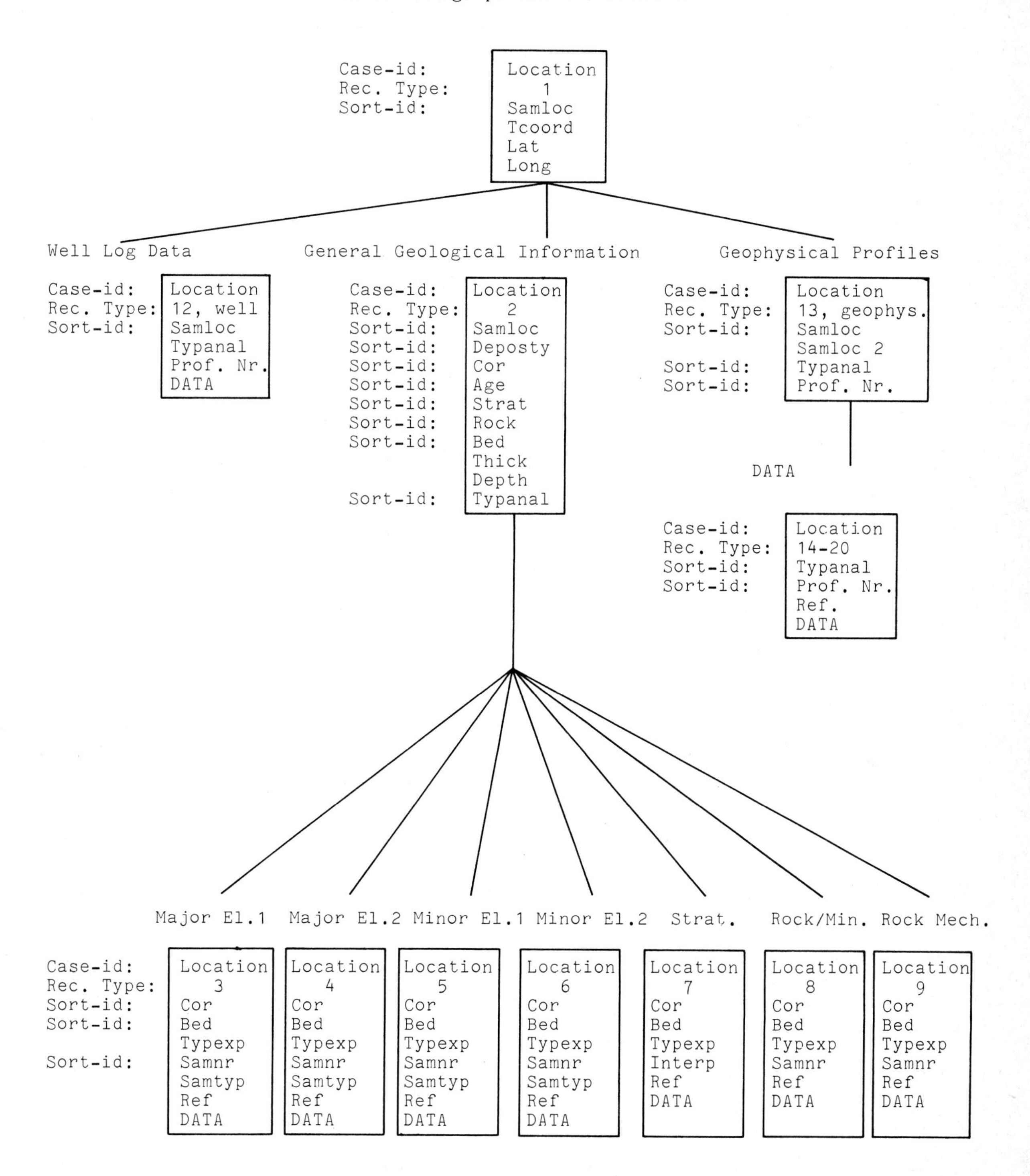

Figure 1: Hierarchical structure of Phosphate DATA Bank

```
. IFNOT            (P2O5 GE 28) Next
                  Record
. MOVE VARS       Bed,...
. PROCESS REC     2, from (1, 101, 3, 281, 102,
                  1, Bed)
                  thru (9999, 101, 3, 281, 102,
                  99999999, Bed)
. MOVE VARS       ....
. END PROCESS REC
. PROCESS REC     5, with (3, Bed)
. MOVE VARS       ....
. END PROCESS REC
. PROCESS REC     7, with (3, Bed)
. MOVE VARS       ....
. END PROCESS REC
. PROCESS REC     8, with (3, Bed)
. MOVE VARS       .....
. END PROCESS REC
. COMPUTE         X1=....
. COMPUTE         X2=....
  .
  .
  .
. PERFORM PROCS
END PROCESS REC
SPSS SAVE FILE    Filename=disc/
                  Variable=X1, X2 ,..
END RETRIEVAL
```

An alternative to the above retrieval could be formulated with the record types being nested within record type 2. In this case, only three sort-ids are necessary, however, the number of records that are assessed is greater. The beginning of such a retrieval is:

```
RETRIEVAL
CASE IS           2014
PROCESS REC       2, from (1, 101, 3)
                  thru (9999, 101, 3)
. MOVE VARS       Bed,....
. PROCESS REC     3, with (3, Bed)
. IFNOT              (P2O5 GE 28)
                  Jump out
  .
  .
  .
OUT:
  END PROCESS REC
  .
  .
```

Both retrievals are simple to write and enable the user to manipulate data from all records.

3.3. Regional Reserve Calculations

The phosphate and trace element reserves should be estimated for selected areas in the Western United States Phosphate Field for the two phosphate bearing members of the Phospharia Formation. Within the two members, the Meade Peak and the Retort, only possible ore beds should be considered, e.g. a P_2O_5 greater than 24%. The areas chosen cover in some cases different counties within a state or areas bordering two or more states. Retrieving data from the Meade Peak Member for the area bordering Idaho, Montana, Wyoming, and Utah can be accomplished by the following procedure:

```
RETRIEVAL
CASE IS               2014
PROCESS REC           1
. IFNOT               (Lat GT ....OR LT...)
                      and (Long GT...OR LT...)
                      Next Record
. MOVE VARS           Samloc, Lat, Long
. PROCESS REC         2, with (Samloc, 101, 1,
                      281, 102)
. MOVE VARS           Bed,...
. PROCESS REC         3, with (1, Bed)
. IFNOT                 (P2O5 GE 24)
                      Jump out
. MOVE VARS           ....
. END PROCESS REC
. PROCESS REC         5, with (1, Bed)
. MOVE VARS           ....
. WRITE               ....
. END PROCESS REC
OUT:
   END PROCESS REC
. END PROCESS REC
END PROCESS REC
END RETRIEVAL
```

Substitution of sort-id 104 for 102 in the above furnishes data on the Retort Member. The data is then used as input to calculate the reserves using geostatistical methods. The results of these calculations are applied in the exploration strategies to help define cell dimensions of areas that are to be investigated.

4. CONCLUSION

Designing a hierarchical data bank is best based on both geological and retrieval considerations. Although each data bank that is used for mineral exploration is problem oriented, some possible pitfalls that are common to all hierarchical data banks can be delineated:
1. Assigning values to variables (=bookkeeping),
2. Order of sort-ids
3. Minimizing sparsely occupied records, etc.
Additional problems such as stratigraphy (=correlation), etc. that reflect the geological complexity of the mineral deposit type or types must also be considered. The hierarchical data bank structure formulated for phosphate exploration and research has proven to be flexible and can be readily adapted to other types of mineral deposits.

5. REFERENCES

[1] Jacob, K.D., Uses of Phosphates, Min. and Met. 25 (1944) 488-491.

[2] Sheldon, R.P., Exploration for Phosphorite in Turkey - A Case History, Econ. Geol. 59 (1964) 1159-1175.

[3] Howard, P.F., Exploration for Phosphorite in Australia - A Case History, Econ. Geol. 67 (1972) 1180-1192.

[4] Robinson, B.N. et al., SIR, Scientific Information Retrieval User's Manual Version 2 (Sir Inc., Evanston, IL., U.S.A. 1980).

[5] Omara, S., Phosphatic Deposits in Syria and Safaga District, Egypt, Econ. Geol. 60 (1965) 214-217.

[6] Sheldon, R.P., Physical Stratigraphy and Mineral Resources of Permian Rocks in Western Wyoming, U.S.G.S. Prof. Paper 313-B (1963) 49-273.

Data for Science and Technology, P.S. Glaeser (ed.)
North-Holland Publishing Company

A METHOD FOR SHADOW REMOVAL FROM MULTISPECTRAL SCANNER DATA AND ITS USE IN MINERAL EXPLORATION

R. Sinding-Larsen

Department of Geology, The Norwegian Institute of Technology
7034 Trondheim-NTH, Norway

A multidimensional scaling method is used on airborne MSS data to compensate for a proportional variation in spectral reflectivity between thematically similar pixels. This method allows one to correct the raw MSS data for shadow and other proportional effects, thereby permitting more powerful use of the MSS data in mineral exploration.

INTRODUCTION

In travelling from its initial source to the remote sensor, the electromagnetic radiation in the visible and near infrared undergoes absorption, reradiation, reflection, scattering, polarization and spectrum redistribution. The interpretation of the relationship between multispectral scanner (MSS) data and geological features of interest for mineral exploration is greatly complicated both by the number and the magnitude of the factors which affect the recorded signal. Spatial variability in the intensity of spectral reflection from pixels that are thematically similar can be caused by shadow effects or effects due to varying topographic inclination. The removal of these effects is consequently a useful preprocessing step in the interpretation of multispectral scanner data for mineral exploration or thematic mapping in general.

MULTIDIMENSIONAL SCALING OF REMOTELY SENSED MSS DATA

Let us consider the $\underline{n}$ pixels within our remote sensing study area as a "cloud" of $\underline{n}$ points in $\underline{p}$-dimensional vector space where $\underline{p}$ is the number of spectral bands measured within each pixel. Similarity in spectral intensities between pixels in this p-dimensional vector space will be represented by the interdistance between pixel points.

Two pixels that are thematically similar and with shadow falling on one of them will show a proportional relation between the intensities measured in each of the spectral bands. We will therefore try to define a multidimensional scaling procedure which adjusts pixels to equality if they have proportional spectral intensities. Such a vector space will automatically compensate for any proportional reduction in the reflected intensities e.g. due to a shadow effect.

A multidimensional scaling procedure with these properties (1) is equivalent to correspondence analysis (2). This procedure can be described as follows with reference to the MSS data matrix X (nxp), where the n rows represent $\underline{n}$ $\underline{p}$ dimensional pixel vectors. If the elements of this data matrix are divided by the sum of all the elements in the matrix, we obtain a matrix B with element:

$$b_{ij} = x_{ij} \Big/ \sum_{i=1}^{n} \sum_{j=1}^{p} x_{ij}$$

If each element of B is scaled according to

$$b_{ij}/b_{i.}\sqrt{b_{.j}}$$

resulting in a streching of the spectral band axes by a factor of $1/\sqrt{b_{.j}}$

then the distance between pixels q and r in this vector space will be given by:

$$d^2_{qr} = \sum_{j=1}^{p} \left(\frac{b_{qj}}{b_{q.}\sqrt{b_{.j}}} - \frac{b_{rj}}{b_{r.}\sqrt{b_{.j}}}\right)^2$$

- where $b_{r.}$ is the row sum or albedo of the ith pixel relative to the total albedo from the n pixels of the study and $b_{.j}$ is the proportion of the total albedo coming from the reflected intensity in the jth spectral band.

Let us verify the algorithm by computing the distance between pixel q and pixel r where the spectral intensities in pixel q, are proportional by a factor of k to those in pixel r, i.e.
$x_{rj} = kx_{qj}$.
In matrix B the following relations are found:

$$b_{rj} = k\, b_{qj} \qquad b_{r.} = k\, b_{q.}$$

$$d^2_{qr} = \sum_{j=1}^{p} \left(\frac{b_{qj}}{b_{q.}\sqrt{b_{.j}}} - \frac{b_{rj}}{b_{r.}\sqrt{b_{.j}}}\right)^2$$

$$= \sum_{j=1}^{p} \left(\frac{b_{qj}}{b_{q.}\sqrt{b_{.j}}} - \frac{kb_{qj}}{kb_{q}\sqrt{b_{.j}}}\right)^2 \equiv 0$$

Thus if pixel points in this space are close to each other it will signify that the "spectral profiles" of the pixels have similar shapes.

The interdistances between pixels will remain unchanged when pixels or spectral bands are referred to new coordinate axes defined by principal components.

The next step in the processing is the computation of a cross-products matrix from which principal components are extracted. The pixel points are then projected onto these axes. Pixels with proportional spectral intensities having zero interdistances are equal in this scaled multidimensional space and can be projecting them onto the first principal component.

In figure 1 the albedo from raw MSS data is shown within a 3 km^2 area in Bamble, south of Norway. We can observe shadows from trees falling onto the highway (E18) near the upper left border. In Figure 2 the same area is shown with MSS data processed according to the multidimensional scaling procedure described. This picture shows clearly that reduced spectral reflection due to shadow affecting the highway is compensated for and it can further be seen that shadows on the flanks of small hills in Figure 1 have equally been removed in Figure 2.

Fig.1 MSS albedo.

Fig.2 1st principle component.

APPLICATION TO MINERAL EXPLORATION

The multispectral 10 band (Daedalus) air-borne survey over Bamble in south of Norway was conducted to test the efficiency of the MSS method for identifying limonitic alteration related to mineralization.

A study area was chosen near an old nickel mine where limonitic alteration can be observed. In order to compensate for proportional effects, the spectral axes were scaled as described. A principal component analysis was then performed on a model area selected so that half of the pixels would fall within the zone of limonitic alteration. The first principal component will in this case be computed so that there is maximum separation between the projections of limonitic and non-limonitic pixels.

The effect of this linear classifier is shown in Figure 2 where limonitic alteration shows up in light near the lower right corner. By field checking, the efficiency of this method, in separating limonitically altered from unaltered rock exposures, was verified. This procedure (3) allowed the recognition of limonite altered pixels (2.5 x 5 m) within a densely forested area and thereby helps speed up field work.

CONCLUSIONS

The use of a multidimensional scaling procedure which stretches differentially the spectral reflectivity vectors by the reciprocal of the square root of their pixel sums and stretch, differentially, the pixel vectors by the reciprocal of the square root of their albedo permits the compensation of effects causing proportional spectral intensities between pixels. By processing the data in this way and performing principal component analysis, an efficient linear classifier was obtained separating pixels containing limonitically altered from unaltered rock exposures.

ACKNOWLEDGEMENTS

I hereby acknowledge NTNF the Royal Norwegian Council for Scientific and Industrial Research for supporting the collection of the remotely sensed MSS data. I also thank director J.M. Monget of the Centre de Télédétection et d'Analyse des Milieux Naturels of the Paris School of Mines, for precious guidance during the analysis of the data at his centre.

BIBLIOGRAPHY

[1] Hill, M., Correspondence analysis: a neglected multivariate method, J.R. Statist. Soc., Ser. C, 23 (1974) 340-354.

[2] Benzécri, J.P., L'Analyse des Données 2, l'analyse des correspondances (Dunod, Paris, 1973)

[3] Sinding-Larsen, R., Use of MSS Data in Mineral Exploration. Unpublished report. (1981).

Data for Science and Technology, P.S. Glaeser (ed.)
North-Holland Publishing Company

THE PROBLEMS OF THERMODYNAMICS IN GEOCHEMISTRY AND COSMOCHEMISTRY

I.L. Khodakovsky

Vernadsky Institute of Geochemistry and Analytical Chemistry
U.S.S.R. Academy of Sciences, Moscow, U.S.S.R.

The physico-chemical modelling of a particular natural process is considered as an only way of quantitative checking of conclusions being a final result of geological observations. A number of examples of thermodynamic treatment of natural processes are presented. The tremendous body of current experimental values of inorganic compounds including minerals as well as the constant income of new data stimulate the foundation of the consistent data bank capable of transformation with the input of new data. Different approaches to this problem are discussed. The necessity of the interconnection of the automatized system of consistency and the programming set of the equilibrium calculation of natural systems is argued.

1. INTRODUCTION

At present geology is widely using the knowledge obtained in physics and chemistry which reflects the scholars' endeavors on a qualitative level to resolve the problems of the origin of geological objects - mountain rocks, minerals, natural water, ore deposits, etc. The research in this field has been mostly developed in geochemistry and cosmochemistry.

In the last two decades alongside with the detailing of the statistical picture of the modern distribution of elements and their isotopes in terrestrial and non-terrestrial objects considerable attention has been paid to the creation of physicochemical models of various natural processes, which are based on chemical thermodynamics data.

The present work will deal with only a few of the most important problems of thermodynamics in geochemistry and cosmochemistry.

2. PHYSICO-CHEMICAL SIMULATION - A QUANTITATIVE BASE OF THEORETICAL ANALYSIS OF NATURAL PROCESSES

2.1 Review of the problem

It is generally accepted that experiments in geology are far more difficult than in physics and chemistry because geological objects are characterized by a greater size, geologic time scale largely exceeding the human time scale, a considerable number of state parameters and lineal and non-lineal relations among them. Under these conditions the use of physico-chemical simulation based on the laws and data of thermodynamics of equilibrium and non-equilibrium processes and the use of computers may have a great theoretical and practical importance. Physicochemical simulation makes it feasible, without being confined to complicated and expensive experiments, with a minimum of time to investigate many essential regularities of natural processes. The simulation of various combinations of factors on the computer permits us to consider several ways of processes. This allows us the possibility to analyse systems which are extremely difficult and, in a number of cases, absolutely impossible to reproduce. Thus, for many natural processes physico-chemical simulation is practically the only way of quantitative verification of notions arising from the generalization of geological observations.

An important advantage of chemical thermodynamic methods lies in the fact that they make it possible to represent the most sophisticated natural process as a "totality" of simpler processes down to the level of concrete chemical reaction parameters which are: temperature, pressure and concentration of separate components. Thus, already nowadays this is a way, in principle, to consider fairly complicated geological and cosmochemical processes. Therefore, the problem of a compltete description of complicated natural systems and of revealing the main factors of mineral formation is confined to the consideration of possible interactions between the analysed simple systems which take place during their synthesis.

It is quite natural that the research described is rather long and the solution of many now disputable problems can be achieved only in the far future. However, this direction is notable for a certain strictness and unambiguity of conclusions. If geology were ever to become an exact science, to a considerable extent it would be due to a wide use of chemical thermodynamic methods.

In this context we need to consider on concrete examples what conclusions can be drawn with the help of thermodynamic methods in geochemistry and cosmochemistry.

2.2 Physico-chemical conditions formation of meteoritic minerals

A theory of the solar system origin reveals the

genetic relationships among its components such as Sun, solar nebula, different types of meteorites; asteroids, comets, planets and their satellites.

After Hoyll's and Cameron's hypothesis many scientists suggested that the solar system was originated by the collapse of a large cold cloud of interstellar dust and gas. The collapse was in progress until the solar radius became approximately similar to the present value. At that stage the temperature and pressure of the solar interior were considered to be sufficient for occurrence of spontaneous nuclear reactions. After the Sun was formed, the process of solar nebula cooling started. This process was accompanied by the substantial differentiation of elements. The evidence of differentiation is reflected in the chemical composition of meteorites, comets, asteroids, planets and their satellites.

Urey was the first to note that the considerable elemental fractionation observed in meteorites may be explained by nebular condensation. Larimer, Grosman, Lewis and other scientists carried out the equilibrium thermodynamic calculations of the gas condensation, the former having solar composition. The calculations were applied to the inner zone of the Solar system. The reactions of the primary condensate substitution were also taken into account. This sequence is: 1) condensation of oxygen-bearing compounds of refractory elements such as Ca, Al, Ti, et al; 2) condensation of Fe-Ni alloy; 3) $MgSiO_3$ condensation; 4) formation of alumino-silicates Na and K; 5) condensation of FeS; 6) formation of tremolite; 7) condensation of serpentine; etc.

All described condensates and products resulting from their reaction with nebular gases are minerals of meteorites with the exception of some minerals of enstatite and iron meteorites. As it was shown by J. Larimer and M. Bartholomay (1) and as seen from Fig. 1 borrowed from the work (2), the typomorphic minerals of enstatite meteorites such as osbornite TiN; ololhamite CaS; sinoite Si_2N_2O, et al. cannot be condensed from the nebular gas with the current solar C/O ratio to 0.55. The condensation of these minerals can occur only providing the C/0 ratio > 0.83 at $P_{tot} = 10^{-4}$ atm.

It is notable that condensation temperatures of oxygen bearing refractory compounds containing Ca, Al, Ti under "oxidizing" conditions (C/O = 0.55) are almost similar to the condensation temperatures of nitrides and carbides of these elements under "reducing" conditions (C/O = 0.83-1.0).

The author of the present study suggested (3) that one of the two mentioned high temperature condensates or refractory elements was never formed in the nebular cooling but it was captured during the formation of the Sun - nebula system. Recent isotopic investigations of meteorites give evidence that the oxygen-bearing refractory minerals in the carbonaceous chondrites are considered

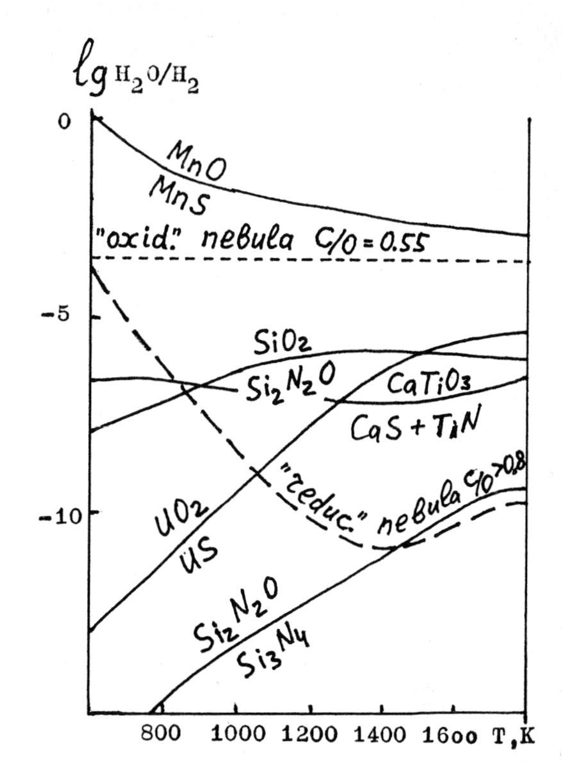

H_2O/H_2 ratio changes:

dashed curve for the C/O ratio > 0.83
dotted curve for the C/O ratio = 0.55

Figure 1: Possible condensates from Solar Nebula gases

as the "alien" material in the Solar system. These minerals are characterized by the anomalous isotopic composition of O, Ti, Mg, Si, etc.

If the hypothesis of "reducing" nebular gases (C/O = 0.83-1.0) should be confirmed in future investigations the observed C/O ratio for the solar gas is to be considered as an evidence of the Solar capture of the nebula in favour of Schmidt's hypothesis or alternatively of the temporal variation of the solar C/O ratio in terms of the simultaneous formation of the Sun and the nebula. In this respect determining the C/O ratio in the atmospheres of giant planets - Jupiter and Saturn - is of great interest since they must have preserved the C/O ratio for nebular gases.

2.3 Thermodynamic prediction of mineral composition of Venus surface rocks

To show another example of the analysis of natural processes by methods of chemical thermodynamics we present here the calculations of the mineral composition of Venus surface rocks.

The assumption of complete chemical equilibrium in the system, Venus troposphere-surface rocks, has been considered as the starting point in all geochemical concepts of Venus atmosphere-surface interaction.

In March 1982 the Soviet space missions "Venus-13" and "Venus-14" managed to determine the bulk chemical composition of surface rocks. However, the mineral composition in Venus surface rocks may be theoretically predicted only by analysing the minerals' stability under the conditions of the planet surface (T = 475 °C, P = 96 atm).

Thus thermodynamic calculations are useful for the estimation of undetected tropospheric gas concentrations as well as for the evaluation of surface rock mineral composition. The latter is of special interest as the return of a Venusian soil sample using a space probe is undoubtedly a complicated engineering problem. Our preliminary predictions of surface rock mineral composition were carried out using the Gibbs free energy minimization method. The equilibrium composition of the 16-component multisystem open with respect to water vapor, carbon dioxide, carbon monoxide, sulfur dioxide, hydrogen chloride, and hydrogen fluoride was calculated (4)(5). Magnetite and tremolite are predicted as main rock-forming minerals provided the sulfur abundance in Venus rocks is not above 2 wt.%. Pyrite is predicted as a dominant mineral in highlands whereas anhydrite is a typical mineral of lowlands (Table 1).

Mineral	h, km		
	0	+5	+10
Forsterite, Mg_2SiO_4	-	0.3	-
Clinoenstatite, $MgSiO_3$	1.3	-	-
Anorthite, $CaAl_2Si_2O_8$	26.3	26.4	26.2
Albite, $NaAlSi_3O_8$	29.9	30.1	29.0
Microcline, $KAlSi_3O_8$	10.4	10.4	10.4
Quartz, SiO_2	0.3	-	-
Tremolite, $Ca_2Mg_5[Si_4O_{11}]_2\cdot(OH)_2$	22.6	24.2	24.9
Marialite, $Na_8[AlSi_3O_8]_6\cdot Cl_2$	-	-	0.8
Sphene, $CaTi[SiO_4]O$	-	0.2	-
Rutile, TiO_2	1.0	0.9	0.95
Tephroite, Mg_2SiO_4	0.3	0.3	0.3
F-apatite, $Ca_5[PO_4]_3F$	0.2	0.2	0.2
Herzinite, $FeAl_2O_4$	-	-	0.1
Pyrite, FeS_2	-	0.2	0.2
Anhydrite, $CaSO_4$	0.7	-	-
Magnetite, Fe_3O_4	6.9	6.8	6.8

Table 1: Mineral composition of Venus surface rocks (T=475 °C, P=96 atm)

Notes

Initial values: $X_{H_2O} = 1.3\cdot 10^{-3}$

$X_{CO} = 2.10^{-5}$

$X_{HF} = 10^{-8}$

$X_{CO_2} = 0.96$

the chemical composition, of basalt in (5); S - contents 0.16 mass %.

The variations of silica, sulfur, and total iron content in the initial matrix rocks (basalts, rhyolites) are found not to effect the stability of the pyrite-magnetite-anhydrite assemblage, assuming the current tropospheric measurments of H_2O, CO_2 and CO.

This buffering system evidently maintains the sulfur and oxygen fugacities of the near-surface Venus troposphere.

2.4 Physico-chemical cassiterite-forming conditions in hydrothermal deposits

Our last example of thermodynamic analysis of natural processes is concerned with revealing the conditions of Sn mineral - cassiterite-forming, the latter being the most important from the economic point of view. Real ore-forming solutions contain many dissolved particles (Ca^{2+}, Mg^{2+}, Fe^{2+}, $Si(OH)^0_4$, HS^-, et al.) which under certain conditions can deposit tin in the form of such minerals as: malayaite - $CaSn(SO_4)O$, shenphlesite $MgSn(OH)_6$, stokesite - $CaSnSi_3O_9\cdot 4H_2O$, stannite - Cu_2FeSnS_4, etc. Though the named minerals are not so widely spread as cassiterite, if their forming conditions are known it is possible to find the range of SnO_2 stability depending on the values of such important physico-chemical parameters of hydrothermal solution as temperature, pH and the activity of Ca^{2+}, Mg^{2+}, Fe^{2+}, $Si(OH)^0_4$, HS^-, etc.

Thus, for example, the substitution reaction of a wide-spread cassiterite-quartz-malayaite association

$$SnO_{2(k)} + SiO_{2(k)} + Ca^{2+}_{(p-p)} + H_2O_{(l)} =$$

$$= CaSn(SiO_4)O_{(k)} + 2H^+_{(p-p)}$$

is dependent on the activity ratio of ions H^+ and Ca^{2+}. As shown in Fig. 2 (6) acid solutions are more favorable for cassiterite-forming in hydrothermal deposits.

The points on Figure 2 describe the solutions of gas-liquid inclusions in quartz from quartz-cassiterite veins, calcium ion activity and pH

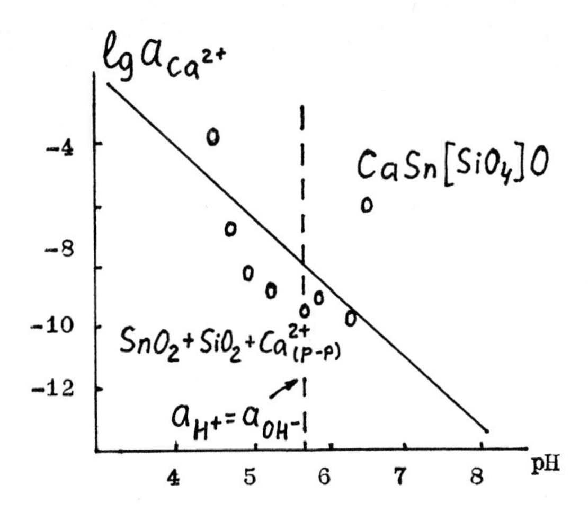

O- compositions of tin-bearing solutions according to the data (7)

Figure 2: Stability fields of minerals in system CaO-SnO_2-SiO_2-H_2O at 300 °C

solution values are calculated taking into account the possibility of complex-formation (7). It can be seen that thermodynamic calculations are consistent with the data obtained from the investigation of mineral-forming solutions.

3. PROBLEMS OF COMPLETION AND RELIABILITY OF THERMODYNAMIC DATA

3.1 Main trends of experimental thermodynamic investigation in geology

The efficiency of use of thermodynamic methods for the solution of problems in geochemistry and cosmochemistry largely depends on the completion and quality of thermodynamic data. Very often it is absolutely impossible to resolve a lot of problems for lack of or minimum reliability of thermodynamic data. Therefore, at present systematic research work is being intensively carried out with a view to determine thermodynamic properties of minerals and aqueous solutions over a wide range of temperatures and pressures. Since it is not our concern to give a comprehensive review of experimental studies aimed at obtaining thermodynamic data for geology we shall only dwell on the basic trends in this field:

- Calorimetric determinations of thermodynamic properties of minerals (measurements of low temperature and high temperature heat capacity, enthalpy of solutions in water, acid solutions and in melts);
- Studies on the solubility of minerals in water and aqueous solutions of salts, acids and bases over a wide range of temperatures and pressures with the aim to determine species existing in aqueous solutions and equilibrium constant of heterogenic reactions;
- Experimental studies of dissociation, hydrolysis and complex formation equilibria in aqueous solutions by potentiometric, conductivity, spectrophotometric and other methods;
- Thermochemical determinations of thermodynamic properties of substances dissolved in water over a wide range of temperatures (measurements of enthalpy of reactions and of heat capacity of solutions);
- Calorimetric and thermodynamic investigation of mixing functions of crystalline solutions and processes of crystalline disorders in minerals;
- Experimental studies on the hydration and carbonization reactions of silicates and other minerals;
- PVT-data determination for minerals, water, gases (CO_2, CH_4, H_2, H_2S, etc.), gas mixes and aqueous solutions;
- Investigation of phase transformations in a wide range of pressures to the level, characteristic for the depth zones of Earth;
- Experimental studies on the isotopic exchange reactions in mineral systems over a wide range of temperatures.

Geology is like no other scientific discipline since it deals with such a large variety of investigated systems and conditions of their existence. Nowadays it seems irrelevant to obtain all thermodynamic data necessary for geology.

Therefore, methods of evaluation of thermodynamic properties of minerals and aqueous solutions which have an accuracy equal to that of an experimental method are now being widely developed.

3.2 Basic requirements for a thermodynamic data bank and their computer based correlation program

An increasing flow of experimental data on thermodynamic properties of inorganic substances (including minerals and aqueous solutions) makes it extremely necessary to create a computer-based thermodynamic data bank able to quickly assume and transform new incoming information. Requirements for thermodynamic data banks have already been discussed in the scientific literature.

Thermodynamic data banks are based on the experimentally measured thermodynamic properties (C_p, etc.) of substances and their changes in reactions. All the experimental data must be analysed for systematic errors.

Those with fragment systematic errors are excluded from the estimation of selected values of thermodynamic properties. In many investigations authors present corrections of measured values which may be due to admixtures or complexity of analysed processes (for example concommitant hydrolysis and ionic association in aqueous solutions, etc.). In this case the correlation process may result in a discrepancy between the values of thermodynamic properties of substances (or their changes in reactions) obtained in investigations and currently accepted "best" values. Therefore, it becomes indispensable to improve the values of corresponding corrections. Thus, the algorithm of the correlation program must provide for some corrections of the primary thermodynamic information.

An ideal correlation should be made for all the thermodynamic properties of inorganic substances using all the literature data available characterizing the state and changes of natural systems depending on temperature, pressure and their composition. Since at present this problem cannot be resolved with the help of a single "super"-program, it is necessary to use a complex-program, consisting of a number of interconnected programs, fundamental service.

Service programs deal with correlation of all the data through one, two or three properties (e.g. C_p°, V, etc.) for a certain substance or reaction.

They include:
- calculations of coefficients of the equation of heat-capacity temperature dependence for gases, solids, liquids and substances dissolved in water;
- calculations of coefficients of equations of state for gases, liquids and solids;
- calculations of partial molal (molar) values ($C^\circ_{p_2}$, etc.) for substances in aqueous (crystalline) solutions;
- calculations of enthalpy of dilution of aqueous solutions;
- calculations of coefficients for equations of equilibrium constant dependence on ionic strength of aqueous solutions, etc;
- calculations of activity solution coefficients, etc.

The values of thermodynamic properties for all investigated substances obtained with the help of preliminary evaluation are then correlated simultaneously for all (or for groups) of substances using the fundamental programs. These programs are aimed at estimating such values of thermodynamic properties which:
- should be based on the overall complex of experimental data and characterized by the least uncertainty;
- should conform to thermodynamic laws, i.e. $\Delta G_T^\circ = \Delta H^\circ - T\Delta S^\circ$;
- should conform to assumptions accepted in calculations of thermodynamic values, concerned with determination of standard state (e.g. additivity of thermodynamic properties of ions in aqueous solutions).

The correlation of basic thermodynamic functions (ΔG_f°, ΔH_f°, S°) is made in such a way that in most cases standard entropies of substances, determined by the measurements of the low temperature heat capacity and having the least uncertainty are not subject to changes in the correlation process. The latter may also concern CODATA Key Values.

The dialogue scientific complex "DIANIK" has been recently developed by A.I. Shapkin, V.A. Dorophееva, A.V. Garanin and the author of the present work in the Institute of Geochemistry and Analytical Chemistry named after V.I. Vernadsky, Academy of Sciences, U.S.S.R. (8). This complex meets all the requirements for a thermodynamic data bank and correlation programs. In the process of its establishment it was found relevant to utilize the experience acquired at the Institute for High Temperatures, Academy of Sciences (U.S.S.R.), National Bureau of Standards (U.S.A.), the Institute of Geochemistry, Siberian Department Academy of Sciences U.S.S.R. (U.S.S.R.) and the Moscow State University, Department of Geology, in the development of updated tables of thermodynamic values and mathematical simulation of chemical systems.

The mathematical potentials of "DIANIK" in solving the task of thermodynamic data correlation are illustrated by calculated results of the correlation of ΔH_f° values for ions in aqueous solutions at 25 °C (Table 2). In our calculations we utilized experimental data on enthalpies of 54 reactions borrowed from the hand-book of the Chemical Thermodynamics Division, NBS, U.S.A. A similar calculation was carried out at our request by Dr. D. Garvin and Dr. R.L. Nuttal according to the NBS programs. An obvious agreement of the values obtained by these two independent methods is clearly seen from Table 2.

Ions	Results of Vernadsky Institute	Results of NBS
	ΔH_f° kcal·mol^{-1}	ΔH_f° kcal·mol^{-1}
OH^-	-54.976±0.011	-54.979±0.014
F^-	-80.137±0.075	-80.140±0.092
Cl^-	-39.917±0.040	-39.913±0.039
Br^-	-28.986±0.032	-28.991±0.04
I^-	-13.528±0.034	-13.528±0.04
Li^-	-66.542±0.07	-66.546±0.071
Na^+	-57.459±0.065	-57.456±0.071
K^+	-60.288±0.063	-60.290±0.071
Rb^+	-60.014±0.062	-60.016±0.071
Cs^+	-61.707±0.066	-61.706±0.071

Table 2: Results of the correlation of $\Delta H^\circ_{f,298.15}$ values for aqueous ions

The essential peculiarity of our computing complex, which differs from the one previously set up, is the use of the a feed-back-principle in the solution of two main tasks of chemical thermodynamics, i.e. the tasks of correlation of thermodynamic data and mathematical simulation of the physical chemistry systems.

Algorithms of mathematical simulation of the chemical systems (worked out by V.A. Doropheeva (9)) permit us to calculate chemical compositions of different physico-chemical systems using the Gibbs free energy minimization method. This part of the structure includes the algorithm for analysis of simulation results which are determined by the errors of initial thermodynamic data (8).

The accuracy of a numerical solution of physico-chemical simulation tasks is primarily determined by the initial data errors. In this context it seems relevant that the selection of reference substances (simple substances, oxides, ions in aqueous solutions, gases, atoms and molecules, etc.) should not be made by computer depending on the primary thermodynamic data and composition of the system. Thus, there exists a direct feed-back between correlation algorithms and the algorithms of equilibrium composition calculations, Table 3 shows the results of calculations of solubility of calcite (1 mol $CaCO_3$ + 1 kg H_2O), using the Gibbs free energy minimization method. In the first case a set of reference substances consists of simple substances, in the second case - they are ions in aqueous solutions. As shown in Table 2, in the second case, errors in calculations of concentrations are on the average 1.5 times less. "DIANIK" automatically chooses an optimal set of reference substances by a special interaction procedure. In fact, the suggested method of thermodynamic data correlation and their calculations based on the equilibrium composition of physico-chemical systems makes it feasible to avoid accumulation of errors in thermodynamic values used in calculations of equilibrium composition of physico-chemical systems. Besides, it provides for the use of Gibbs free energy minimization method for calculations of equilibrium composition of systems in cases where initial thermodynamic data are limited by reaction equilibrium constants (9).

Species	REFERENCES $O_{2(g)}$ $H_{2(g)}$ $C_{(s)}$ $Ca_{(s)}$			SUBSTANCES H^+ OH^- CO_3^{2-} Ca^{2+}		
	ΔG_f^o, kcal/mol	m	δm,%	ΔG_f^o, kcal/mol	m	δm,%
Ca^{2+}	-132.11±0.07	$2.06x10^{-4}$	3.1	0±0	$2.09x10^{-4}$	2.5
H_2CO_3	-148.86±0.12	$1.88x10^{-8}$	4.0	-22.64±0.09	$1.84x10^{-8}$	1.8
HCO_3^-	-140.30±0.04	$1.28x10^{-4}$	3.2	-14.08±0.06	$1.28x10^{-4}$	2.6
CO_3^{2-}	-126.22±0.05	$7.83x10^{-5}$	3.3	0±0	$7.82x10^{-5}$	2.2
OH^-	-37.59±0.06	$1.28x10^{-4}$	2.6	0±0	$1.31x10^{-4}$	1.9
H^+	0±0	$7.94x10^{-11}$	4.4	0±0	$7.98x10^{-11}$	3.8
H_2O (ℓ)	-56.69±0.06	55.51	0.0	-19.07±0.04	55.51	0.0
$CaCO_{3(s)}$	-268.97±0.2	0.9998	0.3	-10.64±0.02	0.9997	0.15

Table 3: Calculation of equilibrium concentrations in the system 1 mol $CaCO_{3(s)}$ - 1 kg H_2O at 298.15

ACKNOWLEDGEMENTS

I especially appreciate the help of Dr. A.I. Shapkin, V.A. Doropheeva, A.V. Garanin for obtaining and discussing the results of the work. I am very grateful to Dr. D. Garvin and Dr. R.L. Nuttal for the calculations so kindly provided at our request and for discussing the problems of thermodynamic data correlation. I also thank E.L. Sergeeva, Y.V. Semyonov, and I.O. Bobkova for their help in the preparation of the paper.

REFERENCES

[1] Larimer, J.W., and Bartolomay, M., The Role of carbon and cosmic gases: some applications to the chemistry and mineralogy of enstatite chondrites, Geochim. Cosmochim. Acta, 43 (1979) 1455-1466.

[2] Doropheeva, V.A., Petaev, M.I., and Khodakovsky, I.L., On the influence of nebular gas chemistry on the condensate compositions, Abstr. of papers submitted to the Thirteenth Lunar and Planetary Science Conference, Houston, March 15-19, Part I (1982) 180-181.

[3] Khodakovsky, I.L., On the carbon to oxygen ration in solar nebular gases, Abstr. of papers submitted to the Thirteenth Lunar and Planetary Science Conference, Houston, March 15-19, Part I (1982) 385-386.

[4] Barsukov, V.L., Volkov, V.P., and Khodakovsky I.L., The mineral composition of Venus surface rocks: A preliminary prediction, Proc. Lunar Planet. Sci. Conf. 11th (1980) 765-773.

[5] Khodakovsky, I.L., Volkov, V.P., Sidorov, Yu.I., and Borisov, M.V., Venus: preliminary prediction of the mineral composition of

surface rocks, Icarus, 39 (1979) 352-363.

[6] Khodakovsky, I.L. Barth, A., Semenov, Yu.V., Klintsova, A.P., Volosov, A.G., Zhdanov, V.M., Turdakin, V.A., and Barsukova, M.L., Thermodynamic properties of malayaite and conditions of its formation in nature, Geokhimiya N 9 (1982) 1298-1306.

[7] Suschevskaya, T.M., and Ryzhenko, B.N., Calculation of the tin-bearing hydrothermal solutions, Geokhimiya N 7 (1977) 1091-1095.

[8] Khodakovsky, I.L., Garanin, A.V., and Shapkin, A.I., Fundamentals of the system of computer-based correlation of thermodynamic data, Thesis of the IX-th All-Union Conference of Calorimetry and Chemical Thermodynamics, Tblisi (1982) 58.

[9] Doropheeva, V.A., and Khodakovsky, I.L., "Minimization Method" calculation of equilibrium composition of multicomponent systems using the equilibrium constants, Geokhimiya N 1 (1981) 129-135.

Data for Science and Technology, P.S. Glaeser (ed.)
North-Holland Publishing Company

JAFOV: DATA BASE ON THE JAPANESE FOSSIL VERTEBRATES

Niichi Nishiwaki, Kaichiro Yamamoto and Tadeo Kamei

Department of Geology and Mineralogy, Faculty of Science, Kyoto University
Kitashirakawa, Sakyo Ward, Kyoto, 606, Japan

A project has started to generate a data base on fossil vertebrates deposited in Japan in the Data Processing Center of Kyoto University, which can be easily called from any place in Japan through the intra-university computer network system. It is composed of two parts. The one is the reference data on specimens such as taxonomic name, age, locality, portion and bibliography which were reported in scientific papers. The other is the numeric data of measurements of specimens which were collected with a digitizer. The former is used for quick retrieval of specific specimens, the latter is used for statistical analysis, and both are connected through a common sample ID number. The system can manage fossil specimens scattered in many museums and institutions, and it helps paleontologists for faster and more detailed research.

1. Introduction

With development of personal computer systems in recent years it has become popular to use them for scientific research not only to analyze data but also to manage information, and there are many examples of computer applications in the geological sciences (1). Many kinds of geologic data bases have been generated in many institutions, which contain reference and/or numerical data, and each of which has its own structure and format. The authors have been continuing efforts to generate geological data bases by using a small data base management system, which was developed mainly for a specific research project (2) (3) (4).

It is now necessary to manage these data bases as a whole because it has become clear that the activities on geological data processing overlap with each other, and because more detailed information is required for the progress of science.

In vertebrate paleontology it is necessary to observe all the specimens, including those which were described in olden times with a modern point of view. Unfortunately, specimens of fossil vertebrates are scattered in many museums and institutions, and many of them have not yet been described. It is possible but not easy to find all the specimens required for the study only by using scientific reports and catalogues. Of course, there is a registration system in several museums which is convenient to search specific specimens, but it generally contains only the information in the museum.

The authors have organized a project to generate a data base system to manage all the specimens of fossil vertebrates deposited in Japan, and many paleontologists have cooperated with us in collecting and updating data for our data base. The data base has been created in the Data Processing Center of Kyoto University by using a general DBMS since it is easily used by any researcher through the intra-university computer network.

The aim of the project is to enter and supply all the information on the specimens, but only a small part of the data have been entered until now since it has just begun.

It is not sufficient for vertebrate paleontology to study only the specimens deposited in Japan. The authors deeply appreciate suggestions from other countries.

2. JAFOV System

2.1. General content

The JAFOV (JApanese FOssil Vertebrates) system manages all the information on the specimens of fossil vertebrates deposited in museums and institutions in Japan. In the system the data on the specimens without scientific description are also included, but in the first step it

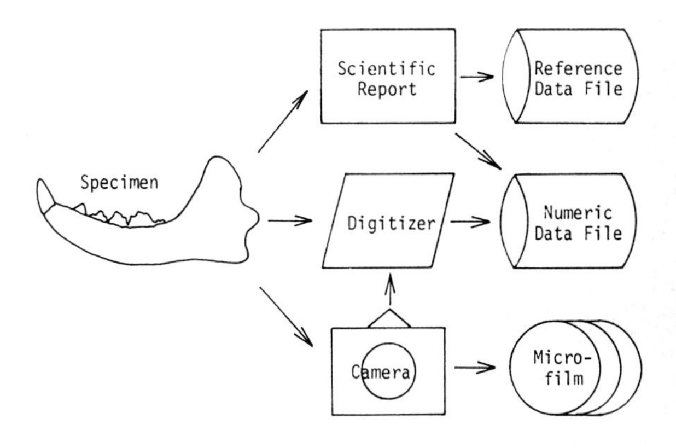

Figure 1. Contents of the JAFOV system

treats only specimens which were reported by paleontologists.

The system is divided into three parts according to the type of data to be managed (Figure 1). One manages reference data on the specimen, such as taxonomic name, locality, age, depository, bibliography, and so on, which are generally described in the scientific report.

The second manages numeric data on the specimen, which are mainly composed of measurements of the specimen, such as height, width, angle, number of items, and so on. Part of the measurement data is described in the scientific report, but the rest must be measured by our groups using a digitizer or some other instruments.

The third manages imagery data on the specimen which are in the form of photos or illustrations. Because of the efficiency of storage space and retrieval, microfilms are selected as storage media for imagery data.

The three parts mentioned above have a common sample identification number (sample ID) for a specimen, and they can be easily connected with each other.

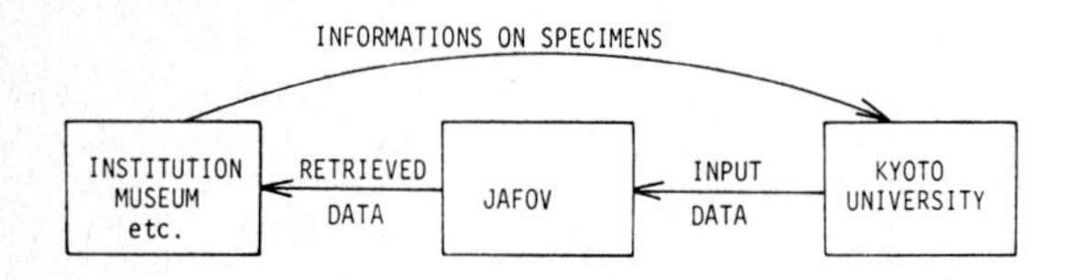

Figure 2. General flow of informations in the JAFOV system

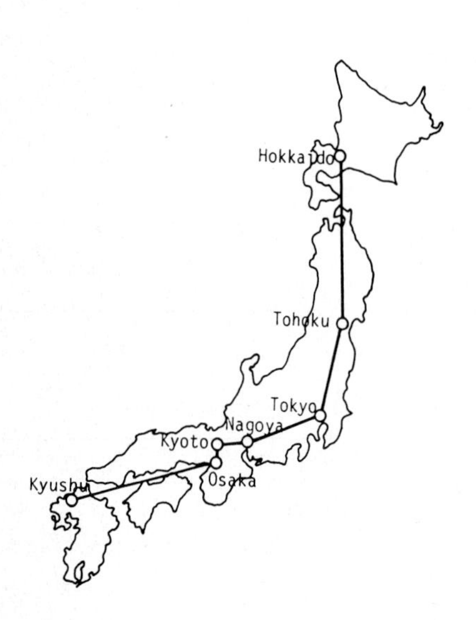

Figure 3. Intra-university computer network in Japan

2.2. Information flow

The JAFOV system was created and is managed by the paleontological group of Kyoto University, and data files are in the Data Processing Center of Kyoto University. But the specimens of fossil vertebrates are scattered in many museums and institutions in Japan, and it is impossible for our group alone to collect all the information. The authors have sent letters to museums and institutions which have specimens of fossil vertebrates requesting their help for our project. The information given by researchers outside of our group is checked and transformed into the standard format, and then entered into the system (Figure 2).

On the other hand, the data in the system can be used by any researcher in Japan. Information is supplied generally through the data base management system, but it is also possible to supply it on magnetic tape or as a printed list.

2.3. Network system

The data files of the JAFOV system in the Data Processing Center of Kyoto University can be easily called from any point in Japan through the intra-university computer network system which connects the seven large computer centers (Figure 3). The researcher who wishes to use the JAFOV system may connect his terminal with the nearest computer by a local call, and then call Kyoto University using the network system. A local call is cheap and the speed of data transformation by the network system is very high. After this procedure he can use the JAFOV system with standard TSS procedures as if he was in our department (Figure 4).

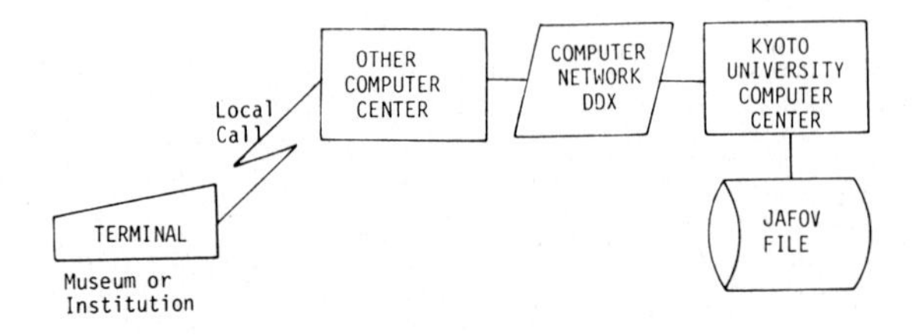

Figure 4. Connection between a researcher and the JAFOV system

3. Reference Data File

3.1. Construction of data

The reference data file has many keys of data on the specimens such as type, taxonomic position, locality, horizon, geologic age, portion, depository and bibliography (Figure 5). Some of them should be written on the JAFOV data sheet by the researcher himself who studied the specimen, though others should be added by our groups. The reason why such procedures are adopted is partly because it is very troublesome for each research-

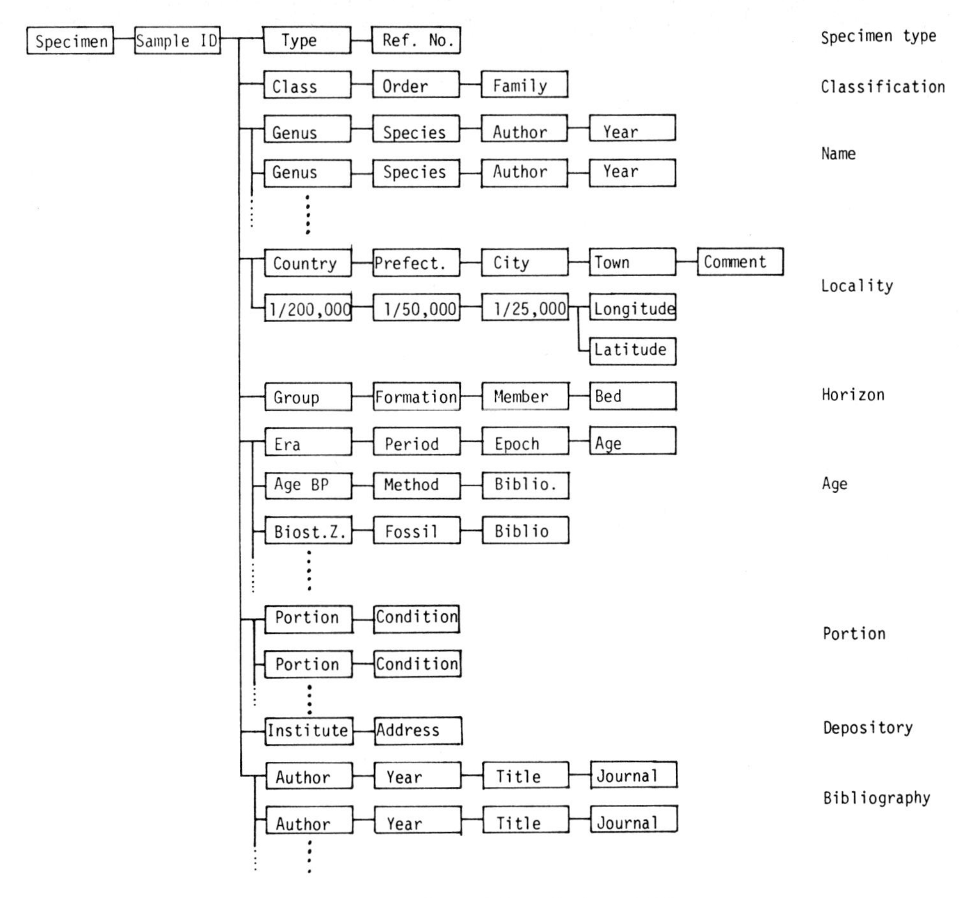

Figure 5. Keys in the reference data file of the JAFOV system

er to fill out all the items, and partly because it is necessary to convert some data by using the code tables.

The FAIRS of the Data Processing Center of Kyoto University is used as the DBMS of the reference data of the JAFOV system (5) (6). The input data for the FAIRS must be specified with the data description data, and the transformation of data into the specified format is performed by using a data transformation system which was developed for this project.

The procedure of data construction for the JAFOV system is summarized in Figure 6.

3.2. Code table

For the efficiency of storage space and to avoid misspelling of data, the following code tables were prepared for the data transformation system which creates the input data for the FAIRS system by modifying original data and adding codes.

a. Code of specimen type
b. Taxonomic code (Table 1)
c. Country and prefecture code
d. Map code of 1/200 000, 1/50 000 and 1/25 000 (Figure7, Table 2)
e. Code of radiometric dating method

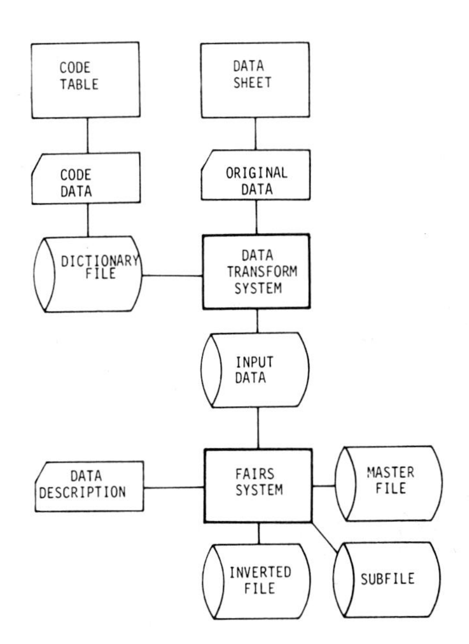

Figure 6. Transformation from original data into the input data for the FAIRS

Table 1. Example of taxonomic code of the JAFOV system

102300 ORDER PROBOSCIDEA		103201 FAMILY DIACODECTIDAE		103613	PHOCAENIDAE
102301 FAMILY MOERITHERIIDAE		103202	LEPTOCHOERIDAE	103614	PHYSETERIDAE
102302	GOMPHOTHERIIDAE	103203	HOMACODONTIDAE	103615	PATRIOCETIDAE
102303	MAMMUTIDAE	103204	DICHOBUNIDAE	103616	CETOTHERIIDAE
102304	STEGODONTIDAE	103205	ACHAENODONTIDAE	103617	ESCHRICHTIIDAE
102305	ELEPHANTIDAE	103206	CHOEROPOTAMIDAE	103618	BALAENOPTERIDAE
102306	DEINOTHERIIDAE	103207	CEBOCHOERIDAE	103619	BALAENIDAE
102307	BARYTHERIIDAE	103208	ENTELODONTIDAE	103700 ORDER RODENTIA	
102400 ORDER SIRENIA		103209	SUIDAE	103701 FAMILY PARAMYIDAE	
102401 FAMILY DUGONGIDAE		103210	TAYASSUIDAE	103702	SCIURAVIDAE
102402	TRICHECHIDAE	103211	ANTHRACOTHERIIDAE	103703	CYLINDRODONTIDAE
102403	PRORASTOMIDAE	103212	HIPPOPOTAMIDAE	103704	PROTOPTYCHIDAE
102404	PROTOSIRENIDAE	103213	CAINOTHERIIDAE	103705	ISCHYROMYIDAE
102500 ORDER DESMOSTYLIA		103214	ANOPLOTHERIIDAE	103706	APLODONTIDAE
102501 FAMILY DESMOSTYLIDAE		103215	XIPHODONTIDAE	103707	MYLAGAULIDAE
102502	CORNWALLIIDAE	103216	AMPHIMERYCIDAE	103708	SCIURIDAE
102600 ORDER HYRACOIDEA		103217	AGRIOCHOCERIDAE	103709	OCTODONTIDAE
102601 FAMILY GENIOHYIDAE		103218	MERYCOIDODONTIDAE	103710	ECHIMYIDAE
102602	HYRACIDAE	103219	OROMERYCIDAE	103711	CTENOMYIDAE

Table 2. Example of map code of the JAFOV system

1/200,000 MAP = NAYORO			
13 KYOWA	9 ONNENAI	5 NIUPU	1 OUMU
14 FUKINODAI	10 NAYORO	6 SANRU	2 KAMIOKOPPE
15 SOEUSHINAI	11 SHIBETSU	7 SHIMOKAWA	3 NISHIOKOPPE
16 HOROKANAI	12 KENBUCHI	8 OKUSHIBETSU	4 SHOKOTSUDAKE
1/200,000 MAP = ASAHIKAWA			
13 TAKADOMARI	9 PIPPU	5 AIBETSU	1 KAMIKAWA
14 FUKAGAWA	10 ASAHIKAWA	6 TOUMA	2 DAISETSUZAN
15 AKABIRA	11 BIEI	7 SHIBINAI	3 ASAHIDAKE
16 KAMIASHIBETSU	12 FURANO	8 TOKACHIDAKE	4 TOKACHIGAWAJOURYUU
1/200,000 MAP = YUUBARIDAKE			
13 IKUSHUNBETSUDAKE	9 YAMABE	5 NISHITAPPU	1 SAHORODAKE
14 ISHIKARIKASHIMA	10 ISHIKARIKANAYAMA	6 OCHIAI	2 SHINTOKU
15 MOMIJIYAMA	11 HIDAKA	7 CHISAKA	3 MIKAGE
16 HOBETSU	12 IWACHISHI	8 HOROSHIRIDAKE	4 SATSUNAIDAKE

Table 3. Portion code for tetrapods and fishes

CODE FOR TETRAPOD	CODE FOR FISH
100 SKULL & LOWER JAW	100 CRANIAL PART
110 SKULL	110 SKULL
120 LOWER JAW	120 OTOLITH
130 TOOTH	130 TOOTH
200 AXIAL SKELETON	200 ENDOSKELETON
210 VERTEBRA	210 VERTEBRA
220 RIB	220 PECTORAL GIRDLE
230 STERNUM	230 PELVIC GIRDLE
300 APPENDICULAR SKELETON	240 PTERYGIOPHORE
310 PECTORAL GIRDLE	300 EXOSKELETON
320 PELVIC GIRDLE	310 BONY PLATE
330 LIMB	320 SPINE
400 BONY PLATE	330 SCALE

f. Era and period codes
g. Code for biostratigraphic classification
h. Code of portion (Table 3)
The last one has been prepared separately for tetrapods and fish since both of them are too different to use a same division.

The map code is prepared for easy correlation of locality name and geographical coordinates, that is, a map name is given from a set of longitude and latitude by using the map code, and vice versa. The map code of the Geographic Survey of Japan is adopted in the JAFOV system (Figure 7).

3.3. Inverted file

For the quick search of data, several kinds of inverted files were created. The higher speed of search is performed by preparing an inverted file on the more detailed key, but it requires larger file space. Inverted files in the JAFOV system, therefore, are limited to the keys which are used in the first step of the search procedure. Consequently search on the more detailed keys should be performed by using a master file after the search on inverted files. The following keys were selected for creating inverted files.
a. Family name (code)
b. Prefectural name (code)
c. 1/25 000 map name (code)
d. Formation name
e. Period (code)

4. Numeric Data File

4.1. Data format

It is not easy to define a common format which is used to record measurement of data of speci-

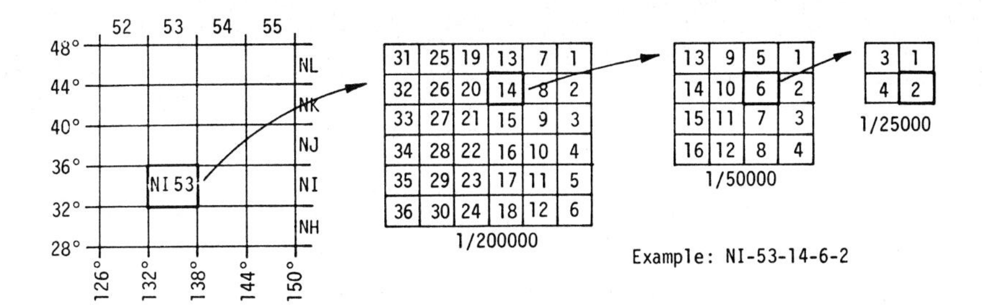

Figure 7. Relationship between the geographic coordinates and map codes

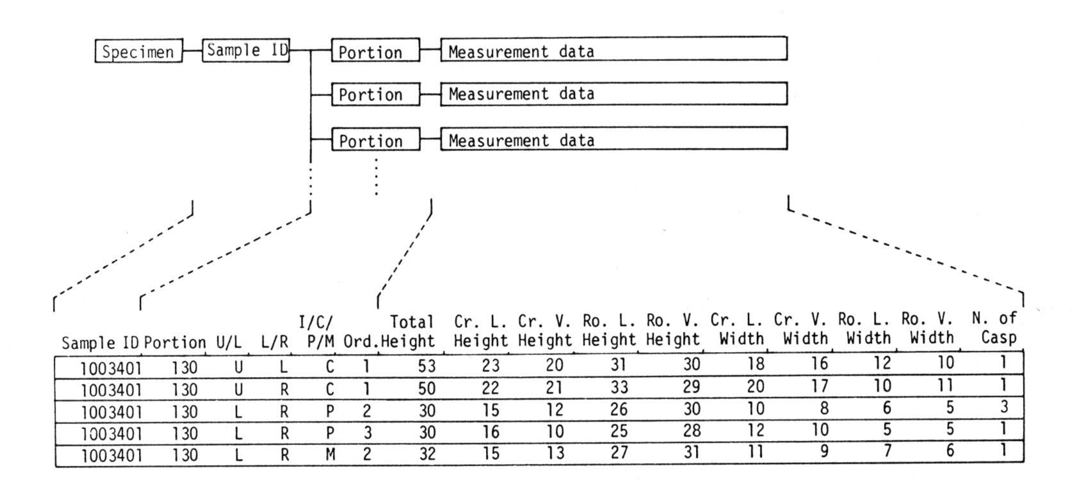

Sample ID	Portion	U/L	L/R	I/C/P/M	Ord.	Total Height	Cr. L. Height	Cr. V. Height	Ro. L. Height	Ro. V. Height	Cr. L. Width	Cr. V. Width	Ro. L. Width	Ro. V. Width	N. of Casp
1003401	130	U	L	C	1	53	23	20	31	30	18	16	12	10	1
1003401	130	U	R	C	1	50	22	21	33	29	20	17	10	11	1
1003401	130	L	R	P	2	30	15	12	26	30	10	8	6	5	3
1003401	130	L	R	P	3	30	16	10	25	28	12	10	5	5	1
1003401	130	L	R	M	2	32	15	13	27	31	11	9	7	6	1

Figure 8. Example of the numeric data in the JAFOV system

mens of all kinds, because the method and variables of measurement differ according to the portion and taxonomic position of the specimen. In the JAFOV system a common format is defined for each portion in Table 3. Codes of subdivision of each portion are included in the record when they are needed. As a result, a portion which is subdivided into many parts has many number records, and a specimen with more than one portion has records of different data format. It is easy to select data for statistical analysis by using codes of portion and its subdivisions. An example of the specimen 1003401 shows that it has five teeth (portion code = 130) of different positions which are specified by the third to sixth variables in a record, and the remaining ten variables are measurement data (Figure 8).

4.2. Connection with reference data

Since the two data files have different data structures from each other and are managed by using different DBMSs, it is difficult to connect them directly. But is is possible to connect them by using a common sample identification number. That is, after selection with specified age and area by using the reference data file, the sample ID numbers retrieved will be passed to the numeric data system to perform selection with measurement values, and statistical analysis by using statistical packages. On the other hand, it is possible by using the reference data file to obtain the locality and horizon of the samples which were detected in the course of statistical analysis. Furthermore, it is possible to pass the data to a graphics program (Figure 9).

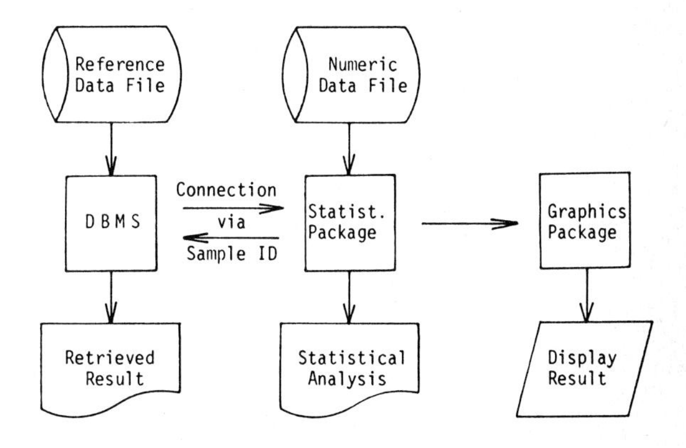

Figure 9. Connection between reference data and numeric data files

5. Discussion

The data in the JAFOV system can be easily used by any researcher through the intra-university computer network and a local call.

The project is supported by many paleontologists who cooperate in our project collecting and updating the input data.

The JAFOV system manages three different kinds of data concerning specimens of fossil vertebrates deposited in Japan. They are managed by using systems different from each other. They have a common key of the sample ID number, and it is possible to connect them by using this key.

The JAFOV should include data of specimens which are deposited in foreign countries because it is impossible to study vertebrate paleontology only by using specimens in Japan. An international computer network of low cost will be expected to proceed with such a project.

In the near future it might be possible to send imagery data through telecommunications at a low cost. Then the JAFOV will become a total data management system for specimens of fossil vertebrates.

6. Acknowledgement

This project is supported by the grant in aid for scientific research of the Ministry of Education of Japan (Project No. 56480023). The FACOM M-200/382 system in the Data Processing Center of Kyoto University and the HWP-2647A terminal system in our department were used. The authors wish to thank Professor S. Ishida, Dr. H. Kamiya, Messrs. H. Nakaya, N. Kuga, A. Takemura, K. Hirota, R. Hirayama, Y. Ito, H. Saegusa and M. Watabe who have been cooperating in this project.

7. References

[1] Wadatsumi, K. ed., Symposium on Information Earth Science, The Earth Monthly 3 (1981) 276-371.

[2] Nishiwaki, N. Yamamoto, K. and Wadatsumi, K., The BIWA System as a prototype data base to manage various kinds of data in geology and related sciences, in Glaeser, P. (ed), Data for Science and Technology (Pergamon Press, Oxford, 1981) 156-160.

[3] Nishiwaki, N., Yamamoto, K. and Wadatsumi, K., A small data base management system for researchers and its significance in geological sciences, Geol. Data Proc. 5 (1980) 63-85.

[4] Yamamoto, K, Harada, K., Okazaki, Y, Yasuda, S. and Wadatsumi, K., Retrieval system for private data of references and specimens, Geol. Data Proc. (1977) 17-22.

[5] Yamamoto, K., Nishiwaki, N. and Kamei, T., JAFOV: Data base on the Japanese fossil vertebrates (1), Geol. Data Proc. 7 (1982) 21-30.

[6] Murao, Y., Hiroike, H., Ozawa, Y., Watanabe, T. and Hishino, S., Construction manual for the FAIRS data base (Kyoto Univ. Data Proc. Cent., 1982).

Data for Science and Technology, P.S. Glaeser (ed.)
North-Holland Publishing Company

DATA FOR MATERIALS SELECTION AND SUBSTITUTION

N.A. Waterman

Director, Michael Neale and Associates Ltd.
43 Downing Street, Farnham, Surrey GU9 7PH, U.K.

The relationship between engineering materials, product design and manufacture is examined in the context of the product life cycle. The influences of engineering materials on the stages of product invention, innovation, commercial manufacture, evolution and obsolescence are illustrated by reference to past and current engineering components, mechanisms and systems. Progress through these stages depends on the provision of data for the selection and substitution of materials and manufacturing methods. Current shortcomings of available data are discussed and recommendations for improvements are made.

1. INTRODUCTION

Civilisation depends on the provision of materials. Materials in the form of food and drink, drugs and medicine, fuel, clothing, furniture, housing, tools, machines and equipment. The extraction of raw materials from the earth, their processing into the forms required by mankind and ultimately their disposal is illustrated in Figure 1. Many branches of engineering, eg mining, agricultural, chemical, mechanical, civil, electrical, etc. are involved in this cycle of creation, application and decay.

At the creation stage of the cycle all branches of engineering are involved in the conversion of materials from one form into another with the aid of the forces of nature. These forces are contained or created and subsequently controlled by other materials viz the engineering materials which are the substances of the components and products which make up the tools, machines and equipment of engineering systems.

In the application stage, the engineering materials must contain or create and control the forces necessary for the successful operation of the product and at the same time withstand the harmful effects of these forces and the surrounding environment.

At the destruction phase the engineering materials must either be convertible to a recyclable form or disposable in an acceptable manner.

The principle role of engineering materials is therefore as the media in which the 'forces of nature' are contained, created, manipulated and displayed. The 'forces' can be useful when properly controlled eg the transmission of load, the conversion of mechanical force into electrical voltage and vice-versa, the conversion of voltage into optical effects; or damaging when uncontrolled eg fatigue, corrosion and wear.

As the working media of engineering devices, materials play primary and secondary roles. Primary roles may be fulfilled by gases, liquids and solids.

eg as steam in a water pumping steam engine

as a liquid in crystal display devices

or as a solid eg piezoelectric polyvinylidene film in a microphone.

In this primary role materials are the hosts of potentially useful scientific phenomena.

In their secondary or supporting role materials are used in order to connect, control, contain or display the useful phenomena. The forms are usually (although not necessarily) solid eg connecting rods, cylinders, pistons, linkages, electrical connectors or simply space containers which protect and locate the elements of an engineering device. Calling this a secondary role does not imply any diminished importance. Indeed the success of a product depends critically on the correct use of materials not least in that the appearance and hence the first reaction of a potential customer will depend on which materials are chosen for a secondary role.

A distinction between primary and secondary roles is that relatively few types of materials can fulfil the primary roles whereas a relatively large number can, at least in principle, fulfil the secondary role for many products.

The skill and ingenuity of the engineer involves the creation of components, mechanisms and systems which utilise the primary and secondary roles of materials with the maximum efficiency and economy.

Engineering materials are the focus of this paper; in particular the data required for the selection and substitution of engineering materials.

FIGURE 1 MATERIALS CYCLE

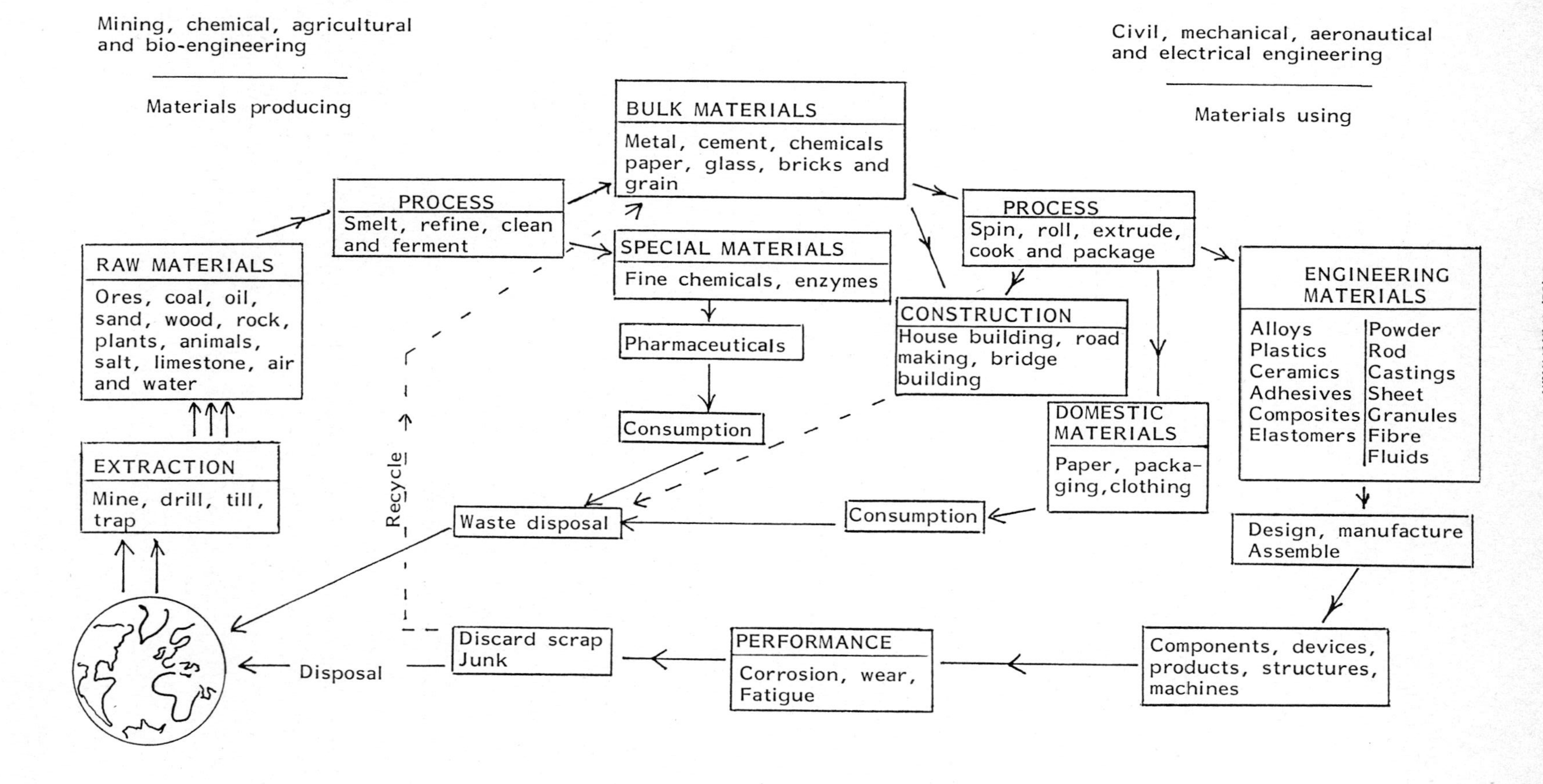

2. THE NATURE OF ENGINEERING

All engineering products must pass through a cycle similar to that of materials viz creation, application and decay. The invention on which they are based may be defined as the first practical control of the forces of nature by the creation of an assembly of components to perform a function which satisfies a specific need. Eg Savery invented the first practical steam engine by creating a system of cylinders, pipes and valves which utilised the suction force, created by the partial vacuum caused by condensing steam, to raise water from coal mines or to supply towns and villages.

The need for greater efficiency and economy stimulates improvements which may radically change the original invention by introducing a different engineering mechanism or even utilising a different scientific phenomenon (force of nature). Eg Newcomen introduced a piston into the steam engine which in addition to utilising the vacuum also harnesses the 'force' of atmospheric pressure. Watt greatly improved the efficiency of Newcomen's engine by introducing a separate condenser.

When the invention plus improvements plus specific need (market) are such that commercial manufacture is justified a design specification is drawn up. This activity may be defined as design for manufacture.

As experience is gained with the manufacture and operation in service of a product certain small improvements will be made which add up to the evolution of the product.

While these improvements may be small, and therefore evolutionary in terms of the end product, radical or innovative improvements may be necessary by the component or materials manufacturer. For example the replacement of light emitting diode displays by liquid crystal displays may be regarded by the digital watch or calculator manufacturer as an evolutionary development; however it required fundamental or innovative improvements in liquid crystal formulation and manufacture and a radical change of display device manufacture by others, before such end-product evolution was possible.

Obsolescence occurs when either a company's product has failed to evolve to match its competitors or the product itself is made obsolescent by another based on a different engineering principle or scientific phenomenon eg electronic calculators replacing mechanical adding machines, hydrocarbon fuelled internal combustion engines replacing reciprocating steam engines, etc.

At the end of its life cycle a product or a particular model may be resurrected if the external conditions which caused obsolescence are subject to significant changes or a new customer emerges. Eg there is a renewed interest in steam engines for third world countries which have no oil supplies; sails have recently been employed on large ships as an addition to conventional power systems; the Peoples Republic of China ordered Viscount aircraft after they had become obsolescent in Western countries. Evidently when a product is resurrected after a considerable period of time there is scope for evolutionary and innovative development.

A schematic life cycle of engineering products which illustrates these points is given in Figure 2. The original invention which satisfies a specific need is radically improved under the dual influence of changing requirements (market pull) and scientific and engineering developments (technology push). When the developed product reaches a stage where a market demand is sufficiently well identified to justify commercial manufacture a design specification will be defined and further major technical developments must await the launch of a new model before finding application. Once the product is being manufactured and service experience has been accumulated small, evolutionary improvements will be made to increase customer satisfaction or (more usually) to reduce manufacturing costs.

After a period of time, the continuation of changing market demands, and radical improvements in the technology concerned with the design or manufacture of the product will justify the launch of a new model. This model will in time undergo evolution and so-on around the cycle again until the product reaches maturity and finally becomes obsolescent. The obsolescence may be caused by a new invention based on different scientific phenomena or engineering principles which offers great improvements of efficiency or economy over existing products or may be due to failure to evolve to match market requirements or the improved technology or low price of a competitor's product.

In developed economies most new needs will not normally be satisfied in the first instance by an original invention. The natural tendency of manufacturing industries will be to try to satisfy the need with an existing product. If this is not possible evolution will be tried first as this offers the quickest and least disruptive route to a solution and hence is, initially at least, likely to be the most profitable. When this fails innovations or eventually new inventions may be necessary. Hence a new need may be satisfied by entry into any part of the product life cycle. This paper is concerned with how the selection of engineering materials and product manufacturing processes contributes to each stage of the life cycle of products and how substitution of new materials and processes can influence the progression from stage to stage.

3. INVENTION

At the invention stage of a new product or

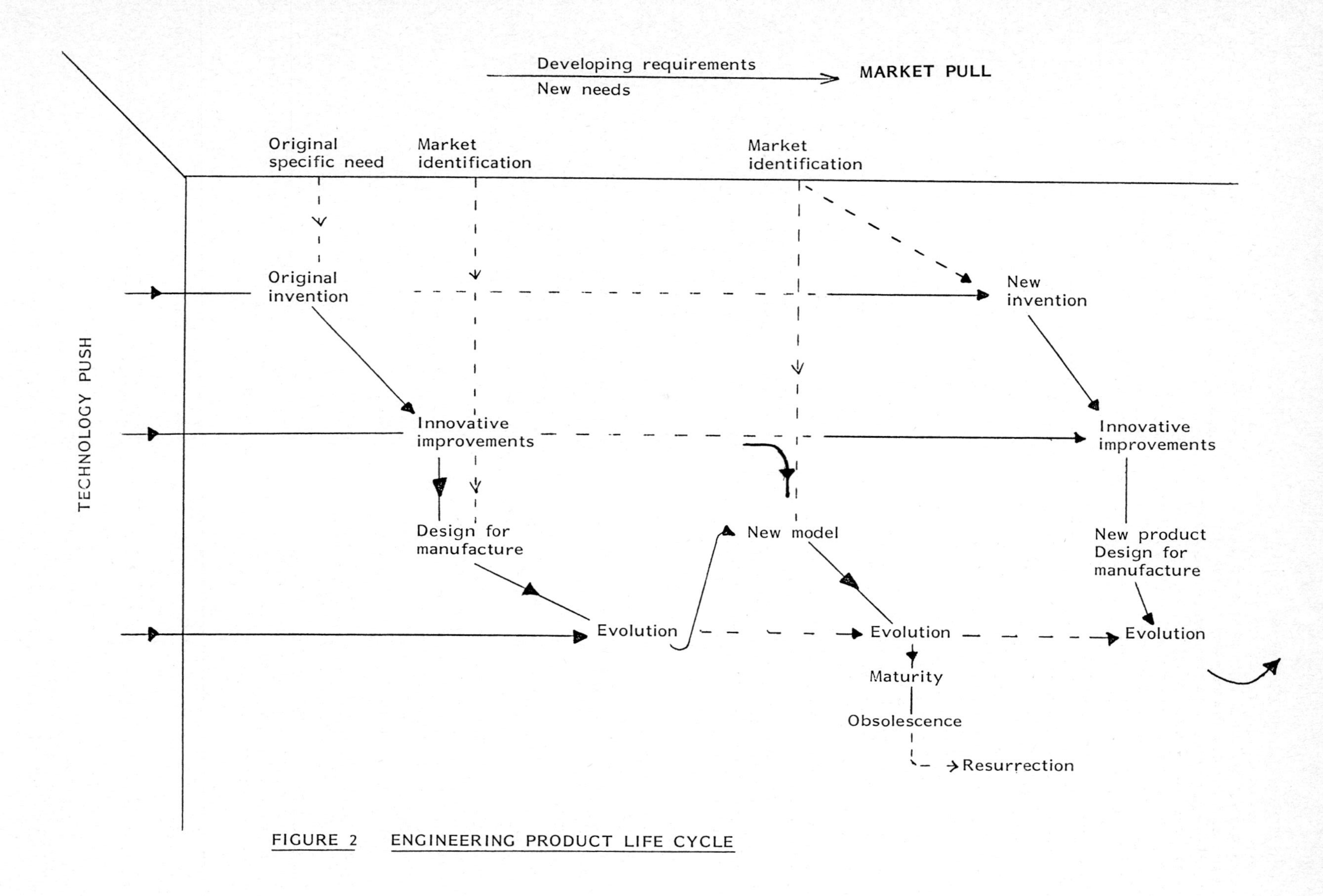

FIGURE 2 ENGINEERING PRODUCT LIFE CYCLE

device the contribution of materials depends on whether the primary or secondary role properties are being employed. In the example of liquid crystals in a display device, doped silicon in a transistor, ferrites in an electrically insulating magnet, polyvinylidene film in a telephone head set and zinc selenide in a dry copier the material and its properties are the working medium of the invention and play a primary role.

In their secondary role materials are much less important to the process of invention. Good examples of this are Newcomen's steam engine and Whittle's jet engine.

Newcomen's atmospheric steam engine made minimal demands on the properties of the rather poor engineering materials available in his day. Whittle demonstrated the feasibility of the jet engine before the development of high strength, high temperature and nickel alloys (although of course the availability of these materials allowed great improvements in the efficiency of the engine).

Hence it is mainly in the primary role that the new materials have the potential to stimulate the invention of new engineering devices. To stimulate new inventions data is required on potentially useful scientific phenomena in a form which informs engineers of

(i) The principle eg conversion of mechanical stress into voltage.

(ii) The quantitative relationships and boundaries of operation.

(iii) The current limitations which might be removed by appropriate research and development.

While textbooks and review articles may contain some of this information it is rarely presented with the needs of the engineer as the prime motivation for publication.

4. INNOVATIVE IMPROVEMENTS

The contribution of materials to the innovative stage of the product life cycle is mainly in a secondary role as the working media of cylinders, connectors, etc. in engineering mechanisms. It is important to note that different mechanisms which achieve the same end result can make very different demands on materials. Eg compression of gas may be obtained by a piston in a cylinder which can be arranged to require very little of materials in the way of strength and durability but sacrifices efficiency by allowing some leakage. An alternative mechanism is a diaphragm which may demand good fatigue resistance from the material.

The potential of new materials depends on the scientific phenomena and engineering mechanisms chosen to perform required functions and how many functions are combined. For example in a vehicle suspension system the requirements are springing, energy damping and anti-roll. The options are:

(i) Interconnected fluid springs.
Eg Moulton's 'Hydrolastic' and 'Hydragas' suspension systems. (1)

(ii) Mechanical springs which may be helical or leaf type, plus an anti-roll bar.

The system chosen should maximise, through ingenious design, the useful properties of the primary working material and the secondary, containing and transmitting, materials and minimise their limitations. Moulton's use of fluids, rubber, steel and reinforced hose materials in the Hydrolastic and Hydragas suspension systems for the BMC 1100, Mini, 1800 and later BL vehicles is an excellent example of innovative improvement which does just this.

Developments of composite leaf springs which are a close imitation of metal counterparts is an example of how by constraining the scope of component development the maximum advantages of new materials will not be obtained.

Hence the potential for new materials substitution is great at the innovation improvement stage of a product life cycle if the potentially useful properties and shortcomings of a material are appreciated and the chosen mechanism is designed accordingly.

5. DESIGN FOR MANUFACTURE

The activities involved in the design for commercial manufacture of a new product, based on an invention plus innovative improvements, are illustrated schematically in Figure 3. At this stage in the product life cycle it is unlikely that materials performing the primary role of the product or engineering system will be changed from those of the original invention. However it is possible that the materials selected for the secondary role in the invention will not be the optimum for commercial manufacture.

The requirements of the selected engineering materials are that they are stable, definable substances which can satisfy the design functions when converted into the components of the product by the chosen manufacturing methods. The choice of manufacturing methods may be constrained by the need to employ existing production facilities.

The potential for new material at this stage is limited by the necessity to demonstrate that their properties are better or at least equal to those already proven in the prototype developed previously and to be more easily converted into the desired components. As the performance of engineering materials in service or their processability cannot be predicted with confidence from laboratory tests on standard samples, product testing to prove the suitability of new

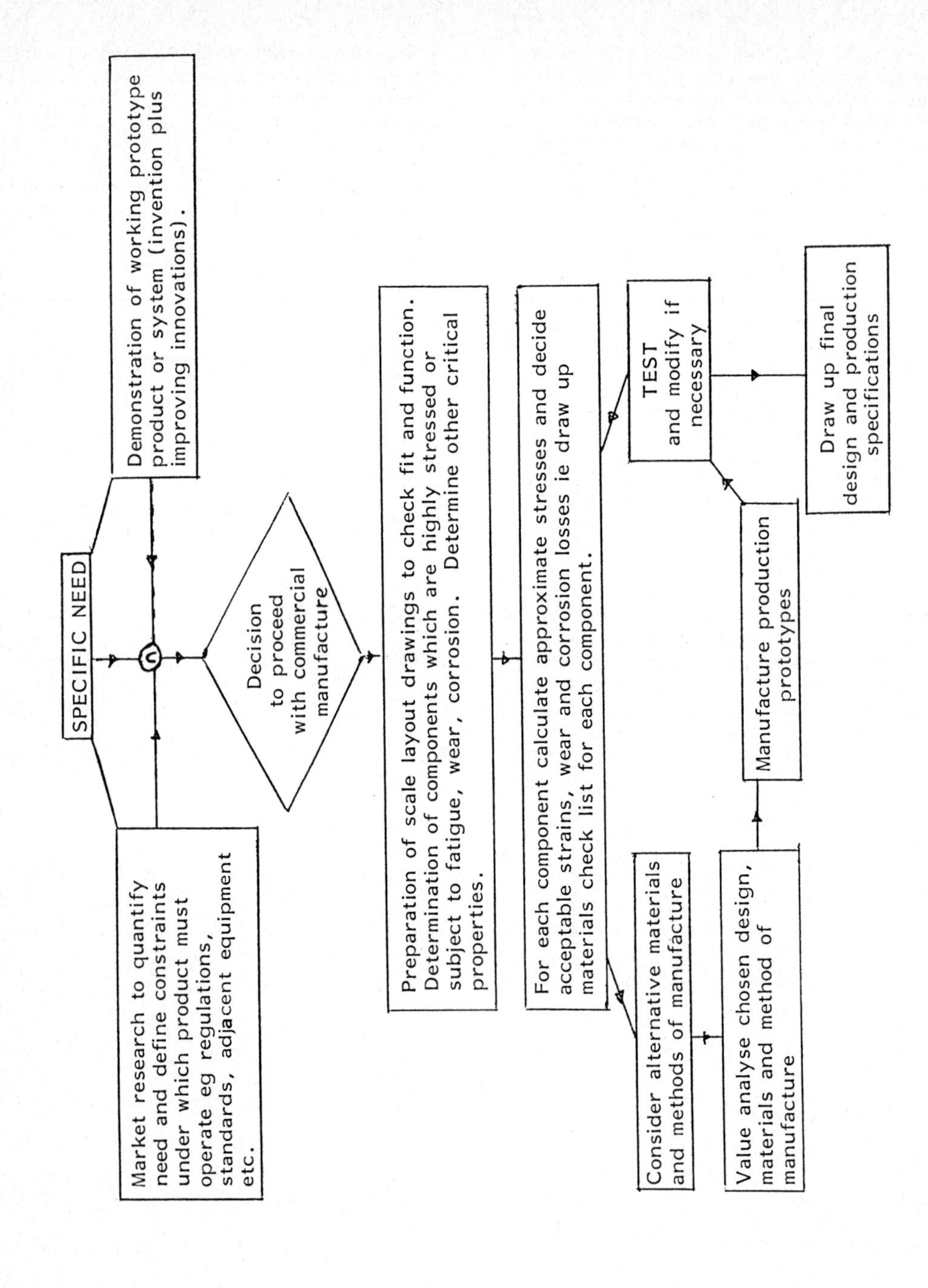

FIGURE 3 DESIGN FOR MANUFACTURE

materials is necessary as well as manufacturing trials. Information on these tests and trials is rarely published. Much of the data that is published on materials is contained in trade literature - which has a number of shortcomings. If the material is a metal the literature will normally say that it conforms to one or another national standard. This means only that a bar of metal has been subjected to various laboratory tests and surpassed certain specified minimum levels of performance. The situation is similar for plastics where specially moulded test bars and other geometrically simple specimens are subjected to standard laboratory tests which are defined by national standards. Where the intended end-use is more critical, especially in terms of safety, eg aircraft materials, the appropriate regulatory body will lay down more stringent requirements and may even define acceptable levels of impurities of metals, and required fire resistance behaviour of plastics and such information is then included in trade literature. However all this information demonstrates is that the material has been prepared in such a way that it is of a consistent quality level. For example impact test results on simple plastic bars will not predict what will happen if a spanner is dropped on to an instrument casing made of the same plastic. Such tests do not even guarantee that the winners of 'league table' of impact resistant plastic materials as defined by the test will necessarily have the best resistance to the spanner when the plastic is moulded into the casing.

The net result is that, based on the so called property data (actually test results) contained in the trade literature of materials suppliers, the would-be purchaser will conclude that all the materials of the same type from different suppliers are more or less the same. A plastics producer may, by clever formulation, (polymer + additives) achieve a slightly higher value of a standard test result (eg heat deflection temperature) than the competitors and this fact may be used as the basis of a marketing campaign which gives that producer a larger market share.

This may be all important to the production economics for the plastics company but the engineer who wishes to select a plastic for the product under design has other needs. To the product engineer, it is not the differences between one manufacturer's heat resistant nylon and another which is significant; rather it is the differences between heat resistant nylon and ordinary nylon, eg how much extra does the improved material cost, what disadvantages, if any, come with the improved heat resistance and most important of all, will the heat resistance be retained when the plastic granules (raw material) are moulded into the product and when the product is in service? Would a plastic based on a different and fundamentally more heat resistant polymer eg polysulphone be cost effective? and so on. To answer these questions the engineer would have to collect literature and additional information from a large number of suppliers.

It was to fill this need for engineering information that the Fulmer Materials Optimizer was conceived and produced (2). This continuously updated loose leaf publication covers not only plastics but all commercially available engineering materials and components manufacturing methods. The Fulmer Materials Optimizer is not simply a collection of trade literature appropriately catalogued, it is a critically analysed digest of the available information (ie more than just trade literature) and also includes a materials selection system which puts it in the category of a 'how-to-do-it' type information.

Trade literature on materials has two major omissions. One that has already been alluded to above is that information is not given on how the material will perform when converted into a real component (ie not a simple test bar). The other is the almost total absence of cost data. When information is provided, in response to specific requests, it refers only to raw material costs.

Material suppliers attempt to cover the former omission by including photographs and some details of components and products where their materials have been employed. However the information rarely includes such details as, how long those products have been in service, what conditions they operate under and how many have been produced over what time scale.

6. EVOLUTION

Small changes in the materials specification which improve product performance or more usually ease manufacture will contribute to product competitiveness and fight off obsolescence. This requires a detailed understanding of materials properties, how to obtain them and their effect on component manufacturing efficiency and product service performance.

The manufacturing process contribution to product evolution is similar to the above with small changes aimed at reducing costs without impairing end product properties or improving the end product without increasing costs. The overall aim should be to approach as near as possible, the Boston Learning curve which sets a target of 20 - 25% reduction in costs for each cumulative doubling of production.

The potential for the substitution of new materials and processes to assist product evolution is constrained by proven record of established materials and need to employ existing production equipment. Components made of new material must fit within space permitted by adjacent components. Opportunities normally require emergence of a problem with current materials (eg failure in service) or process (eg an environmental constraint) or substantial potential savings, in total, on materials and production method.

Successful product evolution requires continuous monitoring of product performance and manufacturing productivity and demands an effective internal information system. However most products will experience difficulties during manufacture or service which require external information inputs. Unlike scientific experiments, where conditions are closely controlled in order to investigate the relationship between two variables and check a theory or establish an empirical law, the working conditions of engineering products, mechanisms and systems cannot be so precisely controlled. Product testing should identify major faults but problems of fatigue, creep, corrosion and wear which are time dependent may not be accurately simulated in these tests. It is also impossible to anticipate the precise application and range of conditions under which a product or mechanism may be used. In addition when a product is part of a complex engineering system (as in process plant for example) or is large or expensive (eg a shaft for a very large electricity generating set) testing in order to find inherent design faults is not practical or economically possible.

All of the above uncertainties point to the need to monitor the performance of engineering products in service while they are in operation and examine them after they have failed or been withdrawn from service for whatever reason. Current work on fatigue performance monitoring of automotive components is a good example of this (3).

In the case of large process plant (refineries, power stations) aeroplanes, etc. ie where breakdowns can be dangerous as well as very expensive, monitoring is a well established activity requiring information on diagnostic techniques and non-destructive testing equipment, electronic signal processing instruments, recorders etc. Breakdowns are thoroughly investigated to determine their cause and measures taken to prevent their re-occurrence. When safety is involved the results of such investigations are communicated to other interested parties. As a consequence of these activities of monitoring and failure investigation, experience is accumulated which should result in safer and more efficient operation of plant and (although this is less certain), improved design of the next generation of plant and equipment.

For most engineering products the threat of major accidents or catastrophic losses does not exist, however this does not mean that performance monitoring or failure investigation should be neglected. Every product in service, whether operating adequately or otherwise, may be regarded as a generator of information which should improve the next generation of products ie those which will incorporate evolutionary developments of the current products.

Experience of the performance and limitations of materials is often accumulated in certain individuals and this will be lost with the departure of these individuals from the organisation. Surveys of product and materials experience would seem an obvious safeguard but the need for such surveys is not widely recognised. The problems associated with compiling such experience surveys is generally that there is not too little information but that there is too much anecdotal data which needs to be analysed and incorporated into a logical scheme or rejected if unreliable. The activity requires an input from someone with a thorough understanding of the product and materials technology but not so involved that personal experience prejudices judgement.

When a product design has reached the evolutionary stage there may still be opportunities for significant advances in production technology. For example cold forging or powder metallurgy methods may replace machining from bar stock when component production totals reach levels which allow amortisation of the increased tooling costs of these newer manufacturing techniques. For the product making company which makes such a change in production techniques the step may be described as innovative if they embark on the manufacture themselves or evolutionary if the components are brought in. Whichever way the decision arises the engineers involved will require information on alternative production techniques and since these changes are made on the basis of improved production economics, comparative costing data is also required. However manufacturing costs are made up of several factors - capital investment charges, labour rates, productivity, consumables, scrap production, etc., etc. hence it is not surprising that authoritative comparative information is not available.

Evolutionary development of mass production techniques can also be effected by modifications to materials specifications. Most materials specifications are originally written to satisfy design performance requirements and optimisation for production may not be possible in the short time available in which to get a product ready for sale. Modifications which ease production and do not impair performance must be carefully thought out from an understanding of basic materials technology allied to production trials.

During the evolution stage of a product life cycle new models will be launched from time to time which incorporate the latest results of research and development on design methods, materials and manufacturing techniques combined with the lessons learned from the experience surveys. The required information is a distillation of all that is known about a product, its design, manufacture and operation. This most important activity of retrieval, analysis and presentation of the quintessential information is necessary if an engineering product is to make significant advance in performance, efficiency or economy. The need for the activity has been stressed elsewhere (4) and the absence of such information for most engineering components, mechanisms and products is a serious omission.

One of the main reasons for lack of such distilled

information is the enormous volume of disparate information which must be collated and critically analysed. For example it has been shown (4) that to produce a design method for journal bearings requires the combined output of around 70 research projects, each of which may be deemed critical, in order to provide engineers with one piece of directly applicable information.

7. SUMMARY AND RECOMMENDATIONS

The contribution of engineering materials to each stage of the product life cycle, the consequent materials data requirements, currently available data sources and their limitations are summarised in Table 1. It can be seen that the major limitation is the inapplicability of the available data to the actual operating conditions on the end product. Three possible means of removing this limitation are:

(i) To attempt to relate the results of short term, laboratory controlled, tests on standard samples to real products through research on the properties and behaviour of materials and the construction of appropriate theories, formulae and models.

(ii) To conduct surveys on the state of materials in products which have been in service for a meaningful time-span; particularly those products which have been operating near the limits of a particular material's characteristics.

(iii) To monitor, through the deployment of sensors, the actual conditions under which components and products operate and reproduce these conditions for laboratory tests.

Materials scientists and engineers have traditionally devoted considerable resources to the first option. It is recommended that the problem of supplying data for materials selection and substitution might be more quickly and satisfactorily solved by devoting more resources to the other two options.

Finally if we are to capture the opportunities for evolution of existing engineering products and the creation and innovative development of new products which new or improved materials offer then much greater collaboration is necessary between materials producers and materials users. When such collaboration is achieved on materials developments for specific end products the data necessary for materials selection or substitution is generated automatically.

REFERENCES

1. Moulton A.E. & Best A. (1979) 'From Hydrolastic to Hydragas Suspension'. Proc. Instn. Mech. Engrs. Vol 193 P.15.

2. The Fulmer Materials Optimizer edited by N. A. Waterman & A.M. Pye. First published in 1974 by the Fulmer Research Institute, Stoke Poges, UK and continuously updated.

3. Dabell B. J. & Hill S.J. - (1979) 'The use of Computers in Fatigue Design and Development' Int. Jnl of Fatigue Vol 1 No. 1

4. Neale M. J. (1977) 'Technology Transfer - The Application in Industry of the Results of Research'. Proc. Instn. Mech. Engrs. Vol 191 P.333.

TABLE 1 PRODUCT LIFE CYCLE, MATERIALS CONTRIBUTION, DATA REQUIREMENTS, SOURCES AND LIMITATIONS.

STAGE IN PRODUCT LIFE CYCLE	MATERIALS CONTRIBUTION	CONSEQUENT DATA REQUIREMENTS	DATA SOURCES	LIMITATIONS
Invention and innovation	Primary role as host of potentially useful scientific phenomenon.	Quantified details of phenomenon and limits of stable operation.	Text books. Popular articles.	Rarely written with needs of engineer in mind.
	Secondary role as containment, control or display in engineering device which uses phenomenon.	Data on interaction of secondary materials with primary materials and details of any deleterious effects.	Generally data is derived during proto-type trials.	Results often not published.
		Conventional property data on strength, modulus, heat resistance etc. related to conditions in service.	Trade literature. Previous experience of well tried materials.	Data often not applicable to new conditions in service.
Design for manufacture	As stable substances with defined properties which fulfil primary and secondary roles and permit manufacture within cost targets.	Primary - as above + more information on criteria for stable quality.	Raw material suppliers.	Objectivity. Alternative sources needed.
		Secondary - as above + comparative data on alternative materials properties under in-service conditions.	National standards.	Most materials standards do not cover fitness for purpose.
			Fulmer Materials Optimiser. In-house data from direct experience.	Data may not relate to precise conditions and of product application and environment.
		Secondary - comparative data on alternative manufacturing techniques including total (ie raw material + processing) costs.		
Evolution	As above + means of achieving small improvements in end-product performance and manufacturing efficiency.	Details of possible effects of small changes of composition and processing of production performance.	Research literature. Raw materials suppliers and process operators.	Difficulty of quantifying likely benefits compared with established practice.
		Feedback from product application, experience failure analyses and production monitoring.	Internal.	Rarely published.

Data for Science and Technology, P.S. Glaeser (ed.)
North-Holland Publishing Company

COOPERATION IN DEVELOPING COMPUTERIZED MATERIALS DATA BASES

J.H. Westbrook

General Electric Company, Research and Development Center
120 Erie Boulevard, Schenectady, New York 12305, U.S.A.

Three concurrent activities are described: the DIPPR program sponsored by AIChE; the November '82 Workshop on on-line materials data bases sponsored by CODATA and others; and the feasibility study of an inter-society on-line system sponsored by the (U.S.) Metal Properties Council (MPC) and others. The objective of DIPPR is to compile, evaluate and provide access to data on compounds important to the chemical process industry. Led by AIChE, the program has been successful in securing cooperation and financial support of 56 companies, institutes and federal agencies in pursuing several independent data projects, the largest of which is a computerized data base of physical and thermodynamic data for 1000 chemical compounds of industrial significance. The Workshop will examine the important issues related to development of on-line systems of engineering properties of materials such as: barriers, roles of stakeholders, economics, system definition, user interface, and impact of system application. The MPC Feasibility Study under the author's direction is evaluating the feasibility of an interlinked system of distributed data bases held by individual professional societies and other institutions. The study will focus primarily on the political, economic and management aspects of such a large and complex cooperative effort and will recommend an action plan to MPC.

INTRODUCTION

In this paper are described three data projects which emphasize one of the key themes of CODATA - inter-organizational cooperation in the evaluation of, access to, and dissemination of data. Further, all three share additional common characteristics in that they all are of industrial interest and all are concerned with the engineering properties of materials. It is considered worthwhile to bring them to the attention of the participants in the 8th International Conference of CODATA, not only for the potential interest in the projects per se but also to illustrate the different modes and benefits of cooperative data activities. The three projects to be described are briefly the Design Institute for Physical Properties Research (DIPPR) sponsored by the American Institute of Chemical Engineers, the feasibility study of the Metal Properties Council aimed at an inter-society computer-based materials property data system, and a multi-sponsored materials data workshop to be held later this year. Following these descriptions, a concluding section will summarize the benefits of cooperation exemplified by these projects and urge that these models be followed in other parts of the data field.

THE DESIGN INSTITUTE FOR PHYSICAL PROPERTY RESEARCH (DIPPR)

This program, organized in 1978 and managed by the American Institute of Chemical Engineers, serves the chemical process industry in compiling, measuring and evaluating thermodynamic and physical property data for industrially important compounds. Its work is supported by 56 industrial companies, institutes and federal agencies. Broadly speaking its goals are:

- to search out and evaluate existing physical property and thermodynamic data to eliminate errors in values and to assess reliability
- to generate and compile new data when needed values are not in the existing literature
- to create a reliable, central, and easily accessible data bank.

In pursuing these goals it benefits the chemical process industry by:

- providing a single comprehensive source of data of stated reliability and currency
- eliminating wasteful duplication of individual data efforts thereby lowering costs for data acquisition and evaluation to individual companies
- facilitates the work of chemical plant construction companies, equipment vendors, and process simulation companies, thus enabling them to devise more efficient, safer and more economical process, plant and equipment designs.

- keeps industrial scientists and engineers informed of the latest data relevant to their field and the most powerful data storage and manipulation techniques as well as providing them with personal contacts with the leading practitioners in these fields.

Organization - DIPPR functions as a virtually independent financial entity within AIChE which provides a legal and administrative framework for DIPPR's activities and which ultimately provides an appropriate publication medium for its results. The work of DIPPR is coordinated by two main committees: the Administrative Committee which is responsible for financial, administrative and policy matters, and the Technical Committee which initiates, implements and supervises individual data projects. Further, each data project has its own steering committee which reports to the Technical Committee and whose chairman sits on the Administrative Committee. These relationships are shown in Figure 1.

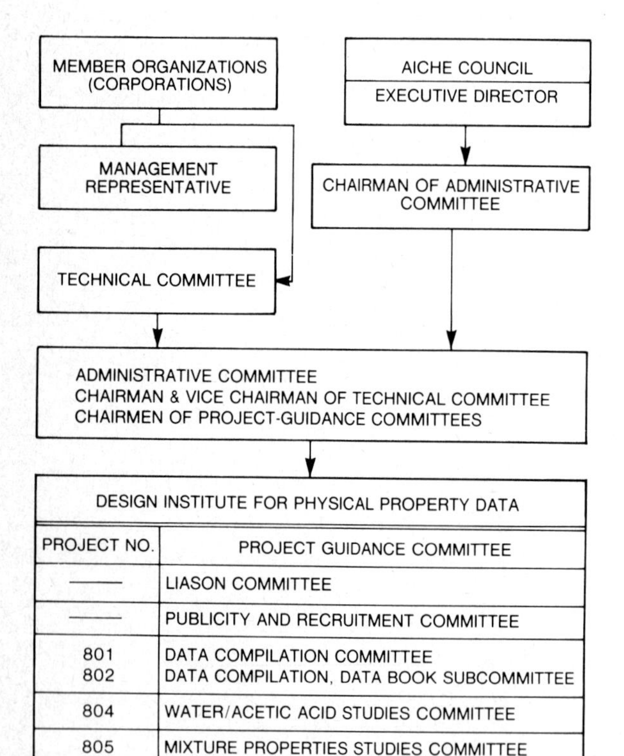

Fig. 1 Organizational Structure of AIChE's DIPPR program

Participants - Industrial companies or technical institutes become members of DIPPR by paying an annual fee (currently $750). They then appoint a Management Representative and a Technical Representative from their own organization to DIPPR. Particularly through the Technical Representative who sits on the Technical Committee, participants can exercise considerable influence in suggesting and planning new projects, can recommend prospective contractors, can assist in screening proposals, and evaluate results. The input from industry is thus quite direct in both formulating and operating the program.

Funding - Each DIPPR project is separately funded. Most are conceived as extending over several years but funding requirements are estimated each year. Each project must be first approved by a majority vote of the Technical Committee who then submit the project to the Management Representatives of all participant organizations for balloting on a weighted basis, taking into account both the size and nature of the participant's business (chemical producer, engineering contractor, process equipment manufacturer or process software firm). Approval of a project by a Management Representative also implies commitment of a pro rata weighted share of the costs. Income from other sources (government grants, contributions, dues or publication income) may be used to reduce required funding to be committed by ballot. Thus both project choice and funding level are directly controlled by sponsor companies. Access to the findings of the project is confined to the particular sponsoring organizations both during the active lifetime of the project and for a full year thereafter. Full public access is permitted only after later publication by AIChE.

Projects - Five projects are currently underway with a total expenditure of over $400 000/yr. These include a Data Compilation Project (1), a Data Prediction Manual Project, an Experimental Data on Mixtures Project (2), an Electrolyte Phase Equilibria Project and a Pure Component Properties Data Project. New projects are proposed each year but some fail to secure ballot approval, some are completed within the year and others are approved for continuation from prior years. The Data Compilation Project, for example, the earliest and largest project, involves compilation and evaluation of physical property and thermodynamic data on 25 properties and three temperature-dependent properties of ~1000 industrially important chemical compounds and is being performed under contract by Pennsylvania State University. The first compounds being considered are those of largest annual dollar volume use in the U.S. The collected data and pertinent descriptive records are being stored on a computer together with the results of evaluation, estimated accuracy and original literature reference. Each datum is characterized as experimentally determined or predicted, of known or unknown origin, evaluated or unevaluated, and assigned an accuracy code on a scale of 1 to 9, ranging from an estimated

0.2% accuracy to >100% probable error. Interactive computer graphics assist in data evaluation, and a database management system known as INTERACT permits selective print-out from the computer file with considerable versatility for both editing and information retrieval purposes. This project is now in its third year of operation and has compiled and evaluated data on more than 200 compounds. A partial file for one substance is reproduced in Figure 2.

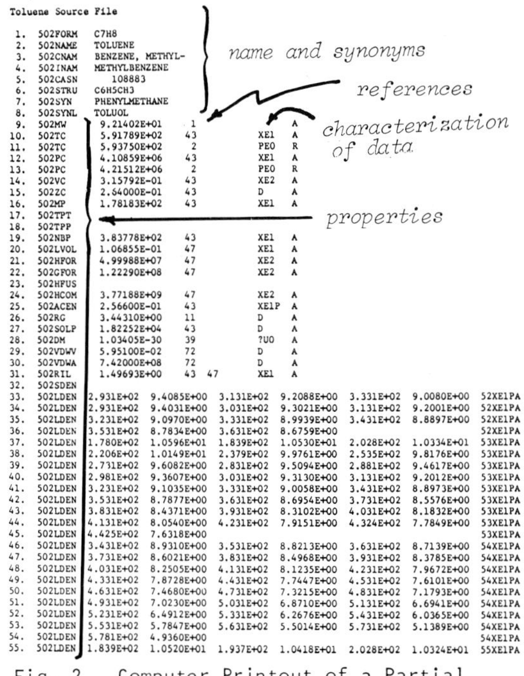

```
Toluene Source File

 1.  502FORM   C7H8
 2.  502NAME   TOLUENE
 3.  502CNAM   BENZENE, METHYL-
 4.  502INAM   METHYLBENZENE
 5.  502CASN      108883
 6.  502STRU   C6H5CH3
 7.  502SYN    PHENYLMETHANE
 8.  502SYNL   TOLUOL
 9.  502MW      9.21402E+01    1               A
10.  502TC      5.91789E+02   43         XE1   A
11.  502TC      5.93750E+02    2         PE0   R
12.  502PC      4.10859E+06   43         XE1   A
13.  502PC      4.21512E+06    2         PE0   R
14.  502VC      3.15792E-01   43         XE2   A
15.  502ZC      2.64000E-01   43         D     A
16.  502MP      1.78183E+02   43         XE1   A
17.  502TPT
18.  502TPP
19.  502NBP     3.83778E+02   43         XE1   A
20.  502LVOL    1.06855E-01   47         XE1   A
21.  502HFOR    4.99988E+07   47         XE2   A
22.  502GFOR    1.22290E+08   47         XE2   A
23.  502HFUS
24.  502HCOM    3.77188E+09   47         XE2   A
25.  502ACEN    2.56600E-01   43         XE1P  A
26.  502RG      3.44310E+00   11         D     A
27.  502SOLP    1.82252E+04   43         D     A
28.  502DM      1.03405E-30   39         ?U0   A
29.  502VDWV    5.95100E-02   72         D     A
30.  502VDWA    7.42000E+08   72         D     A
31.  502RIL     1.49693E+00   43  47     XE1   A
32.  502SDEN
33.  502LDEN  2.931E+02  9.4085E+00  3.131E+02  9.2088E+00  3.331E+02  9.0080E+00  52XE1PA
34.  502LDEN  2.931E+02  9.4031E+00  3.031E+02  9.3021E+00  3.131E+02  9.2001E+00  52XE1PA
35.  502LDEN  3.231E+02  9.0970E+00  3.331E+02  8.9939E+00  3.431E+02  8.8897E+00  52XE1PA
36.  502LDEN  3.531E+02  8.7834E+00  3.631E+02  8.6759E+00                          52XE1PA
37.  502LDEN  1.780E+02  1.0596E+01  1.839E+02  1.0530E+01  2.028E+02  1.0334E+01  53XE1PA
38.  502LDEN  2.206E+02  1.0149E+01  2.379E+02  9.9761E+00  2.535E+02  9.8176E+00  53XE1PA
39.  502LDEN  2.731E+02  9.6082E+00  2.831E+02  9.5094E+00  2.881E+02  9.4617E+00  53XE1PA
40.  502LDEN  2.981E+02  9.3607E+00  3.031E+02  9.3130E+00  3.131E+02  9.2012E+00  53XE1PA
41.  502LDEN  3.231E+02  9.1035E+00  3.331E+02  9.0058E+00  3.431E+02  8.8973E+00  53XE1PA
42.  502LDEN  3.531E+02  8.7877E+00  3.631E+02  8.6954E+00  3.731E+02  8.5576E+00  53XE1PA
43.  502LDEN  3.831E+02  8.4371E+00  3.931E+02  8.3102E+00  4.031E+02  8.1832E+00  53XE1PA
44.  502LDEN  4.131E+02  8.0540E+00  4.231E+02  7.9151E+00  4.324E+02  7.7849E+00  53XE1PA
45.  502LDEN  4.425E+02  7.6318E+00                                                53XE1PA
46.  502LDEN  3.431E+02  8.9310E+00  3.531E+02  8.8213E+00  3.631E+02  8.7139E+00  54XE1PA
47.  502LDEN  3.731E+02  8.6021E+00  3.831E+02  8.4968E+00  3.931E+02  8.3785E+00  54XE1PA
48.  502LDEN  4.031E+02  8.2505E+00  4.131E+02  8.1235E+00  4.231E+02  7.9672E+00  54XE1PA
49.  502LDEN  4.331E+02  7.8728E+00  4.431E+02  7.7447E+00  4.531E+02  7.6101E+00  54XE1PA
50.  502LDEN  4.631E+02  7.4680E+00  4.731E+02  7.3215E+00  4.831E+02  7.1793E+00  54XE1PA
51.  502LDEN  4.931E+02  7.0230E+00  5.031E+02  6.8710E+00  5.131E+02  6.6941E+00  54XE1PA
52.  502LDEN  5.231E+02  6.4912E+00  5.331E+02  6.2676E+00  5.431E+02  6.0365E+00  54XE1PA
53.  502LDEN  5.531E+02  5.7847E+00  5.631E+02  5.5014E+00  5.731E+02  5.1389E+00  54XE1PA
54.  502LDEN  5.781E+02  4.9360E+00                                                54XE1PA
55.  502LDEN  1.839E+02  1.0520E+01  1.937E+02  1.0418E+01  2.028E+02  1.0324E+01  55XE1PA
```


Fig. 2 Computer Printout of a Partial Source File for a Typical Substance (Toluene) in the DIPPR Program

THE METAL PROPERTIES COUNCIL FEASIBILITY STUDY

The Metals Properties Council (MPC) is an entity formed by four professional societies (the American Society for Metals, the American Welding Society, the American Society of Mechanical Engineers, the American Society for Testing and Materials) and the Engineering Foundation and is supported as well by over 50 industrial companies, government agencies and trade associations. The purposes of the MPC are as follows:

a) to identify major unfulfilled needs for reliable data on the engineering properties of metals and alloys,

b) to evolve, plan, and conduct programs for collecting, generating and evaluating such data so they may be useful,

c) to arrange for making such data available promptly by reports, publications, correspondence or other means,

d) to keep informed of and to utilize the results of related activities, both national and international, in order to avoid duplication of effort, and

e) to act as co-sponsor, with the American Society of Mechanical Engineers and the American Society for Testing and Materials, of the Joint Committee on Effect of Temperature on the Properties of Metals, and to raise funds for financing the activities of the Joint Committee.

Since 1964 it has been active in a variety of data projects: data compilation, data evaluation, round-robin testing, test method development etc., concentrating on the engineering properties of commercial metals and alloys. Its programs are supervised by a Technical Advisory Committee overseeing the work of several individual technical sub-committees and task groups such as fatigue, effect of temperature on properties of metals, corrosion and data storage and retrieval systems. These committees propose specific projects for implementation, seek multi-sponsor support for them in terms of either dollars and/or services, and review the progress of the work. Final results of testing and data-analysis are not published directly by the MPC but appear among the publications of its parent societies [e.g. (3)]. Over 100 such reports have appeared thus far (4). Other outlets for the work of MPC are the numerous symposia, panel discussions and workshop sessions it sponsors.

Recently the MPC has become aware of the many advantages that would accrue from the capability for computer access to a comprehensive store of evaluated metals properties data. Since a de novo approach to building such a computer file would be enormously expensive and would duplicate in part machine-readable files already in existence, MPC determined instead to conduct a feasibility study of an information system assembled by interlinking independent computer-based files of materials properties built, or to be formed, by various professional societies or other reputable organizations. This approach would offer numerous advantages as summarized in Table I.

Table I ADVANTAGES OF A COORDINATED LINKED SYSTEM VS A MONOLITHIC COMPREHENSIVE SYSTEM

To the User	To Data Base Proprietor	To Both
Single access point	Broader Market	Useful at Early Stage of Development
Useful Front-end Interface	Shared Inputs	Less Costly to Build
Common Support Packages	Shared Revenues	Decreased Redundancy of Effort
Possible Merger With Secured Private Files	Promotion of Ancilliary Info Services	Better Currency
Quality Control of Data	Improved Access to Own Data	

The writer was engaged to conduct a year-long study of the political, economic, managerial and technical problems and opportunities inherent in a coordinated linked network system and to recommend to the MPC and associated sponsors of the study an action plan for implementation of its findings. Analogous to MPC's usual funding pattern, part of the costs of the study were underwritten by MPC itself from its general receipts and the balance by specific sponsors. While the final report of the study will not be available until the Spring of 1983, the approach to the study can be outlined and some preliminary findings summarized.

The specific tasks laid out for the study included:

- justify need and define the user market
- identify immediate and future participants
- identify databases suitable for inclusion
- recommend plan for management, direction and funding
- identify technical problems to be overcome
    - data selection and evaluation
    - data input
    - computer hardware
    - computer software
    - networking
- define materials-properties matrix suitable for a "Phase I" system
- define search, display, and manipulation capabilities appropriate to a "Phase I" system

Rather than to gather information by questionnaire, it was determined instead to conduct a series of structured interviews with selected institutions with strong interests in the engineering properties of materials. These included professional societies, trade associations, "not-for-profit" institutes, government agencies, private companies and information vendors. Personal visits were supplemented by telephone and correspondence with selected individuals and a literature search. Throughout this survey attempt was made:

- to identify and list all potentially contributory organizations;
- to ascertain the intensity and breadth of interest in the system visualized;
- to identify candidate databases for inclusion;
- to characterize the need for this type of information resource;
- to develop some broad specifications for system design; and
- to elicit other contacts helpful to the study

Thus far, over 40 machine-readable databases of engineering properties of materials have been identified. At present, none of these are publically available, many are tiny in size, intended only as pilots or prototypes, and none have been designed with the thought of compatability with other databases. Furthermore, if these databases were to be displayed in a materials-properties matrix, an alarming state of redundancy and lack of complementarity would be revealed. This result is only to be expected since all these databases were individually designed and constructed. It points up the task that lies ahead if a reasonably comprehensive system of coordinated individual databases is to be achieved.

Let us presume that the problems of scope and compatibility have been resolved and examine how such a system might appear to a user as shown in Figure 3. Following telephone connection via network to host computer(s) and input of passwords, identification numbers and other protocols, the user enters his query at a local terminal (see top center of figure). A software program, here called "query database match", with the help of a "synonym file" directs the user's question to that database or bases that is most appropriate to his field of inquiry (databases A and C in this example). Each individual database is presumed to have its own search and retrieval program which extracts the requested information. These numbers can then be printed, displayed on a screen or stored in a temporary file for later manipulation by a series of standard programs shown in the boxes stacked vertically at the right. Thus, the final result the user obtains might be a single number, a table of values, a graphical plot or a statistical analysis of a set of data. Note here that the updating and expansion of each database is presumed to be carried on independently as shown at the upper left of Figure 3. Conventional online bibliographic search services facilitate this process, but the data retrieved must then be subjected to human evaluation and automatic consistency or plausibility checks, where possible, before actual entry to the database.

It is remarkable that nearly all of the separate pieces shown in the Figure are now available! Virtually all that remains is to extend the collective scope of the database and modify the software so as to achieve the necessary compatibility. The latter, I am told, is a relatively trivial task for an experienced computer

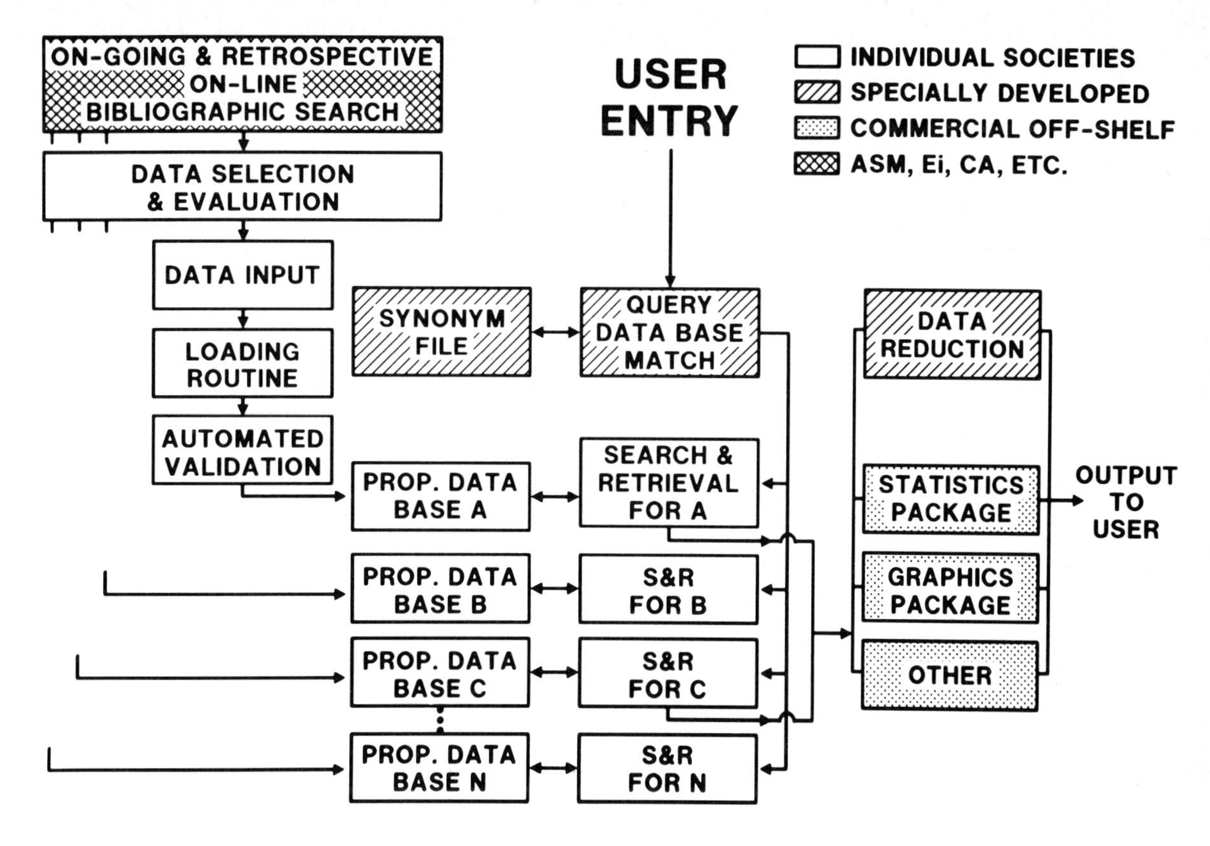

Fig. 3 Schematic Flow Chart for an Interlinked Set of Autonomous Databases Forming a Cooperative On-line Materials Information System

programmer, but, substantial extension of the collective content of the databases is a tremendous intellectual task of data identification, retrieval, and evaluation which will demand a large, coordinated, cooperative effort.

Skipping ahead again to the happy day when all of these problems have been solved, we can imagine the sequence of frames that might be presented a user upon logging on to a linked system of autonomous databases as shown in Figure 4a-4f. *Note that this series of constructions is a mock-up; most of these databases do not exist in fact and no software has been written to effect the interlinkage portrayed.* Figure 4a arrays the databases included in the system and instructs the user how to obtain direct access or automatic query matching. By exercising the options shown in the menu of Figure 4b, the user is led to the database(s) most nearly fitting his needs (Figure 4c). Subsequent frames (Figures 4d and 4e) show progressively finer scales of tables of contents of an exemplary database with the nature of the expansion at the option of the user. Thus the user has some assurance that he has found an appropriate resource and some knowledge of its organizational scheme before he is instructed how to connect with and search the database in question (Figure 4f).

Whether the above or some similar scheme can be realized in the near future depends not so much on development or application of computer technology, but upon the collective dedication of the various interested groups to cooperate, both in sharing the work that must be done and also in effecting the necessary standardization and compatibility.

WELCOME TO CMIS

MPC's COMPUTERIZED MATERIALS INFORMATION SERVICE

THE SERVICE PRESENTLY PROVIDES ACCESS TO THE FOLLOWING DATA BASES:

		CODE
EPRI	SOFT MAGNETIC MATERIALS	EP
DOE	PROPERTIES FOR SOLAR SYSTEMS	SS
ORNL	PROPERTIES FOR ADVANCED REACTORS	AR
CINDAS	PHYSICAL PROPERTIES OF METALS	CN
NAVY	SHIP CONSTRUCTION DESIGN FILE	SC
ASM	METALS HANDBOOK	MH
AA	FRACTURE TOUGHNESS OF ALUMINUM ALLOYS	FT
NBS	STANDARD REFERENCE MATERIALS	SR

TYPING IN THE CODE LETTERS WILL DIRECTLY CONNECT YOU TO THE FILE OF INTEREST

ALTERNATIVELY TYPE IN FS (FILE SELECT) AND YOUR RESPONSES TO THE FOLLOWING MENUS WILL AUTOMATICALLY CONNECT YOU TO THE FILE OR FILES BEST MEETING YOUR NEEDS

a)

MATERIAL CLASSES		APPLICATION AREAS		PROPERTIES GROUPS	
ALLOY STEELS	AS	BEARINGS	BM	CHEMICAL/CORROSION	CH
ALUMINUM + ALLOYS	AL	ELECTRICAL/MAGNETIC	EM	ELECTRICAL/MAGNETIC	EM
CARBON STEELS	CS	ELECTRONICS	EI	MECHANICAL	ME
CAST IRONS	CI	NUCLEAR	NM	THERMAL	TH
COPPER + ALLOYS	CU	STRUCTURAL	SM	OTHER	OT
PRECIOUS METALS	PM	WELDMENTS	WM	ALL	AX
REFRACTORY METALS	RM	TOOL MATERIALS	TM		
STAINLESS STEELS	SS	OTHER	OT		
SUPERALLOYS	SA				
TITANIUM + ALLOYS	TI				
OTHER	OT				
ALL	AX				

ENTER ONE CODE FOR EACH INTEREST AREA

MATERIALS CLASS *AL*

APPLICATION AREA *OT*

PROPERTIES GROUP *TH*

b)

THE FOLLOWING DATA BASES CONTAIN THE MOST INFORMATION PERTINENT TO YOUR NEEDS

		CODE
DOE's	PROPERTIES FOR SOLAR SYSTEMS	SS
CINDAS's	PHYSICAL PROPERTIES OF METALS	CN

TYPE IN CODE FOR DATA BASE YOU WISH TO SEARCH *SS*

c)

DATA BASE SS

THIS DATA BASE FILE IS ORGANIZED BY

COMPONENT GROUP		MATERIALS CLASS	
THERMOELECTRICS	TH	FERROUS	FE
PHOTOELECTRICS	PH	NON-FERROUS	NF
HEAT EXCHANGERS	HE	SEMICONDUCTOR	SC
OTHER	OT	PLASTIC	PL

TO DISPLAY A LISTING OF ALL SUBCOMPONENTS AND PROPERTIES INCLUDED, TYPE IN COMPONENT GROUP *HE*

TO DISPLAY A LISTING OF ALL SPECIFIC MATERIALS INCLUDED, TYPE IN MATERIALS CLASS ____

d)

HEAT EXCHANGERS HE

SUBCOMPONENTS	PROPERTIES
CONDENSER	EMISSIVITY
CONVECTOR	HEAT TRANSFER COEFFICIENT
COOLING SYSTEMS	OXIDATION RESISTANCE
DIFFUSER	SPECIFIC HEAT
FINS	THERMAL CONDUCTIVITY
LIQUIFIER	THERMAL EXPANSION COEFFICIENT
RADIATOR	
RECUPERATOR	
VAPORIZER	

NOTE TOPICS OF INTEREST AND TYPE CARRIAGE RETURN (CR) TO CONTINUE SEARCH

e)

TO EXTEND YOUR SEARCH OF THE SOLAR SYSTEMS DATA BASE PROCEED AS FOLLOWS:

____

____

____

____

____

f)

Fig. 4 Initial Sequence of Frames Obtained by a User upon Interrogation of a Hypothesized Interlinked System of Autonomous Databases

THE MATERIALS DATA WORKSHOP

The prior discussion of the MPC study has emphasized the fragmentary and uncoordinated efforts that have proceeded in recent years toward building on-line systems for data on engineering properties of materials. In recognition of this situation, it was proposed some time ago that a convocation be held of experienced and knowledgeable individuals in a workshop mode to attempt to chart a way out of the dilemma. Funding was finally arranged and a workshop of about 75 persons will be held in November 1982 at a conference site near Crossville, Tennessee. The Workshop is sponsored jointly by CODATA, the National Bureau of Standards, the Oak Ridge National Laboratory and the Fachinformationszentrum of the Federal Republic of Germany. The inclusion in the early registration list of the majority of the world's leaders in building and operating machine-readable materials property databases bodes well for the success of the Workshop.

There is no lack of interest in the subject and the computer technology tools necessary to achieve a comprehensive on-line system are in hand. The magnitude and complexity of the task, however, demand cooperative and coordinated effort by all interested bodies: professional societies, trade associations, industrial users and generators of materials data, universities, research institutions, technical publishers and government. Such an effort by its very nature requires a thorough understanding of all the goals and a common agreement on how to achieve them.

Broadly speaking, the organizers see the Workshop as a unique opportunity for a detailed assessment of the feasibility and timeliness of constructing on-line materials data systems. More specifically it is hoped to identify the critical issues, rank order these problems for solution, publicize the current activities in the field and array the possible options for implementation. To accomplish these goals, it is planned to divide the whole group of participants into small task forces of 6 to 8 persons to examine sets of preformulated questions in such areas of barriers, roles of stakeholders, economics, user identification, scope definition etc. Examples are shown below.

a) Barriers-What problems can be identified that might prevent development of comprehensive on-line materials data systems?

b) Economics-How can we approximate the economic value of ready access to reliable materials property data?

c) User Identification-Which of the following application areas have the greatest need for computerized properties data bases - materials selection, structure and component design, or automation of materials processing?

Each question set will be considered independently for half a day by two or possibly three independent task groups. These task groups then are to report their conclusions and recommendations to a plenary session of the entire Workshop. It is expected that from these reports will emerge points of agreement, points of disagreement, and new issues not contained in the original sets of questions. New task groups will then be constituted from new groupings of individuals to consider unresolved and new issues and after extended independent discussions, will again be reconvened in plenary session for reporting and discussion. A final written report will be prepared by the discussion leaders and organizers summarizing the Workshop and its conclusions and recommendations. By the time these CODATA Conference proceedings appear in print, the Workshop report should also be available. It is presumed that CODATA's Advisory Panel on Data for Industrial Needs will continue to play a role in evaluating and implementing the recommendations of the Workshop.

SUMMARY

The foregoing discussion has described three different data projects on engineering properties of materials, each carried on in a cooperative mode by a grouping of independent organizations. It is to be hoped that they will serve as useful models for future cooperation. In concluding this article, it will be well to review the many benefits of cooperation in such endeavors.

Consensus - In order best to deploy the limited intellectual and financial resources of the materials data community it is essential to develop a consensus on both goals and means of achieving them. Which materials and which properties are the most important? Which applications of materials databases are the most critical? Which computer technologies will be the most useful and cost-effective? ... and so on. The broader the base of groups contributing to the formulation of the consensus, the more sound and actionable it is likely to be.

Cost-sharing - Building and maintaining databases and information systems of the type described herein will be enormously costly enterprises, often beyond the capabilities of individual professional societies, private companies and even nations. Cooperation at any level - professional, national, or international can significantly reduce these costs.

Reduced Redundancy - Part of the survey conducted for the Metal Properties Council revealed a marked redundancy in the machine-readable databases extant for engineering properties of materials. Most of the effort has concentrated on mechanical properties of the common structural metals and alloys with substantial duplication. Not only is this wasteful of funds but more importantly, it is wasteful of the limited intellectual resource of people who are capable of selecting, evaluating and compiling reliable property data. Technological database formation is not a mere clerical function such as might suffice for the bibliographic, sociometric or financial fields but requires the attention of knowledgeable people.

Focussed Expertise - Corollary to the previous point, advantage is also to be gained by fragmenting the total task so that real expertise can be brought to bear on local areas of the whole materials-properties matrix. There really are no experts in materials data per se but rather experts on corrosion behavior of stainless steels, on dielectric properties of polymers, or fracture-toughness of ceramics. The necessary breadth of competence in this sense is rarely within a single organization but must be sought by coordinated cooperative linkages between organizations.

Standardization - The element of standardization is of vastly greater importance in the technological data field as opposed to scientific data and nowhere more so than with reference to materials properties data which are strongly history and structure dependent. (5) If materials data are to be truly "portable" and reliable in diverse application areas, there must be standardization of test methods, property definitions, nomenclature, units, data reporting schemes, and materials characterization. Thus, the terms 18/8 stainless steel, polyvinyl chloride, or silicon carbide ceramic, are not adequate material descriptors, nor do such properties as fracture toughness or thermal shock resistance have the precise definition and lack of ambiguity characteristic of scientific parameters such as density, specific heat or ionization potential. Here too, cooperation is essential.

Critical Mass of Effort - Many data projects are of such character that the necessary effort for successful accomplishment could not be justified by unilateral effort of a single organization. This observation applies not only to specialized data areas, but paradoxically even to those of the broadest general interest. All too often it appears to be that "Everybody's business is nobody's business"! Through cooperation of several groups a sufficient effort can often be accumulated to get a job done which otherwise wouldn't be attempted.

Increased Reliability - Mere accumulation and compilation of data is inadequate for its successful application. Above all, the data must be reliable. By achievement of more comprehensive data files and by more timely realization of them through cooperative activities, attainment of the goal of increased reliability is facilitated by illuminating the gaps, inconsistencies, conflicts, and other inadequacies in our data collections.

Compatibility - While materials, properties and data standardization can contribute to the ease of data transfer, interpretation and application as discussed under Standardization, in the computer age this is not enough for true compatibility. If two machines are to be able to "talk" to each other or if one user at his own terminal is to be able to interrogate a variety of databases, extract and merge results, the computer languages must be the same or interpretable; software packages for retrieval, display or manipulation must not be unduly restricted by the host machine or original database structuring; and networking must not impose additional barriers or constraints. These aspects of compatibility imply a need for cooperation, not so much among data specialists as among hardware manufacturers, system designers, programmers or other software specialists.

Forum for Technical Exchange - Finally the mere existence of cooperative data projects between otherwise independent individuals and organizations both requires and creates opportunity for a variety of forums for technical exchange via workshops, symposia, conferences and informal person-to-person interactions. This very CODATA meeting in a troubled country at a parlous time in world affairs is an outstanding exemplar of how much more remains to be done in technical cooperation in the data field and of the promise which it offers.

References

(1) Danner, R. and Daubert T.E. "The DIPPR Data Projects at the Pennsylvania State University" in A Review of AIChE's Design Institute for Physical Property Data and World-Wide Affiliated Activities, S.A. Newman, ed. AIChE Symposium Series #230 77 (1981)8

(2) Zudkevitch, D. "AIChE's DIPPR Experimental Program on Mixture Properties" in A Review of AIChE's Design Institute for Physical Property Data and World-Wide Affiliated Activities, S.A. Newman, ed. AIChE Symposium Series #230 58 (1981)58

(3) Smith, G.V. "Evaluations of the Elevated Temperature Tensile and Creep Rupture Properties of 12 to 27 Percent Chromium Steels" Data Series Publication DS59 American Society for Testing and Materials (1980) 330 pp.

(4) Annual Report of Metal Properties Council, 1981

(5) Westbrook, J.H. "Materials Standards and Specifications" Kirk-Othmer Encyclopedia of Chemical Technology 3rd ed. v15 (1981) 32

Data for Science and Technology, P.S. Glaeser (ed.)
North-Holland Publishing Company

DATA FOR ALLOY DESIGN

Shuichi Iwata

Department of Nuclear Engineering, Faculty of Engineering
University of Tokyo, 7-3-1 Hongo, Bunkyo-ku, Tokyo, Japan

In designing alloys we take advantage of two sorts of information, i.e. data on experiments and methods for estimation of materials properties. Data on experiments are compiled to set up the method of estimation, for example, multiple regression, microstructural modeling of materials behaviors. Although materials data can be stored and used as a set of numerical data to a practically acceptable level by using commercial DBMSs, the methodology of estimation of materials properties are not used in a systematic manner due to the complexity of alloy data per se. Thus we must use our knowledge completely for the best estimation and consequently the best alloy design. In this paper the way in which various sorts of materials data are used in the process of alloy design are reviewed.

1. INTRODUCTION

Since the attempt of Michael Faraday in "Steel and Alloys", there have been many approaches to alloy design as described in our earlier paper. [1]

Essential logics change, sometimes drastically with time because experience refutes a logic considered essential earlier. However, in the case of alloy design, the first principle, that microstructural design results in various improvements on material properties, has been reconfirmed step by step since the above mentioned first attempt.

General trends to organize individual tactics concerning materials into a systematic strategy first began in the field of organic chemistry in the 1960s. [2] In the field of metallurgy, the first object of many contributions was the design of nickel base super alloys in the first half of the 1970s although several approaches also were carried out concerning newer metals. This period corresponded to the sprouting of systematic alloy design, and almost all activities had some sort of optimistic viewpoint.

Against this optimistic atmosphere about alloy design, introspections during the concluding discussions of the Battelle Materials Science Colloquia "Fundamental Aspects of Structural Alloy Design", included "The field of alloy design is best described by Lewis Carroll's Jabberwocky". [3]

This review begins with "Jabberwocky" in respect to time, i.e. 15-19 September 1975, and sections are arranged roughly in the order of the size of the information, ranging from nuclear data to performance data in a large system.

2. TRENDS OF ALLOY DESIGN

By using the TOOL-IR system in the University of Tokyo Computer Center [4], the general trends of alloy design could be inferred statistically as shown in Table 1, where 52 papers are contained in both categories "Alloy Design" and "Materials Design". What is important in this table is the small number of papers related to predictive design of alloys and materials, which is a reflection of the difficulty in estimating and predicting materials properties.

The contents of the alloy design are listed in Table 2. The most frequent subjects concern materials problems in system design. However, the first two rows, namely, phase diagrams and microstructures, are most important and essential in the case of alloy design.

Table 1. Papers on Materials Design by CAS.

CAS Vol. No.	Alloy Design	Materials Design	Design by Prediction
86	21	28	1
87	10	26	1
88	21	41	0
89	41	48	2
90	18	41	4
91	25	85	1
92	28	95	3
93	22	86	0
94	25	67	1
95	21	65	1
96	21	61	0
Total	253	643	14

Table 2. Contents of Alloy Design

CAS Vol. No. / Contents	86	87	88	89	90	91	92	93	94	95	96	Total
Phase Diagram	1	--	--	--	1	1	1	--	1	--	--	5
Microstructural Design	3	--	2	--	--	1	2	--	4	3	3	18
Corrosion	1	--	1	1	1	4	6	--	--	1	1	16
Fatigue Creep	2	1	--	--	1	1	1	--	3	2	2	13
Fabrication	1	1	1	--	--	2	4	--	1	--	2	12
Design of Experiments	1	1	2	2	--	--	--	3	2	--	--	11
System Design	2	6	13	12	31	7	5	8	8	---	--	92
Correlations	2	1	--	--	--	--	--	1	--	--	1	5
Review	1	--	--	25	10	3	3	4	2	5	1	54
Optimization	--	--	1	--	1	--	1	--	2	1	--	6
Materials Selection	--	1	--	--	--	--	1	3	9	1	5	20
Others	--	1	--	--	5	3	2	--	--	--	3	14

Table 3. Papers on Predictions and Estimations of Alloy Properties

CAS Vol. No. / Contents	86	87	88	89	90	91	92	93	94	95	96	Total
Creep Fatigue	16	7	9	16	13	9	19	15	10	9	20	143
Microstructure Phase Diagram	6	0	1	10	7	10	8	5	5	2	2	56
Corrosion	1	1	1	2	0	2	3	3	4	4	1	22
Modeling	1	1	1	3	0	1	0	0	1	1	2	11
Others	5	5	7	2	7	10	5	9	1	4	5	60
Total	29	14	19	33	27	32	35	32	21	20	30	292

Work on predictions and/or estimations of alloy properties has also been focused mainly on creep and fatigue similar to the case of alloy design as shown in Table 3.

Microstructural changes are usually predicted by taking advantage of such information as phase diagrams, temperature-time-transformation diagrams, models about deformations, phase transitions, interactions of several kinds of defects, precipitates, solute atoms and so on. The numbers in the row of Microstructure and Phase Diagram being relatively small, a great deal of work has been done on this subject and several compilations on binary phase diagrams have been published by M. Hansen, R.P. Elliott, F.A. Shunk and W.G. Moffatt.

In spite of these activities, almost all materials engineers or investigators are not satisfied with these compilations as their general interests concern multi-component alloys, not only equilibrium states but also nonequilibrium states. What is important in alloy design is that compilations of essential and fundamental data should be informative enough for users to compare their own data which may be somewhat different with the adopted data in the compilation enabling them to evaluate their own experiments.

Other categories in Table 3 are not as active as those in the first two rows concerning predictions because of the facility of the relevant experiments.

3. DATA FOR ALLOY DESIGN

3.1 Nuclear Data

Neutron data are used to select materials for components of nuclear reactors, for example, spacer alloys of a lower thermal neutron absorption cross section. As far as alloy design is concerned, nuclear data on neutrons are well compiled except for high energy data, which are important in the evaluation of transmutation effects about fusion reactor structural materials and breeding ratios of tritium.

Compiled data are available in the form of magnetic tapes named JENDL-1,2 (Japanese Evaluated Nuclear Data Library), ENDF/B-IV, KEDAK-3 and so on. Data needs are also edited into a file named WRENDA (World Request List for Nuclear Data). Documents are indexed in CINDA (Computer Index of Neutron Data). Collaboration has been carried out among such data centers as BNL, IAEA-NEA, ORNL, CJD, JAERI, KFK and so on.

Several needs concerning inelastic cross section exist in order to estimate the damage states of alloys more exactly. However, organization of this type of data is well established in data generation, compilation and dissemination, so that alloy designers could become users of nuclear data generally.

As for predictions on the safe service life of reactor components, uneconomically large margins are set up unwillingly due to the inaccuracy of dosimetry data and theories for predictions. This inaccuracy suggests the difficulty of developing a materials data base for nuclear applications.

The characterization of the radiation environment in fission and fusion reactors, ion accelerators, high voltage electron microscopes, and neutron sources is carried out by experiments of dosimetry foils, theoretical neutron transport calculation, current measurements and so on. There are various sorts of uncertainties concerning cross section covariance, multi-group cross sections, and displacement cross sections.

The materials data base which deals with radiation damage should have capabilities of arranging data structures in accordance with the temporal, which may become the final, understanding of experiments. The objective of this data base is to establish correlations between the results of microscopic damage estimations and macroscopic property changes in irradiated materials.

For the correlation of neutron damages and other particle damages, it is important to estimate damage profiles exactly. In the calculations, stopping cross sections, which are compiled in the Handbook of Stopping Cross-Sections for Energetic Ions in All Elements by J.F. Ziegler, are used. To correct deviations between the theory and the experiments, several attempts have been promoted as in Ref. [5].

Examples of materials design using the above information concern ion implantation. Nitrogen ion implantation results in the increase of T_C of niobium annealed at 900 °C after the 3 x $10^{17} N_2^+/cm^2$ implantation. [15] It is essential for obtaining fundamental understanding of the phenomena when new fabrication methods are applied. This example requires the same sort of data as the ion damage experiment.

3.2 Models for Estimation

A great deal of deductive approaches have been performed by theoretical metallurgists. The cohesive energy, bulk modulus and atomic volume were calculated by the local-density theory for third- and fourth-row metals, when the only inputs to the calculations are the atomic number and the crystal structure. [9], [10]

The prediction of crystal structure is one of the most difficult problems due to the too small energy differences in the present quantum mechanical framework. Chelikowsky [8] expanded the data base to which quantum structural pseudopotential theories can be applied, and discussed inducing quantitative rules from chemical bonding in solids as one theoretical extrapolation of the Darken and Gurry plot [1], [16].

Watson and Bennett derived an electron negativity scale for the noble and transition metals on the basis of the electron band theory. [9], [10] They showed structural maps for transition metal -transition metal compounds, using the band-theory estimate of electronegativities and d-band vacancies as illustrated in Figure 1. This sort of data compilation which gives the theoretical background to the method of predictions of σ-phase formation in superalloys, is known as the PHACOMP method.

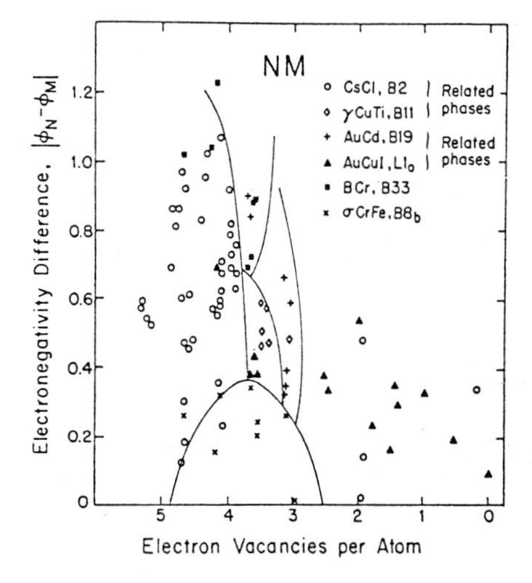

Figure 1. Structural maps for transition-metal transition -metal compounds.(9)

Newer approaches to the prediction of phase stabilities have been proposed by Machlin [11], [12], who reviewed the long time stabilities of stainless steels, superalloys, and proposed a simulation model of reactor alloy embrittlement. He uses a pair-potential calculation to predict phase formation, and his method seems to be successful for various applications.

In the case of the fields where there is little experience, it is necessary to carry out experiments and modeling simultaneously with mutual feedback. Such fields concern microgravity [14], ion implantation, amorphous alloys [17], ultra-high pressure conditions to obtain new materials, and fusion reactor materials. For these fields, the alloy designer is required to be a producer of newer data as well as a mere user of relevant data.

As there is a long time lag between data production, their compilation and dissemination, the activities of materials data centers generally are inclined to fundamental data rather than the current topics concerning such exotic materials as listed above. The continuous-cooling transformation diagram, for example, belongs to the category of fundamental data, even if it was used for the individual purpose of alloy design [13].

Modeling for alloy design needs information on microstructural and microchemical evolution, so that all sorts of phase information, for example, equilibrium phase diagrams of multi-components, TTT diagrams, CCT diagrams, solubilities of minor elements and so on need to be compiled. Theoretical approaches should be combined with experiments, as indispensable complements, especially for low temperature equilibrium states and high temperature states.

3.3 Microstructural Information

Microstructural data are generally represented by microphotographs of optical- and electron-microscopy, pole figures and descriptions of observations. These data might be quantified into a set of parameters, as in Figure 2, for example, in order to design alloys quantitatively.

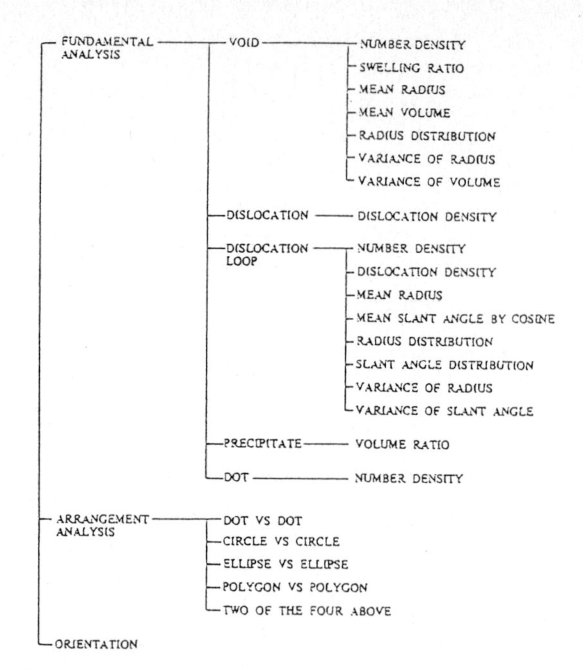

Figure 2. List of Microstructural Parameters

The first review of microstructural design was summarized in the book "Alloy and Microstructural Design" in 1976. [18] - [27]. Strengthening methods for ferrous and nonferrous alloys are well known as a set of methods and underlying theories. However, methods to increase the ductility are not established as well as the strengthening methods. Plastic instability, crack initiation and propagation, DBTT, dislocation channeling, 475 °C embrittlement are not well understood due to the limitations of observations. Environmental effects are also not well understood and not as yet compiled.

In order to obtain clear snapshots of the predominant mechanisms, Ashby proposed the deformation mechanism maps [29] on creep. Crack propagations by fatigue, creep and chemical attack have to be analysed in order to make clear shots of the next step. This type of snapshot has to be multi-dimensional, and computer assisted understanding of such data will be required. The mechanism maps for materials of nuclear plants will be the most sophisticated ones, containing all the data described in this review.

Composite strengthening is a relatively newer technique, and the feeling of design is best satisfied. Materials obtained by this strengthening mechanism are directionally solidified eutectic alloys, fiber reinforced metals of aluminum base matrix alloys and metal filaments or polycrystalline fibers of various materials. [19] In the case of these materials, the interfacial properties have considerable effects on the properties of any composite material, they are difficult to control, and, it is also difficult to predict their nature.

Stoloff indicates the functions of the fiber-matrix interface as:

1. the determination of wetting and bonding;
2. the transfer of stresses from matrix to fibers;
3. the control of fiber spacing;
4. the prevention of fiber-fiber contact;
5. an aid in fiber alignment;
6. service as a diffusion barrier;
7. the protection of fibers from damage;
8. provision as an alternative crack path.

Controllable parameters concerning microstructure are very large, and various approaches have been proposed, for example, the works in Ref. [53].

Yukawa uses DTA data in order to predict the real state of solidifications with stategies similar to the ones of nickel base super alloys.

Attempts to increase creep resistance and stress rupture resistance are focused mainly on nickel and/or iron base super alloys. [20], [21], [34], [39], [49], [50], [51], [52], [53]. They apply the PHACOMP method, which could be theoretically deduced as described earlier, and multiple regression equations which are induced from a large number of relevant data bases. Several fruitful results are mainly due to the compilation of materials data in the form of analytical expressions. As an additional remark, this sort of compilation should be carried out on the basis of theoretical understanding, which is usually summarized in a way as shown in Table 4, so as to restrict relevant data sets to those with no singular data in them.

In general, microstructures are extremely sensitive, not only to alloy chemistry, as shown in Table 4, but also to the thermal-mechanical history, cooling rate, size, welding, chemical environments, etc. A small amount of alloying elements promotes both grain refinement of the austenite and precipitation hardening in the ferrite of high-strength, low-alloy steel, while the cleavage failure, ductile rupture, delamination, Bauschinger effects on flow stress are affected by residual stresses from fabrication, which are the function of the degree and temperature of deformation, the subsequent rate of cooling, alloy contents, shape and distribution of inclusion, etc. [30] Several attempts, however, are induced from complicated results of experiments, by changing the view of alloy designing tactics. Zackay showed an example of using the α-to-γ phase transformation in iron-base alloys strengthened by the Laves phase rather than alloy carbides. [23], [31] Combinations of multiple transformation are proposed by Thomas as tactics for dislocation multiplication. [38] Recrystallization and precipitation are controlled to produce heterogeneous microstructures by thermomechanical treatment. [40] The precise control of alloy contents and the thermomechanical process may produce such fascinating results as superplasticity [41], duplex-precipitation [64], or dual phase [65]. Newer fabrication and/or processing techniques consequently pave the way to newer alloys, namely, metallic glasses [42], amorphous alloys [17], superalloy-powder processing [43], directionally solidified alloys [34], [43], etc.

The limitations of alloy design should be made clear after much consideration. Tien concluded that second phase particles can certainly enhance the flow strength, creep resistance, and stress-rupture life of alloys, but are usually not beneficial alloy-design elements with respect to uniaxial ductility, plane-strain ductility, stress-rupture ductility and toughness. [39] This sort of trade-off is in most cases given to both materials designers and materials users as tentative conclusions of the alloy design or alloy selection. Afterwards the microstructral information available is used for performance modeling as shown in the next section.

As for corrosion resistance, a substantial amount of empirical testing in the environment of interest is required due to the complexity of surface structures and their dynamics as a boundary between environment and substrate. Pettit pointed out that the level of knowledge in this field is sufficient to permit a coupling between the science and art [48], and proposed techniques for the design of corrosion-resistant alloys with qualitative descriptions. [24], [25], [26],

Table 4. Role of Alloying Elements in Superalloys by R.F. Decker et al.

		γ'					Instability			Carbides					Stable Grain
										815 C		980 C			
Element	γss	Vol %	r	γ'ss	Coherency Strains	Ripening	$\gamma' \to \eta'$	$\gamma \to \gamma$bct	$\gamma' \to Ni_3Nb$	MC	$M_{23}C_6$	MC	$M_{23}C_6$	M_6C	Boundaries
Co	↑W	↑M	--	↑W	nil	nil	--	--	--	--	--	--	--	--	↑
Fe	↑M	↑S	--	↑W	↓	nil	--	--	--	↑W	↓W	--	--	--	--
Cr	↑S	↑S	--	↑S	↓	↓	--	--	--	↓M	↑M	↓M	↑S	↓S	--
Mo	↑S	nil	--	↑S	↓	↓	--	--	--	↑S	↓S	--	↓S	↑S	↑
W	↑S	↑M	--	↑S	↓	↓	↓	--	--	↑M	↓M	--	↓S	↑S	--
V	↑W	↑M	--	nil	--	--	--	--	--	--	--	--	--	--	--
Nb	--	↑S	↑	↑M	↑	↓	↑	↑	↑	↑S	↓S	--	--	--	--
Ta	--	↑S	↑	↑M	↑	↓	↑	↑	--	↑M	↓M	↑S	↑S	↓S	--
Ti	--	↑S	↑S	↑S	↑	↑	↑	↓	--	--	--	--	↑S	↓S	--
Al	↑S	↑S	↓	↓	↓	↑	↓	↓	↓	--	--	↓S	↑S	↓S	--
B	--	--	--	--	--	nil	↓	--	--	--	↓*	--	--	--	↑
Zr	--	--	--	--	--	nil	--	--	--	--	↓*	--	--	--	↑
C	--	--	--	nil	↑	--	--	--	--	↑	↑	--	--	--	↓
Mg	--	--	--	--	--	--	--	--	--	↑	↓*	--	--	--	↑

*at grain boundaries ss=solid-solution hardening
W = weak S = strong ↓ = decrease
M = medium ↑ = increase

[48] Very few parameters have been developed from first principles and knowledge about corrosion is generally specific. Thus experience remains the best indicator of alloy-environment behavior. General guidelines for alloy design are at first given as a set of control variables and their qualitative effects. Quantitative predictions are carried out by coupling the general guidelines with the experimental results in almost all cases. Stress corrosion cracking, however, requires predictions and estimations of aggressive environments around cracks and stress states, as well as well-designed experiments. The cases of fuel elements in light water reactors provide good examples that predictions are indispensable rather than effective in materials design.

Microstructural information ranges from the data on point defects (e.g. formation energy of vacancy, its migration energy, those of divacancies, and so on), dislocations, grain boundaries, phase boundaries, etc., to the textures, microstructural anisotropy, segregations, etc. Almost all theories are limited in range and scope, while structure sensitive properties are literally sensitive to each microstructural element. Dominant factors should be extracted in accordance with each application of a material so as to obtain a clear understanding of the material behavior. Here, experience plays an important role and could be represented as an objective knowledge base related to parts of data in the materials data base. One of the approaches to the organization of individual knowledge is the modeling, where various sorts and levels of data are rescaled and arranged according to a scenario. [67], [68]

Modeling is not equal to prediction. It is performed on the basis of the methods of prediction, but it is also based on the analysis of requirements to materials. [33] - [37], [45], [54], [57] Available data are used tentatively for extrapolation and/or interpolation in modeling, whereas microstructural information on the various levels should play the key role in order to maintain the physical base of logical continuity. For example, the dynamic changes of shape with temperature and applied stress could not be used as functional materials for engineering applications without microscopic understanding of the underlying microstructural changes. [56] - [61]

The history of alloy development is also described as the one of microstructural modifications. [62], [63] Further developments may be performed along these lines [3], establishing correlations between control parameters of fabrication and service condition, microstructures, and properties of interest.

Requirements for alloy design must be defined through performance modeling of the designed alloy under the service conditions, even if tentative. Therefore, the relation between the materials data and the models is complementary.

The mechanical designer wants to know the results of more realistic tests rather than fundamental data, for example, of plane-strain ductility and of cracked plates suggestive of various weld defects rather than of uniaxial tensile test data. In such cases of fatigue tests at high temperature ranges it is better to plan experiments effectively by taking advantage of analytical methods, and as a result many predictive methods have been developed. [27], [32], [44], [66], [70] Amost all works on Creep and Fatigue in Table 3 belong to this category.

The alloy designing process in general starts with several reference alloys, the selection of which can be carried out by taking advantage of a materials data base or handbooks, e.g. [70]. The improvement of materials is oriented by requirements, e.g. [69], tactics being selected from deliberate reviews of available knowledge, e.g., [1] - [3], [5] - [72], and finally by trial and error techniques. Analysis follows, including that of microstructures. The process is repeated until the goal of alloy development is reached.

4. CONCLUSION

(i) Data for alloy design have several levels corresponding to one's view, ranging from nuclear data to performance data.

(ii) The process of alloy design is divided into two phases, qualitative design and quantitative design.

(iii) Qualitative design is carried out based on our understanding of each event, while quantitative design should be performed with the materials data base and modeling.

(iv) The role of data is twofold, snapshots to provide better understanding, and input data for models.

(v) The data for the selection of reference alloys for alloy design are available as handbooks, review papers, or materials data bases, but our knowledge on qualitative design is not available in a well-characterized form.

(vi) The first step in alloy design is to represent our understanding on alloys sufficiently so as to induce and/or deduce tactics for alloy design by a set of numerical data and their relations, even if they are only implicit representations.

ACKNOWLEDGEMENT

The author would like to express his gratitude to Emeritus Prof. Mishima and Prof. Ishino for their discussions on alloy design for over more than a decade. Also, support for the presentation of this paper by CODATA is gratefully acknowledged.

REFERENCES:

[1] Iwata, S., Ishino, S. and Mishima, Y., Alloy Design by Automatic Modeling and Estimation of Values from Experimental Data, J. of Fac. of Eng'g, The Univ. of Tokyo(B) 33(1976)545-610.

[2] Corey, E.J. and Wipke, W.T., Computer-Assisted Design of Complex Organic Syntheses, Science 166 (3902) (1969) 178-92.

[3] Hirth, J.P. and Clauer, A.H., Concluding Agenda Discussion-Critical Issues, in Jaffee, R.I. and Wilcox, B.A. (eds.), Fundamental Aspects of Structural Alloy Design (Plenum Press, New York, 1977).

[4] Yamamoto, T. et al., TOOL-IR: An On-Line Information Retrieval System at An Inter-University Computer Center, in Proc. 2nd UJCC (Japan Convention Services, 1975).

[5] Brandt, W. and Kitagawa, M., Effective stopping-power charges of swift ions in condensed matter, Phys. Rev. B 25 (1982) 5631-5637.

[6] Moruzzi, V.L., Williams, A.R., and Janak, J.F., Local density theory of metallic cohesion, Phys. Rev. B 15 (1977)2854-2856.

[7] Moriarty, J.A., Simplified local-density theory of the cohesive energy of metals, Phys. Rev. B 19 (1979)609-619.

[8] Chelikowsky, J.R. and Phillips, J.C., Quantum-defect theory of heats of formation and structural transition energies of liquid and solid simple metal alloys and compounds, Phys. Rev. B 17 (1978)2453-2477.

[9] Watson, R.E. and Bennett, L.H., Transition metals : d-band hybridization, electronegativities and structural stability of intermetallic compounds, Phys.Rev. B 18 (1978)6439-6449.

[10] Watson, R.E. and Bennett, L.H., Electron Factors in the Occurrence of Sigma and structurally Related Transition Metal Alloy Phases, Scripta Metallurgica 12 (1978) 1165-1170.

[11] Machlin, E.S., and Shao, J., Alloy Design for Long Time Stability. Final Report, EPRI-FP-1068.

[12] Machlin, E.S., Correction Terms to Pair Potential Model Values of the Energy of Formation for Transition Element-Polyvalent Element Phases, CALPHAD. 5(1981)1-7.

[13] Coldren, A.P. and Eldis, G.T., Using CCT Diagrams to Optimize the Composition of an As-Rolled Dual-Phase Steel, J. of Metals, March(1980)41-48.

[14] Carberg, T. and Fredriksson, H., The Influence of Microgravity on the Solidification of Zn-Bi Immiscible Alloys, Met. Trans. 11A(1980)1665-1676.

[15] Gamo, K. et al., Control of Tc for Niobium by Nitrogen Ion Implantation, JPN. J. appl. Phys. 16(1977)1853-7.

[16] Ogawa, K., Alloy Design from the Viewpoint of Physical Metallurgy, J. Mat. Sci. Soc. of JPN. 16(1979)204-209.

[17] Masumoto, T., Designing the Composition and Heat Treatment of Magnetic Amorphous Alloys, Mater. Sci. & Eng. 48(1981)147-165.

[18] Copley, S.M. and Williams, J.C., High-Strength Nonferrous Alloys, in Tien,J.K. and Ansell, G.S.(eds.), Alloy and Microstructural Design (Academic Press, New York, 1976).

[19] Stoloff, N.S., Composite Stregthening, ibid.

[20] Tien,J.K., Malu, M. and Purushothaman, Creep Resistance, ibid.

[21] Freche, J.C., Stress Rupture Resistance, ibid.

[22] Laird, C., Fatigue Resistance, ibid.

[23] Zackay, V.F. and Parker, E.R., Fracture Toughness, ibid.

[24] Duquette, D.J., Aqueous and Stress Corrosion Resistance, ibid.

[25] Berstein, I.M. and Thompson, A.W., Resisting Hydrogen Embrittlement, ibid.

[26] Pettit, F.S., Giggins, C.S., Goebel, J.A. and Felten, E.J., Oxidation and Hot Corrosion Resistance, ibid.

[27] Li, C.Y., Ellis, F.V. and Huang, F.H., Mechanical Equations of State, ibid.

[28] Haasen, P., Mechanical Properties of Solid Solutions, in Ref.(3).

[29] Frost, H.J. and Ashby, M.F., Deformation-Mechanism Maps for Pure Iron, Two Austenitic Stainless Steels and a Low-Alloy Ferritic Steel, ibid.

[30] Embury, J.D., Evensen, J.D. and Filipovic, A., The Mechanical Properties of High-Strength Low-Alloy Steels, ibid.

[31] Zackay, V.F. and Parker, E.R., Progress in Ferrous Alloy Design, ibid.

[32] MaClintock, F.A., Mechanics in Alloy Design. ibid.

[33] Shewmon, P.G., Radiation Environments and Alloy Behavior, ibid.

[34] Versnyder, F.L. and Gell, M., New Directions in Alloy Design for Gas Turbines, ibid.

[35] Simenz, R.F. and Steinberg, A., Alloy Needs and Design, ibid.

[36] Bement, A.L., Needs for Alloy Design for Nuclear Applications, ibid.

[37] McMahon, C.J.Jr., Problems of Alloy Design in Pressure Vessel Steels, ibid.

[38] Thomas, G., Utilization and Limitations of Phase Transformations and Microstructures

in Alloy Design for Strength and Toughness, ibid.

[39] Tien, J.K., Alloy Design With Oxide Dispersions and Precipitates, ibid.

[40] Hornbogen, E., Design of Heterogeneous Microstructures by Recrystallization, ibid.

[41] Alden, T.H., Processing and Properties of Superplastic Alloys, ibid.

[42] Davis, L.A., Metallic Glasses, ibid.

[43] Kear, B.A. and Giamei, A.F., The Manipulation of Superalloy Microstructures by Advanced Processing Techniques, ibid.

[44] Sanders, T.H.Jr., Mauney, D.A. and Staley, J.T., Strain Control Fatigue as a Tool to Interpret Fatigue Initiation of Aluminum Alloys, ibid.

[45] Frost, B.R.T., Radiation Damage Considerations, ibid.

[46] Campbell, J.D., Eleiche, A.M. and Tsao, M.C.C., Strength of Metals and Alloys at High Strains and Strain Rate, ibid.

[47] Suh, N.P., Microstructural Effects in Wear of Metals, ibid.

[48] Pettit, F.S., Design of Structural Alloys With High Temperature Corrosion Resistance, ibid.

[49] Watanabe, R., Alloy Design of Solid Solution Strengthened Nickel-Chromium-Tungsten Super-Alloys, Tetsu to Hagane 63(1977)118-24.

[50] Kodaira, T., Material Design Data of 2.25 Cr-1Mo Steel and Hastelloy-X for the Experimental Multi-purpose Very-High Temperature Gas-Cooled Reactor, JAERI-M-6213 Report, Japan Atomic Research Institute (1975).

[51] Harada, H. et al., A Series of Nickel-base Superalloys on $\gamma-\gamma'$ Tie Line of Alloy Inconel 713, Tetsu to Hagane 65(1979)1049-1058.

[52] Harada, H. and Yamazaki, M. Alloy Design for γ' Precipitation Hardening Nickel-base Superalloys Containing Ti, Ta, and W, ibid. (1979)1059-1068.

[53] Yukawa, N., Correlation of Microstructures and Properties of Directionally Solidified (Ni,Cr)-Cr_7C_3 Eutectic Alloy, Japan Society for the Promotion of Science, 123rd Committee(Heat Resisting Metals and Alloys) Report Vol.19 No.7(1978).

[54] German, R.M. et al., Color and Color Stability as Alloy Design Criteria, J. Metals March(1980)20-27.

[55] Thompson, A.W. and Wonsiewicz, B.C., The Role of Texture in Work Hardening of Copper, Met. Trans., 12A(1981)531-534.

[56] Wayman, C.M., Deformation, Mechanisms and Other Characteristics of Shape Memory Alloys, in Perkins, J.(eds.), Shape Memory Effects in Alloys (Prenum press, New York, 1975).

[58] Falk, F., Model Free Energy, Mechanics, and Thermodynamics of Shape Memory Alloys, Acta Metall. 28(1980)1773-1780.

[59] Bricknell, R.H. and Melton, K.N., Thin Foil Electron Microscope Observations on NiTiCu Shape Memory Alloys, Met. Trans. 11A(1980) 1541-1546.

[60] Shapiro, S.M., Neutron Scattering Studies of Pretransitional Phenomena in Structural Phase Transformations, Met. Trans. 12A (1981)567-573.

[61] Tirbonod, B. and Koshimizu, S., Dislocation Relaxation in the Martensitic Phase of the Thermoelastic NiTi and NiTiCu Alloys, J. de Physique (1981)C5-1043-47.

[62] Inman, P.F., Understanding Copper Alloys (John Wiley & Sons, New York, 1980)

[63] Pickering, F.B., Physical Metallurgy and the Design of Steels (Applied Science, London, 1978).

[64] Ohashi, T. and Ichikawa, R., Duplex-Precipitation Hardening in Al-Zn-Mg Alloys Highly Supersaturated with Zr, Met. Trans. 12A(1981)546-549.

[65] Kim, N.J. and Thomas, G., Effects of Morphology on the Mechanical Behavior of a Dual Phase Fe/2Si/0.1C Steel, Met. Trans. 12A(1981)483-489.

[66] Wilson, D.A., and Hoeppner, D.W., A Statistical Investigation of Microstructure and Crack Growth in Titanium Alloys, J. Eng. Power, Trans. ASME103(1981)180-7.

[67] Odette, G.R., Modeling of Microstructural Evolution Under Irradiation, J. Nucl. Mater. 85&86(1979)533-543.

[68] Stoller, R.E. and Odette, G.R., A Model Based Fission-Fusion Correlation of Cavity Swelling in Stainless Steel, J. Nucl. Mater. 103&104(1981)1361-1365.

[69] Stein, C. (eds.), Critical Materials Problems in Energy Production (Academic Press, New York, 1976).

[70] Damage Tolerant Design Handbook, MCIC-HB-01, (Battelle Columbus Laboratories, 1975).

[71] Structural Alloys Handbook, MPDC, A DOD Materials Imformation Center, (Belfour Stulen, Michigan, 1977).

[72] Fusion Reactor Materials Data, Supported by the Grant-in-Aid for Fusion Research, The Ministry of Education, Science and Culture, (1982).

*[57] Halbach, K., Design of Permanent Multipole Magnets with Oriented Rare Earth Cobalt Material, Nucl. Instr. and Meth., 169(1980) 1-10.

Data for Science and Technology, P.S. Glaeser (ed.)
North-Holland Publishing Company

ESTIMATION OF SOME MATERIALS DATA BY ARTIFICIAL INTELLIGENCE

Zdzislaw Hippe

The I. Łukasiewicz Technical University
35-959 Rzeszów, Poland

The estimation of structural data of organic materials might be done by means of heuristic computer programs that mimic the sequence of operations of human beings while interpreting empirically various types of molecular spectra. The paper describes briefly a general strategy of computer-assisted structure elucidation and gives some real examples obtained by means of originally developed software SEAC-2.

1. INTRODUCTION

Sophisticated computer program systems for information storage and retrieval of data for materials science are important topices in artificial intelligence (AI). However, the basic principle of operation of such systems - accessing encyclopedic data bases of well-indexed facts - is simultaneously the most limiting factor of their application (1). This is because data are now available for only a very negligible number of materials, substances, mixtures, etc. Hence, an entirely new approach seems to be a more promising solution to the problem stated: an "intelligent" base with a limited number of universal data (constants) and facts about relevant theories, to provide the internal possibility of an automatic prediction of any required property of materials. This was successfully realized for the automatic accessing of various materials data, mostly physical or physico-chemical, for elements, metals, alloys, pure substances or even mixtures (2-5). However, there are many instances where some other types of data besides numerical ones are required. One of the most important types of materials data - for chemical substances - is the molecular structure, elucidated from other data, usually obtained by empirical interpretation of molecular spectra, like mass-spectrum (MS), nuclear magnetic resonance spectrum (proton or ^{13}carbon, ^{1}H-NMR or ^{13}CNMR, respectively), or infra-red spectrum (IR).

2. GENERAL BACKGROUND OF COMPUTER-ASSISTED STRUCTURE ELUCIDATION

The first time the computer was applied to automatize structural investigations was in the United States at Stanford University for the interpretation of mass-spectra (6); the original computer program system DENDRAL has been the subject of constant development (7-11). Somewhat later, from other research groups, investigations on so-called integrated computer program systems for structure elucidation were also reported, primarily those from Arizona State University (12,13), Toyohashi University of Technology (14,15), Technical University of Rzeszow (16-18) and Institute of Analytical Chemistry and Geochemistry of the Academy of Sciences of U.S.S.R. (19,20).

The overall strategy of computer-assisted elucidation of chemical structures consists of five formal steps (21):

(a) correlation, inferring the possible structural fragments by means of one or any combination of spectral methods: MS, ^{13}C-NMR, ^{1}H-NMR, IR, Raman, UV,
(b) consistency test, selection from the set of structural fragments found in step (a) those subsets that are internally consistent,
(c) structural assembly, combination of partial structures (substructures) found in step (b) into a meaningful total structure (tentative or candidate structure),
(d) spectrum prediction, prediction of selected spectral features (or the whole spectrum) for the candidate structure arrived at in step (c), and experimental spectra are compared. If they agree, the candidate structure may be correct. Go back to step (c) to generate another tentative structure; if no other candidate structures can be generated, then stop.

The versatility of such a strategy depends strongly on the efficiency of the algorithms (applied in subsequent steps); they usually reflect the skill and experience of the algorithm designer. It is believed that the actual performance of computer program systems for structure elucidation by AI may be described as follows: they operate generally at the level of a post-doctoral analyst; their performance is good not because they know any more than experienced spectroscopists, but because:

(a) they use most of the rules applied by an analyst to solve structure determination problems,
(b) they apply the same set of rules in every task, even in routine problems,
(c) they apply systematically the whole set of rules each time, without mistakes and loss of memory (22).

3. AN EXAMPLE OF ESTIMATION OF MOLECULAR DATA BY AN AI-PROGRAM

The general algorithm for computer-assisted elucidation of structural data may be illustrated by real results obtained by means of the AI-program system called SEAC-2. This user-oriented software may be exploited on one of six realization levels dependent on the information entropy of the input data. Results of estimation for an organic material

$$\left(\begin{matrix}CH_3\\ CH_3\end{matrix}\!\!>N{-}CH_2{-}CH_2{-}\underset{\underset{O}{\|}}{C}{-}O{-}CH_3\right)$$

throughout all the realization levels are exemplified by computer printouts.

At level I (only IR-spectrum is available), the set of substructures detected contains all substructures which are included in the molecule, but it also contains groups with atoms which are not present in the compound. This fact is due to the inherent ambiguity of any spectral technique used. (See computer printout below.)

```
          INPUT DATA
          **********

QUALITATIVE COMPOSITION: UNKNOWN
IR-SPECTRUM:
W(1/CM)     T(%)      D(1/CM)
**************************
2959         31         60
2778         23         80
1754          3          0
1460         30          0
1370         23          0
1227          1          0
1163         66          0
1099         54          0
1036         33          0
 966         60         40

           RESULTS:
           ********

REALIZATION LEVEL I
===================

STRUCTURAL GROUPS FOUND (IR-ANALYSIS)

SET NO      GROUP         PROBABILITY, %
************************************
   1        CH3               40
   2        CH2               35
   3        CO-O-CO           20
   4        CO-O              20
   5        CO                20
   6        O-CO-O            20
   7        N=O(N)            20
            N=O               20
   8        SO                20
   9        C-SO2             20
  10        C-SO2-O           20
```

When information about qualitative composition is used (level II), any substructure containing elements that are not present in the test substance are rejected. (See computer printout below.)

```
          INPUT DATA
          **********

QUALITATIVE COMPOSITION: CHNO

              RESULTS:
              ********

REALIZATION LEVEL II
====================

STRUCTURAL GROUPS FOUND (IR-ANALYSIS)

SET NO      GROUP           PROBABILITY, %
*************************************
   1        CH3                 40
   2        CH2                 35
   3        CO-O-CO             20
   4        CO-O                20
   5        CO                  20
   6        C-CO-O              20
   7        C*NO                20
   8        N=O(N)              20
            N=O                 20
```

At level III, apart from the detected structural fragments, the computer prints sets of substructures which supplement the empiric formula. (See computer printout below.)

```
            INPUT DATA
            **********

EMPIRIC FORMULA: C 6  H 15  N 1  O 2
DEGREE OF UNSATURATION = 1

               RESULTS:
               ********

REALIZATION LEVEL III
=====================

STRUCTURAL GROUPS FOUND (IR-ANALYSIS)

SET NO      GROUP           PROBABILITY; %
*************************************
   1        CH3                 40
   2        CH2                 35
   3        CO-O                20
   4        CO                  20
   5        N=O(N)              20

SET OF SUBSTRUCTURES FULFILLING EMPIRIC
FORMULA

SET NO      GROUP           PROBABILITY, %
*************************************
   1
          4 CH3                 40
            CO-O                20
            N
            CH
   2
          3 CH3                 40
          2 CH2                 35
            CO-O                20
            N
```

Identification by IR and ^{1}H-NMR, when the qualitative composition is known (level IV), increases the identification probability for groups detected by IR-spectrometry and supplements the set by substructures with specified molecular environments. (See computer printout on following page.)

INPUT DATA
****

QUALITATIVE COMPOSITION: CHNO

IR-SPECTRUM:

W(1/CM)	T(%)	D(1/CM)
2959	31	60
2778	23	80
1754	3	0
1460	30	0
1370	23	0
1227	1	0
1163	66	0
1099	54	0
1036	33	0
966	60	40

NMR-SPECTRUM:

TAU(PPM)	INT.(H)	MULT./C.CONST.
5.95	2	3.07
7.55	2	3.07
7.80	6	1.00
8.05	3	1.00

RESULTS:
****

REALIZATION LEVEL IV

STRUCTURAL GROUPS FOUND (IR-ANALYSIS)

SET NO	GROUP	PROBABILITY, %
1	CH3	40
2	CH2	35
3	CO-O-CO	20
4	CO-O	20
5	CO	20
6	C-CO-O	20
7	C*NO	20
8	N=O(N)	20
	N=O	20

STRUCTURAL GROUPS FOUND (NMR-ANALYSIS)

SET NO	GROUP	PROBABILITY, %
1	CH2(CH2,N)	22
	CH2	22
	CH2(CH2,O)	22
	CH2	22
2	CH2(CH2,CO)	15
	CH2	15
	CH2(CH2,N)	15
	CH2	15
3	CH3(N)	15
	CH3	15
	CH3(CO)	15
	CH3	15
4	CH3(CO)	22
	CH3	22
	CH3(C)	22
	CH3	22

At level V, when the empirical formula of the material being analyzed is known, and both spectra (IR+^{1}H-NMR) are given, the computer creates the connectivity matrix of the structure examined. Here any column vector contains the information about connectivities of the given substructure: 1 stands for single-bond connection, whereas 0 points out no connectivity between considered species. For instance, 1 in the third row of the first column informs us that substructure no. 1 (N) is connected to substructure no. 3 (CH2) by a single bond. Thus any structure may easily be redrawn from the matrix. (See computer printout below.)

INPUT DATA
****

EMPIRICAL FOMULA: C 6 H 15 N 1 0 2
DEGREE OF UNSATURATION = 1

RESULTS:
****

REALIZATION LEVEL V

STRUCTURAL GROUPS FOUND (IR+NMR-ANALYSIS)

SET NO	GROUP	PROBABILITY, %
1	CH3	40
2	CH2	35
3	CO-O	20
4	CO	20
5	N=O(N)	20
6	CH2(CH2,O)	22
	CH2(CH2,N)	22
7	CH2(CH2,CO)	22
	CH2(CH2,N)	22
8	CH3(CO)	22
	CH3(N)	22
9	CH3(CO)	22
	CH3(C)	22

CHEMICAL FORMULA
****

CONNECTION MATRIX NO 1

NO	GROUP	1	2	3	4
1	N	0	0	1	0
2	CH2	0	0	1	1
3	CH2	1	1	0	0
4	CO-O	0	1	0	0
5	CH3	0	0	0	1
6	CH3	1	0	0	0
7	CH3	1	0	0	0

At level VI, which yields the same results but in much shorter machine-time, the user has the possibility of an interactive change of the content of the substructure set used as input data for structure generation. In other words, the user may formulate some constraints, for instance - based on his (or her) own knowledge about the substance being analyzed - to join some of the detected substructures into a larger macrofragment. The results presented here are only a small example of recent solutions and possibilities in the application of artificial intelligence in accessing selected materials data. It may well be predicted that future developments in this field may be supercomputers (fast computers with large main-line memories) and by simultaneous processing of spectral data from more than two or three spectral techniques.

REFERENCES

[1] Nilsson, N.J. Problem-solving in Artificial Intelligence, McGraw-Hill, Inc., New York, 1971, p.1.

[2] Heller, R.S., Milne, G.W.A., Feldman, R.J. and Heller, S.R., J. Chem. Inf. Comp. Sci. 16 (1976) 176.

[3] Gurvich, L.V., Iorish, V.S. and Yungman, V.S., 8th CODATA Conference, Jachranka (October 1982).

[4] Yamazaki, M., Yoshida, S., Ihara, H. and Gonda, S., ibid.

[5] Maczynski, A., ibid.

[6] Lederberg, J., Sutherland, G.L., Buchanan, B.G., Feigenbaum, E.A., Robertson, A.U., Duffield, A.M. and Djerassi, C., J. Am. Chem. Soc., 91 (1969) 2973.

[7] Delfino, A.B. and Buchs, A., Fortschr. d. chem. Forsch., 39 (1979).

[8] Michie, D. and Buchanan, G.B. in Carrington, R.A.G. (Ed.), Computer for Spectroscopists, A. Hilger, Ltd., London, 1974, p. 114.

[9] Smith, D.H., Masinter, L.M. and Sridharan, N.S., in Wipke, W.T., Heller, S.R., Feldman, R.J. and Hyde, E., Computer Representation and Manipulation of Chemical Information, J. Wiley, New York, 1974, p. 287.

[10] Lindsay, R.K., Buchanan, G.B., Feigenbaum, E.A. and Lederberg, J., Applications of artificial intelligence for organic chemistry - The Dendral Project. McGraw-Hill, Inc., New York, 1980.

[11] Smith, D.H., Gray, N.A.B., Nourse, J.C. and Crandell, C.W., Anal. Chim. Acta, Comp. Techn. and Optim., 133 (1981) 471.

[12] Shelley, C.A., Woodruff, H.B., Snelling, C.R. and Munk, M.E., in Smith, D.H. (Ed.), Computer-Assisted Structure Elucidation, ACS Symposium Series 54, American Chemical Society, Washington D.C., 1977, p. 92.

[13] Shelley, C.A. and Munk, M.E. Anal. Chim. Acta, Comp. Techn. and Optim., 133 (1981) 507.

[14] Yamasaki, T., Abe, H., Kudo, Y. and Sasaki, S., in Smith, D.H. (Ed.), Computer Assisted Structure Elucidation, ACS Symposium Series 54, American Chemical Society, Washington, D.C., 1977, p. 108.

[15] Sasaki, S., Abe, H., Fujiwara, I. and Yamasaki, T., in Hippe, Z. (Ed.), Data Processing in Chemistry, PWN-Elsevier, Warsaw, 1981, p. 186.

[16] Hippe, Z., Hippe, R., Duliban, J. and Debska, B., Techn. Univ. Res. Project, Rzeszow, 1981.

[17] Debska, B., Guzowska-Swider, B. and Duliban, J., Proc. Int. Conf. Data Processing in Chemistry, Techn. Univ. Publ. Rzeszow, 1979, p. 71.

[18] Debska, B., Duliban, J., Guzowska-Swider, B. and Hippe, Z., Anal. Chim. Acta, Comp. Techn. and Optim., 133 (1981) 303.

[19] Gribov, L.A., Elyashberg, M.E. and Serov, V.V., ibid. 95 (1977) 75.

[20] Gribov, L.A., ibid., 122 (1980) 249.

[21] Clerc, J.T. and Koenitzer, H., in Hippe, Z. (Ed.), Data Processing in Chemistry, PWN-Elsevier, Warsaw, 1981, p. 151.

[22] Slagle, J.R., Artificial Intelligence. The Heuristic Programming Approach, McGraw-Hill, Inc., New York, 1971.

Data for Science and Technology, P.S. Glaeser (ed.)
North-Holland Publishing Company

A SEMICONDUCTOR DATABASE AND ITS APPLICATION TO HETEROSTRUCTURE DESIGN

Masato Yamazaki, Sadafumi Yoshida, Hideo Ihara and Shun-ichi Gonda

Electrotechnical Laboratory
1-1-4 Umezono, Sakura-mura, Ibaraki, 305, Japan

This paper discusses a semiconductor database for III-V compounds and a computer-based design-aid system using the database. A materials modelling concept that systematizes various properties by interpolation procedures is introduced for the design-oriented database. A LISP language is conveniently used to embed such procedures in each data record.
Input/output capabilities of graph data using a graphic tablet and a graphic display terminal enable easy compilation and complete verification of the data.
Some examples of semiconductor materials design using the design-aid system called CAMS are shown, and the materials modelling concept is proved to be important for the task.

1. INTRODUCTION

Recent advances in semiconductor physics and related device technologies have shown us extended possibilities in this field. However, to fully develop and utilize them, a variety of data including crystal structures, thermal properties, electronic band strucutres, opto-electronic properties, etc. become necessary and a materials database has been awaited, especially for compound (including solid solution) semiconductors.

We have developed a semiconductor database and are now developing a computer-based design-aid system called CAMS for III-V group materials.
A materials modelling concept has been introduced for the design-oriented database. A materials model is defined as the complete description of certain properties using interpolation procedures including a set of continuous formulae. The materials model has been constructed for the lattice parameters and the energy band gaps.

Using this model, materials design, especially for ternary and quaternary solid solutions, becomes realistic. The design of lattice matched heterostructures such as laser diodes and superlattices has also been greatly supported.

2. SEMICONDUCTOR DATABASE

2.1 Database structure

To develop a computer-aided materials design system, we need a systematized collection of data for the target materials. We think a design oriented database ought to consist of two parts, one is a collection of raw data and the other is an elaborated or systematized one which we call a materials model.

Figure 1 shows the relationships among a raw database, a materials model and design-aid programs. The former preferably should contain as much experimental data as possible and forms the

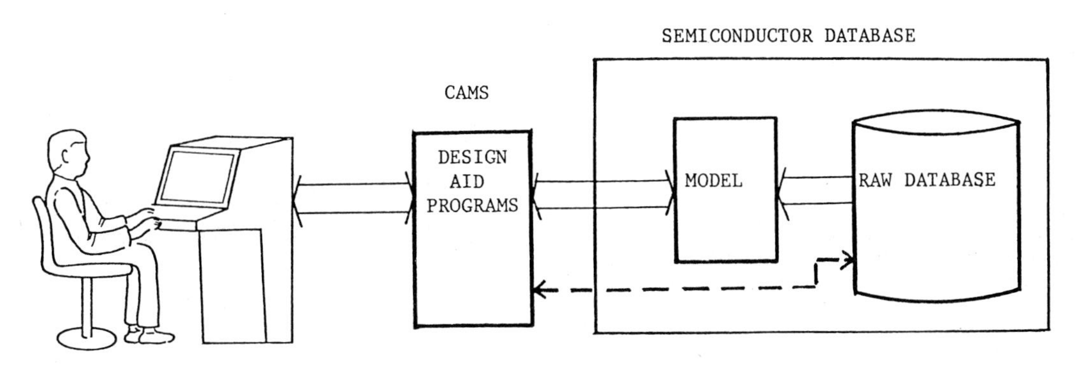

Figure 1 : Structure of the semiconductor database and the design-aid system CAMS

basis of the latter. It may contain large deviations and usually does not cover all materials and conditions we want to consider. On the other hand, the latter should be constructed from a set of highly evaluated data and should cover the whole range. It preferably should be a set of continuous representations of the properties of every material.

Whether or not a good materials model can be constructed depends on the quality of the data and the estimation theory of the field. If the materials model is to be satisfactory, the design-aid programs need not access the raw database, so they should be made with an elaborated structure and have many possibilities. But if, on the contrary, the model is poor, the design-aid programs would be difficult to construct and if they could be made, their ability would be low. In general, whether good theoretical estimation is available or not decides the level of the model and decides the possible level of the design-aid programs.

2.2 Raw semiconductor database

In the construction of the database, we restricted the materials into III-V compound semiconductors. The materials set was chosen not only because it is important for many fields of application but also because it has been investigated extensively and a reasonable amount of data are expected to be available. The materials included in this database are listed in Table 1, where underlined names indicate that very few data could be obtained for them.

Accumulated properties are crystal types, lattice parameters, their thermal properties, energy band gaps, electron affinities, electron effective masses, electron mobilities, refractive indices, etc. These are collected mainly from the data books by JEIDA (The Japan Electronics Industry Development Association) (1)(2) and other references (3)(4).

The graph input facility which we have developed for the purpose makes it possible to easily compile many graph data with sufficient precision. A calibration method of graph input also has been developed. The complete verification of the input data can be accomplished by reproducing them on the graphic display immediately after the input. The graphic tablet and the monitor display used in our system are shown in Figure 2.

Figure 3 shows an example of the graph data input, (a) is the original graph, (b) a part of the record created on a disc file, and (c) a reproduced graph. In the database, the scales can be changed arbitrarily and data of the same kind can be merged at ease, so that comparisons among such data can be performed conveniently.

2.3 Semiconductor model

A materials model has been constructed for the III-V group binary, ternary and quaternary semiconductors. The lattice parameters and the energy band gaps have already been modelled.

The modelling process is as follows. Firstly, an appropriate consistent set of data is selected from the raw database. At this stage, facilities of merging and graphical output of the data help us very much. Secondly, formal representations are made for the data. Afterwards, these representations are embedded in the data model in LISP language. The following are examples of the modelling processes.

GROUP	TYPE	MATERIALS
BINARY	III - V	Al-P Al-As Al-Sb Ga-P Ga-As Ga-Sb In-P In-As In-Sb
TERNARY	III - III - V	AlGa-N AlGa-P AlGa-As AlGa-Sb AlIn-P AlIn-As AlIn-Sb GaIn-P GaIn-As GaIn-Sb
	III - V - V	Al-PAs Al-PSb Al-AsSb Ga-PAs Ga-PSb Ga-AsSb In-PAs In-PSb In-AsSb
QUATERNARY	III - III - V - V	AlGa-PAs AlGa-PSb AlGa-AsSb AlIn-PAs AlIn-PSb AlIn-AsSb GaIn-PAs GaIn-PSb GaIn-AsSb
	III - III - III - V	AlGaIn-P AlGaIn-As AlGaIn-Sb
	III - V - V - V	Al-PAsSb Ga-PAsSb In-PAsSb

Table 1 : Semiconductors in the database

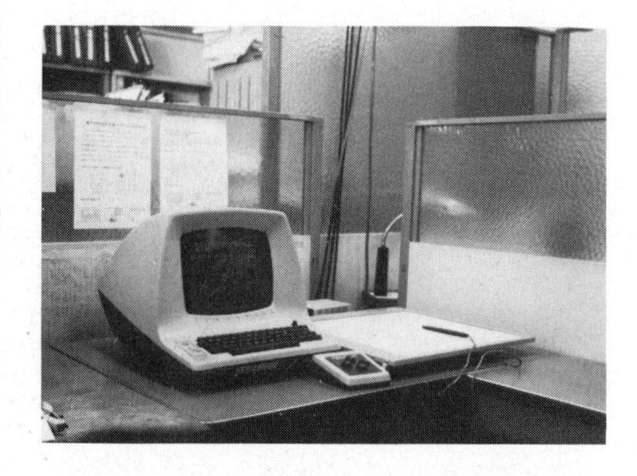

Figure 2 : Graph input/ output equipment

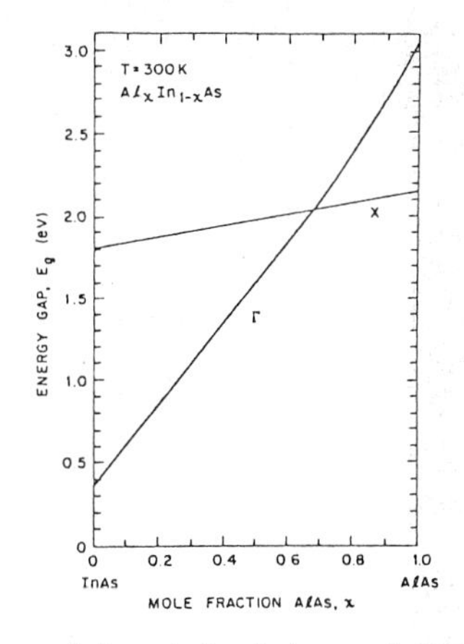

(a) original data (Ref.3)

```
FIG#= 100 FIGID=CP5.3.3          BGN= 105000 END
 TITLE=COMPOSITIONAL DEPENDENCE OF THE DIRECT
AL>X<IN>1-X<AS>
TGCOMPOSITIONAL DEPENDENCE OF THE DIRECT AND
IN>1-X<AS>
KWM=<AL>X<IN>1-X<AS>,X=MOLE FRACTION <AL><AS>
TXMOLE FRACTION <AL><AS>
AX .000000E+00, .100000E+01,LIN,NON
TYENERGY GAP EG (EV)
AY .000000E+00, .400000E+01,LIN,EV
LN 1, .458089E+00, .137029E+01,GAMMA DIRECT
NP  25
P1 .325767E-02, .354143E+00
P1 .321121E-01, .442205E+00
P1 .793275E-01, .550042E+00
P1 .122441E+00, .653994E+00
P1 .171659E+00, .765936E+00
P1 .210669E+00, .870083E+00
P1 .259873E+00, .990221E+00
P1 .311094E+00, .111442E+01
P1 .362302E+00, .124274E+01
```

(b) digitized

```
FIG. #100      FIG. ID=CP5.3.3      DATE =6,24,82
TITLE  =COMPOSITIONAL DEPENDENCE OF THE DIRECT AND THE INDIRECT ENERGY G
        APS FOR <AL>X<IN>1-X<AS>
KEYWDS =M=<AL>X<IN>1-X<AS>,X=MOLE FRACTION <AL><AS>,Y=ENERGY GAP,T=300K
UNIT-X =NON
UNIT-Y =EV
```

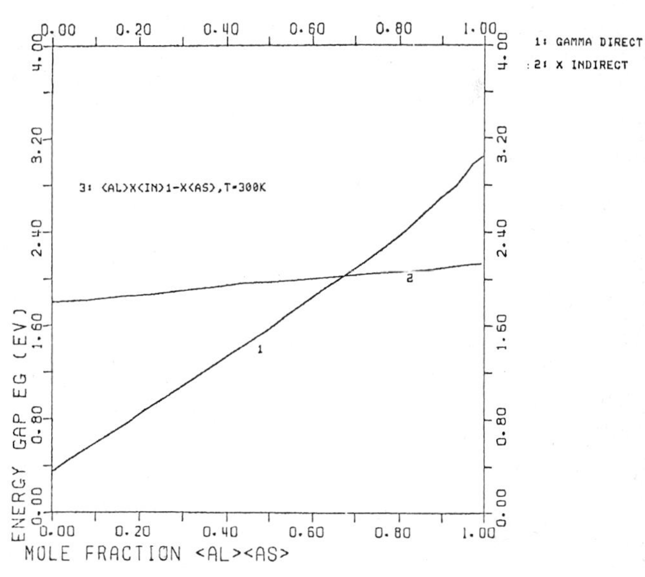

(c) reproduced

Figure 3 : An example of graph data input and reproduction

Modelling of lattice parameters

All the materials in Table 1 except for AlGa-N are cubic zincblende structures. The lattice constants of the binary materials are given in Table 2. In addition, thermal expansion properties are also given for some of the materials. For example, the thermal expansion for GaAs is given as;

$$\frac{\Delta L}{L_0}(\%) = -0.151 + 4.239\times10^{-4}T + 2.916\times10^{-7}T^2 - 9.36\times10^{-11}T^3 \quad (200\text{–}1000\ K) \qquad \text{--- (1)}$$

$$= -0.0894 - 1.135\times10^{-4}T + 1.535\times10^{-6}T^2 \quad (0 - 200\ K) \qquad \text{--- (2)}$$

MATERIAL	LATTICE CONSTANT at 300 k (Å)
Al-P	5.451
Al-As	5.6613
Al-Sb	6.1358
Ga-P	5.45122
Ga-As	5.65325
Ga-Sb	6.09601
In-P	5.86899
In-As	6.0587
In-Sb	6.47944

Table 2: Lattice constants of the binary materials

The modelized representation of the lattice constant of Ga-As are shown in Figure 4. This means its crystal structure is the cubic zincblende and the lattice constant 'a' is given as 5.65325 angstroms at 300 K and at other temperatures calculated by the LISP function equivalent to formulae (1) and (2).

```
12.000   [GA-AS [A ( ZINCBLENDE
12.100    (A [COND
12.200       ((= TMP 300.)(LIST 5.65325 1))
12.300       ((AND (>= TMP 200.)(<= TMP 1000.))(LIST (* 5.65325 (+ 1.
12.400          (- .00151) (* 4.239E-6 TMP)(* 2.916E-9 TMP TMP)
12.500          (- (* 9.360E-13 TMP TMP TMP)) )) 2))
12.600       ((AND (>= TMP 0.)(< TMP 200.))(LIST (* 5.65325 (+ 1.
12.700          (- 8.54E-4)(- (* 1.135E-6 TMP))(* 1.535E-8 TMP TMP))) 3))
12.800       (T 'DATA_NOT_AVAILABLE)]) ) ] ]
```

Figure 4 : Lattice parameter model of Ga-As

For ternary and quaternary systems, a linear interpolation formula called Vegard's law is thought to hold as the first approximation. Thus for $A_xB_{1-x}C$ type ternary materials, the lattice constant is calculated as;

$$Aabc = Aac \cdot x + Abc \cdot (1-x), \quad \text{--- (3)}$$

for $A_xB_{1-x}C_yD_{1-y}$ quaternary materials

$$Aabcd = Aac \cdot xy + Abc \cdot (1-x)y + Aad \cdot x(1-y) + Acd \cdot (1-x) \cdot (1-y), \quad \text{--- (4)}$$

and for $(A_xB_{1-x})_yC_{1-y}D$ type quaternary materials

$$Aabcd = Aad \cdot xy + Abd \cdot (1-x)y + Acd \cdot (1-y). \quad \text{--- (5)}$$

Figure 5 shows the cases of $Al_xGa_{1-x}As$, $Al_xIn_{1-x}P_ySb_{1-y}$ and $(Al_xGa_{1-x})_yIn_{1-y}As$. In this figure, functions 'V3-VEGARD','V4a-VEGARD' and 'V4B-VEGARD' correspond to formulae (3),(4) and (5), respectively.

Modelling of energy band gaps

The energy band gap is the most important parameter that directly relates to the electronic and opto-electronic properties of a semiconductor. It is defined as the separation between the energy of the lowest conduction band and that of the highest valence band. Considering the energy-momentum (E-K) space, the upper edge of the valence band always appears at K=0 (Gamma point), while the bottom of the conduction band can appear either at K=0 (Gamma point), or along the <111>axes (L point), or along the <100>axes (X point). If these extreme points are at the same position, an electron can make a transition from the valence band to the conduction band (or vice versa) directly, so such semiconductors are called direct. Otherwise the transition must involve the emission or absorption of a phonon and the transition is not direct, so these are called indirect. This categorization is particularly important for opto-electronic applications.

Table 3 shows the band gap values of every binary material at Gamma, X and L positions, and their temperature dependencies at the minimum gap points. A part of the energy gap model for binary materials is shown in Figure 6. Band gaps of a ternary material $A_xB_{1-x}C$ can be estimated theoretically as;

$$E_gabc = E_gac \cdot x + Ebc_g \cdot (1-x) - \frac{\alpha abc}{\sqrt{\frac{E_gac+E_gbc}{2}}} \cdot x(1-x) \quad (6)$$

where αabc is called a sag parameter and the value is determined either experimentally or theoretically. Figure 7 shows experimental data of the band gap energies of $Al_xGa_{1-x}Sb$ that are stored in the raw database and experimental formulae for the energy gaps can be obtained from the data by least mean square fitting method. These formulae are;

$$Eg_\Gamma(x) = 0.721+1.22x+0.257x^2 \quad \text{-- (7)}$$

$$Egx(x) = 1.02+0.477x+0.103x^2 \quad \text{-- (8)}$$

$$Eg_L(x) = 0.821+0.727x+0.052x^2 \quad \text{-- (9)}$$

```
39.000  [AL-GA-AS [A (ZINCBLENDE
40.000      (A (V3-VEGARD 1 'AL-AS 'GA-AS 'ZINCBLENDE TMP X Y))) ] ]

60.000  [AL-IN-P-SB [A (ZINCBLENDE
61.000      (A (V4A-VEGARD 2 'AL-P 'AL-SB 'IN-P 'IN-SB 'ZINCBLENDE TMP
62.000       X Y))) ] ]

58.000  [AL-GA-IN-AS [A (ZINCBLENDE
59.000      (A (V4B-VEGARD 2 'AL-AS 'GA-AS 'IN-AS 'ZINCBLENDE TMP X Y))) ] ]
```

Figure 5 : Lattice parameters for AlGa-As, AlIn-PSb and AlGaIn-As systems

MATERIAL	Eg (eV) at 300 K			EgT (ev)	
	Gamma	X	L	POSITION	FORMULA
Al-P	3.58	2.46	3.6	X	$2.49-3.18\times10^{-4}T^2/(T+588)$
Al-As	2.95	2.16	2.3	X	$2.24-6.0\times10^{-4}T^2/(T+408)$
Al-Sb	2.20	1.60	1.60	X,L	$1.69-4.97\times10^{-4}T^2/(T+213)$
Ga-P	2.78	2.26	2.67	X	$2.34-5.771\times10^{-4}T^2/(T+372)$
Ga-As	1.43	1.86	1.72	Gamma	$1.53-5.405\times10^{-4}T^2/(T+204)$
Ga-Sb	0.721	1.02	0.821	Gamma	$0.807-3.78\times10^{-4}T^2/(T+94)$
In-P	1.34	2.24	1.86	Gamma	$1.41-3.63\times10^{-4}T^2/(T+162)$
In-As	0.359	1.80	1.46	Gamma	$0.419-2.50\times10^{-4}T^2/(T+75)$
In-Sb	0.175	0.955	0.775	Gamma	$0.236-2.99\times10^{-4}T^2/(T+140)$

Table 3 : Energy gaps and their temperature dependencies of binary materials

```
10.000   [AL-AS [EG
11.000     (GAMMA [CASE TMP ((300.)(LIST 2.95 1))(T (EG1MOD .787 2
12.000              (EG1CAL  'AL-AS 'X TMP)))])
13.000     (X (LIST (- 2.239 (* 6.0E-4 (/ (* TMP TMP)(+  TMP 488.)))) 1))
14.000     (L [CASE TMP ((300.)(LIST 2.93 1))(T (EG1MOD .767 2
15.000              (EG1CAL  'AL-AS 'X TMP)))]) ] ]
```

```
28.000   [GA-AS [EG
29.000     (GAMMA (LIST (- 1.519 (* 5.405E-4 (/ (* TMP TMP)(+  TMP 204)))) 1))
30.000     (X [CASE TMP ((300.)(LIST 1.86 1))(T (EG1MOD .436 2
31.000        (EG1CAL  'GA-AS 'GAMMA TMP)))])
32.000     (L [CASE TMP ((300.)(LIST 1.72 1))(T (EG1MOD .296 2
33.000        (EG1CAL  'GA-AS 'GAMMA TMP)))]) ] ]
```

Figure 6 : Part of energy gap model for binary materials

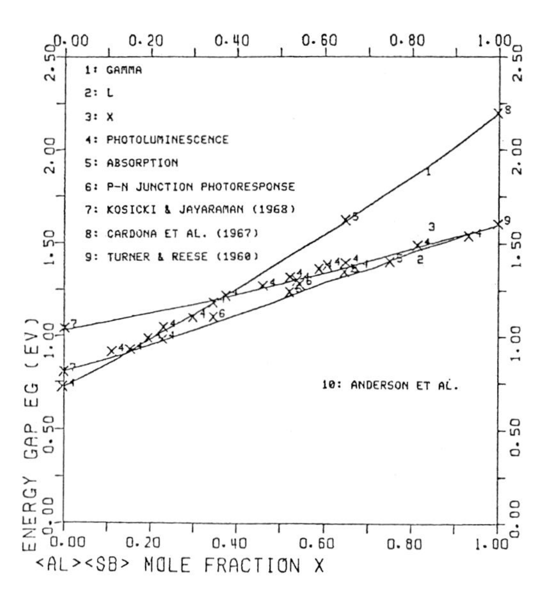

Figure 7 : Band gap energies of $Al_xGa_{1-x}Sb$

(experimental)

Energy gap model for $Al_xGa_{1-x}Sb$ is shown in Figure 8. Eg values other than 300 K are assumed to be interpolated by those of the component binary materials (Al-Sb and Ga-Sb this case). The interpolation formula is 'EG1INTQUAD' in this figure. The interpolation method can be extended for quaternary materials (4). The formulae for $A_xB_{1-x}C_yD_{1-y}$ materials are as follows;

$$E_g\,abcd = E_g abc \cdot y + E_g abd \cdot (1-y) - \frac{\alpha\,abcd}{\sqrt{\frac{E_g abc + E_g abd}{2}}} \cdot y(1-y) \quad (10)$$

where

$$E_g abc = \text{Eg of } A_xB_{1-x}C, \quad \text{--(11)}$$

$$E_g abd = \text{Eg of } A_xB_{1-x}D, \text{ and} \quad \text{--(12)}$$

$$\alpha\,abcd = \alpha\,acd \cdot x + \alpha\,bcd \cdot (1-x) \quad \text{--(13)}$$

Error estimation facilities

To evaluate the estimation error in the materials model, we attached an error factor to each datum and an interpolation formula. Sum of the factors is calculated along the estimation path of the required property, and it shows the quality of the estimated value.

```
76.000   [AL-GA-SB [EG
77.000     (GAMMA [CASE TMP
78.000       ((300.)(LIST (+ .726 (* 1.004 X)(* .459 X X)) 1))
79.000       (T (EG1INTQUAD .459 2 'AL-SB 'GA-SB 'GAMMA TMP X))])
80.000     (X [CASE TMP
81.000       ((300.)(LIST (+ .998 (* .464 X)(* .118 X X)) 1))
82.000       (T (EG1INTQUAD .118 2 'AL-SB 'GA-SB 'X TMP X))])
83.000     (L [CASE TMP
84.000       ((300.)(LIST (+ .779 (* .729 X)(* .342 X X)) 1))
85.000       (T (EG1INTQUAD .342 2 'AL-SB 'GA-SB 'L TMP X))]) ] ]
```

Figure 8 : Energy gap model of $Al_xGa_{1-x}Sb$

3. DESIGN-AID SYSTEM CAMS

3.1 Retrieval of database

Retrieval of a property of a given material, and inversely that of a material of a given property are the most basic facilities. Figure 9(a) shows some examples: the lattice constants of Ga-As at 300 K and 77 K, that of Al_xGa_{1-x}-$P_y As_{1-y}$ at 300 K which closely lattice matches with Ga-As, and the energy gap of these materials. Figure 9(b) shows examples of inverse queries where the materials with given properties are obtained as outputs; the ternary materials with energy gaps between 0.3 and 1.3 eV, between 0.15 and 0.20 eV (narrow gap), and between 6.0 and 6.1 eV (extra wide gap), respectively.

3.2 Design chart of quaternary semiconductors

There are two degrees of freedom in quaternary solid solutions and these freedoms are very important for materials design. For example, two parameters such as a lattice constant and an energy gap can be adjusted simultaneously. On the other hand, a computerized design-aid system becomes necessary to utilize such freedoms. Figure 11 shows part of a design chart for the energy gaps of quaternary solid solutions.

3.3 Heterostructure design

Heterostructure design is becoming very important for the development of heterojunction devices such as laser diodes and superlattices.

The typical structure of a heterostructure laser is illustrated in Figure 12(a). To confine the carriers (electrons and holes) and the emitting light to the active layer, which is necessary for the laser oscillation, the profiles of the energy bands, and the refractive indices among the active layer and the confining layers should be as illustrated in Figure 12(b). In addition, these layers should be in good lattice match with the substrate.

```
 1 - COMMAND_PLEASE
 2 - CALC
 3 - MATERIAL:PROPERTY_PLEASE
 4 - GA-AS:A
 5 - TMP_PLEASE
 6 -       300.000
 7 - ((ZINCBLENDE       5.65325     1))

 8 - COMMAND_PLEASE
 9 - CALC
10 - MATERIAL:PROPERTY_PLEASE
11 - GA-AS:A
12 - TMP_PLEASE
13 -         77.000
14 - ((ZINCBLENDE       5.64799     3))

15 - COMMAND_PLEASE
16 - CALC
17 - MATERIAL:PROPERTY_PLEASE
18 - AL-GA-P-AS:A
19 - X_PLEASE
20 -    7.000E-01
21 - Y_PLEASE
22 -    3.000E-02
23 - TMP_PLEASE
24 -       300.000
25 - ((ZINCBLENDE       5.65324     2))

26 - COMMAND_PLEASE
27 - CALC
28 - MATERIAL:PROPERTY_PLEASE
29 - AL-GA-P-AS:EG
30 - X_PLEASE
31 -    7.000E-01
32 - Y_PLEASE
33 -    3.000E-02
34 - TMP_PLEASE
35 -       300.000
36 - ((X          2.091      2))
```

(a) Material -> Property

```
 1 - COMMAND_PLEASE
 2 - INQ
 3 - GROUP:PROPERTY_PLEASE
 4 - TERNARY:EG
 5 - RANGE_PLEASE
 6 - (1.3   1.31)
 7 - MATERIAL(S)_FOUND
 8 - (AL-GA-SB(X= .570  .580)(TYPE INDIRECT-X))
 9 - (GA-IN-AS (X= .890 .890)  (TYPE DIRECT))
10 - (IN-P-AS (X= .960 .960)  (TYPE DIRECT))

11 - COMMAND_PLEASE
12 - INQ
13 - GROUP:PROPERTY_PLEASE
14 - TERNARY:EG
15 - RANGE_PLEASE
16 - (.15 .2)
17 - MATERIAL(S)_FOUND
18 - (AL-IN-SB (X= .000 .010)  (TYPE DIRECT))
19 - (GA-IN-SB (X= .000 .160)  (TYPE DIRECT))
20 - (IN-AS-SB (X= .000 .050)  (TYPE DIRECT))
21 - (IN-AS-SB (X= .630 .740)  (TYPE DIRECT))

22 - COMMAND_PLEASE
23 - INQ
24 - GROUP:PROPERTY_PLEASE
25 - TERNARY:EG
26 - RANGE_PLEASE
27 - (6.0 6.1)
28 - MATERIAL(S)_FOUND
29 - (AL-GA-N (X= .910 .950)  (TYPE DIRECT))
30 - COMMAND_PLEASE
31 - END
```

(b) Property -> Material

Figure 9 : Examples of the database retrieval

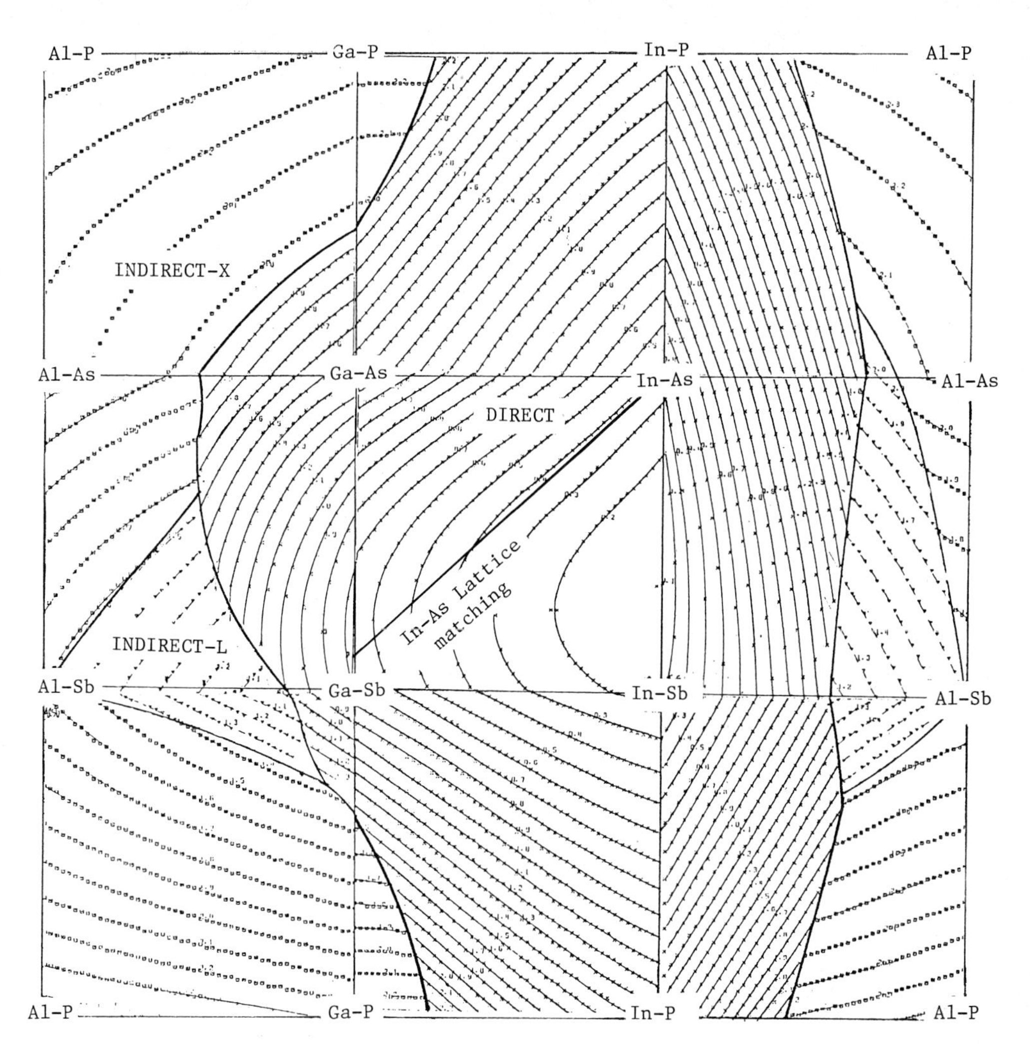

Figure 10 : Design chart of quaternary materials

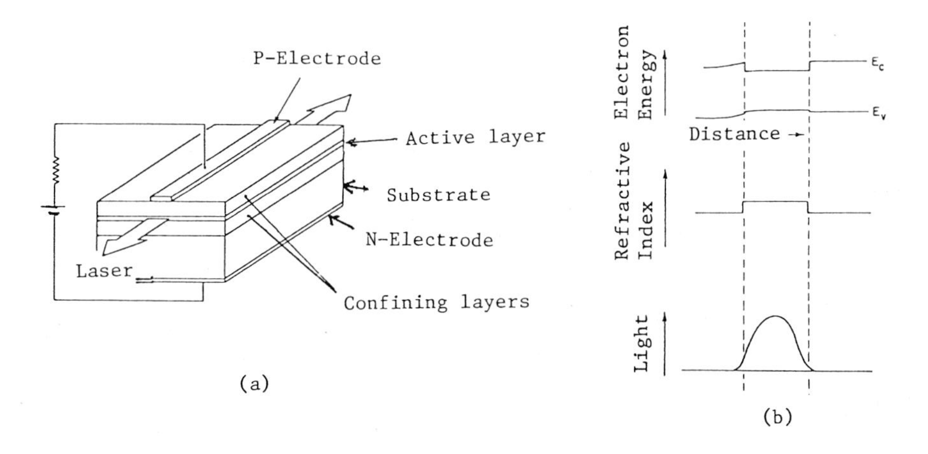

Figure 11 : Structure of a double-hetero structure laser

Given the wavelength of a laser, Eg of the active layer should be designed as:

$$Eg\ (eV) = 1.2398/L\ (\mu m) \qquad --(13)$$

Considering a laser using $Ga_xIn_{1-x}As_ySb_{1-y}$, on In-As substrate, possible Eg values are obtained between 0.38 to 0.67 eV along the lattice matching line with In-As from Figure 11, so the possible wavelength of this system is between 1.9 and 3.3 microns. The confining layers can be designed so as to satisfy the above mentioned Eg and refractive index profiles and good lattice matching.

In the design of superlattices, given an energy band gap lineup where we want to determine a pair of materials that meet the condition, an energy lineup is given as in Figure 12(a). For example, the conduction band barrier height Hb, and the valence band barrier height are given as

$$0.5 \leq Hb \leq 1.0 \text{ and} \qquad --(14)$$

$$0.0 \leq Lb \leq 0.5. \qquad --(15)$$

A pair of binary materials Ga-Sb : In-As is found to satisfy the conditions and has good lattice matching.

4. CONCLUSION

We have discussed the semiconductor database and its application to materials design. The materials modelling concept that we have proposed has revealed itself to be very effective for the application. On the other hand, to construct such a model, an adequate theoretical basis for interpolation and evaluated data are required. As for lattice parameters and energy band gaps including the temperature dependencies of III-V group binary, ternary and quaternary semiconductors, the primary model can be implemented.

Although our model contains several uncertain data and some assumptions in interpolation procedures, it is still quite useful, and expected to be the base of the evaluation of both the data and the theory. The use of PETL language (5) which was developed in our laboratory, a dialect of LISP, enables us to embed arbitrary formulae and estimation procedures directly in each data record, thus the model is very easy to maintain and improve.

We have also presented a design-aid system called CAMS based on the model, and have introduced several important facilities with some design examples.

ACKNOWLEDGEMENT

The authors would like to express their gratitude to Dr. K. Sato, the Vice-President of the ETL, Mr. S. Wakamatsu and Dr. H. Kawakatsu, our division chiefs, and Mr. M. Kakikura, the section chief, for their continuing support of the present work. Thanks are also given to Mr. M. Tsukamoto who has developed the PETL language, and Miss S. Naito for implementing programs.

REFERENCES

[1] Data book for III-V group semiconductor solid solutions, JEIDA, 1981

[2] Data book for III-V group binary semiconductors, JEIDA, 1982

[3] Casey, H.C. and Panish, M.B., Heterostructure Lasers Part A & B, Academic Press, 1978.

[4] Sasaki, A., Nishiuma, M. and Takeda, Y., Energy Band Structure and Lattice Constant Chart of III-V Mixed Semiconductors, and AlGaSb/AlGaAsSb Semiconductor Lasers on GaSb Substrate, Jap. Jnl. App. Phys., vol. 19, No. 9, 1980.

[5] PETL Reference Manual, Electrotechnical Lab, 1982.

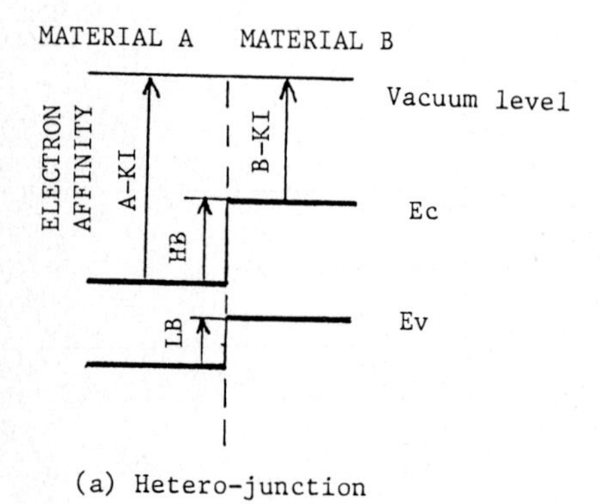

(a) Hetero-junction

```
 1 - COMMAND_PLEASE
 2 - SRCH_BINARY_PAIR
 3 - CONDITION_PLEASE
 4 - (AND (HB 0.5 1.0) (LB 0.0 0.5))
 5 - PAIR(S)_FOUND

 6 - IN-P:AL-AS
 7 - (A=IN-P A-EG=1.35 A-KI=4.40 A-A=5.8688
 8 -  B=AL-AS B-EG=2.17 B-KI=3.50 B-A=5.6612)

 9 - IN-AS:GA-SB
10 - (A=IN-AS A-EG=0.36 A-KI=4.90 A-A=6.0584
11 -  B=GA-SB B-EG=0.72 B-KI=4.06 B-A=6.0959)
```

(b) Design Example

Figure 12 : An example of a hetero-junction

Data for Science and Technology, P.S. Glaeser (ed.)
North-Holland Publishing Company

MATERIALS DATA BASE FOR ENERGY APPLICATIONS – II

Shuichi Iwata, Atsushi Nogami, Shiori Ishino and Yoshitsugu Mishima

Department of Nuclear Engineering, Faculty of Engineering
University of Tokyo, 7-3-1 Hongo, Bunkyo-ku, Tokyo, Japan

Materials data range from nuclear cross sections to performances in a system, and they are described by many variables and parameters exceeding one thousand, which require, at first, a data base of graceful interfaces for general users, especially for materials scientists and engineers. In each description of materials data we then take into account the scientific understanding about the materials data as one of "relations", which become implicit representations of our knowledge on materials. Stored data now consist of fundamental data for all materials, performance data of energy systems mainly about metals and alloys, and typical data used for materials design, for example that on semiconductors, superconducting materials, polymers, rubbers, ceramics and superalloys.

1. INTRODUCTION

The importance of materials in many energy systems increases in respect to safety problems, as well as economical aspects, and materials must be selected and designed adequately to attain the expected performance in energy systems.

Materials data are widely spread, ranging from the field which relates fundamental understanding of the behavior of electrons, atoms, vacancies, dislocations, etc., to the performance of components, devices, machines and systems. Moreover, there are various viewpoints: those of the academic scientists, namely, analytical and theoretical; synthetic viewpoints from materials engineers, and investigators; and those of materials users on the selection of the most suitable materials, etc.

Requirements for good and reliable data which can be used in system designs have been duplicated by materials users. As a result, a general trend to share the experimental data on hand mutually has been followed not only among materials investigators, but also between materials engineers and system engineers who want to see materials used to their fullest potential.

Under these conditions, a materials data base has been developed for use as a common authoritative source of materials data as well as a mighty tool for materials selection and design. Technological developments in this area are dynamic and materials are required to maintain sound performance under very severe conditions. Continuing reassessment of materials data is necessary to provide readily usable materials data for the system designer, analyst and materials engineer, and so special emphasis has been put on the development of iterative tools so as to refine data quality easily at the levels of data entry, rearrangement and retrieval.

Stored data comprise materials data for fusion reactors and other energy systems. These data have been selected and evaluated from critical reviews of existing data and from our own experimental data. In this paper, the current status of the systems reviewed and future developemnt plans are explained.

2. DATA COMPILATION AND EVALUATION

In many ways, the projects of developing materials for energy systems have become related dynamically with system design and development of energy systems - in more ways than most materials engineers have ever experienced in other systems.

Materials have been developed and/or improved gradually in accordance with the accumulation of related experiences and knowledge, so that the lore and methodology on materials development leads one to say that experimentation in practice is the best of all instructors. However, it is impractical to expect sufficient amounts of tests and experiments under real working conditions of large scale energy systems as was the case with those energy systems already established. Materials engineers must learn more and more from the available data and establish a methodology, namely, establish a materials data base and the models on materials performances, so as to carry out materials development effectively.

The indispensable prerequisite for the development of a materials data base is an active research group as it is in other fields based on factual data. Therefore, it is extremely important for the development of a data base to rearrange data compilations into a simple and convenient format so as to allow many materials researchers to join in the data compilation, and to design the DBMS for materials data with graceful interfaces for both colleagues and general users.

Methods to establish the graceful interface could be classified into two categories. One is to develop a knowledge base system which is able to understand our queries with minimum designations.

The other one is to design a materials data base whose data structures can easily be understood by users who may be beginners in the area of data base management systems, but who are specialists in materials science and engineering. As for the knowledge system, no commercial software is available now, so that we focused our research on the development of such a materials data base, that is, a relational data base which enables implicit representation of our understanding on materials, relational operations to select and evaluate materials, to set trade-off parameters for system designers, and to establish design criteria as illustrated in Figure 1.

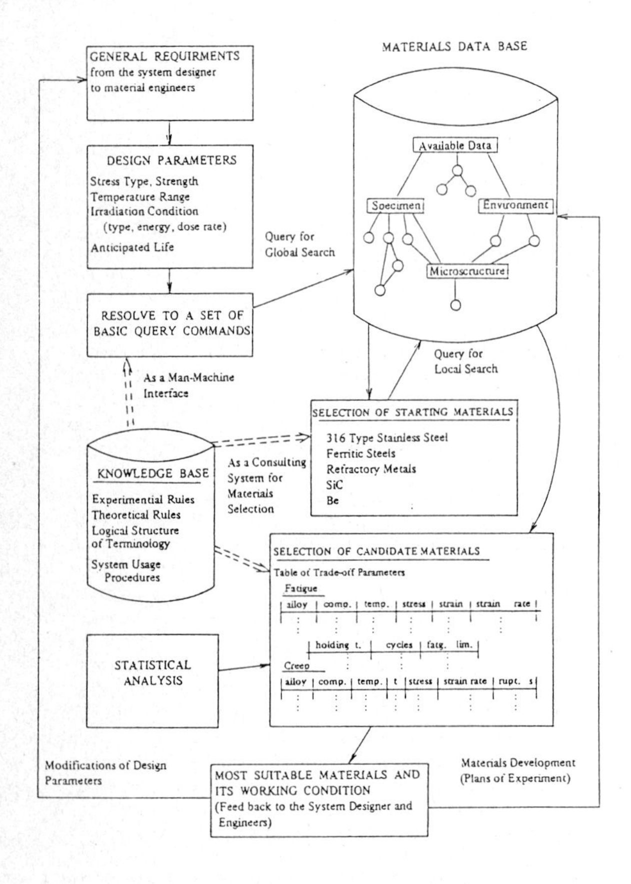

Figure 1 : A procedure of materials design using the materials data base.

The objective of a materials information system is twofold for materials engineers. One is to establish the design criteria for each application of energy systems on the basis of available information. The other is to design and/or develop materials by inducing effective rules through dynamic manipulation of stored information. As to the well-compiled data, we could expect that they will be represented by a set of tables. The above two problems could be solved gradually through deliberate iterations and classified into eight subproblems;

(i) resolution of materials requirements by system designers to a set of queries that could be answered by materials scientists and engineers without sophisticated inferences,

(ii) global and local search for the selection of starting materials through interactive operations of the materials data base,

(iii) creation of temporary data sets which consist of trade-off parameters as operating conditions and corresponding lifetimes,

(iv) synthetic decisions of design parameters from the viewpoint of materials engineers,

(v) data base management for retrieving and displaying materials data,

(vi) consulting, so as to use the data base completely including implicit representations of our understanding on materials properties,

(vii) synthetic decisions on strategies of materials development and system designs,

(viii) continuing improvements of predictive procedures and modeling in accordance with the addition of new data on materials and improvements on system designs.

These subproblems require a total information system with two sorts of capability which can help both materials engineers and system designers as a software tool. Thus the information system first needs "graceful interfaces" for the evaluation of complex materials data at any level of our knowledge on materials, namely, nuclear data, defects data, grain boundaries, precipitates, textures, properties, fabrication methods, total performances. Microstructural information should be dealt with here as key information to relate all the data in a physico-based manner for materials scientists and engineers, but may as well remain hidden for materials users. These differences of views of materials data will be supported by the knowledge base systems, e.g. system designers consulting about materials usage.

Other important aspects of the evaluation of materials data are portability and integrity. The differences in data on the same property should be analyzed and dealt with as a set of sample data for statistical analysis or issues on the fact per se.

The portability of materials data files is checked from two viewpoints: data base management systems and logical aspects. Except for the user interface, there is no problem about the portability of a materials data base if we compile the available data in the form of flat files, and only if we do not need new relations. Products from manufacturers are generally well-characterized by standards, e.g. AISI, DIN, JIS, and we only need the identification number and properties for the conventional uses of the materials.

However, in the case of newer materials and/or newer applications, we need dynamic manipulations of materials data. Relational-like manipulations by "predetermined" relations are essentially not adequate for this case. The materials data are used for materials selection, materials design, for performance modeling, and for establishing design criteria. We sometimes rescale tentatively the data for the correlation or make linear combinations of several fields. Materials data are referred repeatedly upward and downward as shown in Figure 2. Here the data are evaluated through the evaluation of predictive methods, performance models, and theoretical understanding. Figure 3 shows an example that needs predictions due to the difficulty of the experiments.

3. DATA STRUCTURE ON MEASURED DATA

The data structure reported in the former paper (1) has been decided upon on the basis of compiled data. It is convenient for materials selection, design, modeling, and establishment of design criteria. However, this approach could not check the duplication of stored data with the apparent means, though logical checks within the frame of the relational data base management system are possible. Therefore we decompose the preceding data structure to a more simplified one defined on the basis of experiments.

Materials data are regarded as a set of bibliographical data, textual data on experiments, photographs of microstructures, knowledge, tabulated data, figures and image data. These data are converted into a set of relations, and ad hoc files for textual, graphic and image data. Image data are analyzed and summarized into a set of quantified parameters, textual information and image data and voices on the image. Storage devices are not necessarily restricted to digital computers, and laser discs and video tapes and photograph files also may be used as well for the effective usage of image information.

In Figure 4 the process of data entry is shown schematically, where image data are included in the item Descriptions and Tabulated Data. Editing of materials data is left to the software tools - user commands, command editors, command analyzers etc. Then the data structure on experiments becomes very simple as shown in Figure 5.

4. INFORMATION SYSTEM FOR ENERGY APPLICATION

The conceptual procedure for the usage of the

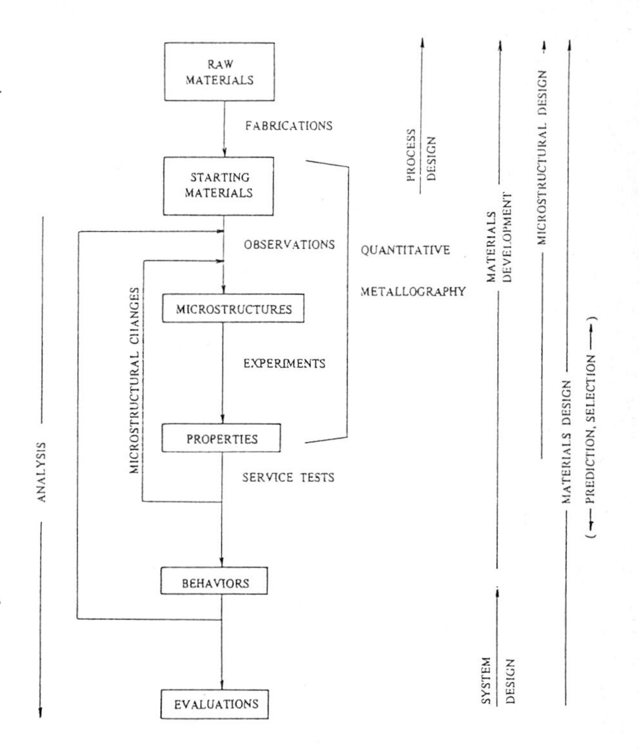

Figure 2 : Map of materials design and analysis connected with microstructural data.

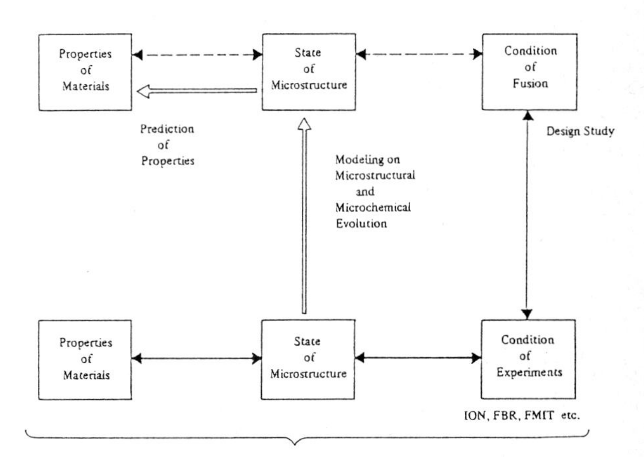

Figure 3 : Materials design in case of no direct data. Simulation experiments are to be compiled and used on the basis of microstructural modeling.

numerical data base of the data structure in Figure 5 is shown in Figure 6. The retrieved results are at first stored in the work file and edited to a set of ad hoc files of private users. We need not expect frequent update of stored data, but must expect various applications of stored data. Therefore, more powerful capabilities are eagerly required concerning the manipulations of stored data, more than conventional DBMSs can provide, even for such capabilities as updating, processing of transaction concurrence, etc. Energy applications are specific with respect to fields, but the needs for the dynamic manipulation of data are essential on materials data handling. Moreover, the needs for such capabilities are universal for numerical data manipulations.

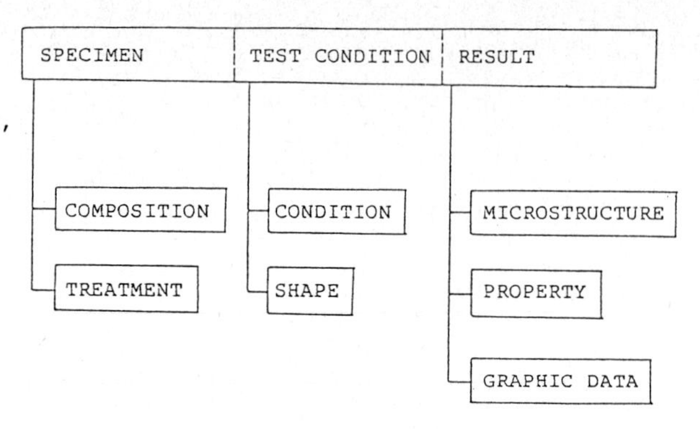

Figure 5 : Data Structure on Materials.

REFERENCE

(1) Iwata, S. et al., Materials Data Base for Energy Applications, in Glaeser, P.S.(eds.) Data for Science & Technology (Pergamon Press, Oxford, 1981).

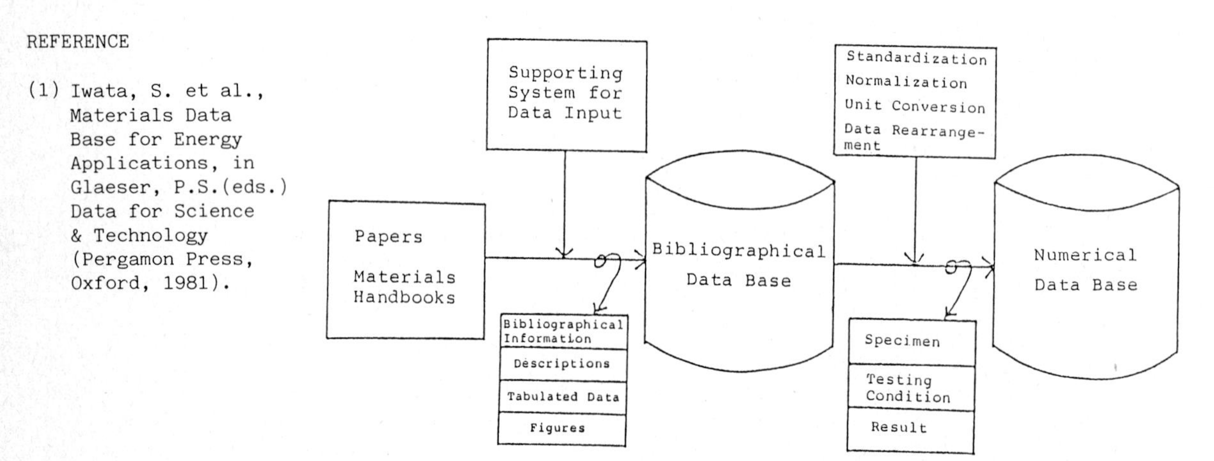

Figure 4 : Procedure for the Construction of Materials Data Base.

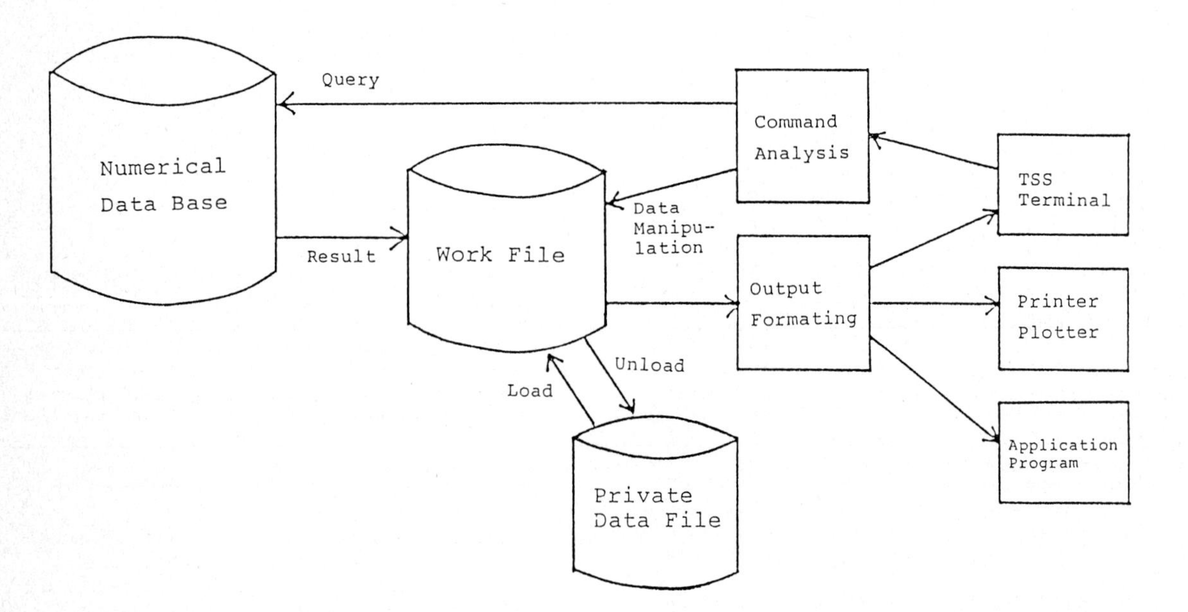

Figure 6 : Conceptual Procedure to use the materials data base of simplified data structure.

Data for Science and Technology, P.S. Glaeser (ed.)
North-Holland Publishing Company

MECHANICAL PROPERTIES: A REVIEW OF SOME RECENT DEVELOPMENTS

G.C. Carter

Numerical Data Advisory Board

National Research Council/National Academy of Sciences, Washington, D.C., U.S.A.

Mechanical properties differ from other properties covered by CODATA in two basic ways: 1) their measurements are reproducible but strongly dependent on exact details of preparation and processing; 2) users are primarily in the private sector for which success depends inseparably on the reliability of the marketed product. Failures can be catastrophic both financially and in terms of quality of life, morbidity, and mortality (e.g. failures of nuclear plants, air and other transportation). CODATA's mechanical properties program needs to suit this engineering and development area, for which government and private sectors play different roles than for other CODATA subjects.
Possible CODATA roles: 1) international standardization (in coordination with ISO, OIML); 2) stimulating computer compatibility (data banking and modeling); 3) stimulating international cooperation (for subjects in which data sharing is agreed to be mutually profitable).

I. Introduction

The subject addressed in this paper is that of mechanical properties. With that is meant those properties that pertain to strength of materials as they perform under various conditions as building materials and parts in technological equipment. Among the properties are, for example, yield strengths, tensile strengths, moduli of elasticity, impact strengths, creep-fatigue, and fracture toughness. These properties are highly structure sensitive. Local structures on the atomic level, dislocations, inclusions, grain boundaries, precipitates, and flaws contribute largely to properties (and failures) of materials. These are introduced and the amount of their presence often controlled during the preparation of the material. Heat treatments, rates of cooling, rolling, presence of ambient gases and other processing parameters all help determine the mechanical properties of the material. Thus, in order to prepare reliable data tables, all these "characterization" parameters are important. In the past, some very important data compilation work has been done for practical engineering applications.

A compilation, "Data Sources for Material Scientists and Engineers" by Westbrook and Desai[1], indicates existence of several valuable resources, although most resources listed therein deal with material properties other than mechanical. A survey presented as part of a larger study by a Panel on Mechanical Properties Data (PMPD) for Metals and Alloys, of the Numerical Data Advisory Board[2], identifies more specifically those relevant to the current subject. More will be said in the following Sections of this paper on the findings of this Panel.

Unlike certain basic physical and chemical properties, such as fundamental constants, or boiling points of pure substances, a mechanical property of a material cannot be considered to have a single value that is valid for all heats (batches) of that material. Rather, a range of values must be considered to be valid with a "minimum safe value" associated with the material. In other words, "design allowables" are of interest, rather than single "best values." In the absence of reliable data, the design engineer must "overdesign," i.e., use excess material, to prevent failure. Today, with growing scarcity of materials, and reduced margins of profit, overdesigning is becoming evermore undesirable. Consequently, the need for more accurate knowledge of materials properties is ever increasing. Many companies are establishing specific data banks for their own needs, but the need for good data generally exceeds by far a company's capabilities to develop the data banks.

It is interesting to observe in this field the growth of a common need for data, while on the other hand competition in the market of products may depend on unique applications of specialty materials. Thus, the roles of the government and non-governmental sectors, at least in the United States, appear to apply differently in this area than in areas more relevant to academic environments. There is a substantial amount of data for commonly available materials that are of common need. To meet these needs, in some isolated instances, private data enterprises have emerged, such as the Engineering Science Data Unit in the United Kingdom[3], that sell their data packages to users from all sectors. Propriety of company data is clearly not a factor limiting data bank development for these data. Rather, cost of data bank development, user orientation, marketing, and cost of data evaluation, or "appraisal," are some of the primary considerations in making a commercial data system feasible. (See Section III, below, and a paper by Westbrook elsewhere in this conference on a Metal Properties Council study of this subject.)

"Appraisal" is the term used in the PMPD report to signify the process of assessing the validity of the data, and their errors and limitations. In other disciplines this process may be referred to as evaluation or validation. This choice of term is because of the complexity of this process for engineering data, and the wide number of variables and specifications that must be taken into consideration, while incorporating a knowledge of the limits of applicability. Only highly qualified experts can perform this task, and of this group, only those with gift for analytical thinking and extreme patience are bound to be amenable to the job. Because of these stringent requirements, not many evaluators are eager for the job. Ironically, not many data base developers, funding agencies, or private concerns comprehend the necessity or expense of this effort either. Thus, this step in data base development continues to be a chronic problem.

II. Recently Completed Studies

A preliminary study done by the U.S. Office of Standard Reference Data (OSRD), National Bureau of Standards (NBS), done a decade ago, concluded that certain mechanical properties should be covered by that program. For various reasons, it was not possible to follow through on this finding.

In the Autumn of 1978, a two-day meeting was held at the American Society for Testing and Materials, again to address the subject of the development of a publicly available data system for engineering data. The meeting brought together a wide variety of interested parties and it can be concluded that at that time both a need for the data was expressed and several "pockets" of on-going data activities were identified.

Shortly after this meeting, the OSRD/NBS requested the Numerical Data Advisory Board (NDAB) to study whether, and if so, to what extent, this office should be engaged in a data program for mechanical properties for metals and alloys. To respond to this request, a separate Panel of the NDAB was convened to reassess the situation for the OSRD in the context of 1979 needs.

The Panel on Mechanical Properties Data (PMPD) for Metals and Alloys was a panel chosen by the U.S. National Research Council and consisted of representatives from various sectors: nuclear design, nuclear materials, non-nuclear energy, heavy moving equipment, the Metal Properties Council, American Society for Testing and Materials, American Society of Mechanical Engineers, American Society for Metals, Federation of Materials Societies, Society for Automotive Engineers, the steel industry, rail industry, and academia. Furthermore, at its first meeting, a large number of representatives attended from other parts of the engineering community, including federal agencies and academic and industrial representation other than that serving directly on the Panel.

The Panel wrote a comprehensive report[2] with recommendations to a broad audience because the Panel realized that the total subject is broad and diffuse and only with the help of this broad audience can all elements of a solution be assembled. International considerations are also included in the report (see Section IV, below).

The Panel considered a totally NBS-based system not to be the solution, but rather, recommended strong NBS participation in the following specific problem areas:

a. Enhanced uniformity of test methods and reporting;
b. Developing definitions of nomenclature and properties;
c. Developing data appraisal procedures;
d. Organizing symposia, conferences, workshops, and other coordinating activities.

These findings are repeated here because it is this type of activity in which CODATA could also be of considerable value on an international level with obvious benefits to member nations.

III. Studies Under Way

After the completion and distribution of the NDAB's Panel report, the subject was taken up by the Metal Properties Council (MPC). An MPC committee was established and is continuing to meet regularly, attempting to give further definition to a realistic plan to develop a self-sustaining computerized data base. A one-year feasibility study to come to grips with details has been contracted out by MPC. The MPC itself is sponsored by the American Society of Mechanical Engineers, American Society for Metals, American Society for Testing and Materials, American Welding Society, and Engineering Foundation. The MPC itself was established to deal with property evaluation, or appraisal, of selected materials and properties as needed by the communities served by these professional societies. The outcome of the feasibility study is therefore of considerably more than academic interest. The study must consider state-of-the-art preparedness (computer and scientific), needs of the sponsoring societies, and needs of the vast and varied user audiences, plus financial marketing details. A large number of groups in the United States look with great interest to the outcome of the study (for further details of the outset of this study see paper in this conference by Westbrook), and a number of significant decisions will be based on the report.

Concurrently with this study an international Materials Data Workshop is being organized, to be held in Crossville, Tennessee, 7-11 November, 1982. The organizing committee is representative of a number of the groups involved in the earlier studies. CODATA is cosponsoring the Workshop together with NBS, the German Fachinformationszentrum, and Oak Ridge National Laboratory. The Workshop will examine the feasibility, desirability, and timeliness of "on-line" materials information systems covering commercially available metals, ceramics, polymers, and composites. The plan is to invite a limited number of technically highly qualified representatives of the various sectors (e.g. engineers and other users, data base developers, evaluators, vendors, professional societies and others). In one week of intensified discourse, study, and writing, the group will prepare a document that addresses: barriers, economics, user identification, scope definition, data base development, data base specifications, the user interface, and data applications. The relevant understandings gained in this activity will be carried over and integrated with the MPC study.

A third study is being done by the U.S. National Research Council's National Materials Advisory Board for the U.S. Department of Defense and the National Aeronautics Space Administration. That Board has established a Committee on Materials Information used in Computerized Design and Manufacturing Processes, to define the information needs of the designers with respect to form of property data presentation, accompanying materials identification, and variability and reliability. Implications of computer-aided design and manufacture (CAD/CAM) are to be recognized by the Committee. If appropriate, a demonstration project (a property data base system incorporating a limited number of alloys) intended to transcend any uncovered deficiencies would be outlined. The Committee is performing its study in full cognizance of the MPC and other activities. It has held several meetings to initiate a dialogue involving the design and the materials communities in order to assess the situation in this developing field.

IV. International Aspects

In order to be precise, the first two recommendations of the NDAB's Panel on Mechanical Properties Data to the general scientific and technical community are quoted here:

"1. There is a continuing need for the organizations concerned with mechanical properties cooperatively

(a) To develop and improve the uniformity of test methods and data reporting;

(b) To develop definitions of nomenclature and properties;

(c) To develop property data appraisal procedures;

(d) To disseminate data including those evaluated in the process of developing test methods;

(e) To provide lists and disseminate information on test methods, data reporting, measurement standards, definitions, nomenclature, and appraisal procedures;

(f) To coordinate data efforts and assess data needs through organizing active committees, symposia, conferences, workshops, and other methods;

"2. The above recommendations are a result of U.S. national considerations. However, the situation is even more urgent on an international level. Large amounts of data that are being generated in other countries are, in effect, lost to the U.S. community because of a lack of uniformity of definitions of nomenclature and property, of test methods, of material identification and other criteria for data reporting, and of appraisal and data analysis procedures. The eventual development of a broad international data system for mechanical properties of metals and alloys will not become feasible in the future if such international uniformity is not developed first. It is recommended that an international coordinating body be established to address the development of international uniformity upon which an international system could be built. Technical and professional societies and trade associations, such as the International Standards Organization, could play a significant role in the international development of a worldwide system."

CODATA, and its Advisory Panel on Data for Industrial Needs, should consider these recommendations and decide in which way it could assist in this subject.

V. Conclusions

Interest in the subject of mechanical property data bases is increasing in the United States. Especially computerized data bases are receiving a considerable amount of consideration. It would seem that, if the time isn't ripe now to establish self-supporting automated data bases, such a time would develop in the not-too-distant future.

Based on recent studies and reports, CODATA could be, and should consider to become, active in several areas, as an international organization with a unique mandate. Among such areas are the following:

1. International standardization efforts. This should be carried out in coordination with other international bodies such as the International Standard Organization and the International Organization for Legal Metrology (of

which CODATA is an observer member). Standardization is needed in definitions of nomenclature, notation, properties, test methods, data reporting. Improved uniformity is needed in data appraisal (evaluation, validation) methods, and minimum materials characterization practices and documentation.

2. Data base compatibility. Enhancement of compatibility will allow better cross-use of software packages with data bases, if items under 1, above, are satisfied. Both modeling procedures and computer interfaces are an aspect of total exchange and access.

3. Cooperation in data base development. International fora for discussions among parties with similar needs could stimulate joint international efforts (either directly among parties, or as CODATA projects) in which data base design, or evaluation, or model development, and data bank costs as well as benefits could be shared.

All three points require international attention, and none are impeded by real constraints resulting from proprietary interests. In talking to a number of scientists or managers of a number of institutions in a number of countries, I have noticed that the barrier in initiating international cooperation stems from a lack of personal contact, interaction, and understanding, rather than from the often quoted, but infrequently verified excuse of privacy for a large number of mechanical property data. It is true that there are always those data that cannot and should not be shared, but they are only in a finite number of readily identifiable high-technology areas. The existence of these isolated cases should not prevent engineers from finding solutions in the many other areas.

References

1. Westbrook, J.H., and Desai, J.D., "Data Sources for Materials Scientists and Engineers," Am. Rev. Mater. Sci. 8, 359-422, 1978. The continuation of this work is presented elsewhere in this conference.

2. Mechanical Properties Data for Metals and Alloys: Status of Data Reporting, Collecting, Appraising, and Disseminating. Panel on Mechanical Properties Data for Metals and Alloys, Numerical Data Advisory Board, National Research Council (National Academy Press, Washington, D.C. 1980).

3. Engineering Sciences Data Unit LTD, 251-259 Regent Street, London, WIR7AD, UK. This is a private company owned by the Royal Aeronautical Society, and operated in association with the Institutions of Chemical, Mechanical, and Structural Engineers. The work is sponsored by appropriate professional organizations.

Data for Science and Technology, P.S. Glaeser (ed.)
North-Holland Publishing Company

A DATA BASE SYSTEM FOR PROPERTIES OF IRON AND STEEL

Hans Rohloff

Betriebsforschungsinstitut, VDEh-Institut für angewandte Forschung GmbH
Düsseldorf, Federal Republic of Germany

The data base system comprises standard data and measured material characteristics using two different data base management systems. The compilation of the mechanical, technological, and physical properties and the extent of data volumes are outlined. The query language to select materials and the evaluation by means of programs for the graphic representation involving the computation of regression functions and frequency distributions are described. Various possibilities for the output by listing and plotting the property values are illustrated. An outlook is given on further development activities.

The Betriebsforschungsinstitut maintains its data base for iron and steel materials since 1970. The entire system with programs for data acquisition, data base management, and data evaluation was developed by our own staff. The development work is sponsored by the German Organisation Gesellschaft für Information und Dokumentation with grants from the Federal Ministry Bundesministerium für Forschung und Technologie. It is intended to describe the present state of development and to give an outlook to the further development plans.

The data base for iron and steel materials has four main fields of application:

1. Selecting materials that meet a certain demand profile as specified by a query. In this way, materials for specific applications or equivalent materials can be selected.

2. Informing about the characteristics of specified materials.

3. Supporting investigations concerning the dependence of material properties on influence factors such as the chemical composition, heat treatment, form and dimensions of the product.

4. Calculating cumulative frequency curves and regression analyses in order to fix guaranteed values for standardization, for technical specifications and for quality control.

Figures 1 to 4 show examples of evaluations.

Fig. 1 shows the formulation of a demand profile which is to be met by a given material. For this purpose, ranges of property values and influencing factors are input line by line (center of Fig. 1, grey: MN LE 2). The complete demand profile is formed by linking the lines with operators such as OR, AND, WHIL, WITH (lower part of Fig. 1, grey: 1 WHIL 3 OR 1 WHIL 4 AND 6 AND 5). The steel grades found are output by their identifiers in the FIND result table (bottom of Fig. 1).

```
KOMMANDO EINGEBEN      FIND READ READ,S READ,D FETCH PRINT DELETE END
>FIND

DB SZ ANWENDERTEXT SKT
** * ************ ***
>01 W  HB30+SI+MN     10

ZNR NAME VOP1 WERT1                    VOP2 WERT2                DIMENSION
----------------------------------------------------------------------------
001 HB30 GE   210                      LE   230
002 RM   GE   700                      LE   800                  N/MM**2
003 BHLZ EQ   ABGESCHRECKT
004 BHLZ EQ   ABG
    **** ***  ******************** ***  ********************
>   MN   LE   2

    **** ***  ******************** ***  ********************
>   SI   NGE  1.1

    **** ***  ******************** ***  ********************
>

ZNR NAME VOP1 WERT1                    VOP2 WERT2                DIMENSION
----------------------------------------------------------------------------
001 HB30 GE   210                      LE   230
002 RM   GE   700                      LE   800                  N/MM**2
003 BHLZ EQ   ABGESCHRECKT
004 BHLZ EQ   ABG
005 MN   LE   2                                                  %
006 SI   NGE  1.1                                                %
    **** ***  ******************** ***  ********************
>

BITTE SUCHKRITERIUM VORGEBEN      ( OR, AND, WHIL, WITH )
>( 1 WHIL 3 OR 1 WHIL 4 ) AND 6 AND 5

>

S U C H E R G E B N I S - T A B E L L E   (FIND - ERGEBNIS)
ZDFT ANWENDERTEXT DB SZ ERG TF   BED SKZ LID LESEPROFIL        ANZ TR   TSU
011  HB30+SI+MN   01 W  004 0100 006 001                       000 0000 0000
----------------------------------------------------------------------------
LNR WERKSTOFFNUMMER
-------------------
001 1.4310
002 1.4439
003 1.4460
004 1.4582

KOMMANDO EINGEBEN      FIND READ READ,S READ,D FETCH PRINT DELETE END
>
```

Fig. 1: Query by a demand profile selecting steel grades

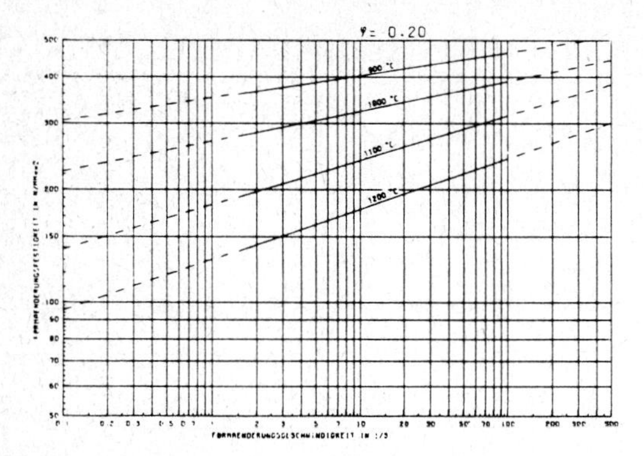

Fig. 2: Hot yield strength for a steel grade as a function of the strain rate at several temperatures and a strain of $\varphi = 0.20$

As an example for the demonstration of property values, Fig. 2 shows the graphic representation of the hot yield strength in dependence on the strain rate which is required to calculate hot working processes such as rolling.

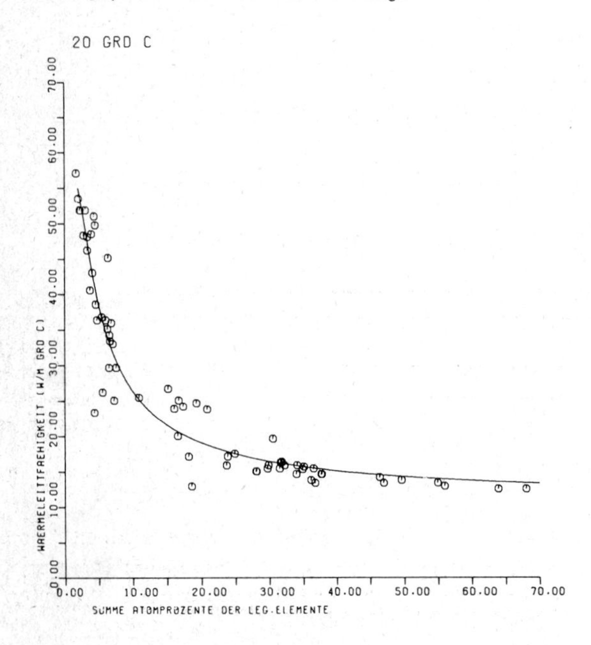

Fig. 3: Heat conductivity at 20 °C as a function of alloy content for steels

Fig. 3 presents the result of the investigation on the influence of the chemical composition on heat conductivity. There is a hyperbolic relation between the heat conductivity at ambient temperature and the sum of alloying elements in the steel if the alloy concentration is expressed in atom-%.

Fig. 4 shows the representation of the cumulative frequency for the tensile strength of a steel grade at 20 °C. In analogy with this representation it is possible to present measured values of any properties, for example, in order to fix limit values for certain failure probabilities in this way.

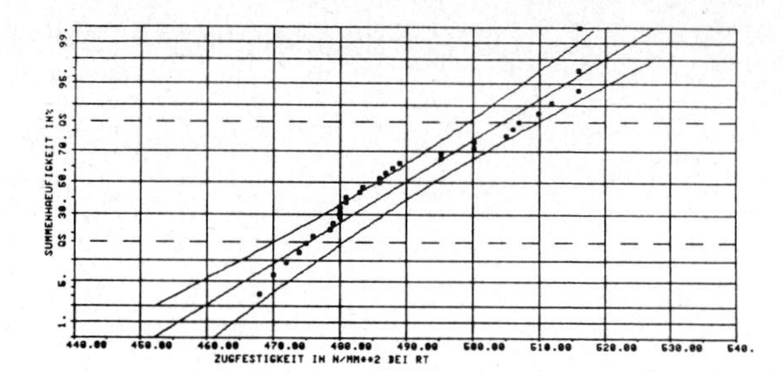

Fig. 4: Cumulative frequency of tensile strength at ambient temperature for a steel grade

For such evaluations, two types of data must be distinguished in the data base:

- standard data
- measured material characteristics

Consequently, the material data base consists of two subsystems for standard data and measured material characteristics, respectively. Both subsystems will now be described separately.

Data base for standard data

The subsystem for standard data so far contains information about 1000 steel grades according to DIN and Stahl-Eisen-Werkstoffblätter, material specifications of the Verein Deutscher Eisenhüttenleute. The acquisition includes the values for a great number of properties given in the table units of the standards. In many cases, parameters such as temperature, range of dimensions, heat treatment, etc. are assigned to these properties, for example to the chemical composition, the yield strength, the heat conductivity. The number and type of properties and of the parameters assigned to them vary considerably from steel grade to steel grade. Both discrete values and ranges of values can be indicated for properties and parameters. All this is reflected in the query language.

An example is shown in Fig. 1. Input lines are marked grey. The final result is at the bottom. The following relational operators are used to describe the demand ranges of values: GE = greater equal, EQ = equal, LE = less equal, NGE = not greater equal. In the demand profile, the operators WHIL or WITH are used for formulating the linkage of a property with a parameter.

As Fig 5 shows, there may be different possibilities for the position of the required interval in relation to the intervals of the standard data. In the case of the inquiry with GE and LE for an interval, the cases a, b and c meet the individual criterion. However, if a criterion is formulated by using NLE and NGE operators no case but b will meet the demand.

With a WHIL linkage, the system is instructed to consider the individual criterion fulfilled for these steel grades only for which the demanded values for the property as well as for the parameter are met. By a WITH linkage, all those cases are additionally included for which the property does not depend on the respective parameter.

For the output of values there are two format types:

- Tabulated summaries for several materials. As an example, Fig. 6 shows the reference of steel grades to standards.
- Data relating to an individual steel grade, indicating all values for properties and parameters.

  The output can be limited to certain groups of properties or individual properties by the user. Fig. 7 shows an example for the chemical composition.

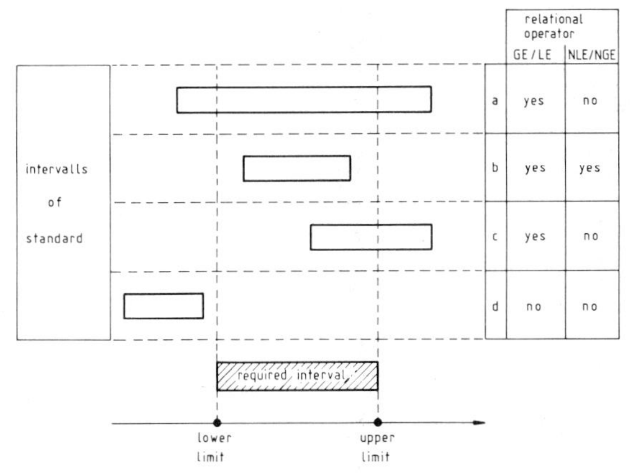

Fig. 5: Find results using different relational operators for the four possible positions of intervals of property values in relation to the required interval

The output of larger tables takes place in batch operation. Fig. 8 shows an example for the chemical composition. In this table fixed columns are assigned to the five most important elements. Further elements are shown in the last column under "Sonstige". Further tables are defined for the following groups of properties:

- Mechanical properties at ambient temperature
- Mechanical properties at higher temperatures
- Long time creep rupture values
- Physical properties

```
KOMMANDO EINGEBEN      FIND READ READ S READ,D FETCH PRINT DELETE END
>READ

LID LESEPROFIL
*** ****************
>ALL NORM

L E S E E R G E B N I S - T A B E L L E    (READ-ERGEBNIS)
ZDFT ANWENDERTEXT DB SZ ERG TF   BED SKZ LID LESEPROFIL        ANZ TR   TSU
011  HB30+SI+MN   01 W  004 0180 006 001 ALL NORM              0010 0161 0000
-----------------------------------------------------------   ------------
ZNR WERKSTOFF  (-NUMMER, -NAME)                     NORM          DATUM
ZNR VARIABLE                      PARAMETER
--------------------------------------------------------------------------
001 1 4310 X12CRNI177                               SEW400-73      1973
002                                                 DIN17224       1968
003                                                 DIN17225V      1955
004                                                 SEW400-73      1973
005                                                 DIN17442       1977
006 1 4439 X3CRNIMON17135                           SEW400-73      1973
007        X3CRNIMON1713                            SEW400-73      1973
008        X3CRNIMON17135                           SEW400-73      1973
009 1 4460 X8CRNIMO275                              SEW400-73      1973
010 1 4582 X4CRNIMONB257                            SEW400-73      1973

KOMMANDO EINGEBEN      FIND READ READ,S READ,D FETCH PRINT DELETE END
```

Fig. 6: Output of the references of steel grades

```
L E S E E R G E B N I S - T A B E L L E    (READ-ERGEBNIS)
ZDFT ANWENDERTEXT DB SZ ERG TF   BED SKZ LID LESEPROFIL        ANZ TR   TSU
011  HB30+SI+MN   01 W  004 0180 006 001 001 ALL               0545 0246 0000
--------------------------------------------------------------------------
ZNR WERKSTOFF  (-NUMMER, -NAME)                     NORM          DATUM
ZNR VARIABLE                      PARAMETER
--------------------------------------------------------------------------
001 1.4310 X12CRNI177                               SEW400-73      1973
003 C       )=0 00, (=  12        AART SCHMELZANALYSE
004 MN      (= 2 00               AART SCHMELZANALYSE
005 CR      )=16 00, (=18 00      AART SCHMELZANALYSE
006 NI      )= 7 00, (= 9 00      AART SCHMELZANALYSE
007 SI      (=1 00                AART SCHMELZANALYSE
008 P       (=  045               AART SCHMELZANALYSE
009 S       )=0 000, (=  030      AART SCHMELZANALYSE
083 1.4310 X12CRNI177                               DIN17224       1968
085 C       (=  12                AART SCHMELZANALYSE
086 SI      (=1 00                AART SCHMELZANALYSE
087 MN      (= 2 00               AART SCHMELZANALYSE
088 CR      )=16 00, (=18 00      AART SCHMELZANALYSE
089 NI      )= 7 00, (= 9 00      AART SCHMELZANALYSE
090 P       (=  045               AART SCHMELZANALYSE
091 S       (=  030               AART SCHMELZANALYSE
522 1.4310 X12CRNI177                               DIN17225V      1955
530 C       )=0 00, (=  15
531 ERSTE NICHT GEPRUEFTE ZEILE

KOMMANDO EINGEBEN      FIND READ READ,S READ,D FETCH NEXT PRINT DELETE END
```

Fig. 7: Output of the chemical composition of steel grades

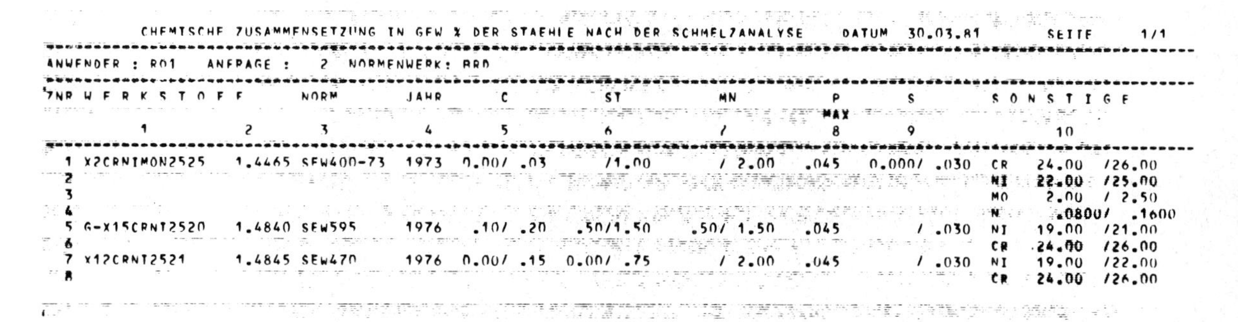

CHEMISCHE ZUSAMMENSETZUNG IN GEW % DER STAEHLE NACH DER SCHMELZANALYSE DATUM 30.03.81 SEITE 1/1

ANWENDER : R01 ANFRAGE : 2 NORMENWERK: BRD

ZNR	W E R K S T O F F 1	2	NORM 3	JAHR 4	C 5	SI 6	MN 7	P MAX 8	S 9	S O N S T I G E 10		
1	X2CRNIMON2525	1.4465	SEW400-73	1973	0.00/ .03	/1.00	/ 2.00	.045	0.000/ .030	CR	24.00	/26.00
2										NI	22.00	/25.00
3										MO	2.00	/ 2.50
4										N	.0800/	.1600
5	G-X15CRNI2520	1.4840	SEW595	1976	.10/ .20	.50/1.50	.50/ 1.50	.045	/ .030	NI	19.00	/21.00
6										CR	24.00	/26.00
7	X12CRNI2521	1.4845	SEW470	1976	0.00/ .15	0.00/ .75	/ 2.00	.045	/ .030	NI	19.00	/22.00
8										CR	24.00	/26.00

Fig. 8: Tabulated chemical composition of steel grades, output in batch operation

Data base for measured material characteristics

The subsystem for measured material characteristics serves to store measured values of test objects. For example, a test object may be a bar or a plate having a specific processing history. On such an object various tests may be carried out. All data relating to one test object usually are stored in one record. Each record contains so-called

- identifying values and
- test values.

These are described in more detail in Fig. 9. The identifying values give a description of the material, its origin and manufacturing history. These data are necessary for the correct interpretation of test values. The test values are the results from the different material tests. The right-hand column of Fig. 9 which concerns the example of a creep rupture test shows all data which ought to be collected.

Identifying Values	Test Values from:	
date of collection or publication	tensile test	
origin	notched bar impact test	
steel grade	hardening test	
material composition	hardenability test by end quenching	particulars relating to creep rupture test:
product		
condition of processing and treatment	creep rupture test	- sample position
melting process	physical tests	- sample type
	hot yield strength tests	- sample size
	fatigue test	- test temperature
		- stress
		- time to rupture
		- elongation after fracture in creep rupture test
		- reduction of area after fracture in creep rupture test
CHARACTERISTICS OF WHICH VALUES MAY BE STORED FOR A MATERIAL		

Fig. 9: Measured material characteristics compiled for the data base

Identifying values appear only once in one record, while test values may occur repeatedly or miss completely, depending on the number and extent of the tests carried out.

There are the following possibilities for data acquisition:

- data acquisition sheets
- special forms
- data mediums
- interactive mode (dialogue)
- connection of test devices

Data acquisition sheets are mainly used to collect values from the literature. Special forms are used for larger data volumes having the same format. A rather universal transformation program is available for the conversion of the data structure of such special forms or of files on data mediums into the data base format.

The interactive acquisition by dialogue is designed for the input of data immediately by material specialists. The acquisition of data by the connection of test devices is under development. A considerable organizational expenditure is necessary here in order to make data accumulating at different times and different places flow together properly in one record.

The data base for measured material characteristics contains a public data volume and various data volumes that are confidential. The public data volume comprises 600 000 values for 350 steel grades. All data come from the literature. Almost the entire literature has been evaluated since 1978 by the documentation department of the Verein Deutscher Eisenhüttenleute.

Two journals that have proved particularly rich data sources were evaluated back to 1970 and 1961, respectively. The special fields of fracture mechanics and hot yield strength were traced back to 1970 and 1957, respectively. The major non-public data volume contains 3 000 000 values concerning 500 steel grades. It is collected by member companies of the Verein Deutscher Eisenhüttenleute (VDEh). It serves to verify and evaluate the German and international material specifications. Intensive use is made of the data volumes for information and evaluation purposes.

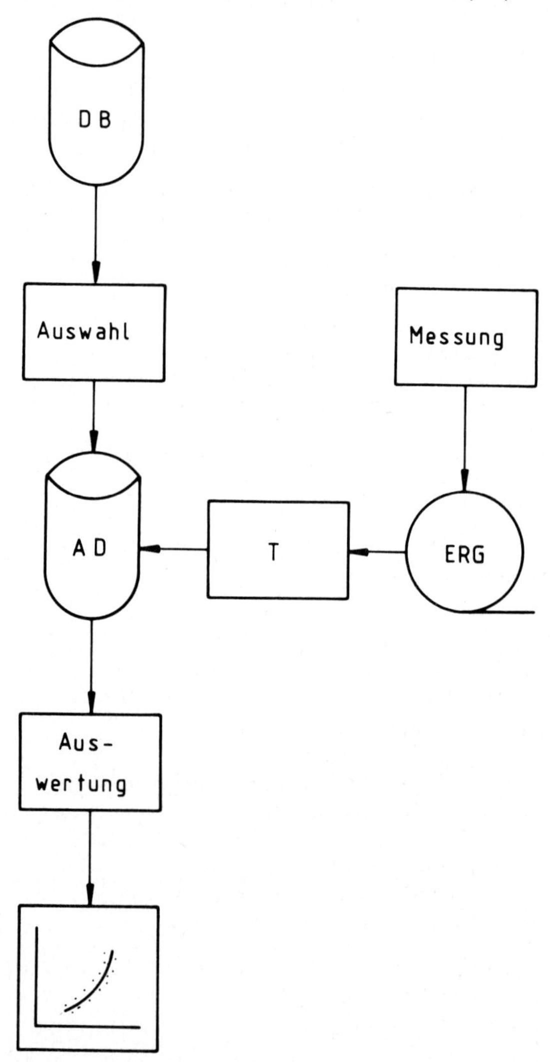

Fig. 10: Evaluation procedure from the data base for measured material characteristics

An evaluation procedure is shown in Fig. 10. In a first step, records are selected from the data base and requested property values from these records are put into a work file (AD).

In a second step, the evaluation proper takes place. Data from measurements can also be transferred into the work file without passing through the data base, by transformation from the data medium. The comfortable and comprehensive evaluation programs can thus be utilized in a flexible way.

The typical result of an evaluation of measured values are graphs, Fig. 4. The evaluations are carried out in an interactive mode with numerous possibilities of intervention with regard to the mathematical methods and the type of presentation. Fig. 4 is a hard copy from a graphics terminal. For the purpose of documentation or for improved resolution, the graph can also be produced on a plotter as, for example, in Fig. 2.

Further development

The planned further development mainly concerns the extension of the data volume for standard data and the on-line utilization.

To extend the volume of standard data, it is intended to evaluate the material standards of the major industrial nations. This will improve the possibilities for the comparison of steel grades in different standards.

Up till now, the ports for on-line connections via telephone lines are limited. This will be improved by conversions of the data base system to another computer and by connecting that computer to DATEX-P, the German packet switched network. DATEX-P is connected to a variety of other networks like Telenet, Tymnet, Euronet, etc. More will follow.

Summary

The main fields of application for the data base for iron and steel materials are:

- Selecting materials in accordance with a given demand profile
- Demonstrating properties of given materials
- Investigating the dependence of properties on the conditions of production
- Data evalation and support of the establishment of guaranteed values.

This is explained by means of examples.

Standard data and measured material characteristics are stored in two data bases with different data base management systems.

The data base for standard data contains the data of about 1000 German steel grades. The peculiarities of properties and influencing parameters in the standards and their handling in the selection procedure are explained. There are different table formats for the output of data on the selected steel grades.

The data base for measured material characteristics stores data of test objects. The data of one test object constitute one record which comprises identifying values and test values. The possibilities for the acquisition of measured material characteristics are described.

Typical evaluation results are graphic representations that are carried out in an interactive conversational mode and output on graphics terminals and plotters.

The futher development of the data base for steel and iron materials concerns the extension of data volumes and the connection to packet-switched data communication networks.

Data for Science and Technology, P.S. Glaeser (ed.)
North-Holland Publishing Company

THE ASM/NBS BINARY PHASE DIAGRAM EVALUATION PROGRAM

T.B. Massalski *(Editor-in-Chief)*
Carnegie-Mellon University, Pittsburgh, PA, U.S.A.

H. Baker *(ASM Manager, Phase Diagram Program)*
American Society for Metals, Metals Park, Ohio, U.S.A.

J. Rumble, Jr. *(Program Coordinator)*
Office for Standard Reference Data, National Bureau of Standards, Washington, D.C., U.S.A.

The availability of compiled and evaluated information on phase diagrams fulfills many needs in scientific, technological and commercial applications. The well-known compilations published during the late 1950's and 1960's have been in constant use, but a formal compilation and evaluation program which would update and extend these compilations has been lacking, with the result of an ever-increasing gap between published and evaluated data. The Binary Phase Diagram Evaluation Program is intended to rectify this situation during the next several years.

The task of the joint American Society for Metals - National Bureau of Standards Phase Diagram Program is to develop a new set of critically evaluated phase diagrams and supplementary information by the middle of this decade.

The well-known compilations of phase diagrams published during the late 1950's and 60's have been in constant use, but a formal compilation and evaluation program which would update and extend these compilations has been lacking, resulting in an ever-increasing gap between published and evaluated data.

When Hansen (1) attempted his first alphabetical compilation during the 1930's, the job must have appeared to be quite manageable. Hansen's original German volume published in 1936 contained some 450 binary phase diagrams in evaluated reviews of more than 800 systems. The major emphasis was on phase boundary information, accompanied by some crystal structure data. Metastable solubilities and phases were not included (except perhaps for Fe-C), nor was any thermodynamic data. Hansen made his own authoritative decisions about phase nomenclatures and crystal structure designations, and probably many other features as well, such as whether or not to use the centigrade scale, atomic percentages, stoichiometric ratios, etc. The postwar English language enlarged edition of Hansen, published in collaboration with Kurt Anderko (2) in 1958, and subsequent volumes by R.P. Elliott (3) and F.A. Shunk (4), which followed Hansen's general scheme are probably the most widely used publications of evaluated binary phase diagrams today. In addition to these volumes, however, a large number of other publications have now appeared, or are being produced; these present partially evaluated binary diagrams to a varying degree of detail and scope, from a mere recording of phase boundaries to attempted calculated predictions of diagrams that have not yet been explored experimentally. Such compilations, taken together, now include thousands of systems. (5)(6)

The first major step that affected plans for phase diagram compilation under the ASM/NBS program was the recognition that, with the mushrooming amount of phase diagram information, an alphabetical, single volume scheme, under the control of one author, or even several authors, was no longer feasible. A single author may be willing (and have time and resources) to review in depth perhaps 20-to-100 diagrams, but not thousands. He or she may be familiar with systems based on a single metal or a group of metals, but not familiar, or interested, in a completely different group. And, of course, support for compilation and evaluation work may be available for systems based on a particular metal, but not all systems alphabetically.

Accordingly, the best solution was to divide the systems into suitable categories, and to arrange for each category to be compiled and evaluated under the guidance and responsibility of a specific Category Editor, each for a given finite set of systems.

The task of coordinating the different categories into a whole, deciding how they should be grouped and published, appointing category editors, and of assuming the overall quality and standards of the binary evaluation program rests with the Editor-in-Chief. He, in turn, is able to seek advice and guidance from individual members of the International Council, from Category Editors, the Editor of the Bulletin and, most important, from the Committee on Constitution of Binary Alloys. In this way not only does the job become manageable again, but also it becomes possible to gain several extra benefits. Category Editors can be attracted from speciality areas, where they are already quite familiar with many of the diagrams based on a given metal. Within each category, the nature and scope of the information to be evaluated and recorded can be expanded. One can separate, or supplement, the bibliographic search needed for each category operation, by a computerized and centralized bibliographic search program. The compiled and evaluated data can be processed in a central organization possessing extensive facilities for computer storing of the information, and for computer graphics display of the resulting phase diagrams. As an added benefit, the Category Editors can be located in different parts of the world, in which activities pertaining to a given category can be coupled to the already existing and related research in a given area.

At present, it is envisaged that, whenever possible, phase diagrams in each category will be evaluated with respect to the following features:

Phase Diagram. The principal line drawing of the phase diagram itself (with possible inserts), which will be produced on a computer graphics display system. This allows each diagram to be displayed, if needed, with different scales, reference axes, etc.

Text. A verbal description of various boundaries, references, compositions, temperature, etc. Also included here can be possible drawings of additional details, partial diagrams with data points, or enlarged sections.

Structures. Tabulated information about crystal structure and lattice parameters, with accompanying verbal comments and references (if needed).

Thermodynamics. A brief analytical (thermodynamic) description of phase boundaries that can be calculated (if such information is available in the literature). Also included can be additional thermodynamic information, and thermodynamics of phase formation in specific important systems.

Metastable. Information about metastable structures and/or metastable phase boundaries, including information about martensites, massive transformations, spinodal decomposition and metallic glass.

References. A complete list of references.

Only the diagrams for which an extensive literature exists are expected to include all of the above steps.

To facilitate compilation of the most precise diagrams, a special Committee on Constitution of Binary Alloys has been appointed. This committee recommends and assures the high quality, scope, content and presentation standards connected with the evaluation of binary systems. It also assists ASM and NBS in various editorial matters and provides advice on worldwide cooperation in phase diagram evaluation and research.

Present members include:

Prof. T.B. Massalski, Committee Chairman, Carnegie-Mellon University, USA.
Dr. I. Ansara, Domaine Universitaire, Saint Martin d'Heres, France.
Mr. H. Baker, Alloy Phase Diagram Program Manager, ASM, USA.
Dr. L.H. Bennett, National Bureau of Standards, USA.
Prof. L. Brewer, University of California, USA.
Dr. J. Cahn, National Bureau of Standards, USA.
Dr. J.R. Cuthill, National Bureau of Standards, USA.
Dr. R.P. Elliott, Cleveland State University, USA.
Prof. K.A. Geschneidner, Jr., Iowa State University, USA.
Prof. M. Hillert, Royal Institute of Technology, Sweden.
Dr. O. Kubaschewski-von Goldbeck, Technische Hochschule Aachen, Federal Republic of Germany.
Prof. J. Nutting, Leeds University, England.
Prof. W.B. Pearson, University of Waterloo, Canada.
Prof. A. Pelton, University of Montreal, Canada
Prof. J.H. Perepezko, University of Wisconsin, USA.
Mr. A. Prince, General Electric Co. Ltd., England.
Prof. F.N. Rhines, University of Florida, USA.
Dr. I.S. Servi, metallurgical consultant, USA.
Dr. R.M. Waterstrat, National Bureau of Standards, USA.
Dr. J.H. Westbrook, General Electric Co., USA.

In various meetings, the binary committee has considered the schemes of phase designations, the degree to which thermodynamic and metastability data is to be included, the most effective ways of displaying the evaluated information, and of maintaining a working liaison with the parallel Committee on Constitution of Higher-Order Alloys. Another task of the binary committee was to consider how to stimulate research on phase diagrams in the future, particularly when controversies in data interpretation must be resolved before a correct diagram can be produced.

One of the objectives of the ASM/NBS program is the preparation of detailed phase diagrams which will be drawn by computer-generated graphics. This involves numerous technical problems which are being solved at NBS. The method has the advantage of great flexibility. Once stored in the computer, the diagrams can be displayed in a variety of ways, as well as in different temperature and composition scales. The evaluated diagrams produced by the various categories are published in the Bulletin of Alloy Phase Diagrams produced by ASM and edited by Dr. L. Bennett at NBS. These evaluations have already been subjected to a review process by both internal and external reviewers: however, they may undergo further modifications through additional input. Hence, they are labeled as "provisional" at this stage. A response to the published evaluations from the readers of the Bulletin is both expected and encouraged between the "provisional" and "final" publications. The latter may take the form of individual monographs pertaining to a given category, in addition to the overall publication of an alphabetical listing of all binary phase diagrams (drawn with computer graphics) in a special ASM volume.

At present, there are some 26 categories operating in several countries. With a few more categories appointed, all the periodic chart will be covered and divided into suitable groups that will impinge on one another until all needed binary systems are "on line". Of course, the essential idea is never to have to do this job again "ab initio". Hence, our task will be to develop a system that will update and extend beyond the existing evaluations with minimum effort and maximum usefulness. In the very near future computer programs must be developed to handle the job on a continuous basis. If we are clever now and lucky later, perhaps in the distant future, computers will be able to take over the job altogether.

REFERENCES

[1] Hansen, M., Der Aufbau der Zweistofflegierungen (Springer-Verlag, Berlin, 1936).

[2] Hansen, M., and Anderko, K., Constitution of Binary Alloys (McGraw-Hill, New York, 1958).

[3] Elliott, R.P., Constitution of Binary Alloys, First Supplement (McGraw-Hill, New York, 1965).

[4] Shunk, F.A., Constitution of Binary Alloys, Second Supplement (McGraw-Hill, New York, 1969).

[5] Bennett, L.H., Kahan, D.J., and Carter G.C., Alloy phase diagram data, Mater Sci. Eng., 24(1) (1976).

[6] Drits, M.E., Bochvar, N.R., Guzei, L.S., Lysova, E.V., Padezhnova, E.M., Rokhlin, L.L., and Turkina, N.I., Binary and Multicomponent Copper-Based Systems, Abrikosov (ed.) Nauka (1979) in Russian.

Data for Science and Technology, P.S. Glaeser (ed.)
North-Holland Publishing Company

THE NATIONAL SYSTEM ON NUMERICAL DATA CONCERNING THE PROPERTIES OF MATERIALS IN CZECHOSLOVAKIA

Margita Šulcová

Czechoslovak Institute of Standardization and Quality
Bratislava, Czechoslovakia

One of the factors assuring the present and prospective development of the national economy is the maximum utilization of factual data on the properties of materials in science, research and in practice. With this in mind, a national system of factual data concerning the properties of materials has been constructed in Czechoslovakia.

The work in this sphere can be divided into:

a) organizational-juridical work which has provided for the elaboration of a legal document (decree) on the operation of the information system, determining the duties of the user and of the owner of data bases on the properties of materials.

b) the actual construction since 1978 of a national system of factual data on the properties of materials in the sphere of metals, plastics and elastomers.

The aim of the Information Data System on the Properties of Materials is to rapidly introduce the newest most efficient materials into all branches of the national economy as well as to use rationally the types of materials employed in a traditional way, ensuring complete and reliable information on their properties in all spheres of science, technology, production, and consumption.

Of the organizational-juridical documents, the most important is the "Decree on the Information Data System on the Properties of Materials" (SID) ensuring the construction, operation and development of the system in Czechoslovakia.

The Information Data System is included in the System of Scientific, Technical and Economic Information which was constructed as a specialized, decentralized information system of factual information on the properties of selected materials, noted for its reliability, completeness and operative supplying of data needed for the national economy.

The factual data information system is centrally coordinated and methodically managed by the system's national center, i.e. the Czechoslovak Institute of Standardization and Quality in Bratislava. The system is of a branch nature. The branches manage the acquisition and utilization of technical and technological data for selected materials to be used in the fabrication of their own products. These data are equally supplied to the National Centre of the System of Scientific, Technical and Economic Information.

The main users of the system are, first of all, organizations dealing with research, development, manufacturing preparation, material production or organizations employing materials.

In addition, the system is used by:

- the Office for Inventions and Discoveries for giving national expertise on inventions, as well as by the inventors and innovators themselves,
- organizations charged with the elaboration of drafts of standards and organizations authorized to approve these drafts,
- organizations checking and verifying the properties of materials,
- state testing laboratories,
- institutions of the System of Scientific, Technical and Economic Information in the framework of their information activity for research, development and for the steering of scientific and technological development.

Branch centers (possessors of bases), as specialized components of the system, are obliged to effect measurements in view of determining accurate values of physical constants, and of obtaining factual numerical data on material properties:

- to build up specialized sets of information to a limited extent,
- to check the reliability of information entering the sets and to put them into corresponding categories,
- to publish general information on the content of information sets and to supply information on:
  a) materials corresponding to the properties requested,
  b) numerical values of properties for required materials,
  c) changeability of materials,
  d) materials on the basis of producer requirements,
  e) values of material properties under conditions specified by the user.

The construction of the international Standard Information Data System on the Properties of Materials (SSID CMEA) and its structural component - the national system (SID) stemmed from the Complex Programme of Further Development and Improvement of Collaboration and Development of Socialist and Economic Integration of Member Countries of the CMEA, in the section concerning the SSID CMEA, section 9, points 8.1, 8.3 and 8.5.

Construction of the system and its application in practice were planned in three stages:

a) organization and methodical preparation of the system's construction,

b) construction of single thematic data bases and their experimental verification,

c) functioning, further development and improvement of the system.

From the point of view of the requirements of the national economy, it was decided to construct primarily thematic data bases for metals, plastics and elastomers.

In order to ascertain the actual state of data bases for metals, plastics and elastomers in Czechoslovakia, we conducted a public inquiry about the "Standard Information Data System on Properties of Materials" which contained 31 questions. We sent the inquiry forms to 120 organizations of a scientific, research or production nature and obtained 85 answers which represented 71% of the total. The public inquiry led to the conclusion that in Czechoslovakia there was no information system having as its aim the one referred to, which is why the respondants asked that a centrally coordinated data base on materials properties be established using the international data base SSID CMEA. The first regional data center in Czechoslovakia is the National Research Institute of Materials (NRIM) in Prague which, so far possesses three specialized thematic data bases on the properties of tool steels, refractory steels, and plastics.

A very ample data base containing more than 250 000 industrial data on the composition and test results of properties of Czechoslovak and foreign produced tool materials has been compiled over a period of many years of research and testing. At the beginning the data were only destined for research purposes. However, these data have permitted the introduction in Czechoslovakia of steels of higher quality and have allowed us to stop using outdated steel types whose performances were not commensurate with the degree of alloying. They have facilitated the continuous modification of steel types in order to conform to production requirements. The gradual introduction of further data, especially on the properties of cemented carbides, is underway.

The files on the properties of tool steels are arranged in three groups: data from standards, data from producers, and data from testing laboratories. In such an implicit form they represent, in a certain way, the degree of reliability of the stored and disseminated information. In the beginning, primarily information on composition and basic properties of materials was stored in the data base. Later on the data base was completed with the collection and storing of data concerning technological properties and data on the thermal treatment of tool materials.

The Information System for the Properties of Refractory Materials is likewise, the former information system which was located at the National Research Institute of Materials branch. It is based on uniform principles of construction. The system structure reflects the logical sequence from production and treatment of the material to be tested to the results of tests carried out on particular samples.

In the sphere of the refractory materials there is a small variety of material types and information. Each material type is, however, tested in detail for heat resistance. This represents the testing over a large range of temperatures, covering all the intervals of admitted compositions and treatments of the material type. As each material type is used in a broad range of temperatures, a multitude of necessary tests are required to cover this range. The long-term characteristics of heat resistance adequate for these materials are obtained only by processing the test results. As heat resistance tests are relatively very expensive, and the processing methods are still being developed, in principle the data from all the tests are preserved. The system structure for information items also includes data on geometry, welding technology and other factors identifying the conditions and the welding methods inasmuch as refractory materials are often used in a welded state.

The Information System on Refractory Materials is at present operational and supplies data on selected materials of domestic and foreign production which are presently in use. Data on more than 10 000 creep tests are stored here. Currently work is concentrated on the completion of the data base by adding further materials and test results.

Construction of the material information system is, from the point of view of the program, practically at an end. The program equipment allows, for example, the drawing of column diagrams with Digigraph equipment of certain properties for a selected group of materials. In drawing functional dependencies, these dependencies, stored by means of points, are converted to a graphic form in a rectangular coordinate system. The transmission module used between the stored data and the user's program is sufficiently versatile to enable one, together with the stored data, to reproduce graphically the parameters of functional dependencies and, in this way, to cover the whole spectrum of material tests. It also permits the use of retrieved data to determine the theoretical or approximate curves. It is self-evident that it differentiates between standard data and other data, which represents a particularly important element with respect to the complexity of the information supplied.

An important realization in the use of the information system in the field of construction was the experimental use of the limited data base on the properties of tool and refractory materials in the framework of the CMEA Standard Information Data System on Properties of Materials (SSID CMEA), in conformity with the Permanent Commission of the CMEA's plan for collaboration in the field of standardization which began in early 1980. The participating delegations from five member countries of the CMEA were informed about the conception, extent and possibilities of the use of thematic data bases during the experiment. In the opinion of the participants, the experimental utilization of the data bases attained its goal and, following this verification, it was recommended to include the information system among the corresponding thematical data bases of the SSID CMEA as a constituent part.

The third thematical data base is the one on the properties of plastics which is based on the collection and classification of factual data on the properties and behaviour of plastics of both Czechoslovak and foreign origin. At present the data base covers 590 materials with about 200 000 data. The data base contains general data, data on physical-chemical, mechanical, optical, electric and heat properties, data on chemical resistance, aging, appropriate treatment technology, and application possibilities.

By 1983 the data base on the properties of plastics will be completed and will be experimentally tested within the framework of the CMEA.

At present we are preparing a future plan aimed at completing the information base with further thematical data bases for materials important from the point of view of the national economy, such as glass, wood and hide.

REFERENCES:

[1] The draft of the decree of Federal Ministry for Technical and Investment Development. Internal material. To be found at the Czechoslovak Institute of Standardization and Quality in Bratislava.

[2] Kraus, V., Base of data on tool materials, Standardization and Quality Bulletin 4, Bratislava, Alfa (1979).

[3] Kraus, V., The results and perspectives of the development of tool steels and alloys, Machine Construction 9, Prague, SNTL (1979).

[4] The information system of material utilization for the assessment of properties, Proceedings for the 30th anniversary of foundation of the NRIM, Prague, NRIM (1979).

[5] Record of the 19th session of the Working Group for the Standard Information Data System of CMEA on the properties of materials and matters (SSID of CMEA) Yerevan (1982).

Data for Science and Technology, P.S. Glaeser (ed.)
North-Holland Publishing Company

STRUCTURAL MATERIALS CORROSION DATA BANK

Marek Dobrucki*, Jerzy M. Kasprzyk**, Eugeniusz Krop**, Jarosław Liskowacki***
Jacek Rakowski** and Krzysztof Wróblewski**

**Institute of Ferrous Metallurgy, Gliwice, Poland*
***Engineering a. Contractors Co. for Heavy Chemicals Synthesis Industry, PROSYNCHEM, Gliwice, Poland*
****Institute of Materials Economy, Katowice, Poland*

A data base was established for computer retrieval of the information concerning corrosion properties of structural materials and coatings. The data are assembled from elementary units describing individual corrosion cases. The following variables and parameters are used: material data, environment data, protection data, corrosion resistance, type of corrosion, informal comments, source and uses (industrial application, devices). A software specially written for the system makes it possible to create and up-date the files and to answer queries written in a special problem-oriented language. Important features of the system include the use of controlled retrieval vocabulary and automatic query enhancement by means of term networks.

1. INTRODUCTION

Causes of losses due to corrosion can be attributed to a certain extent to insufficient information. Data concerning corrosion behaviour of specific materials in well defined environments should be freely accessible to all concerned i.e. designers, manufacturers and users of various products. It seems that the availability of corrosion data is limited not so much by lack of research or publications but rather by dispersion of information and difficult interpretation of existing data in terms of user's conditions. The difficulties are often associated with identification of structural and coating materials which are frequently designated with names conveying little information about their nature, at least for the user who is seldom expert in comparative materials science. Queries often cannot be answered because no data can be found for a material under the exact name quoted by the user and there is no system which could make use of data for equivalent or similar materials. Another problem is associated with oversimplification of corrosion data in generally accessible corrosion tables. Very often they neglect the effects of accessory components of corrosion environments. Numerous failures can be traced to negligence of such accessory components as chlorides in water or sulphur compounds in fuels.

An analysis of a number of real corrosion data demonstrated the following facts:

- in the majority of cases corrosion data can be presented in the form of definite and quantitative parameters which can be recorded in a formalized way;
- those data which are not easily formalized usually are not needed as criteria for retrieval and serve as an additional accompanying information at best;
- there are numerous corrosion data concerning whole classes of materials and environments. It is possible to obtain indirect information about behaviour of specific materials or environments on the basis of data for the respective classes and knowledge of contents of the classes. In a similar situation it is possible to synthesize information about a class by retrieving all accessible data concerning its representatives. Such operations are easily realized by means of predefined networks of hierarchic relations between terms (hierarchic thesauri);
- in the case of materials it is possible to transfer information concerning behaviour of one specific material to all equivalent materials i.e. those with close enough or identical characteristics (esp. composition and structure). This is important because equivalent materials are manufactured in various countries or by various producers under quite different names;
- corrosion data are often divergent or contradictory and no decision about their reliability can be taken without expert evaluation of each individual case. The divergence of data may be due to poor definition of the material e.g. its purity or surface finish , corrosion environment or testing conditions. This is one of the reasons why corrosion resistance data can hardly be used as criteria for retrieval in any computer system.

Such character of data and their logical interrelation suggest that computer processing may produce results or at least perform operations that are impossible to obtain by conventional manual retrieval.

It seems that the feasibility of operations which are impracticable in manual systems makes computer processing attractive in spite of its inherent high costs. It must be noted that a single successful implementation of results obtained from the system in a large industrial installation or a mass-produced consumer product may repay all costs of the system. It must be admitted, however, that the problem of creating and maintaining channels for recycling even a small part of the profits remains to be solved.

Taking into account the above considerations, in 1973 an initial study was taken up to develop a computer bank of corrosion data. In 1978-1982 funds provided by the Institute of Materials Economy and PROSYNCHEM made it possible to implement the system in a computer and to establish an initial data base.

2. DATA BANK STRUCTURE

Data base, the principal part of the Bank, contains data grouped in several specialized files. The most important files (Main Files) are CORROSION CASE FILE (CCF) and MATERIALS FILE (MF).

Each record of CCF contains all selected data concerning a single corrosion case. Corrosion case is defined as an information describing behaviour of an identifiable material in specific environment and exposure conditions.

Records of CCF contain the following fields:
Material data
- material identification code (thesaurus term acting also as an access key to MF),
- material form data (plate, pipe, weld, cladding, etc.),
- material treatment data (brazed, welded, heat-treated, precipitation hardened, sensitized, etc.),

Environment data
- thesaurus terms,
- chemical composition,
- state of aggregation (solid, liquid, boiling liquid, gas, aerosol, suspension, fume, etc.),
- parameters (ranges of temperature, pressure and pH).

Protection data
- thesaurus terms,
- unformalized description.

Types of corrosion (stress corrosion cracking, pitting, crevice, etc.).
Types of equipment.
Branches of industry.
Unformalized comment (important data which cannot be accommodated in other fields).
Source bibliography.
Corrosion data
- corrosion rate unit code or corrosion scale type,
- range of corrosion rate or scale points.

Materials file contains data inherent to a given substance without reference to characteristics depending on treatment or fabrication. Only specific grades of materials are covered by the file and no data for classes of materials are included.

Records of the file contain the following fields:
thesaurus term,
reference to national or industrial standard,
producer country,
elemental composition,
unformalized description.

Some of the above fields in both files are optional.

Compositions of materials and environments are recorded in the form of lists of component specifications. Each specification contains:
component thesaurus term (chemical element or substance for materials and environments, respectively),
unit of concentration,
range of concentrations,
indicator of significance.

Indicator of significance shows whether a component is significant with respect to its influence on general characteristics of the material or corrosive action of the environment. The indicator is used in operations involving matching of compositions.

Descriptions of corrosion cases and materials are prepared in a strictly defined format. Specially developed nomenclature is used for names of materials, environments, substances, etc. Terms allowed in the system are collected in specialized files, the so-called thesauri.

The standardized nomenclature makes possible unequivocal recording of terms and allows introduction of logical relations between terms in hierarchic thesauri. The system uses hierarchic thesauri for materials, environments, substances, and protective means. There are also simple thesauri without hierarchic relations for forms of materials, treatments of materials, states of aggregation, types of equipment, branches of industry, types of corrosion, chemical elements, producer countries, corrosion resistance units, temperature scales, pressure units, and concentration units.

Various types of codes are used for notation of thesaurus terms. All numeric codes contain check digits.

In order to facilitate processing, a number of inverted files is used. The files consist of lists of addresses of records containing given thesaurus terms.

All records in data base files use formats specially designed for the purpose.

In addition to the data base the Bank contains programs for creation and updating of all data base files, translation of queries written in query language, and retrieval of information as well as GEORGE-3 operating system macrocommands for control of the programs and overall processing.

3. DATA BANK OPERATION

Data for the Bank can be taken from any relevant publication, research report, corrosion table, etc. Input data are written in coded form on special data sheets which make it possible to avoid repetition of those fragments of data which are to be used more than once in a batch of data. During final assembling of corrosion case data, reference numbers of predefined materials, environments, etc. are used in combination with the respective parameters.

The data consist of terms, parameters and comments in natural language. The terms must be taken from the appropriate thesauri which in addition to control of the vocabulary also determine the depth of indexing. During indexing by means of hierarchic thesauri, terms from the lowest level of the hierarchy must be used except in cases where the data pertain to the whole class of terms as defined by the thesaurus.

Input data are checked by the appropriate programs and, after conversion, stored in CCF. If necessary, MF and thesauri are first updated with new material data and terms used in the current batch of data for CCF.

The block diagram in Fig. 1 illustrates in a much simplified form the course of information retrieval.

The initial stage of processing of a query written in query language involves formal checking and conversion to Polish notation.

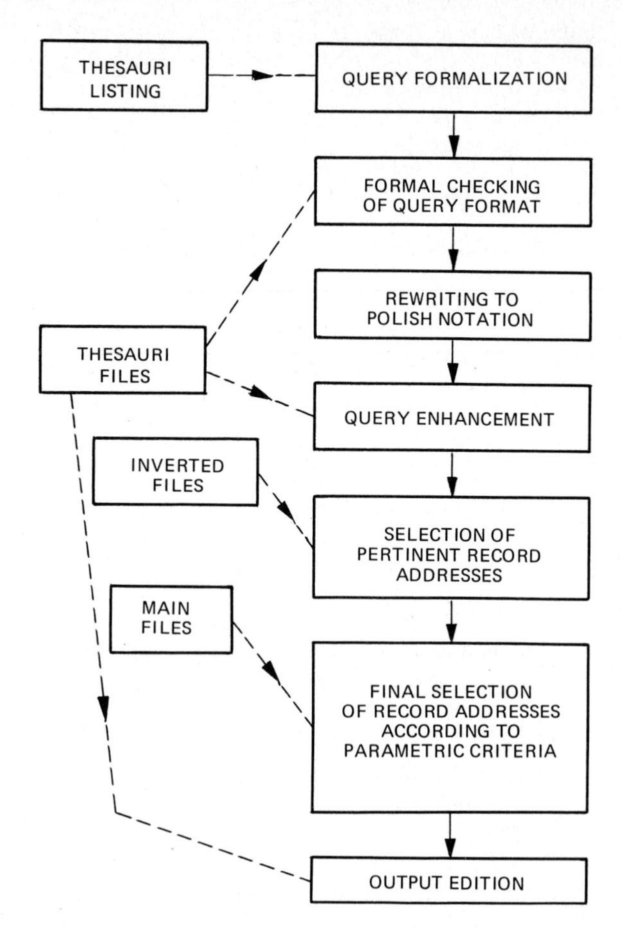

Fig. 1 Simplified block diagram of information retrieval

The next operation (optional) is automatic query enhancement by means of hierarchic thesauri. All query terms concerning materials, environments, substances, and protective means can be expanded i.e. substituted by lists of terms selected according to thesaurus logic. The principle of query enhancement is best illustrated on the following examples involving materials and a simplified fragment of the relevant thesaurus (Fig. 2):

Let a query contain the term "316L(AISI)" (a widely used Cr-Ni-Mo austenitic stainless steel with extra-low carbon content). After enhancement the term will be substituted by the following list of terms:
316L(AISI) or EXTRA-LOW CARBON STEEL or LOW CARBON STEEL or STEEL or Cr-Ni-Mo HIGH ALLOY STEEL or HIGH ALLOY STEEL or ALLOY STEEL or AUSTENITIC STEEL or STAINLESS STEEL[1].

[1] The use of more general terms makes sense only if such terms are really describing properties of whole classes of terms.

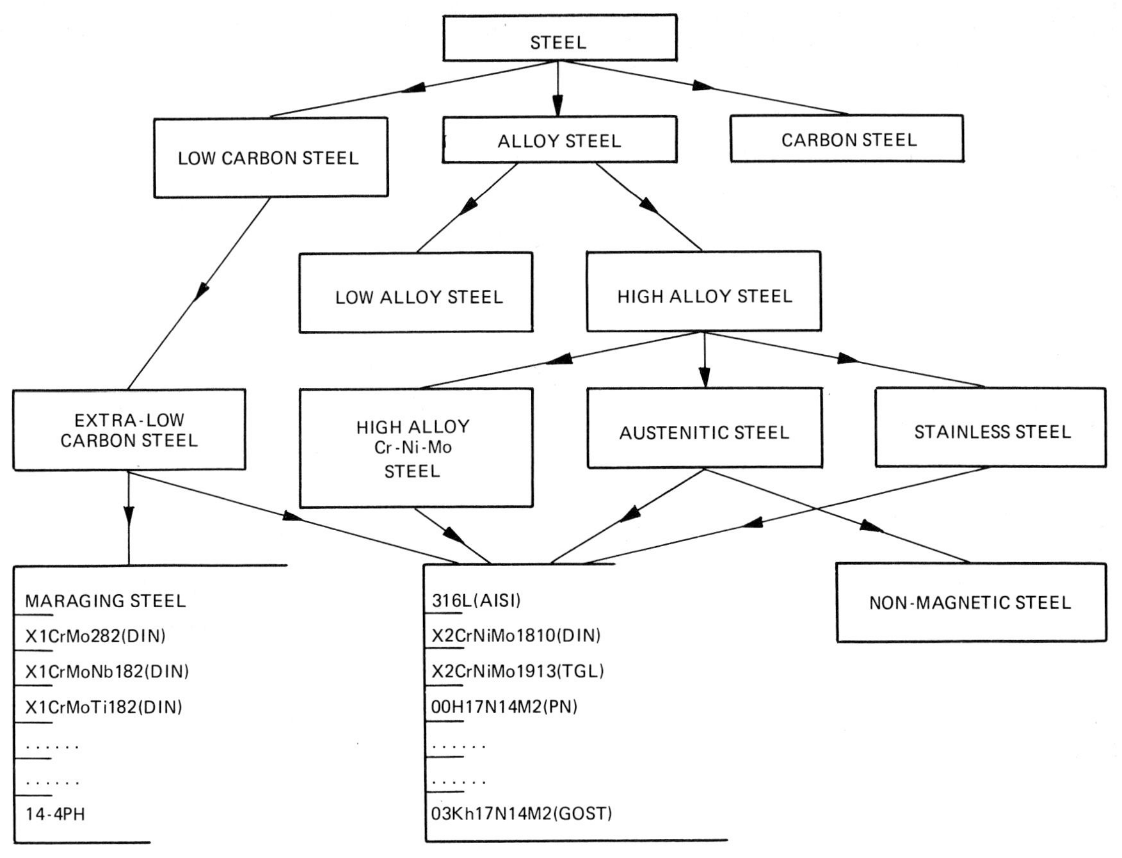

Fig. 2 Fragment of Thesaurus of Materials

This list will be further expanded by adding all materials with equivalent compositions i.e. materials containing the same principal components (according to indicators of significance) in at least overlapping concentrations and containing no other principal components. Selection of such equivalents is performed automatically.

The term "EXTRA-LOW CARBON STEEL" in another query will be substituted by the following list:
EXTRA-LOW CARBON STEEL or LOW CARBON STEEL or STEEL or MARAGING STEEL or X1CrMo282(DIN) or X1CrMoNb182 (DIN) or X1CrMoTi182(DIN) or .... or .... or 14-4PH or 316L(AISI) or X2CrNiMo1810 DIN) or X2CrNiMo1913(TGL) or 00H17N14M2(PN) or .... or .... or 03Kh17N14M2(GOST).

Terms from other thesauri are expanded in a similar way except that equivalents according to composition are not applicable.

The next stage of retrieval involves selection of records fulfilling criteria defined by logical expressions containing thesaurus terms as the operands. Each thesaurus term is substituted by a list of addresses of records containing this term. The list is taken from the appropriate inverted file. After substitution, the specified operations are performed on the lists. This will be illustrated by the following example:

Let a query contain terms combined in the expression:

D1 or (D2 and D3)

and the relevant portion of the inverted file be:

D1 - A2, A3, A5, A8, A10
D2 - A1, A3, A4, A5, A9
D3 - A1, A3, A5, A9, A11, A20

where A1, A2 ... are addresses of records in the main file.

After substitution the expression takes the form:
(A2 or A3 or A5 or A8 or A10) or ((A1 or A3 or A4 or A5 or A9) and (A1 or A3 or A5 or A9 or A11 or A20)).
Execution of logical operations gives the following list of record addresses:

A1, A2, A3, A5, A8, A9, A10

In the consecutive step the selected records are examined in order to eliminate those which fail to satisfy additional parametric criteria such as temperature or pressure ranges defined in the query. As the result a final file of records satisfying all query criteria is obtained.

In the final operation output is edited in the form specified in the query. Output listing contains full decoded descriptions of corrosion cases or materials. The listings can be supplied with alphabetic subject indexes of environments and/or materials.

4. PROBLEM-ORIENTED QUERY LANGUAGE

In order to facilitate formulation of complex logical criteria a special query language was developed. The language makes it possible to define step-wise retrieval criteria by means of thesaurus terms, logical operators, numerical parameters and criterion identifiers.

The following simple criteria can be defined:

material according to elemental composition (presence of given concentration or absence of elements) and/or thesaurus terms;

environment according to chemical composition (presence of given concentration or absence of substances) and/or thesaurus terms;

protective means, type of corrosion, type of equipment, industry according to thesaurus terms.

Complex criteria are defined by means of logical expressions combining identifiers of predefined simple criteria.

The following logical operators are permitted: AND, OR, NOT (in certain types of criteria), <,> , =, (,) , ↑ . The operator '↑' is used for control of automatic query enhancement.

The following example illustrates application of the language.

Query: Corrosion behaviour of austenitic steels containing 18 % Cr, 9 % Ni, max. 0.1 % C and Ti or Nb in concentrated phosphoric acid.

Query in coded form:

```
MD1↑: 014255:
MS1:  CR+[=18] AND NI+[=9] AND C+[< 0.1]
      AND (TI+[ ]OR NB+[ ]) AND FE+[ ]:
M: MD1 AND MS1:
SD1↑ : 002346 AND 005072 AND 006337:
S: SD1:
P: M AND S:
```

where:

MD1 - criterion identifier for material defined by thesaurus terms,
014255 - austenitic steel (thesaurus term),
MS1 - criterion identifier for material defined by composition,
M - final definition of material,
SD1 - criterion identifier for environment defined by thesaurus terms,
002346 - phosphoric acid,
005072 - aqueous solution of acid,
006337 - concentrated solution,
S - final definition of environment,
P - final definition of query.

5. IMPLEMENTATION

At present (Oct. 1982) the system is fully programmed and tested. Implementation is in ODRA 1305 computer with GEORGE-3 Operation System (ICL compatible). System software uses mostly COBOL.

For testing purposes real data for phosphoric acid, ammonia, sulphur dioxide, methyl alcohol, atmosphere, monosodium carbonate, (and sulphuric acid were collected and stored (approximately 10 000 corrosion cases). Initial data for Materials File were taken from selected US, West German, USSR, GDR, Japanese, Czechoslovak, and Polish standards. These material data were complemented by information obtained during collection of corrosion data. At present MF contains data for approximately 3000 materials. It is estimated that some 70 % of the materials encountered during current collection of corrosion data is already present in MF. It seems that for normal operation of the Bank about 100 000 corrosion cases would be needed. This requires appropriate financial and organizational decisions.

The system was designed with room for future extension to other properties (e.g. mechanical). Owing to full autonomy of materials data this is possible without the need of any major modification of the Materials File.

Data for Science and Technology, P.S. Glaeser (ed.)
North-Holland Publishing Company

THE COMPUTER AIDED SYSTEM OF MATERIALS SELECTION

Tadeusz Florczak

Research and Development Center for Machine Technology and Design 'Tekoma'
Warsaw, Poland

Difficulties and disadvantages of processes of materials selection in the engineering industry are discussed and examples of the effective application of materials are presented. General features of a materials information system design for improving the processing of materials selection are included. Possibilities and profits which can be obtained by using the system are noted.

1. Problems of Engineering Materials Selection

Every machine element may be described by the following characteristics:

- features of shape
- properties of inner structure
- properties of surface layer
- other physical features

The necessary condition for obtaining the desired useful properties of such an element excluding (partially) the features of shape, is the use of a suitable material.

In mechanical engineering there exist, on the one hand, great quantities of machine parts (applied materials) working under various conditions, but on the other hand, there exists a wide range of materials, in various forms, the properties of which may be shaped within certain limits. These properties undergo changes with temperature, time and the influence of the environment.

It has been estimated, that several thousand grades of materials used for making machine parts exist in the world. In the Polish engineering industry the quantity of grades of materials used exceeds 10 000, including about 3000 grades of construction materials (1).

Almost every construction material, after a specific treatment, acquires an average of 3 to 7 sets of properties, varying to a great degree in their characteristics and practical applications. The variety of media influencing the materials in differing degrees is quite large.

"The Chemical Abstracts", for example, informs us that the number of chemical compounds increases weekly by about 6000 individual ones which may potentially form components of a medium in which a construction material operates.

It was, therefore, out of the question to apply computers to the process of materials selection for machine elements. "In today's high-cost business climate it is important that the most suitable, economical, and easily processable material be selected the first time" (2).

The process of materials selection is a decision regarding not only the technological and economical criteria, but also the social, legal and political ones.

The decision takes into account the technological characteristics, and availability of materials (particularly the critical ones), prognosis of future costs, influence on the natural environment, recycling problems, energy needs and others, as summarized in Table 1.

Table 1

FUNDAMENTAL PROPERTIES AND INDEXES FOR ANALYSIS IN SELECTION OR SUBSTITUTION OF BUILDING MATERIALS

I. Technological

1) Mechanical properties at working temperature

At static loading

a) yield strength in tension (strain resistance),
b) tensile strength,
c) Young's modulus (deflection resistance),
d) elongation, reduction of area,
e) Poisson's ratio,
f) hardness,
g) yield strength for other types of loads (compression, bending, shear torsion),
h) immediate strength for other types of load,
i) creep limit in tension,
j) creep strength,
k) creep limits for other types of loads,
l) creep strength for other types of loads,

At rapid load or strain

m) impact resistance,
n) impact energy,
o) fracture toughness Klc,
p) crack opening displacement COD,

At cyclic changing load

q) fatigue strength in rotational bending,
r) fatigue strength for other types of loads, fluctuating or alternating loads,
s) fatigue strength exponent n,
t) coefficient of fatigue deformation f,

2) Atmospheric corrosion resistance and/or chemical substances resistance at working temperature

3) Thermal properties

a) melting point,
b) softening point,
c) thermal conductivity,
d) thermal expansion,
e) resistance for rapid changes in temperature,
f) fire resistance,

4) Electric and magnetic properties

a) electric conductivity,
b) electric resistivity,
c) dielectric strength,
d) superficial resistivity,
e) magnetic permeability,
f) magnetic saturation induction,
g) coercive force,
h) maximum magnetic energy,
i) dielectric loss,

5) Other properties at working temperature

a) wear resistance,
b) color, lustre,
c) other,

6) Structure features

a) chemical composition,
b) structure.

II. Manufacturing

1) Limitations of overall dimensions and shape

2) Accuracy of dimensions

3) Accuracy of machining

a) shaping (forming ability, workability, drawability, castability, pressability),
b) joining (weldability, soldering ability),
c) finishing (painting, plating and so on),
d) manufacturing facilities
- ease of displacement (lightness),
- ease of storing and protection during storing,

e) productivity capability being at disposal,

4) Minimization of allowances, use of materials in rough state

III. Production

1) Easy availability of material for starting and maintaining production

2) Scale of losses and wastes

3) Possibility of re-using wastes

4) Productivity

5) Stability of production quality (repeatability)

IV. Utilization

1) Durability/reliability

a) reliability of constructional joints,
b) reliability of electric contacts,
c) environment resistance,
d) adaption to mating with other materials,
e) ease of maintenance and protection,
f) possibility of repairs,

2) Safety of utilization (legal responsibility)

3) Watt-hour efficiency, energy and fuel consumption

4) Comfort and beauty

5) Recycling of wastes, ease of segregation and scrap materials recovery

V. Economical aspects

1) Cost of production

a) purchase costs (discount related to quantity),
b) cost of processing
- consumable materials costs,
- labor costs,
- instrumentation costs,
- equipment costs,
- power consumption costs,
- transferable waste materials,
- profit,
c) inspection costs,
- input inspection costs,
- process inspection costs,
- final inspection costs,
d) storing costs and costs of protection during storing,
e) operating costs and maintenance costs,

(service, guaranties, maintenance, repairs).

2) Costs of obtaining required set of properties and features of the shape of an element

3) Quantity of energy necessary for obtaining required set of properties and features of the shape of an element

4) Investment outlay requirements

5) Relation of a profit to an outlay for production of a part

6) Decrease (increase) in cost by substitution of:

a) single part,
b) set of elements,
c) in the whole cost of a product,
d) in the global cost for user of product.

VI. General economic aspects

1) Reliability of deliveries

a) quantity and distribution of resources in geopolitical situation,
b) possibility of build-up of strategic reserves,
c) possibility of implementing substitutes or alternative solutions,
d) strategy of managing own reserves to assure deliveries in the future,

2) Factors of outlay necessary for production of 1 Mg of various types of elements

3) Factors of energy consumption in the phases of:

a) mining and production of material,
b) processing material,
c) operating a product,

4) Losses of material in the phases of:

a) mining and production of material,
b) processing of material,
c) operating a product.

5) Safety and convenience to the natural environment

The appropriate selection of materials is of particular importance for long-term undertakings, in which the acceptance of the definite material solution means the involvement of an undertaking in the investment process for a period of 10 to 15 years.

The choice of materials is especially important in the following 5 groups of problems (3):

- new product development (devices with superconductors, supermagnets, biotechnology, computer cores, coherent flexible fibre bundles, hovercraft skirts, oil rig structures),

- modification of the existing products to various working conditions (modern aircraft engines, turbines, industrial robots, lasers, solar batteries, durable telephone booths),

- decrease in costs of the existing products (metal replacement in cars, thin walled die castings, transmission cables made of aluminum),

- modification of products to meet the legal security and natural environment protection requirements (replacement of asbestos-cement pipes by cast iron and plastic ones, safe buffers in cars, introduction of plastic in ship and submarine construction, limitations in the use of asbestos, mercury, cadmium and other harmful materials),

- avoidance of premature wear (resulting from abrasive action, corrosion and other factors).

"Although materials in themselves are important to a designer it is their ability to provide the required physical and chemical characteristics which really matters to him" (4).

The optimum selection of a material means taking all possible full advantages of its properties in real applications, and , we add, losing none of its precious merits.

Many attempts have been made to determine numerically the degree of correctness of material selection. P.M. Apoo and W.O. Alexander (5) introduced a Weighted Property Factor combining the set of properties into one parameter determining the material according to weighted factors for particular properties.

Weighted Property Factor WPF has the formula:

$$WPF = \frac{\sum_{i=1}^{NC} P_i W_i}{\sum_{i=1}^{NC} W_i}$$

where:

P_i - value of "i" property of material

W_i - weighting of "i" property

NC - number of specified characteristics of material

Example of calculation WPF

Material	Tensile Strength (MPa)	Conductivity (IACS)
A	500	50
B	1000	10
weighting	1	10

$$WPF_A = \frac{(500 \times 1) + (50 \times 10)}{(1 + 10)} = 91$$

$$WPF_B = \frac{(1000 \times 1) + (10 \times 10)}{(1 + 10)} = 100$$

A higher value of the WPF factor indicates a more advantageous material taking into account the characteristics specified.

The Weighted Property Factor regarding the Scaling Factors was also introduced. The Scaling Factor SF equals 100 divided by the maximum value of the property taken from the list of considered materials. Then the Scaled Weighted Property Factor WPF_{sc_i} has the following formula:

$$WPF_{SC_i} = \frac{\sum_{j=1}^{NC} P_{ji} W_j SF_j}{\sum_{j=1}^{NC} W_j}$$

where:

- P_{ji} - value of "j" property for "i" material
- W - weighting of "j" property
- SF_j - scaling factor for "j" property
- NC - number of specified material characteristics

Example of calculation of WPF_{sc_i}

Material	Tensile Strength (MPa)	Max value	SF	Conductivity	Max value	SF
A	500	1000	$\frac{100}{1000}$	50	50	$\frac{100}{50}$
B	1000	1000	$\frac{100}{1000}$	10	50	$\frac{100}{50}$
weighting	1			10		
L (basic)	1000	1000	$\frac{100}{1000}$	50	50	$\frac{100}{50}$

$$WPF_{SC_A} = \frac{(500 \times 1 \times 0.1) + (50 \times 10 \times 2)}{(1 + 10)} = 95$$

$$WPF_{SC_B} = \frac{(1000 \times 1 \times 0.1) + (10 \times 10 \times 2)}{(1 + 10)} = 27$$

$$WPF_{SC_i} = \frac{(1000 \times 1 \times 0.1) + (50 \times 10 \times 2)}{(1 + 10)} = 100$$

Then the basic material "l" is selected from the list, and the other construction materials are compared to it. In this manner a Relative Weighted Property Factor is calculated.

$$RWPF_i = \frac{WPF_i}{WPF_l}$$

The higher the value of the $RWPF_i$, the better the material, because it shows the higher degree of utilization of the properties in the use of the material under consideration.

The author thinks that it is sometimes advisable to introduce the Factor of Exploitation of Material Property in the determined use as defined by the formula below:

$$WPZ = \frac{\sum_{i=1}^{NW} \left(\frac{P_i}{P_k} \right)}{NW}$$

where:

- P_i - value of "i" property.
- P_k - the most advantageous value of "i" property which has the best existing construction materials (taking into account this property).
- NW - number of characteristics considered.

Example of calculation of WPZ

Material	Tensile Strength (MPa)	Conductivity ($Sm \cdot 10^6$)	Density ($g \cdot cm^{-3}$)
A	400	58	8.8
B	120	37	2.7
L "ideal"	3000	63	1.5

$$WPZ_A = \frac{\frac{400}{3000} + \frac{58}{63} + \frac{1.5}{8.8}}{3} = 0.40$$

$$WPZ_B = \frac{\frac{120}{3000} + \frac{37}{63} + \frac{1.5}{2.7}}{3} = 0.39$$

The highest value of the Factor of Exploitation of Material Property in the determined use is what the "ideal" material could have, i.e. all the properties of the existing construction materials which are the most advantageous under the conditions of utilization (generally at standard pressure and temperature).

In practice such a possibility is of small probability and the value of this factor determines a "degree of compromise", which has been (or will be) attained by the selection of a material (or the designing of a new one).

It is advisable when determining the WPZ factor to prepare bar charts with values of equivalent properties for various grades of construction materials.

It is frequently found in practice, that only some of the various important advantageous properties of a material are used in the determined application. Therefore, the General Factor of Exploitation of Material Properties WPO may be introduced:

$$WPO = WPZ - WPS$$

where:

WPZ - Factor of Material Properties used in the desired application.

WPS - Factor of Material Properties not used in the application.

$$WPZ = \frac{\sum_{i=1}^{NW}\left(\frac{P_i}{P_k}\right)}{NW + NN} \qquad WPS = \frac{\sum_{i=1}^{NN}\left(\frac{P_i}{P_k}\right)}{NW + NN}$$

where:

P_i - value of "i" property.

P_k - the most advantageous value of "i" property which has the best existing construction material.

NW - number of characteristics used in the required application.

NN - number of characteristics not used in the required application ("lost characteristics").

Example of calculation of a WPO

Material	Tensile Strength (MPa)	Conductivity ($Sm \cdot 10^6$)	Density ($g \cdot cm^{-3}$)	Thermal conductivity ($W \cdot mK^{-1}$)	Corrosion Resistance
A	400	58	8.8	335	5
B	120	120	2.7	130	4
L	3000	63	1.5	360	10

characteristics used in the required application (Tensile Strength, Conductivity, Density) — "lost characteristics" (Thermal conductivity, Corrosion Resistance)

$$WPO_A = \frac{\frac{400}{3000} + \frac{58}{63} + \frac{1.5}{8.8}}{3 + 2} - \frac{\frac{335}{360} + \frac{5}{10}}{3 + 2} = - 0.05$$

$$WPO_B = \frac{\frac{120}{3000} + \frac{37}{63} + \frac{1.5}{2.7}}{3 + 2} - \frac{\frac{130}{360} + \frac{4}{10}}{3 + 2} = + 0.08$$

If in certain applications the General Factor of Material Exploitation takes a negative value, we may generally speak about misuse (application) of a material property (although it fulfills the technical and economical demands very well).

The necessary condition for proper materials selection is the knowledge of their properties.

On the one hand, we do not have complete enough data necessary for determining potentially possible ways of using materials, but on the other hand, the science of materials is undergoing continuous development. This situation enforces the idea that the system aiding materials selection should be opened to the inflow of information on the features and properties of materials.

Such a system must fulfill the function of an information bank on materials which enables the use of a data base for aiding materials selection.

Unfamiliarity with the data necessary for accomplishing the correct materials selection and manufacturing processes, which in turn assure the obtention of the required end properties by an element, causes this process to be imperfect.

This may be confirmed by the numerous changes of materials which take place during the designing and manufacturing processes of products and which sometimes have considerable technical and economic effects.

Advantages resulting from the introduction of more suitable material solutions in connection with the processes improving their properties may be certified by the following examples:

- Introducing the low-alloy high-speed steel SW3S2N subjected to the selective nitriding process, instead of the SW18 and SW7M steel, resulted in the increase in cutting tool life of up to 50 percent, and in cold working instruments' lives of up to 30 percent.

The new steel contains 12 percent less of scarce components, and its energy consumption in the heat treatment process is lowered by 20 percent, when compared to standard steels.

- Introducing construction of bicycle pedals in the form of a block made of polyamide by an injection moulding process, instead of steel drawpieces riveted with rubber elements, enabled reduction of manufacturing costs by 7 500 000 zl.

- As a result of introducing the zinc-galvanized steel net for air filters in carburetors, instead of the brazed ones, the material costs were lowered by about 90 percent.

- In cases ensuring the proper placing of details in the hard soldering process (with the use of copper) the composition of fine-grained graphite with binder in the form of pressed plates, burned at a temperature of 1300 °C, was introduced instead of heat resisting steels. The new composition with a low coefficient of thermal expansion and good machinability enabled us to extend the durability of the equipment by several times, to lower the costs of the equipment, and to increase the quality of the soldered connections.

- Various non-ferrous metals, mainly bronzes, were used for friction pairs in car mechanisms. As a result of investigations and experiments, it was proved that identical pairs made of low-alloy steels or cast iron and subjected to the holding at a temperature of 560-700 °C in furnaces with the atmosphere of gaseous sulfur and ammonia for a period of 0.5 to 10 hours gave an increase in durability of mating parts by about 30 percent which caused yearly savings of several million zlotys.

Investigations carried out in the period 1977 to 1980 in the Polish engineering industry proved that more important implementation of material and process manufacturing solutions yielded an average saving of 0.8 to 1.2 million zlotys per year.

2. Structure and Functions of the "MATERIALS" System

The following predicted scopes of using the data base were stated at the moment the system was started:

- the improvement of the selection of materials and manufacturing processes for improving properties and the assessment of the correctness of the solutions,

- prediction of scopes of effective implementations of modern material and manufacturing process solutions in the engineering industry,

- aid in designing materials with required properties.

Many attempts have been undertaken to use computer data banks for achieving the targets mentioned above. For a full range of construction materials used for machine elements, however, satisfying results have not yet been attained.

The Molecular Design firm in California, for example, introduced computer aided designing of new chemical compounds predicting their properties and finding new applications for already known compounds. Engineers at General Motors Research Laboratories established, by the use of micromechanics theory for a set of resin-filler-fibers in the SMC plastics, the procedure enabling the prediction of stiffness of compositions of such types of material.

At the present stage of development, the "MATERIALS" system aids the selection of materials and manufacturing processes for the improvement of properties, based on the assumption that a designer determines the required form and set of properties when searching materials.

The designer, therefore, must possess knowledge of the characteristics relating to various groups of materials (differences resulting from different methods of determinations).

The following three elements were the most important for building the system:

- the method of compact but sufficiently precise recording of information on material properties,

- a special software,

- filling the base of data on materials.

In order to fulfill the needs of designing and materials selection, together with improving manufacturing processes, the method of "technological" recording of the value of a material property was chosen.

The value of each property of a certain material is recorded in the system by taking into consideration the following features:

- working (inspection) temperature,

- method of primary and/or plastic forming,

- treatment changing properties in the entire volume of the material,

- treatment changing properties of the surface.

Such a method of recording enables us, among other things, to determine the costs for obtaining the required set of properties of a material (element).

It is possible to place the information for any grades of a material in the system's data base.

It is necessary, prior to that, to have prepared the following dictionaries:

- a dictionary of materials,

- a dictionary of properties,

- a dictionary of temperatures,

- a dictionary of methods of primary and/or plastic forming,

- a dictionary of treatments changing properties in the entire volume of the material,

- a dictionary of treatments changing properties of the surface.

Considering the limited computer equipment, the material selection process is effected by a batch system (which is inconvenient).

The following main types of questions may be answered by using the system:

- What are the properties of the specified material in question?

- What properties does a material have in the

determined form? (defined by its name and the main dimension).

- What are the properties of a given material after subjecting it to the defined property-changing treatments in question?

- What materials have the required set of properties?

- What materials have the required set of properties after particular primary and/or plastics forming?

- What material is a substitute for a material imported from abroad taking into consideration, for example, its chemical composition?

For further development of the system's functioning, the part concerned with determining the influence of aggressive media on material properties was tested. In the process of materials selection the data enabling the assessment of the usefulness of the material solution for working in the expected medium, often of a very complex composition, are of substantial importance. The influence of an aggressive environment has considerable impact on material properties. Examples:

- many types of plastics must be protected against various chemical substances for the stress corrosion effect which develops immediately upon contact,

- the rate of corrosion of "Fe" main-water pipes increases considerably when the temperature of water exceeds a permissible level,

- several materials used for cable and conduit insulation are good nourishment for insects in the tropics,

- a number of materials used in manufacturing processes pose a danger to human health and the natural environment (mercury, lead, cadmium, and others),

- some materials exhibit a total change in their properties in various media, for example: plastic zinc becomes brittle when submerged in a solution of mercuric nitrate.

Information on the impact of a particular medium on a material is usually obtained in the process of carrying out long-term and costly environmental tests. For this reason, it is of particular importance that all available data on this subject be collected and recorded in a way which enables their immediate use by a designer for new designs. This will facilitate better designing of solutions, avoid unexpected failures, and will decrease or eliminate costs of material tests. Searching for the data on material properties in various media is carried out under the assumption that the user determines the working environment by means of its components and features, and makes requests related to the required properties of the designed material solution in this medium. The fundamental condition for the full usefulness of the system is its complete, if possible, file of reliable information on material properties. At present the data base contains about 40 000 pieces of information on almost all grades of Polish construction materials for machine building. Considering the as yet incomplete content of the base, difficulties occur during use with furnishing information on "subtle" questions. Generally, however, the usefulness of the system has been confirmed for the following questions:

- material selection, taking into account required properties of machine elements, designated in the query,

- selection of nationally made substitutes for materials imported from abroad.

REFERENCES

[1] Florczak, T, System "MATERIALY", Gospodarka Materialowa 11 (1979).

[2] Kusy, P.F., Plastic Materials Selection Guide, SAE Paper 760663.

[3] Pye, A., Materials Optimization - the key of product profitability, Materials in Engineering (September 1980).

[4] Dowsing, F.J., Fulmer Research Institute Helping optimize materials selection, Metals and Materials (February 1977).

[5] Apoo, P.M., and Alexander, W.O., Computer analysis speeds of assessment of engineering materials, Metals and Materials (July-August 1976).

Data for Science and Technology, P.S. Glaeser (ed.)
North-Holland Publishing Company

GENERIC REPRESENTATION OF MATERIALS IN A POLYMER DATABASE (PCMRDB)

Y. Fujiwara,[1] T. Nakayama, [2] A. Amada, K. Iida, K. Hatada, [3], Y. Hirose, A. Nishioka, [4], N. Ohobo, I. Suzuki, T. Yasugahira and S. Fujiwara, [5]

[1] *Institute of Information Science and Electronics, The University of Tsukuba, Sakura-mura, Niihari-gun, Ibaraki 305 Japan, where correspondence should be addressed.*

[2] *International Foundation for Advancement of Science*

[3] *Faculty of Engineering Science, The University of Osaka*

[4] *Faculty of Engineering, Fukui College of Technology*

[5] *Department of Chemistry, Chiba University*

Japan

Polymeric materials are usually represented in terms of monomers or repeating units. This is not sufficient to distinguish different structures of polymers with the same generic names such as polyethene because polymers and especially copolymers are actually mixtures. The present paper is to report one of the ways how the sequence of monomers, irregularity of polymerization, chain ends, branches and so on are described and are processed in polymer databases. It was shown that the flexible representation of polymer structures by expandable connectivity information as well as an internal thesaurus consisting of various synonyms, CA registry numbers and molecular formulae is effective.

1. INTRODUCTION

One of the major problems in compiling data is the identification of the entities unless a simple database of small scale is of concern. Various data models have been proposed to increase performance of database management systems, but the restrictions imposed on data are not thoroughly discussed yet. For example, the relational model is very popular because it is easy to use and is theoretically well-founded. It requires that values are atomic and may not be structured. This means that data should be simple and that you may not input data requiring a considerable amount of descriptive information. Another requirement is that a relation hardly allows null values, which results in rejecting a large fraction of available data. The network model is flexible but cannot directly accept many-to-many relationships. The access path in a network model is usually not simple. The hierarchical model has the problem of classification. These restrictions imposed by data models are still rather clear and may be overcome to some extent by hybrid types of file construction.

The present paper focusses on another type of restriction which is caused by using identification numbers as keys and is usually disregarded. A typical example is the registry numbers of CAS (Chemical Abstracts Service) which are widely used for identification of chemical compounds.

2. INFORMATION CONTAINED IN POLYMER DATA

The polymer database (PCMRDB) is taken as an example. This database contains bibliographic, spectral, graphic and text data of C-13 NMR of polymers, and has been compiled for seven years. Although this is not a large database, it has comprehensive data in this particular field and the contents are of high quality in the sense of critical evaluation as well as expert level of information[1].

In order to describe polymeric materials, it is necessary to include not only structures of repeating units but also other structural and non-structural information, as shown in Figure 1; where explicit structural information contains that of repeating units, stereoregularity, irregular structures (e.g., branches, chain ends, head-to-head addition of vinyl monomer, 1/2 or 3/4 addition in 1/4 addition of polyisoprene), degree of polymerization, and dispersion; implicit structural information contains that of production processes (e.g., hydrogenated PIP/EPR), reactions (e.g., chlorinated PE), starting monomer (e.g., polyvinylalcohol from polyvinyl acetate); related structural information contains that of additives, catalysts and so on; and non-structural information of the types of properties (e.g., high density PE), processes (e.g., high pressure polymerization), use (e.g., engineering plastics) [1-3].

Although polymeric materials have many attributes to be described, all of them are not usually obtained. The lack of data is significant and the material cannot be fully specified. Moreover, it is not always required to know the value of every attribute, but the generic classes of materials are of concern. Sometimes, all of these attributes are not sufficient to describe compounds and yet further information is required such as isotope substitution, higher order structures and so on.

On the contrary, minor differences between distinct values may be neglected from a practical

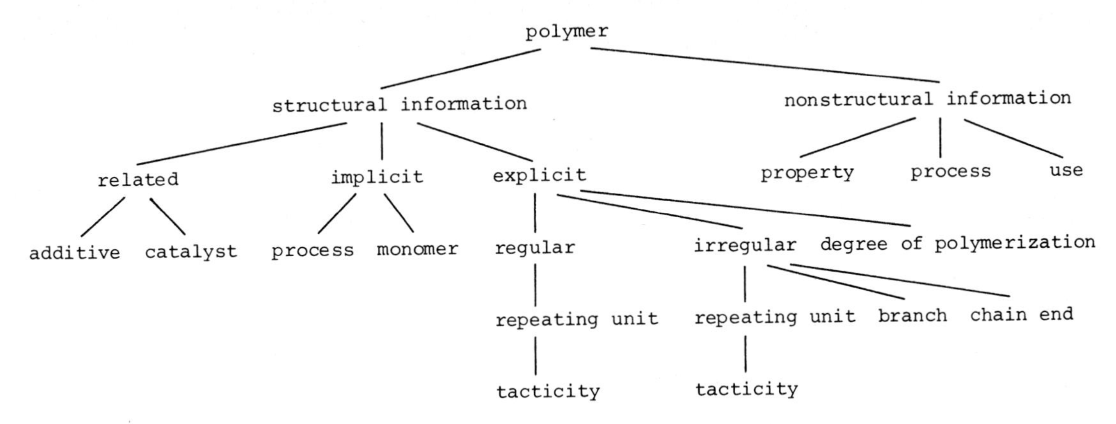

Figure 1. Description of polymer information

point of view. In a word, polymeric materials are mixtures of many kinds of isomers and homologues, and the required specification depends on the view of the user, where the corresponding sets of attributes are combinations of all attributes irrespective of the levels. Therefore, the constitution of key attributes must be dynamic and may not be unique.

3. GENERIC REPRESENTATION OF POLYMER STRUCTURES

A practical way of specification of polymer structures is given in terms of repeating units and extended connectivity matrices as discussed below. The BCT representation of chemical structures was discussed in our previous paper [4,5], where BCT stands for *b*lock-*c*utpoint *t*ree and the principal idea of BCT is to reduce chemical graphs into simple trees by replacing chemically meaningful subgraphs by new nodes. The extended BCT EBCT) has been devised to describe generic chemical structures of low molecular weight such as Markush formulae in patent documents, where a set of substituents gives a descriptive unit (a node) of an EBCT and the definition of connectivity relation between the nodes is extended accordingly.

As for polymers, the structure is represented also by EBCT, where a repeating unit is treated as a node (a descriptive unit). The connectivity relation between the nodes is represented by a connectivity matrix whose elements indicate connection as well as tacticity. Some generic representations of polymers are illustrated by examples:

(1) <u>simple repeating unit</u> Figure 2a shows a main structure of a homopolymer, where R is a simple repeating unit. This is regarded as an EBCT consisting of a single node. The repeating number *n* is collectively taken into account for EBCT representation, and so is the outermost repeating number in a complex repeating unit explained below.

(2) <u>complex repeating unit</u> A complex repeating unit appears in copolymers or heterostructures:

1° copolymer: Figure 2b shows a copolymer and its EBCT, where

$R_1 = \{r_1, r_1\text{-}r_1, r_1\text{-}r_1\text{-}r_1, \ldots\}$; $r_1 = \text{-CH}_2\text{-}\underset{\substack{|\\X}}{\text{CH}}\text{-}$

$R2 = \{r_2, r_2\text{-}r_2, r_2\text{-}r_2\text{-}r_2, \ldots\}$; $r_2 = \text{-CH}_2\text{-}\underset{\substack{|\\Y}}{\text{CH}}\text{-}.$

2° head-to-head, tail-to-tail: Figure 2c shows an irregular structure, head-to-head and tail-to-tail, which is located at the junction point between regular sequence with opposite direction.

3° branch: Figure 2d shows a structure with a branch.

(3) <u>chain end</u> Figure 2e shows a structure with a chain end. The EBCT represents a connectivity relation between a chain end and a simple repeating unit, and node E_1 gives a generic descriptive unit if several candidate chain ends exist.

The pair of elements of a connectivity matrix, $(c_{ij}, c_{ji}) = (h,t)$, indicates a head-to-tail bonding, and the pair $(c_{ij}, c_{ji}) = (t,t)$ indicates a tail-to-tail bonding. c_{ij}=1 means that node i is a fixed substructure and that the substructure is connected to node j at the position of consituent atom 1.

4. DYNAMIC VIEW FOR POLYMER DATA

The representation of polymer structure is given in terms of generic descriptive units and corresponding EBCT as discussed above. While it gives a generic representation of polymer structures, there is another aspect of polymer data to be considered. That is problems of null values and structured values. The real data of polymers are supplied in incomplete form leaving various unknown or inaccurate values. These values are likely to become known some time or other, which results in the update operation of generalization/specialization of entities. This is mainly due to the characteristic feature of polymers, i.e., the fact that polymers are mixtures. Although it is difficult to find all the attributes of polymer entities, it is necessary for the effective data model to consider newly known values, because those values

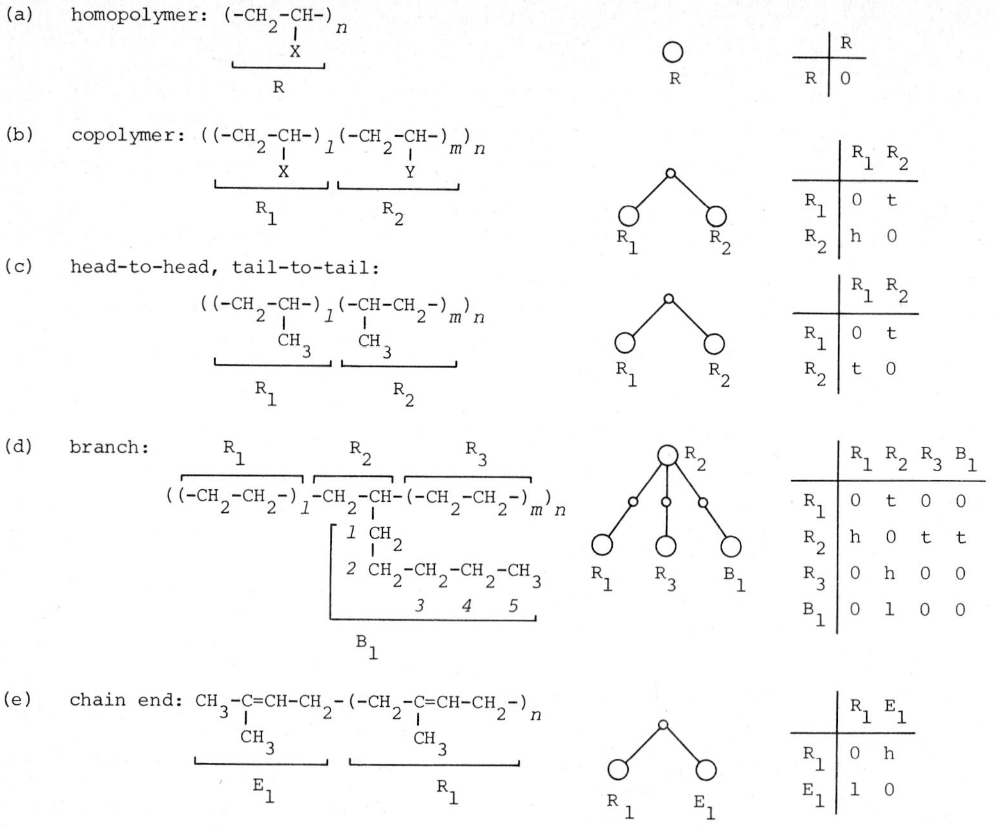

(a) homopolymer: $(\text{-CH}_2\text{-CH(X)-})_n$ — R

	R
R	0

(b) copolymer: $((\text{-CH}_2\text{-CH(X)-})_l(\text{-CH}_2\text{-CH(Y)-})_m)_n$ — R_1, R_2

	R_1	R_2
R_1	0	t
R_2	h	0

(c) head-to-head, tail-to-tail: $((\text{-CH}_2\text{-CH(CH}_3\text{)-})_l(\text{-CH(CH}_3\text{)-CH}_2\text{-})_m)_n$ — R_1, R_2

	R_1	R_2
R_1	0	t
R_2	t	0

(d) branch: $((\text{-CH}_2\text{-CH}_2\text{-})_l\text{-CH}_2\text{-CH(}R'\text{)-(-CH}_2\text{-CH}_2\text{-)}_m)_n$, with R' = $^{1}\text{CH}_2\text{-}^{2}\text{CH}_2\text{-}^{3}\text{CH}_2\text{-}^{4}\text{CH}_2\text{-}^{5}\text{CH}_3$ — R_1, R_2, R_3, B_1

	R_1	R_2	R_3	B_1
R_1	0	t	0	0
R_2	h	0	t	t
R_3	0	h	0	0
B_1	0	1	0	0

(e) chain end: $\text{CH}_3\text{-C(CH}_3\text{)=CH-CH}_2\text{-(-CH}_2\text{-C(CH}_3\text{)=CH-CH}_2\text{-)}_n$ — E_1, R_1

	R_1	E_1
R_1	0	h
E_1	1	0

Figure 2. Sample generic representation of polymers

are usually of high value from a scientific or technical point of view. That is, the effective database should support a dynamic viewing facility according to the transformation of data structure.

As an example of the confusion, polyethylene has more than nine hundred synonyms and all of the popular polymers have synonyms of practically the same order. Consequently, it causes difficulty in retrieving polymer data from a large database. A standard nomenclature appears to overcome this problem, but it is not so effective as expected. Identification numbers are very simple and clear, and appear to be better than names. Here is one of the results as shown in Figure 3. The number 9003-53-6 is the CA

```
REG   9003-53-6
FOR (C8 H8)x
IND Benzene, ethenyl-, homopolymer (9CI)
DEN  55465-00-4,  53986-84-8,  60328-46-3,
     56451-72-0,  58799-53-4,  63849-49-0,
     60120-16-3,  60880-98-0,  58033-91-3,
     61584-90-4,  61584-89-2,  56748-62-0,
     53112-49-5,  51609-83-7,  57657-06-4,
     39470-87-6,  54596-41-7,  62932-49-7,
      9044-64-8,   9055-91-8,  11120-46-0,
     12627-11-1,  40494-15-3,  51609-87-1;
```

Figure 3. Registry numbers for polystyrene

registry number of polystyrene or ethenyl benzene homopolymer according to the 9th collective index of CAS. This is a unique number at present, although there are many deleted numbers which should be replaced. Unfortunately, those deleted numbers were once printed and all the published data scattered over the world cannot be erased or replaced by the revised number instead. In addition, these examples are not exceptional but all popular materials are in similar situations.

One of the methods to solve the problem is to make use of a thesaurus which saves the trouble of preparing search profiles. The polymer database (PCMRDB) with automatic generation facility of the internal thesaurus has been implemented on the ORION database management system. Figure 4 shows the PCMRDB/ORION system configuration. Table I shows statistics of the internal thesaurus, where the number of synonym sets is 95 and

Table I: Statistics of Thesaurus

number of papers	number of nodes	CPU time for compilation	number of synonym sets
100	61	17	9
550	680	744	95

the vocabulary in the sets consists of 680 names. That is, the average number of synonyms in a set is seven, which is much smaller than the order of hundred or thousand.

5. CONCLUSION

It is important to incorporate generic concepts and structured values into database descriptions as well as specific/atomic values. PCMRDB has been organized in that manner, where generic structures are represented by EBCTs and some of the structured values are represented in the form of a thesaurus. Although the implementation presented in this paper is not refined as a general data model, it was shown that the idea of a dynamic view corresponding to the operation of generalization and specialization seems to be substantial, and that the extended BCT is of vital use for structure representation.

REFERENCES:

[1] Fujiwara,Y., Hatada,K., Hirano,T., Kawamura, T., Kondo,S.,Matsuzaki,K., Nishioka,A., Tanaka,Y. and Tomita,B., A carbon-13 database for advanced research in polymers, CODATA Bull. 40 (1981) 35-38.

[2] Haeuser,J. and Herz,M., Searches for polymers in the BASIC files derived from the Chemical Abstracts Service Chemical Registry System, J.Chem.Inf.Comput.Sci. 21 (1981) 180-182.

[3] Fugmann,R., POLIDCASYR: The polymer documentation system of IDC, J.Chem.Inf.Comput.Sci. 19 (1979) 64-68.

[4] Fujiwara,Y. and Nakayama,T., A graph data base for storage of chemical structures organized by the block-cutpoint tree technique, Anal.Chim.Acta 133 (1981) 647-656.

[5] Nakayama,T.and Fujiwara,Y., BCT representation of chemical structures, J.Chem.Inf. Comput.Sci. 20 (1980) 23-28.

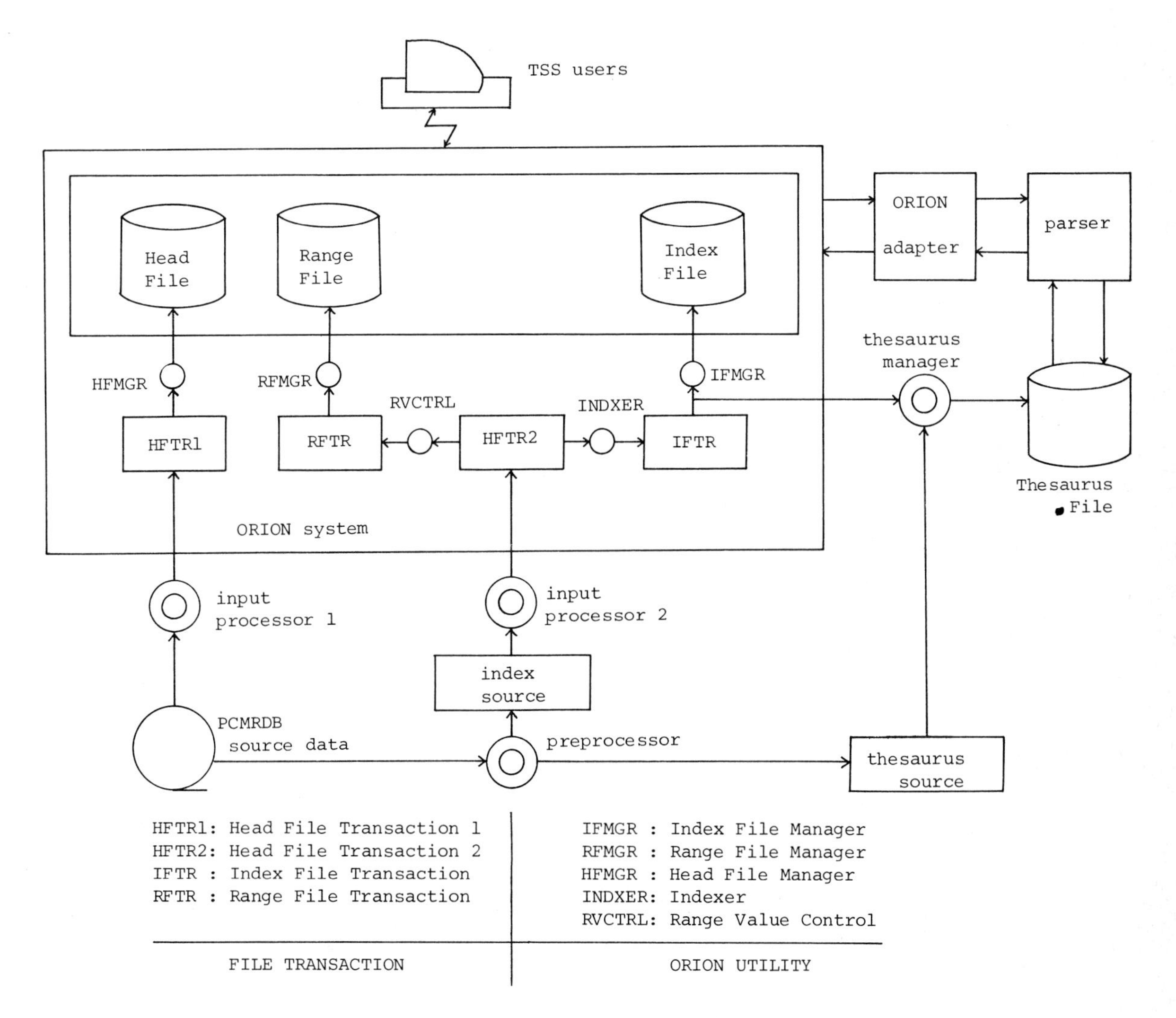

Figure 4. PCMRDB/ORION system configuration

Data for Science and Technology, P.S. Glaeser (ed.)
North-Holland Publishing Company

CONCEPTUAL DESCRIPTION OF SYSTEM TRANSFORMATIONS AND REACTIONS

Jacques-Emile Dubois

ITODYS, Université Paris VII, 1 rue Guy de la Brosse
75005 Paris, France

Transformation between potential states of a system can be described by using formalized procedures expressly derived to describe stages of an evolution. Thus, the successive steps in a puzzle or a combinatorial game can be identified starting from transformation operations and can be handled by syntax. The methodology of production systems used in artificial intelligence is well adapted to handling these dynamic changes. More generally, in order to translate operation sequences and filiations among engendered objects certain transformations defined by progressive and ordered graphs are taken into account. In chemistry, reactions are formalized by coded operators, using construction concepts similar to those which generate graphs of compounds. By generative grammars associated with structural manipulation, one can handle both real and imaginary transformations whose syntax is often linked to the semantic description of states. The search of the best primitives and operators could lead to identification of some general underlying real mechanisms ruling a visible transformation. The dynamic conception of describing evolving situations through graphs of states could thoroughly change the DBMSs handling of transformation phenomena and of static situations.

Solving problems heuristically, isolating the best answers to a problem within a space of possible solutions, - these are ancient goals of science. The past twenty years have seen a methodological renaissance in the study of complex problems. In this talk, I shall attempt to show relationships between different artificial intelligence theories.

COMPLEX SYSTEMS

Cybernetics is the science of maintaining order in complex systems, natural or artificial: its basic aim is to control the evolution of these systems. In this context, information expresses transmission of data associated with the changes which mark a complex evolution. The old association of matter and energy as the basic principles of change is gradually replaced, especially for complex systems, by a trio composed of matter, energy and information. Information, often expressed by special codes, presides over the growing organization and the understanding of evolving complex systems.

This brief introduction attempts to clarify the relationships proposed by Kitagawa (Cybernetics Congress 1976). (Figure 1). One soon realizes the need to go beyond the study of simplified or mutilated systems and to make serious efforts to extend and to find new conceptions both on the level of systems logic and of the linguistics which appears at various points (formal and conventional representation of data and of relationships, message forms, correction of errors). On the logical level, efforts are primarily made in accepting spaces of less restrictive states, in evaluating the field of possible solutions and their actualization, on uncertainty in operations, all this with the aid of exceptional methods of scanning and screening of computer programs.

TRANSFORMATION OF STATES

It is more delicate to describe transformations than static facts. Such a description can be made through the states of a space of states, but also on the level of passages between state representations (Fig. 2). An often conventional appreciation of the state provides a descriptor capable of translating transformations from state 1 to state 2. In Figure 3 the roles of transformation representation are assembled. In order to consider the passage from one state to another in a complex system and to take into account the variety of evolutions, their precision, the automatic correction of errors or the minimizing of their effects, one links to transformation functions essential factors of information redundancy which monitor evolution.

Artificial intelligence has recently handled transformation description by integrating it as a production system. (Fig. 4). In the following study, we apply this methodology and its related terminology. One must first deal with a given initial state, then certain given production rules define the operators which intervene to move the system to another state. These production rules, which I shall give in detail later, must define the precise starting point, the precise arrival point and how you move. The control involved in such a strategy means that sometimes one is overwhelmed by the number of dimensions of one's space and, therefore, one must set boundaries, arbitrarily or logically. These decisions are on a different level and should not be confused with production. They must very often take notice of data evaluation and certain aspects of redundancy.

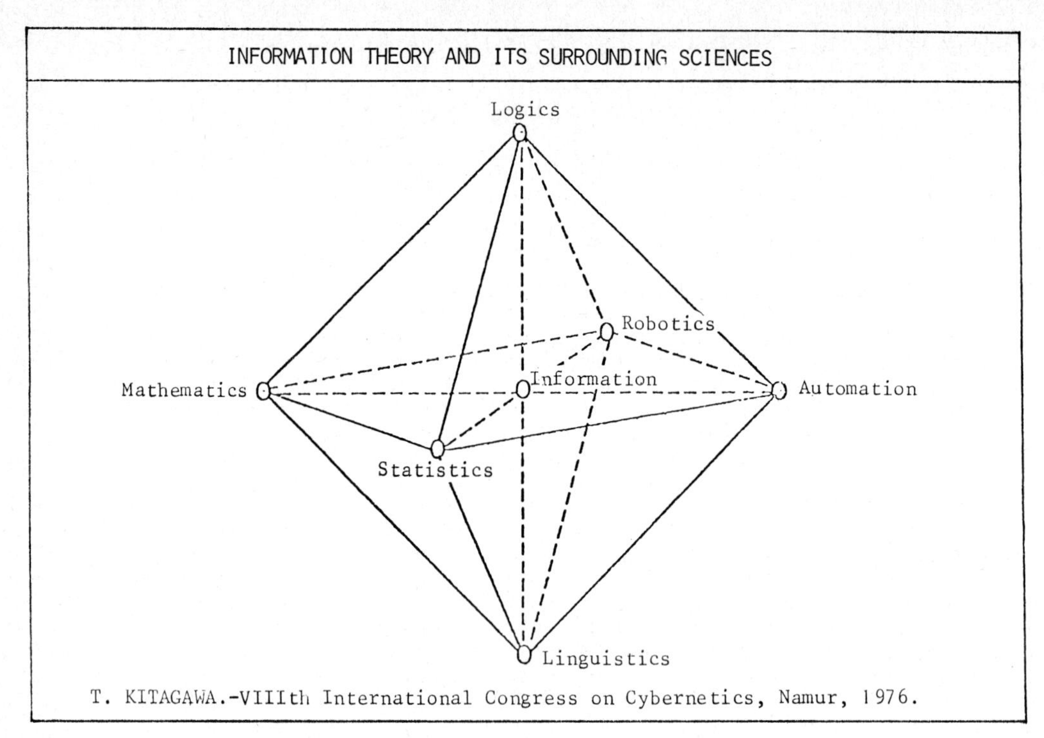

Figure 1

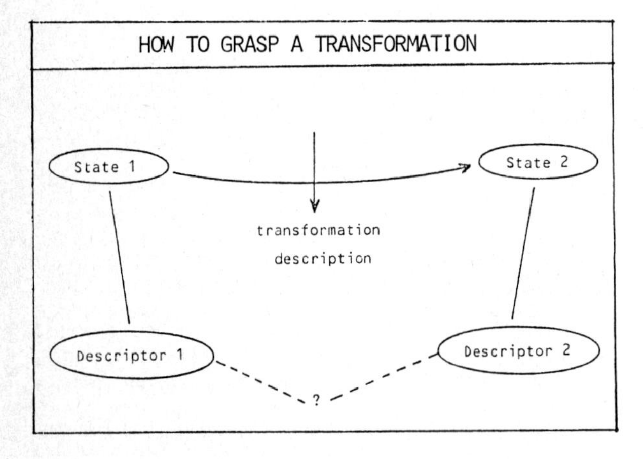

Figure 2

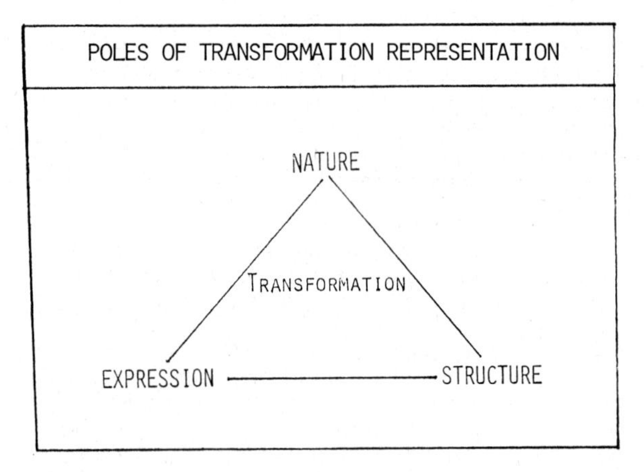

Figure 3

Transformation operations can be of different types, e.g. geometric where state 1, state 2 operations are concerned, or textual to deal with textual representations of states. We shall see that in both cases the translating or the transformation of states is controlled by a set of grammatical instructions. Both logic and linguistics are concerned in most transformations. Moving from state 1 to state 2 can be expressed in terms of fundamental parts of 1 and 2, of their primitives, whose transfer will be controlled by defined operators. The transformation deals mainly with topological arrangements, position of neighboring primitives and their chromatism which defines their intrinsic nature. The simplest production deals with nodes and edges and can be easily handled by graph theories oriented to graph description and their combinatoric aptitude controlled by permutation (displacement of their nodes and edges).

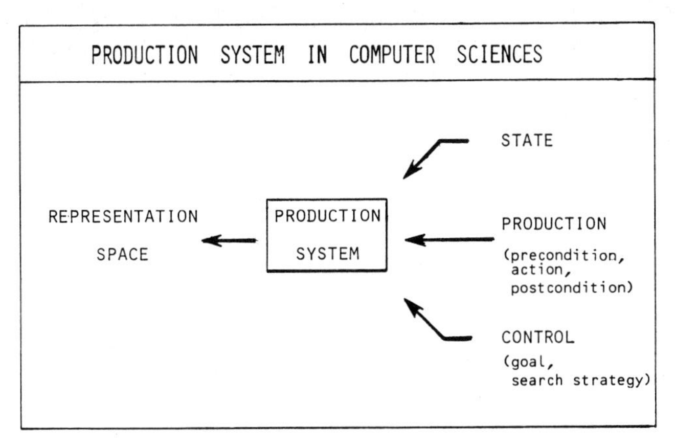

Figure 4

COMBINATORIAL GAMES

Two kinds of evolution are very typical of many different situations: both can be used for a large variety of primitives. They can be considered as typical production games.

The first game (Fig. 5) is typically combinatorial as there is a constant number of sites (edges, nodes), whereas the second one has an increasing number of sites. Both of these can be described by production rules, controlling the production of the second and third states from the starting graphs. In chemistry the first series deals with the isomeric graph of an entity whereas the second one concerns homologous series. Both of these populations need to be ordered before being processed in a production system. In fact, production control implies a given ordering of the primitives and their labelling, and it creates new states labelled in their growing populations. The states are defined, but different ordering can supply different overall organizations of the states. This organization of a population of states by construction rules leads also to ordered graphs whose nodes are occupied by the local states – (these ordered graphs are called hyperstructures). In other words this state space can be considered as a directed graph.

It is clear that transformation of states depends on the primitives, their nature and their combinatorial rules. The more fundamental the primitives and the construction operators, the stronger the production system will be. Of course, a good definition on the level of these choices corresponds to powerful elementary rules whose applications often imply a basic grammar capable of leaving the construction system **open and unbounded.**

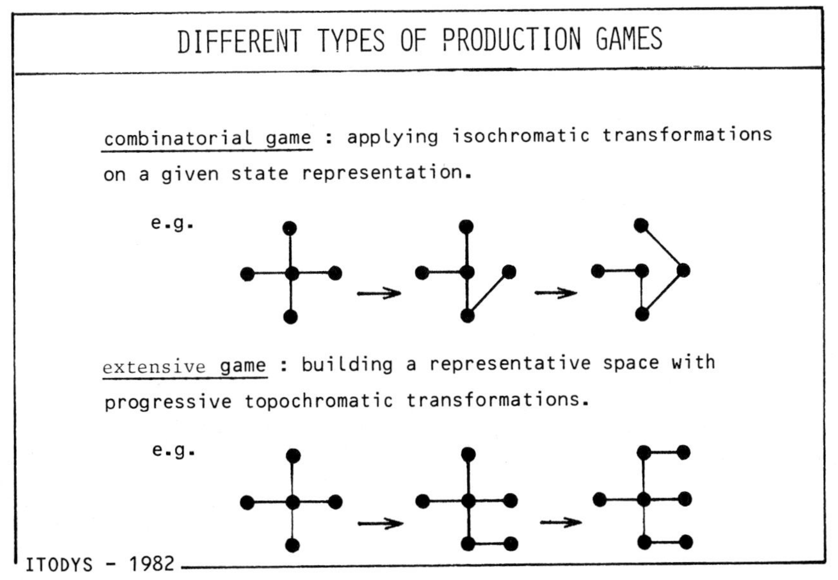

Figure 5

STATE- SPACE SEARCH METHODS

In order to describe the evolution of a complex system, graph language is most useful for its aptitude to describe the pathways followed by the operators from the starting state to the goal state. It allows handling of the states by different strategies. These can be based on the enumeration of pathways, on the creation of redundant states and their selective eliminations. Heuristic strategies tend to produce or generate the different states without redundancy by appropriate rules or admissible functions of potential candidates of the production rules. In Fig. 6 the generation of a four-nodes graph is acquired by different heuristic strategies giving early warning of redundancy.

Rules and primitives are usually acquired by trial and error. Strategies for state-space search can be more or less associated with the grammars used for the canonical production system. In the DARC System the ordering of the molecules (structures) to be generated infers a given ordering of the organization of the state-space considered as an ordered or oriented graph called hyperstructure (HS). In most modelling this synchronism between orders (S and HS) is impossible or not interesting.

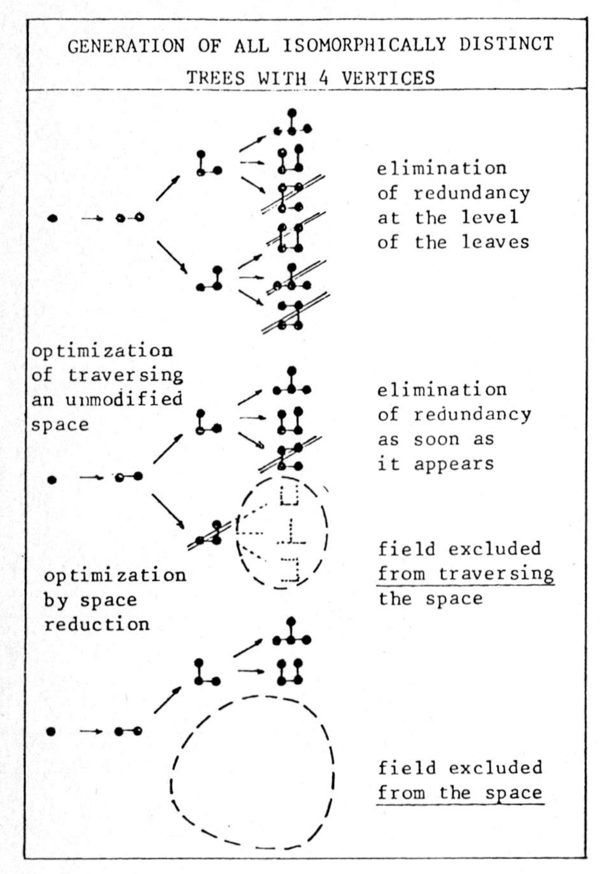

Figure 6

It is important to realize that the generation of graph files by heuristic transform strategies is not conducive to an easy retrieval of some particular information. It is, therefore, necessary to introduce in the file a certain amount of redundancy. This can easily be done by creating subfiles of local topochromatic information centered on certain primitives of the graphs (substructures - SS).

LINGUISTIC APPROACH & TRANSFORMATIONS

A sentence can be generated with a production system. In fact, the concept of generative grammars has been considered independently.(Figure 7). A state (Fig. 8) can be represented by a descriptor, a word or a sentence. This approach has been investigated thoroughly. From examples presented in Fig. 7 it is clear that a sentence can be built through different linguistic trees. The same is true for names of chemical structures and their coding, usually with a metalanguage.

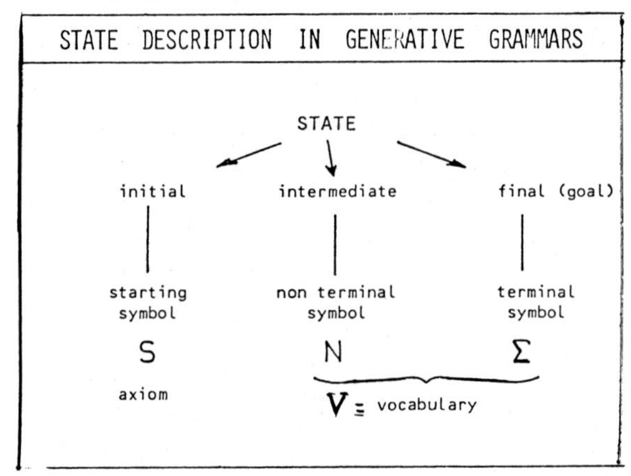

Figure 7

A grammar is defined as G = (V,P,S), where V stands for vocabulary and can be decomposed in terms which are non-terminal symbols and terminal ones; P are productions or deduction rules; and S the starting symbol or axiom. Such a grammar is used to generate a language L(G). L(G) is a set of terminal symbol strings generated from the source by a series of productions. The strings are sentences of this language. The complexity of languages, regular, context-free, context-sensitive, and unrestricted types call for grammars of increasing complexity where symbols and assembly rules become more elaborate. The following grammars can be classified according to their complexity: string, uniform trees, graphs, webs, plex, shapes, chromatic graphs, graphoids and block hyperstructures.

Operations like **translation or structural transcoding** in chemistry can be conducted by artificial intelligence procedures. (Figure 9-first part). The second part of Fig. 9 clearly stresses the identity of the structural diagram of a molecule with two descriptors: it could also be linked to the systematic name of this compound. The trans-formation from one code to another goes through a **pivot-representation** accessible from the source and the target codes.

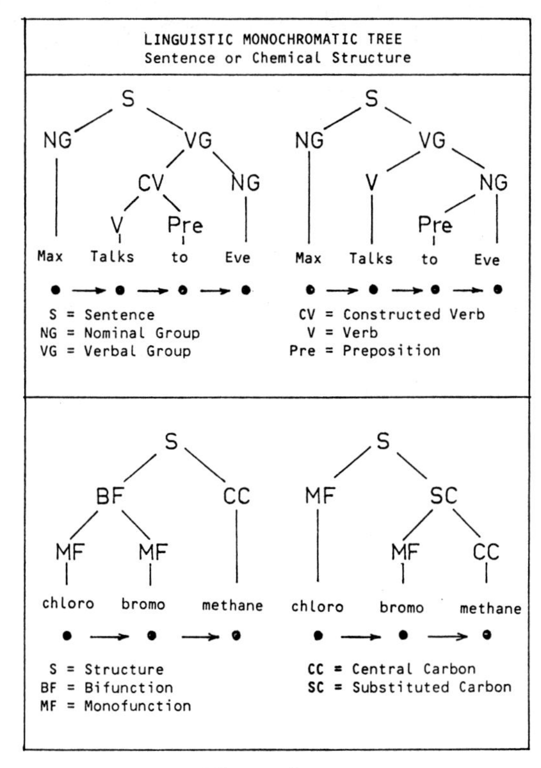

Figure 8

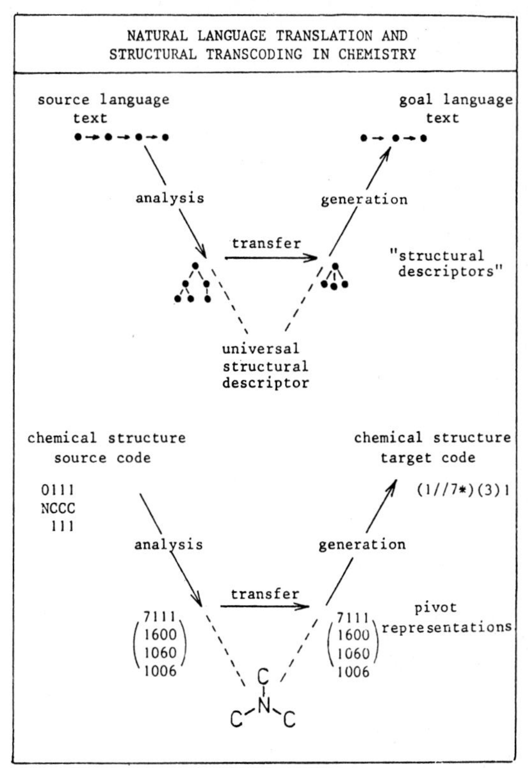

Figure 9

Shapes can be handled automatically as in the combinatorial games shown in Figure 5. The definition of a grammar of shapes makes possible their construction as well as their gradual and controlled reduction. In Figure 10 it is shown how, starting with a line segment as a lineary shape, "C curves" can be plotted by recursive applications of two rules, P_1 and P_2. A simple algorithm controls and describes a multitude of interrelated shapes since they can be located in their constructive hyperstructure.

The complexity of shapes and their topochromatic nature are expressed through different types of generative grammars. Going from one shape to another involves going through some pivot representations. The relations between successive "C curves" are analogous to relations for homologous compounds whose structural diagrams can be considered as shapes or curves.

Complex problems like the simultaneous edition of shapes and their relations within a conversion graph can be considered with generative grammars.

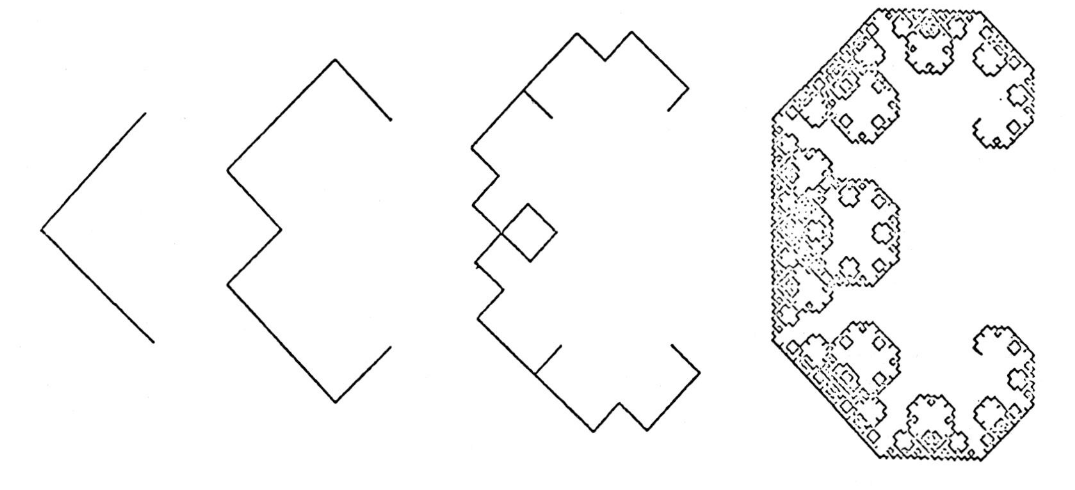

Figure 10 - COMPUTER GENERATION OF "C-CURVES" BY RECURSIVE GRAMMATICAL CONTROL

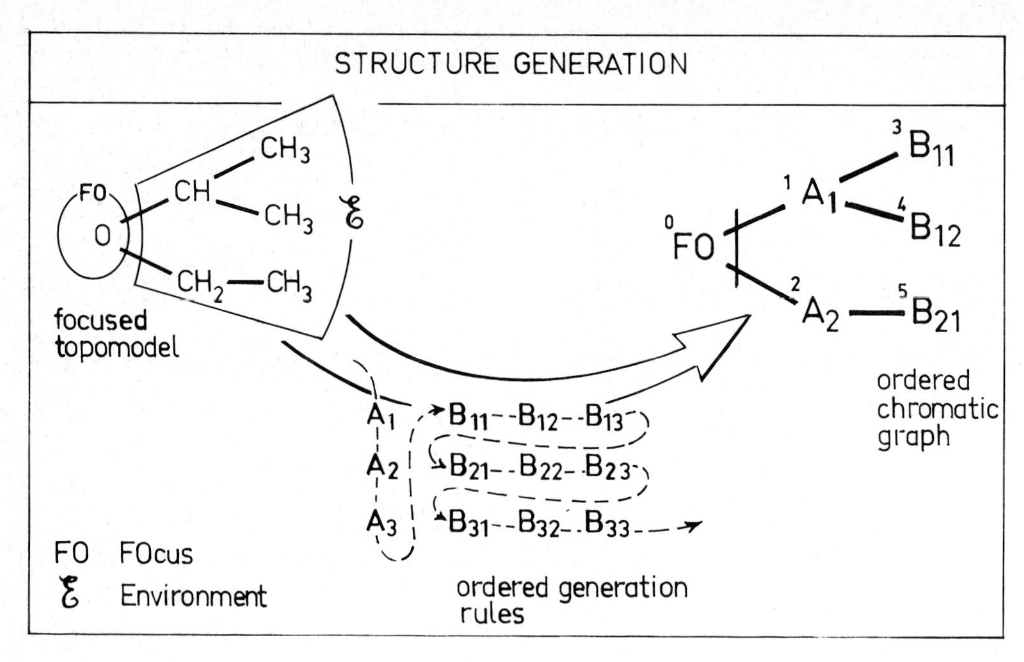

Figure 11

CODING AND PRODUCTION SYSTEMS

We have shown how to deal with real transformations where two states are different and with code transformations of a unique state.

One interesting extension of this methodology is deriving a code to describe a compound. Such a metalanguage has been derived with the DARC language. A compound is virtually built by the progressive generation of all its primitives according to a given grammar. The goal compound is thus ordered by generation rules (Fig. 11). This process associates the compound considered as an ordered goal state with a state space where all its neighboring compounds are defined (Fig. 14).

This construction of a **goal** entity is totally similar to the 8 puzzle problem presented later (Fig. 16). In the state-space of this puzzle, two operators are allowed, whereas in the DARC construction only one operator is allowed (adjunction) to generate a name or descriptor. One canonical pathway to the goal state is considered so as to unify and standardize the codification process.

The production or generation of a code name for a structure implies the generation of its ordered graph and then the obtention of its derived code. If the production goes through formal progressive steps of creation the sites introduced are only defined at certain stages. In between those stages **imaginary states** are considered along the construction pathway. They can be looked at as transient elements perceived as intermediates in the pathways of the corresponding hyperstructures. In certain strategies they can be used for **generic keys** and similarities search. Underlying mechanisms of data structuring may be worked out through those hypothetical entities. These concepts are used in the development of our EURECAS file (more than five million structures) where structure and substructures are handled by such hypothetical chromatic graphs.

Figure 12 and 13 illustrate the generation of imaginary and real structures by a fragmental generative grammar and the DARC topological one.

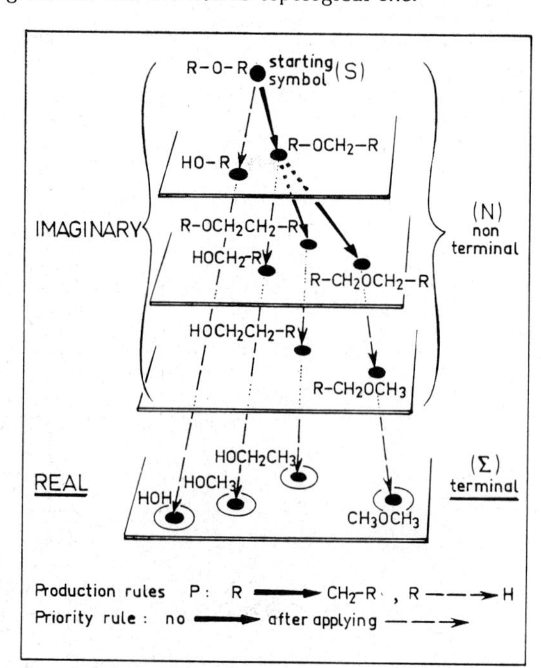

Figure 12 - GENERATIVE GRAMMAR FOR LINEAR ALCOHOLS AND ETHERS

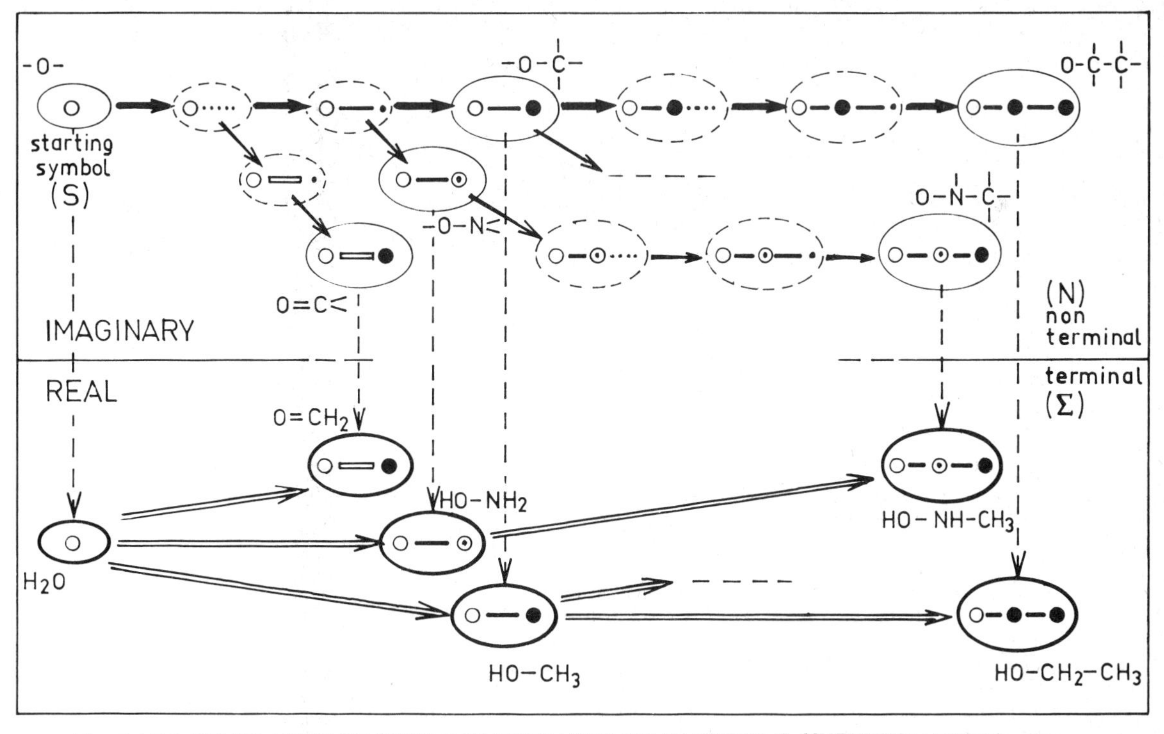

Figure 13 - DARC GENERATIVE GRAMMAR UNDERLYING SYNCHRONOUS HYPERSTRUCTURE GENERATION

POPULATIONS OF DATA STATES

A goal state could be a population of data rather than an isolated object or a chemical entity. In such complex cases the mechanism of the production system can be used with special considerations on the constraints defining the goal state (Fig. 14)

The enumeration of all aliphatic alkanes from methane to dodecane is shown in Fig. 14. The goal state is the population of C_nH_{2n+2} isomers, where n=12. They are located on the outside ellipse. The location of compounds, represented as points, is associated to the DARC grammar used and can be modified to fit certain types of constraints (structural or physical data). Going from the central methane entity to the periphery corresponds to certain homologous pathways. Developments of the structural logic underlying this presentation deals with new anteriority rules of production.

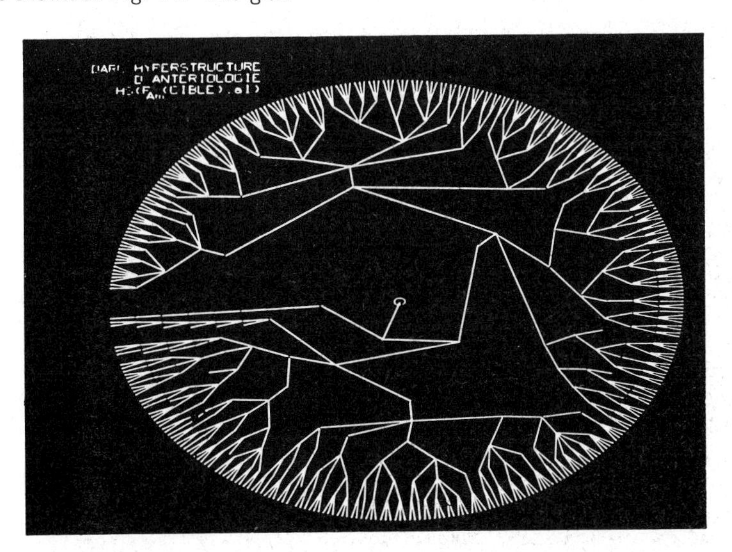

Figure 14 - Exhaustive and ordered population of aliphatic alkanes

But the isolated goal state technique can be used also for enumeration. For instance, to generate the structures imbedded in the first aliphatic E_B alkyl environment, 462 different pathways can link the starting and the final entity -C-t-Bu_3. The entire hyperstructure is covered by a set of five pathways. There are 1600 overlapping such sets.

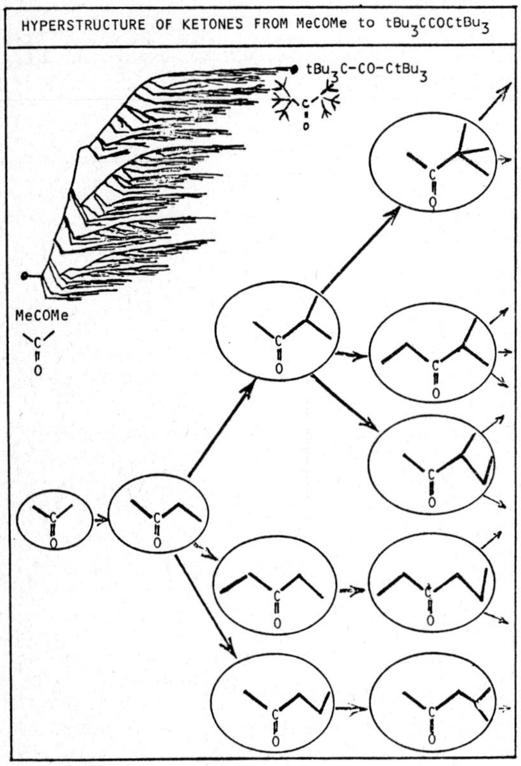

HYPERSTRUCTURE OF KETONES FROM MeCOMe to $tBu_3CCOCtBu_3$

tBu_3C-CO-$CtBu_3$

MeCOMe

Figure 15 (see Figure 18)

This type of consideration could be useful for data treatment where properties change regularly along the hyperstructure (HS) from one state to the next one. (Fig. 17).

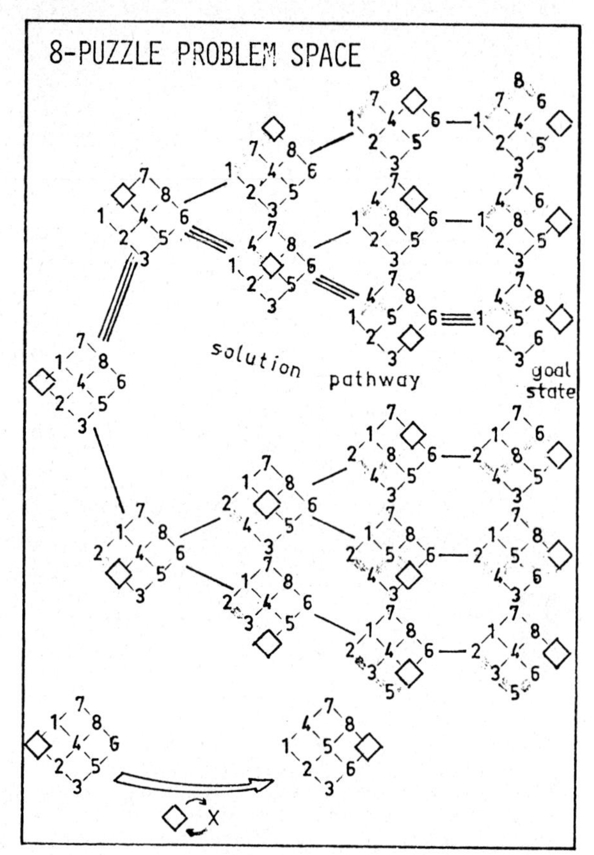

Figure 16

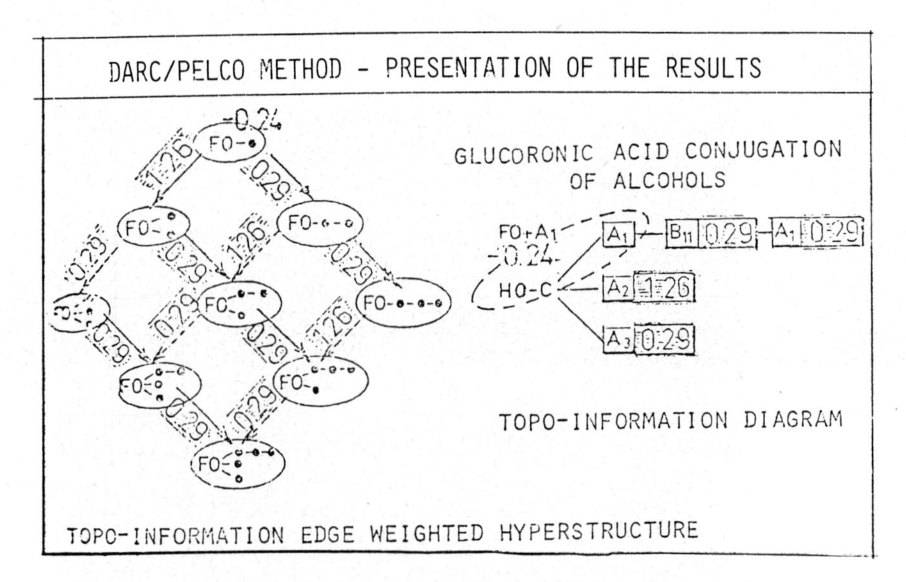

Figure 17 - The computerized validation of nodes and edges shows the regularity of the property evolution of conjugation clearly on the left hand side

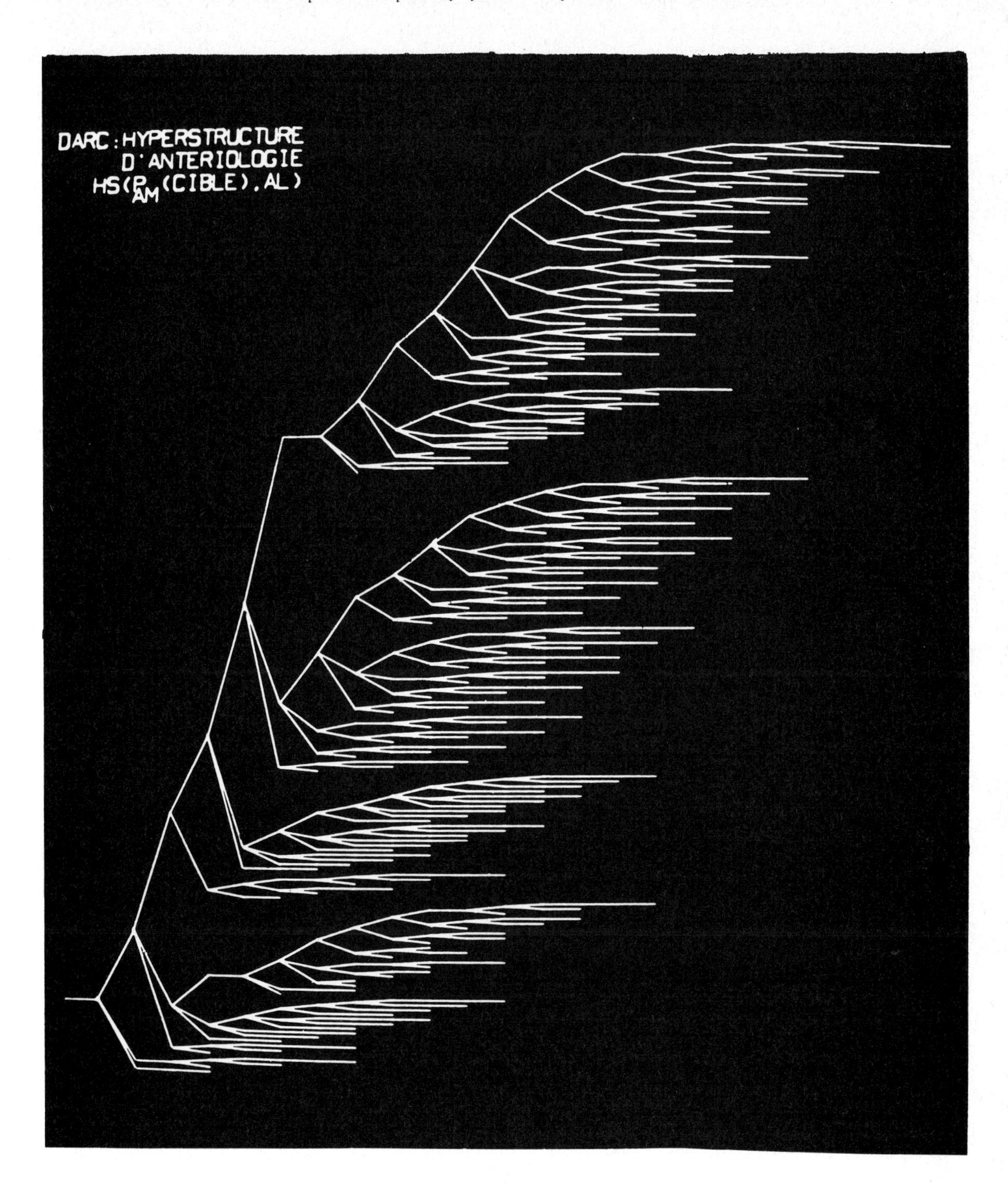

Figure 18 - Generation Hyperstructure. Formal generation of 630 ketones whose environment surrounding the >C-CO-C< focus is made up of at most two ranks of atoms; each graph node corresponds to a ketone from its origin (acetone, lower lefthand side) up to the target (hexatert-butylacetone, upper righthand side; isomers are located on a same vertical axis).

REACTIONS AS TRANSFORMATIONS

The description of a reaction involves three types of information: numerical, structural and textual. We shall mention here only the structural part limited to the essential stoichiometry of the main equation of the reaction. In the naming process of the DARC System one operator has been presented. In reactions to move from state 1 to state 2, it is clear that one needs more basic operators. For efficiency, one should minimize the number of these operators.

This observation brings us back to the search for the necessary primitives to express the details of a reaction where different changes can occur. The problems encountered in such a situation are very complex, and the structural modelling of a reaction has many sides.

One can consider each state of the reaction system as a structural block that undergoes a transformation. Usually the transformation is accessible on the level of substructural blocks where the changes take place. A more complex method consists of searching for the underlying changes that control the reaction in order to relate them to structural mechanisms. These could, in other words, correspond to the real changes and could be classified by types.

With a production system the complex system of a reaction can be tackled through formal and/or practical operators. Certain facts must be considered if one wants to have a coherent treatment of these transformations.

1) Need for two opposite formal operators. With the elementary ablation (ab) and adjunction (ad) operators one can derive a formal classification of the fundamental reactions of chemistry. These formal operators and the reaction types, addition, elimination, substitution, rearrangement, are presented in Fig. 19. Both of these operators concern topological and chromatic operators.

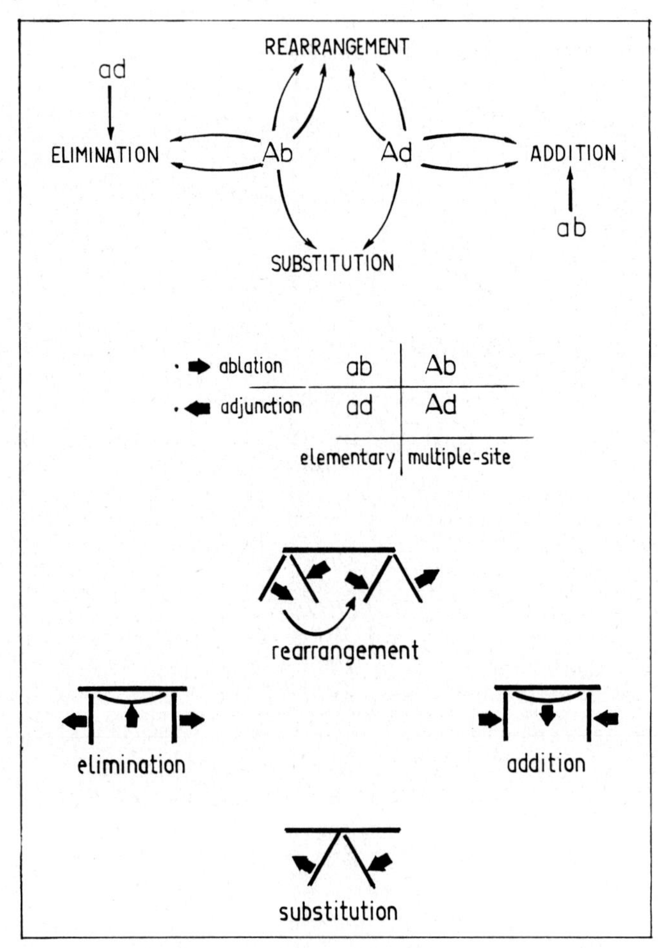

Figure 19 - Expression of the fundamental types of reactions in organic chemistry through formal generative operations

2) Ordering and reaction modelling. In a reaction a site not altered in the transformation of a compound A in B may see its code or ordering changed. The hydration of propene is presented by two different intrinsic orders. An important objective is to handle such changes with a common unique order (Fig. 20). New concepts have recently been derived so as to have a unique representation of the invariant part in a transformation. This is an excellent case where the representation of state 1 of the system is associated with the transformation of state 1 into state 2. These developments are essential to derive certain homomorphism relations in a complex reaction pathway concerning different steps.

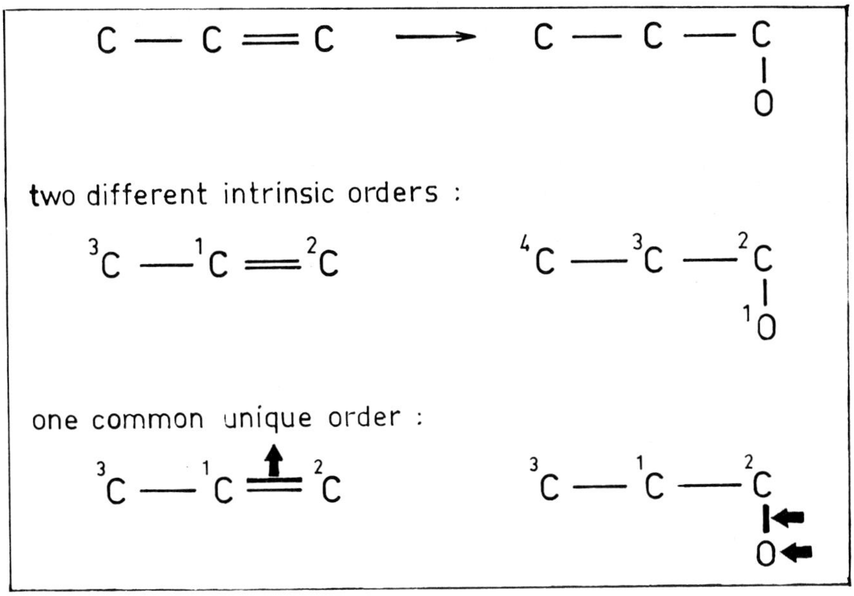

C — C ═ C ⟶ C — C — C (— O)

two different intrinsic orders :

$^{3}C - {}^{1}C = {}^{2}C$ $^{4}C - {}^{3}C - {}^{2}C - {}^{1}O$

one common unique order :

$^{3}C - {}^{1}C = {}^{2}C$ $^{3}C - {}^{1}C - {}^{2}C - O$

Figure 20 - ORDER AND REACTION MODELLING

3) Representation of unitary transformation. More work on graph theories is needed to elucidate real transformations which concern several independent entities. Several disconnected graphs have to be handled as a block (Fig. 21).

In this respect a new category of hyperstructures HS (P, B) has been proposed for the study of simultaneous transformations of several structures. These HS are structured by block relationships generalizing the binary concept R_O. Oriented chromatic hypergraphs or tripartite edged chromatic hypergraphs R_T are useful to characterize the structures of the dead, permanent and newborn entities of referential unitary transformations. These graph concepts allow for transcriptions in accordance with the customary language of any scientific field.

Thus far, parallel efforts in the development of graphic displays for complex networks have been carried on in different fields: enzymatic catalyzed mechanisms, electrical networks, engineering control.... Nonetheless more work is expected in the computerized handling of results. Progress in the area of complex hyperstructures with reference to a richer edition is also needed to improve graphic interaction of scientists with elaborate programs of data handling linked to computer design.

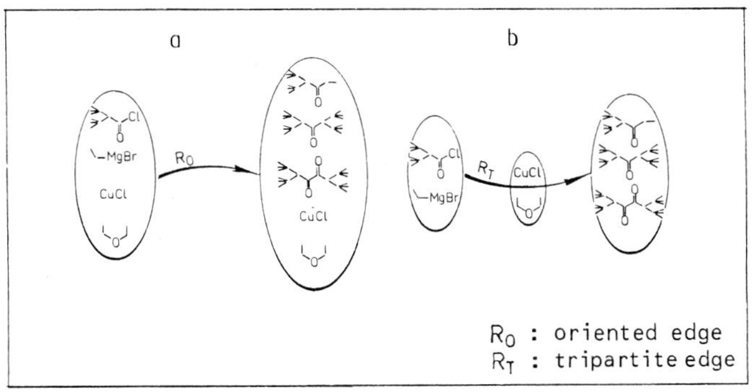

Figure 21 - REPRESENTATION OF UNITARY TRANSFORMATION

CONCLUSION

In this lecture new concepts have been presented for the handling of complex systems and their transformation. It is to be noted that the use of a local order on structures (S) and the state-space (Hyperstructure HS) seem very promising. By playing on these organizations (S and HS) with different ordering functions, it seems possible to discover general underlying real mechanisms. Such assumptions are currently made for natural complex systems (DNA functions), linguistic data (natural grammars) and structure-properties correlations... The importance of the combinatorial problems encountered in these cases is such that these concepts are best handled with the logistics associated with computer-aided design (CAD).

ACKNOWLEDGEMENTS

I would like to thank my collaborators in this domain: Professor A. Panaye, Messrs. G. Georgoulis, R. Picchiottino, G. Sicouri, and Y. Sobel, and also Mme P. Glaeser, for their help and support in the preparation and compression of my lecture into this presentation.

REFERENCES

A. DARC SYSTEM

(1) Dubois, J.E., Structural organic thinking and computer assistance in synthesis and correlation, Isr. J. Chem. 14 (1975) 17-32

(2) Dubois, J.E., Computer-assisted modelling of reactions and reactivity, Pure Appl. Chem. 53 (1981) 1313-1327

(3) Dubois, J.E., Picchiottino, R., Sicouri, G., Sobel, Y., DARC System: block relationships and hyperstructures, C.R. Acad. Sci. Paris, Serie I (1982) 251-256

(4) Dubois, J.E., Panaye, A., Picchiottino, R., Sicouri, G., DARC System: structure of a reaction invariant, C.R. Acad. Sci. Paris Serie II (1982) 1081-1086

(5) Picchiottino, R., Sicouri, G., Dubois, J.E., DARC System: transformational relationships and hyperstructures in chemistry, this volume, 229-234

B. OTHER SYSTEMS

(1) Kitagawa, Tosio, The role of information science in the Unification of Sciences, Research Institute of Fundamental Information Science, Report No. 48 (1974)

(2) Kitagawa, Tosio, Dynamics of reverberation cycle and its implication to linguistics, RIFIS, Report No. 56 (1975)

(3) Nilsson, N.J., Principles of artificial intelligence (Tioga, Palo Alto, 1980)

(4) Winston, P.H., Artificial intelligence (Addison-Wesley, 1979)

(5) Latombe, J.C., (Ed.), Artificial intelligence and pattern recognition in computer aided design, (North Holland, Amsterdam, 1978)

(6) Winston, P.H., Brown, R.H., Artificial intelligence, an M.I.T. Perspective (MIT Press, 1979)

(7) Atlan, H. L'organisation biologique et la theorie de l'imagination (Hermann, Paris, 1972)

(8) Fu, K.S. Syntactic methods in pattern recognition (Academic Press, N.Y., 1974)

(9) Bonnet, A., Applications de l'intelligence artificielle: les systèmes experts, RAIRO No. 4, 15 (1981)

(10) Pinson, S., Representation des connaissances dans les systèmes experts, RAIRO No. 4, 15 (1981

(11) Zadeh, L., A fuzzy set theoretic interpretation of linguistic hedges, J. of Cybernetics, No. 3, 2 (1972)

(12) Lauriere, J.L., Representation et utilisation des connaissances, ISI,1 (1982) 25-42 and 109-133

Data for Science and Technology, P.S. Glaeser (ed.)
North-Holland Publishing Company

RECENT TRENDS IN COMPUTER SYSTEMS ARCHITECTURE AND GENERAL LANGUAGE

Andrzej Janicki

Institute of Mathematical Machines
Warsaw, Poland

The paper presents problems of an influence of computer systems architecture as well as of features of high level languages over an efficiency of extraction of information from scientific experiments. This also concerns acquisition of reference data and planning of target research as well. This paper discusses some non-conventional architectures. Also, some instruments convenient for integrated programming are considered. The features of parallel programming languages ADA and LOGLAN 79 were presented and compared and the results are given. An experimental software environment is proposed. It is equipped with extensive simulation functions and a relation data base for concurrent calculations in multiprocessor and network systems.

1. INTRODUCTION

In many problems of data extraction from research experiments, gathering and describing reference data, planning of target research, etc., the influence of computer systems' architecture and features of high-level programming languages on the effectiveness of adopted solutions, has become more conspicuous. It is indispensable to produce compatible systems of various capacity and problem specialization levels. High computing speed, flexible variation of configurations of hardware and software mechanisms, comfortable means of man-computer communication and high reliability of calculations are required.

The new standard computing speeds of 1 - 2 Megaflops (the abbreviation 'Megaflops' denotes computing speed equal to one million floatingpoint operations per second) have already become insufficient for many experiments and reference data calculations, especially the ones that employ behaviour simulation mechanisms of the tested systems (e.g. atom and molecule movement simulation, weather forecasting, nuclear reactions, etc.).

Up till now, computers that were generally used for scientific applications, have been developing seemingly independently of the development of data processing systems. The outcome of this was the necessity to adapt in practice the structure of applied systems to the equipment and firmware in hand, which was obviously harmful for the applications. The time has come to introduce a different approach which would concentrate on the user's point of view.

This new approach leads first to determining accurately the aims and tasks a given computing system is to execute in practice, and then determine the structure and functions of the technical means and software.

2. SOME ARCHITECTURAL PROPERTIES

The crucial change in computer architecture lies, among others, in the conversion from *layer* structure of the equipment (control processor layer, telecommunication and network processor layer, etc.) to the *functional* structure sub-systems of controlling data and operations, input/output, memory and set service sub-systems, etc. It is being aimed at a far-reaching *transparency* of computer systems, so that the realization of the interesting class of tasks, in which introduction, collection, searching, presentation and transmission of data are of utmost importance, be almost entirely independent of the equipment configuration.

In the hitherto existing systems this transparency was achieved in relation to numerical calculations in which a relatively small part of resources in the usable data processing was used. The use of distributed control in multi-processor systems, different from control in processors based on von Neuman's idea, of their nature intended for individual operation, has increased the efficiency of data exchange between processor, simplified the software controlling computing resources and made the administration of memory resources easier.

The introduction and equipment redundancy enabled the creation of certain system tools for increasing the reliability of computations against interferences, errors and damages. Along with the large integration scale of microelectronic circuits, the reduction of dissipated power and overall dimensions of the equipment, the said tools and measures have considerably reduced the costs of operating computer systems. The far-reaching unification of signal and logic interfaces, and information exchange protocols has simplified the exploitation of these systems. One of the promising ways is the specialization of individual hardware and software mechanisms (the term 'mechanism' is used in the meaning of seclusion of a given fragment of equipment or programs in the computer system), the substitution of more standard-like operations executed by the program with suitable hardware mechanisms, the implementation of new hardware primitives with considerably increased complexity and the formation of multiprocessor configurations. The configurations can be either clusters or spatially distributed ones, owing to the consistently introduced modularity of hardware and software mechanisms interconnected by means of rapid interfaces and tools for separate compiling.

2.1 Parallel computing

The range of implementing into computer systems' architecture such methods of parallel data processing (e.g. multiprocessing developed time-sharing type, matrix type or data flow type) depends substantially on the nature of the computations actually effected.

If the nature of the computations is such that to every program-job of this computation one appointed processor can be allocated, and the data required for all the jobs can be located in the same storage, then the architecture developed for time-sharing becomes convenient.

If a particular computation, associated with the solved problem, is to be made repeatedly in many different ways - thus forming a certain computation matrix - then the most useful architecture is the matrix architecture of the multiprocessor computer system. Each of the basic computers of such a system deals with the operation on a single matrix element.

If it is possible to detach the computations from the particular job, then the application of data flow architecture is advantageous. This type of parallelism of computation does not allow us to increase the number of tasks caused by allocating the tasks to the processors, but permits us to create a combination of many very simple processors performing selected computations at a time when all the information for particular computations becomes useful and the same is directed to the unoccupied processor. This, however, requires special control, which is not complicated.

The programming language of data flow machines is a relatively widely adopted language called VAL, developed at NIT. However, efforts are still being made to find more effective ones. There already exists an experimental high-level language, called LAPSE, worked out by J. Gurd and others at Manchester University, based on a certain syntactic adaptation of the widely known language, PASCAL.

A language of parallel programming called LOGLAN has been worked out in Poland. The fact of possessing the developed coroutine mechanisms qualifies LOGLAN as convenient for programming data flow machines. It is possible to mention the formation of a new style of programming called *functional* programming. It enables a much more sophisticated application of computers, less liable to errors.

Use of mechanisms which create distributed data banks having a common data base, use of the essentially significantly enlarged capacity of storage of all kinds, introduction of highly efficient dialects of concurrent programming languages, as well as use of techniques of direct input and output of data - sound or graphic - are conceptions which together trace the further development of computer architecture (e.g. the fifth generation of computers now being formed).

The first expected practical applications will be in research experiments and engineering projects, as well as in so called *consultant* systems for chemistry, medicine, nuclear physics, geophysics, etc. The necessary facility for the implementation of such systems into practice could be the use of auxiliary methods by so called *conceptual* programming, including elements of automatic task solving.

The results of research carried out in Poland, in applying LOGLAN 79 language for these purposes, allow us to hold the opinion that is has become possible to introduce effective methods of programming and aiding into practice, both by using data flow computers as well as by appropriately improving the present manufactured computers of multiprocessor type organization. It is necessary, however, to equip these computers with the proper programming environment, which will be discussed further.

The optimization of parallel processing, as a method of structural increase of computation capacity, has become basically important. Parallel processing can take place at every level of operations of the computing system, i.e. on the binary level, and, most of all, on the instruction and task levels [the parallel processing on higher operation levels (instruction and task levels) is called 'multiprocessing' as being a term having a narrower sense].

The idea is that many simultaneous operations on data can be effectively executed on these levels. This approach also enables concurrent computing (including the application of data flow processes), i.e. joint execution of various calculations at the same time, possibly as many as are ready to be executed.

We point out that both operations and data here concern at least basic machine language. The introduction of multiprocessing mechanisms into the high level languages is a considerable achievement in the field of computer architecture. Advanced theories of parallel computing have been developed, especially on the basis of algorithmic logic and effective techniques of concurrent programming (the term 'concurrent programming' is used by many authors equally with the term 'parallel programming').

2.2 Cluster and distributed systems

The character of cluster systems is similar to the character of conventional computers. However, the distributed systems are similar to computer networks [1]. There are many known examples of interesting cluster systems with SIMD (Single Instruction Multiple Data) and MIMD (Multiple Instruction Multiple Data) types of architecture (see M.J. Flynn's computer architecture classification [2]). (See Fig. 1).

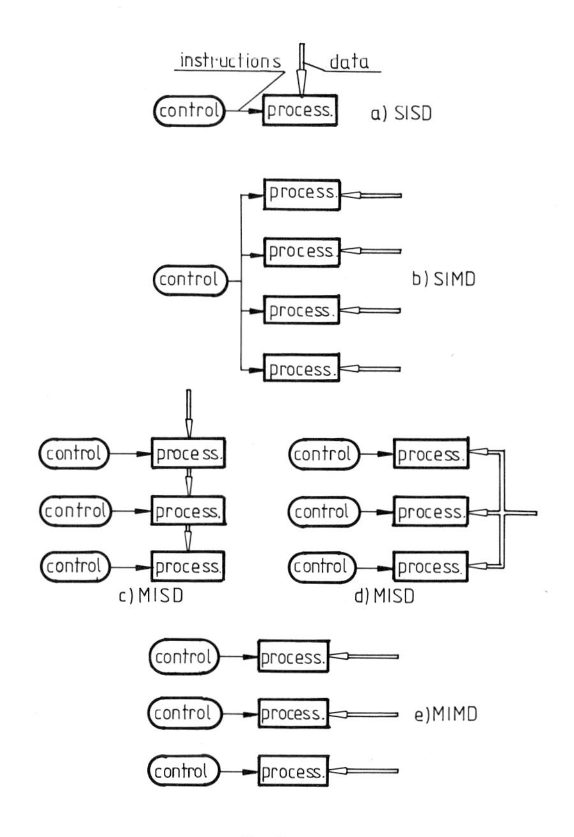

Fig. 1

Nevertheless, computers from the MIMD class are the successors of CDC6600, Univac 1108, and the Borroughs 6700 series, e.g. C.mmp and Prime computers, established a good footing in different fields of applications [3, 4]. However, the products which comprise enhancements, in computer systems existing in today's market, especially in the field of an integrated programming environment (sometimes called a 'software environment'), are focused on by the users. This mainly concerns the minicomputer systems like Level 6, Reality 2000, Vax 11/750, Mera 400, etc. distributed systems [5].

2.3 Instruments for integrated programming

Development of software technology, besides its direct contribution to programming, also provides a new important means for communicating, supervising and managing various project teams.

The everlasting pursuit of more and more convenient programming tools has been inspired by the programmers who demanded that the programming system view, as a link between the abstract algorithm and the computer, be able to catalyze well the contact between the problem to be solved and the possibilities of the hardware.

The present 'state of the art' in programming technology allows us to place programming tools into the following three-class hierarchy (cf. [6]):

a) conventional tools, those which are used to alleviate the task of producing programs during the entire process of their development. The tools in question are text editors, translators, linkers, loaders and debuggers. It seems quite obvious that the properties of the programming language can strongly affect the stages mentioned above.
b) the programming systems provide some tailored collections of tools: the programmable complex structures of the environment, the sophisticated means of data manipulation and communication, and the versatile tools which ameliorate user-system interaction.

   Nowadays, the development of these systems takes in two relatively separate directions. The first leads towards the systems of general purpose; the second towards highly specialized systems. These latter are designed to be user-oriented, however they are usually very closely associated with the unified ideas in the particular domain of the related problem.
c) the programming environment (this term became common along with the ADA programming language [cf. IEE computer magazine, issue June 1982]) which unifies and integrates appropriately designed tools in such a way that their action is directed at one common aim. The objects manipulated by one of these tools are easily and naturally understood by the other. The superior purpose of such a system is a coherent and effective development of tools, maintenance and exploitation of its software resources on the base of underlying high-level programming language. The programs (and their pieces) are uniformly represented at each stage of the program development, in particular at the stage of specification as well as when the program is edited of debugged.

   Some of the principal works in this approach are reported by Teitelbaum [7], Yershov [8], Haberman, Tichy [9], De Remer, Cheatham and others [10].

All these classes are nowadays the subjects of intensive research. For the purpose of this paper, such achievements as the Incremental Programming Environment IPE (cf. [11]), the GANDALF environment (cf. [12]), and most of all the software environments of ADA and LOGLAN (cf. [13])

are particulary interesting. Comparison of the properties of the trench version of ADA language and the Polish programming language LOGLAN 79 is presented in the next sections.

Much attention has been paid recently to such functions of the programming environments as distributed processing and integration of data bases with industrial documents.

Among the highly ambitious projects in the field of programming environments, like the GANDALF integrated environment for software development, or the Cornell program synthesizer, and the French project SOL, a new system called SLOG (programming System in LOGLAN language) is going to appear. At the moment, there are a few experimental installations of the SLOG environment in Poland.

3. SOFTWARE ENVIRONMENT

We will discuss SLOG as representative of integrated software environments for multiprocessor computer systems that is based on compilation technology and on parallel processes and real-time interaction tools, strictly combined with transportable operating systems. Programs are manipulated through a syntax and application directed by a control module instead of a traditional debugger. The central module's commands are provided to the programmer or the user as commands of the editor. Other tools are applied automatically at the appropriate times by SLOG.

These tools and their representations are not visible to the user. The only uniform interface is the user's interface of the editor and the executor. During the construction and manipulation of the programs the user focuses on a small fragment of the program at a time. Choices and changes made cause the system to incrementally compile those fragments and incorporate them within the executable version of the program.

Moreover, SLOG provides among others such programming tools as:
- effective instructions for handling computation, files and interruptions,
- transferrable standard I/0 library inclusive of run-time routines,
- cross-assemblers for other computers having objects formats conforming with host-machine assembler,
- symbolic debugger for alorithmic languages used and for assembler,
- cross-reference text editor equipped with regular form of expressions,
- stream editor used in filters shaped according to text editor,
- formats matching filters,
- language for scanning and conversion of formats,
- lexical analysis generator,
- parser generator extended language,
- text formatter (compatible with Nroff),
- instructions for computer aiding concerning computers,
- table calculator of direct operation,
- statistic program package for fields control and processor utilization,
- personal data information system for controlled access and resources management,
- mechanisms for extensions of programs, languages and basic instructions,
- machine-independent program and implementation solutions.

It was very important to make the right language choice for SLOG. A general language LOGLAN 79 designed for concurrent programming applications, has demonstrated the most promising features, from the software environment point of view, as well.

More precisely, the LOGLAN 79 is a universal high-level programming language. Its syntax is patterned upon PASCAL's. Its rich semantics include the classical constructs and facilities offered by the ALGOL-family programming languages as well as more modern facilities, such as concurrency and exception handling (cf. [14-18]).

The basic constructs and facilities of the LOGLAN 79 include:
1) A convenient set of structured statements,
2) Modularity (with the possibility of module nesting and extension),
3) Procedures and functions (fully recursive; procedures and functions can be used also as formal parameters),
4) Adjustable arrays whose boundaries are determined at run-time in such a way that multidimensional arrays may be of various shapes e.g. triangular, k-diagonal, streaked, etc.,
5) Classes (as a generalization of records) which enable us to define complex structured types, data structures, packages, etc.,
6) Coroutines and semi-coroutines,
7) Prefixing - the facility, borrowed from SIMULA-67, substantially generalized in LOGLAN - which enables us to build up hierarchies of types and data structures, problem-oriented languages, etc.,
8) Formal types treated as a method of module parametrization,
9) Module protection and encapsulation techniques,
10) Programmed deallocator - a tool for efficient and secure garbage collection, which allows the user to implement the optimal strategy of storage management,
11) Exception handling which provides facilities for dealing with run-time errors and other exceptional situations raised by the user,
12) Separate compilation techniques,
13) Concurrency easily adaptable to any operating system kernel and allowing parallel programming in a natural and efficient way.

The language covers system programming, data processing, and numerical computations. Its constructs represent the state-of-the-art and are efficiently implementable. Large systems consisting of many cooperating modules are easily decomposed and assembled, due to class concept and prefixing.

LOGLAN constructs and facilities have appeared and evolved simultaneously with the experiments on the first pilot compiler (working on a Mera-400 16-bit minicomputer). The research on LOGLAN implementation profited by new algorithms for static semantics, context analysis, data structures for storage management, etc.

The LOGLAN compiler provides a keen analysis of syntactic and semantic errors at the compilation stage as well as at run time. The object code is very efficient with respect to time and space. The completeness of error checking guarantees full security and ease of program debugging.

The main problems of data structures in SLOG, which are specification, implementation, and design and verification, can be solved by means of a proper set of algorithmic axioms and a formula which express the property of the corresponding program.

An algorithmic theory of data structures which we used in this paper is defined whenever we are given three elements [17]: an algorithmic language, a logical deductive system, and a set of specific nonlogical axioms, and then we base the theory upon some kind of algorithmic logic.

The kind of logic depends on the collection of program connectives allowed in the construction of programs. In our case the collection of program connectives is:

begin...end	(composit)
if...than...else...fi	(branching)
while...do...od	(iteration)

4. COMPARISON BETWEEN LOGLAN AND ADA

The difference between LOGLAN 79 and other languages lies in different sets of functional and relational symbols and its semantics.

The set can be split into: terms, formulae, programs.

In the case of LOGLAN 79 language:
the structure of the set of terms is as usual,
formulae: are closed with respect to the usual formation rules, contain quantifier-free formulae and fulfill such condition, that,
programs: are built from atomic programs (assignments) by means of program connectives.

A relation of a new programming language to other contemporary languages should be determined. Therefore, the software project LOGLAN 79 (which contains a wide variety of software and theoretical works) also deals with research such as connections between LOGLAN 79 and ADA (ADA is a famous programming language for numerical applications, system programming applications and embedded computer applications with real-time requirements) programming constructs. The latter language introduced many new interesting constructs.

ADA provides tasks - program components which are subject to the following actions: generation, dynamic activation (after which they act concurrently with their fathers) and, finally, termination. A task may only be delayed as a result of synchronization (with other tasks). ADA provides a single synchronizing tool - rendezvous, which is based on C.A.R. Hoare and P. Brinch-Hansen's ideas. LOGLAN 79, in turn, provides processes as program components that are subject to generation, activation, suspension, resumption and termination. Therefore, it is sufficient to know whether rendezvous is expressible by means of LOGLAN 79 synchronizing tools; if so, then the task could be fully expressed in LOGLAN 79.

The details and motivation of correctness are supplied in T. Muldner's papers [18]. LOGLAN's approach to synchronization is essentially different from ADA's approach. LOGLAN 79 provides elementary synchronization tools of the semaphore type which are implemented in the operating system kernel.

More complex synchronizing tools (e.g. those at high level like Hoare's monitor) are created by means of elementary ones and structured functors (prefixing). The other tools, that are necessary for defining a rendezvous, are provided by library classes e.g. QUEUE determining queues. The following LOGLAN 79 feature is of paramount importance: if one provides a dialect, e.g. TASKING which determines tasks and rendezvous, then one can hide auxiliary variables serving for implementation of the corresponding data structure, and, moreover, one can also cover low-level synchronizing tools. Therefore, the properties of the dialect cannot disturbed.

There are no essential differences between the sequential control structure in ADA and LOGLAN 79. Both languages provide tools for a separate compilation of programming modules. LOGLAN's tools are slightly more general than ADA's tools, however. Exception handling introduced in ADA is implemented in LOGLAN 79.

5. TOOLS FOR SOFTWARE ENVIRONMENTS

Proposed tools for SLOG constitute the software environment of modern hardware and contain:
1) LOGLAN 79 compiler (basic and extended version)
2) Language C compiler (source equivalent to programs running under Unix system)
3) Standard languages compilers: PASCAL, FORTRAN IV, FORTRAN 77, BASIC and MUMPS 11
4) Standard operating system (at a level such as the RSX-11 or Unix operating systems)
5) Data base and conceptual programming tools management system (including the LOGLAN Management System)
6) Program designing and manipulating system (through a syntax and application directed editor)

7) Pilot operating system SOM-7 (written in assembler) providing parallel computation facilities on the Mera-400 minicomputer system
8) Pilot operating system SOM-79 (written in LOGLAN 79 or in C language) providing multiprocessor parallel computation with transportability is provided
9) Basic set of programs, utilities and system programs making computer system operation easier and more comfortable.

5.1 Operating systems

The proposed operating systems are examples of well-engineered systems where tools may be used in combination to perform specific applications [19]. Standard operation systems meet the basic specifications generally defined for modern operation systems of the second generation both for controlling part of the system and for handling processors and I/O devices.

Pilot operation systems meet the above-mentioned and other, more sophisticated, requirements. This concerns especially:
- suitable environment creation for development of application programs,
- availability of a wide range of modules for the user and tools to select from depending on the specific project definition and its suitable solution,
- creation of a unified I/O structure for files, peripherals and pipes communication between processors,
- coordination of multiprocessor operations or control over the work of one or two processors in order to provide concurrent computation of processes when the operation system covers the lack of actually processed tasks,
- availability of such properties that can provide the most efficient usage of machine resources and in particular:
- multitasking, multiprogramming and multiuser facilities,
- running processes on foreground and background levels,
- concordance of mechanisms for files, peripherals, processors' communication and for teleprocessing,
- possession of interpreter modified instructions for specific applications in hierarchical, tree-structured, distributed file system,
- pipes type and multiplexed channels of multiple inter-processors communication,
- asynchronous program interruptions,
- overall segmentation (shared data, free records, instruction fields),
- ability to hold programs in memory for real time applications,
- high speed information exchange by means of separated cache storage,
- minimum lock-out time at interruptions for real time applications,
- facility of reliable return after supply errors,
- high speed access to disc storage by means of cache buffer,
- buffered peripherals' controllers computing timers, tracing of program's structure and debugging.

5.2 Relative data base in SLOG

The data base management system in SLOG is coded DBS 79 and it uses strong mechanisms specific to LOGLAN 79 language, such as: classes, hierarchic prefixing, dynamic creation of types, virtual procedures, remote (dotted) access (cf. [20]).

System DBLS 79 has the following data structure:

- DOCUMENT
  is the basic external data unit. This document is of the RECORD type (in this context 'type' corresponds to the formal meanings of this term used in LOGLAN 79) and represents a file of position values. There may be many documents of this type. The name of the document and its set of position data identify its specific type. A document of specific type is linked to a constant set of positions.

- POSITION
  is the basic component of a document and possesses a name defined by the user. The positions are conferred ones of the following types (in the frame of one type of document):
  N - name (8 characters)
  S - chain (72 characters)
  I - integer number
  R - real number

- RECORD
  is a data unit representing a set of documents with a common name which contains this name, ordered position values and some additional information.

  Document type definition

  A set of documents in the DBLS 79 system comprises only such documents which have the same number of positions, the same names of each position and the same type of position value (REAL, INTEGER, CHARACTER).

- DOCUMENT FILE

  comprises documents of the appointed type by realizing records with attributes assigned according to the user's needs.

The DBLS 79 system performs the following basic functions:

data storage	(using procedures NEW, ADD, DEL and auxiliary utility programs INI, COD)
data access	(using procedures GET, NEXT, TPOS, NPOS, POS and auxiliary utility program BASE)
data update	(using procedures ADD, NEW and auxiliary utility program MOD)

Architecture of the DBLS 79 is presented in Figures 2 and 3.

We define the main modules of the DBLS 79 as follows:

LIBRARY OF DATA BASE OPERATION BINARY PROGRAMS

This library is created by compiling the source programs from the sources program library only at the stage of creating a DBMS in LOGLAN 79.

LIBRARY OF BINARY USERS' PROGRAMS

The library contains compiled users' programs together with data base access mechanisms.

SOURCE LIBRARY OF PACKAGES OF DATA BASE OPERATION PROCEDURES FOR USERS' PROGRAMS

Formally each PACKAGE that belongs to this library is a LOGLAN 79 class. Its attributes belong to a respective set of procedures for data base operation.

Procedures and programs presented in Figure 2 are examples of those attributes in PACKAGE.

The library is stored on a disc section. Respective PACKAGES - classes are connected automatically to the users' programs before compilation in case there are modules, e.g. blocks, classes, prefixed by these classes.

SOURCE LIBRARY OF THE USERS' PROGRAMS

The user may apply some base access mechanisms in his program by calling proper procedures contained in the packages of data base operation procedures. However, he should remember that the program unit (block, class), within which procedures are used, should be prefixed by a proper class (PACKAGE) or should contain this class in its prefix sequence. PACKAGE will be automatically connected to the program before compilation through the logic interface module which initiates the user's program compilation.

SOURCE LIBRARY OF PROGRAMMING TOOLS

The library contains a set of classes (in the LOGLAN sense). Each class contains different specialized mechanisms to solve various tasks in the users' programs such as simulation, modelling, numerical calculations, etc.

LOGIC INTERFACE

This interface allows us to maintain the libraries and presents a modular construction (cf. Fig. 3). All modules are based on the properly defined system macrodirectives, which use standard system programs (e.g. selected system processes).

In regard to programming tools, one of them is the class MODELLING.

5.3 System class MODELLING

The MODELLING class is a dialect of the LOGLAN 79 language which gives useful tools for writing programs for adaptive forming of complicated structures and configurations, hierarchical systems, and simulative research. In this class virtual functions of two-staged criteria for quality evaluation of synthesized structures or configurations are declared (cf. [21]).

The MODELLING class is a prefixed class realizing a specific parallelism of computations obtained on a single processor computer due to the mechanisms of parallelism contained in the LOGLAN 79 language. This class is named SYSPP.

The MODELLING class has, as its attribute, procedures which retrieve such data from the data base as is needed for the execution of the application program, as well as the procedures which route the results of computations to the data base. Apart from the above described mechanisms the programmer can use the tools of parallelism derived from the SYSPP class, as well as the tools of data base management DBLS 79.

5.4 Application program ADAPT

The ADAPT program is written using the tools of MODELLING class and is designed for computer aided synthesis of complicated structures using adaptative procedures based on two-staged criteria for evaluating the quality of these structures: inferior criterion and superior criterion (cf. [22]).

The inferior criterion defined in this program and designated as OMEGA is used for estimating from step to step the informative efficiency of the synthesized structure or configuration. The superior criterion designated as KRFI is used to estimate the economical efficiency of this structure (for example the relation of efficiency and costs).

The contents of the synthesis procedures or the efficiency evaluation criteria may be exchanged in the program according to the needs of the user, due to the fact that these functions have been declared virtually in the MODELLING class. The program uses the management procedures of the DBLS 79 relational data base.

The start of the program takes place after the beginning conditions have been defined by the user (Fig. 4). Placing the synthesis procedure in a process formally defined as in LOGLAN 79 allows the activating by the ADAPT program in parallel process activating the corresponding instances of this procedure. They synthesize, in parallel, structures beginning from the sequential annexing to a given beginning configuration of the next element from the set of allowable elements and compare at each step the values of the OMEGA criterion. The proceeding ends according

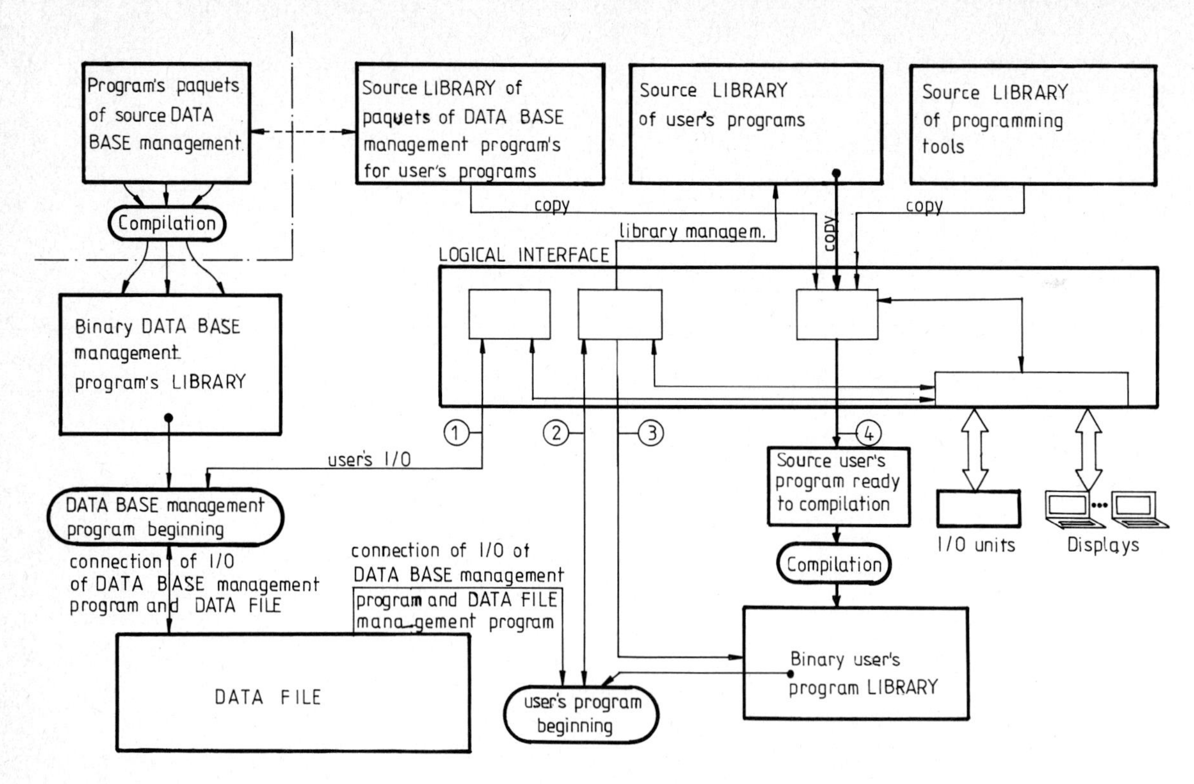
Program's paquets of source DATA BASE management
Source LIBRARY of paquets of DATA BASE management program's for user's programs
Source LIBRARY of user's programs
Source LIBRARY of programming tools
Compilation
copy
library managem.
copy
copy
LOGICAL INTERFACE
Binary DATA BASE management program's LIBRARY
user's I/O
DATA BASE management program beginning
connection of I/O of DATA BASE management program and DATA FILE
connection of I/O of DATA BASE management program and DATA FILE management program
DATA FILE
user's program beginning
Source user's program ready to compilation
Compilation
Binary user's program LIBRARY
I/O units
Displays

Fig. 2

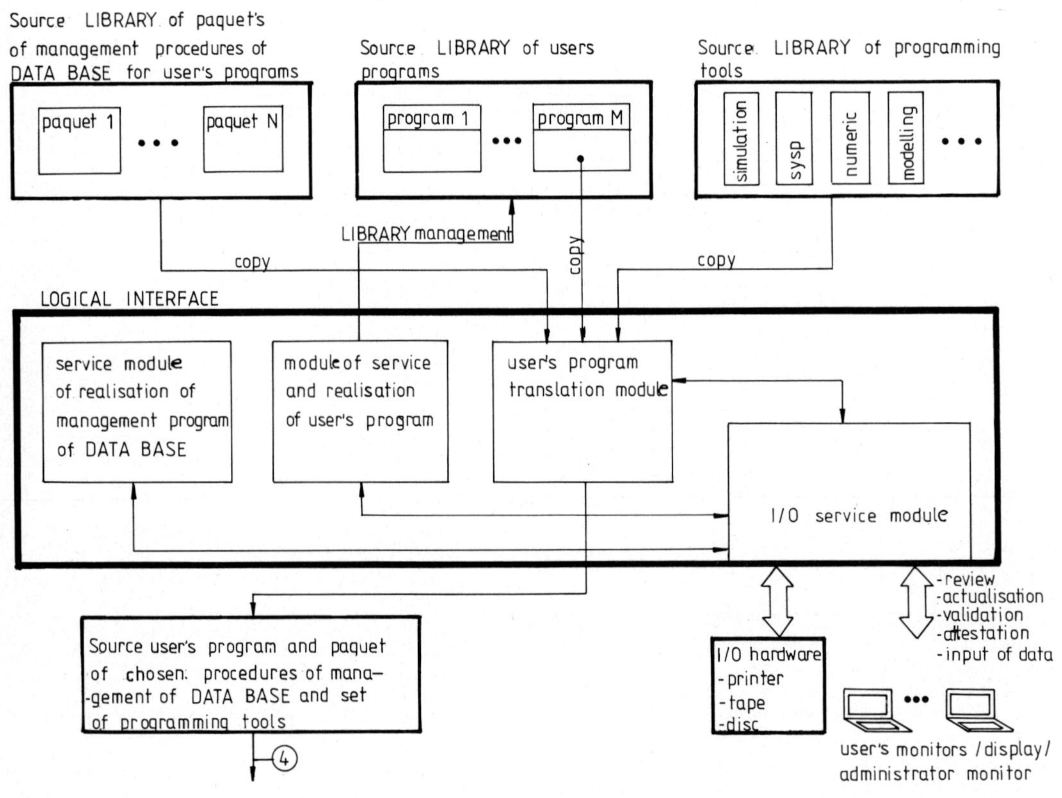
Source LIBRARY of paquet's of management procedures of DATA BASE for user's programs
Source LIBRARY of users programs
Source LIBRARY of programming tools
paquet 1
paquet N
program 1
program M
simulation
sysp
numeric
modelling
LIBRARY management
copy
copy
copy
LOGICAL INTERFACE
service module of realisation of management program of DATA BASE
module of service and realisation of user's program
user's program translation module
I/O service module
-review
-actualisation
-validation
-attestation
-input of data
Source user's program and paquet of chosen procedures of management of DATA BASE and set of programming tools
I/O hardware
-printer
-tape
-disc
user's monitors /display/ administrator monitor

Fig. 3

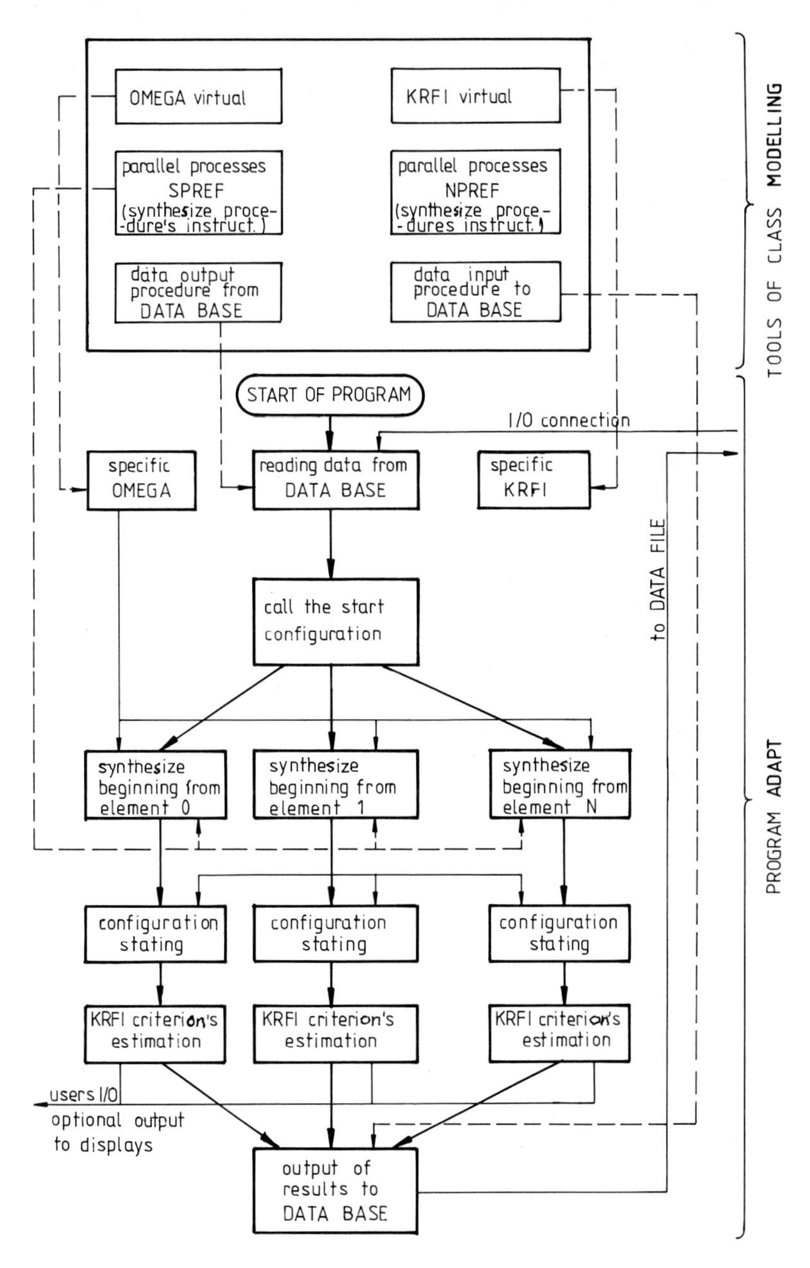

Fig.4

to the accepted rule of stop, for example, on the step after which the valve of function OMEGA does not notably increase in L next steps. The resulting configurations (or structures) obtained in this way are evaluated by means of the URFI criterion and the processes are then ended. The obtained results are placed in the data base by means of management procedures deriving from the MODELLING class. Here the action of the ADAPT program is suspended and the user reviews the resulting structures and their quality appraisal by means of interactive data base management programs (in our case DBLS 79 data base programs). In doing so he uses additional data (standards, altering data, etc.) placed in the data base. In the end the user makes the decision about choosing one of these results or activates the ADAPT program for the next stage of synthesis with different beginning data. This regime of work facilitates the gathering of experience based on several experiments and allows the synthesis of quasi-optimum structures in a condition of insufficient a priori information.

5.5 LOGLAN Management System (LMS)

The LMS comprises a packet of programs facilitating access to source and binary programs written in LOGLAN 79 language and their management. It allows the translation of a program from its source to binary form as well as execution of this program. To this end libraries were founded for retention of programs in their source (LBP) and binary (LBM) form.

Remarks:
- all LMS programs put user's program in stated work section
- the LMS programs automatically include fragments of programs appointed earlier in the source program
- especially they include SIMULATION class if the source program contains a prefixed block with such a class when the version of this class has been deducted by the programmer.

6. FINAL REMARKS

The previous sections have outlined a strictly architectural treatment of programming mechanisms. The aim of functional programming is to design a clear system which includes transparent computer architecture, for a neatly defined language. The LOGLAN 79 (and its next release LOGLAN 82 (cf. [23])) is a suggestion for such a language, where a transparent computer architecture already exists and this paper makes the SLOG software environment available for a user's program organization system.

LOGLAN 79 is a very useful language for both modelling and simulating many theoretical and practical problems. This is also because in the LOGLAN body there exists a special class which is more effective than the well known SIMULA 67 language class called SIMULATION (see Fig. 5).

It is also worth mentioning that LOGLAN 79 has extensive possibilities in the programming of data flow machines. The productive coroutine and process mechanisms and also the prefix sequences make such programming highly efficient.

From the application point of view it is important to stress that there exists a very short distance between some parts of the LOGLAN 79 compiler, operating systems and data base systems written in the language, and the corresponding user-tailored 'silicon software' packages. The term should be used only in reference to programs and subroutines that reside in silicon but are analogous to hardware components. [Significant areas of silicon software are those in which routines cast in silicon offer a modular yet interconnected set of primitives (i.e. operating system primitives).]

It becomes evident that in the proper approach to specifying the properties of a software environment one should forget about the technology

RESULTS OF SQS TEST

Number of objects		10	20	40	60	80	100	120	140	160
Times of computation (in seconds)	I	0,6	1,2	2,8	4	5	7	7,3	9,5	11
	II	2	4,1	9	16	24	33	44	54	

I. In case of SIMULATION tool in Logan class MODELLING.

II. In case of SIMULATION tool in analogically as in „SIMULA 67".

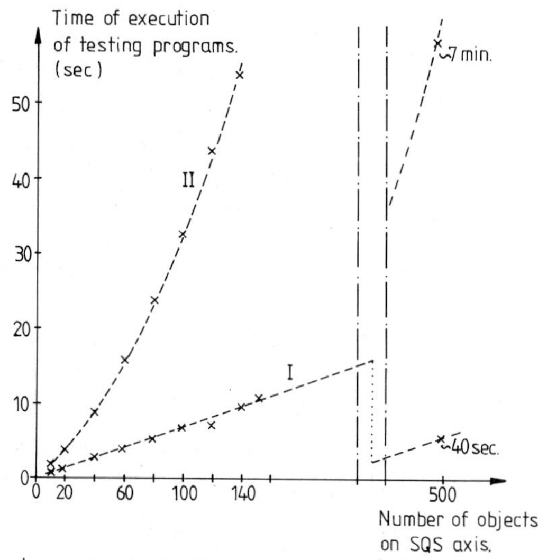

In case I - about 10% of saving of occupied operation memory capacity.

Fig. 5

of its implementation. It springs from the fact that properties of various environments are common, and simultaneously, the user-organization still has the unimpeded choice. However, the choice of tools is strictly defined but there is freedom to choose the way these tools are built up so as to be optimally suited to the application constraints, as well as to the target hardware.

An organization which wants to build up its own software environment on the LOGLAN 79 basis should have a close look at its own procedures and then select the set of tools which is supremely useful for most of their requirements. The whole modularity of the SLOG system allowed us to realize this idea rather easily.

REFERENCES

[1] Liu, I.W.S. and Liu, C.L., Performance analysis of the heterogeneous multiprocessor computing systems, Computer Architecture and Networks. Modelling and Evaluation, ed. Gelende, E. (North-Holland, Amsterdam, 1974, 331-334).

[2] Flynn, M.J., Very high-speed computing systems, Proc. IEEE vol. 54 (1966) 1901-1909.

[3] Wulf, W.A. and Bell, C.G., C mmp - a multi-mini-processor, Proc. AFIPS FICC (1972) 765-777.

[4] Baskin, H.B. et al., Prime - a modular architecture for terminal oriented systems, Proc. AFIPS SICC (1972) 431-437.

[5] Janicki A., Mera 400 - an architecture of modular minicomputer systems, IMM Trans. (1975) preprint (in Polish).

[6] Shyamasundar, R.K., Tools for software development and programming environments, Proc. CSI Con. (1982) 68-84, Madras.

[7] Teitelbaum, T. and Reps, T., The Cornell program synthesizer, a syntax directed programming environment, Cornell Univ. Ithaca (1980) TR 80-421.

[8] Yershov, A.P., The ALPHA programming system, (APIC, Study in processing, Academic Press, New York 1971).

[9] Tichy, W.F., Software development control based on module interconnection. Proc. 4th Int. Con. on Software Engineering (1980) Munich.

[10] Cheatham, T.E. et al., A system for programming environments, Proc. 4th Int. Con. on Software Engineering (1980) Munich.

[11] Mora, R.M. and Feiler, P.H., An incremental programming environment, Proc. 4th Int. Con. on Software Engineering (1980) Munich.

[12] Habermann, A.N. et al., Report on the use of ADA for the development and implementations of part of GANDALF, TR CUM (1979) Pittsburgh.

[13] Bartol, W.M. et al., Semantics and implementation of prefixing at many levels, Inst. of Informatics Report N° 94 (1980) Warsaw Univ.

[14] Wegner, P., Research directions in software technology (The MIT Press, Cambridge, Massachusetts and London, England, 1979).

[15] Janicki, A. and Muldner, T., An approach to formal description of monitor activity, IMM Trans. (1977) - preprint.

[16] Muldner, T. edit., Formal report on the programming language LOGLAN 79, Inst. of Informatics Report (1980) Warsaw Univ.

[17] Salwicki, A., Algorithmic theories of data structures, Lecture Notes in Computer Science (Springer-Verlag, Berlin, Heidelberg, New York, vol. 140 (1982) 458-472).

[18] Muldner, T., Some remarks on new high level programming languages: LOGLAN and ADA, Bull. Mera N° 7,8 (1978) 23-27; 22-31.

[19] Holt, R.C., Structured concurrent programming with operating systems applications (Addison-Wesley, Philippines, 1978).

[20] Dabrowski, J. et al., Relative data base in the SLOG software environment, Bull. Mera (1983) in press.

[21] Janicki, A. and Miziolek, J.K., New software tools for modelling and conceptual programming. Proc. FST TCS Con. (in press).

[22] Janicki, A., An adaptive sequential test. Applicationes Mathematicae, vol. XIII, 3 (1973).

[23] Kreczmar, A. edit., Formal report on the programming language LOGLAN 82, Lecture Notes in Computer Science (Springer-Verlag, Berlin, Heidelberg, New York, 1983) in press.

Data for Science and Technology, P.S. Glaeser (ed.)
North-Holland Publishing Company

QUEUING NETWORK MODELS: A TOOL FOR RELATIONAL DATA BASE DESIGN*

Dario Maio and Claudio Sartori

CIOC - C.N.R. and Istituto di Elettronica, University of Bologna
Viale Risorgimento, 2 - 40136 Bologna, Italy

The operational efficiency of a relational DBMS is greatly dependent not only on the effectiveness of the access structures, but also on the optimizer algorithms and srategies used in solving transactions. Both of these parts of the DBMS must be designed taking into account the hardware and software architecture of the host computer system, as well as the characteristics and use of the information in the data base. The goal of the paper is to present general concepts of a relational data base evaluator currently being developed at CIOC - C.N.R., University of Bologna. The design problem is solved by using at the lowest level queuing network models which are able to capture the most important features of actual systems and to express the complexity of modern configurations by means of mathematically tractable forms.

1. INTRODUCTION

The area of relational data base systems has become a popular source of system research projects. These systems allow the user to manage the data base via high level non-navigational query languages: selection of efficient access paths to the stored data is the task of the query optimizer (1, 2, 3, 4, 5). In most cases one of the most important design considerations is the performance of data base systems. The operational efficiency of a relational DBMS is greatly dependent not only on the effectiveness of the access structures, but also on the optimizer algorithms. Both these parts of the DBMS must be designed taking into account the hardware and software characteristics of the host computer system as well as the kind of applications involved in the DB (6, 7, 8, 9, 10, 11, 12, 13).

While much progress has been made recently in developing strategies for data base design at both logical and physical levels, a review of the literature indicates that the trend towards developing tools for evaluating data base system performance has yet to be fully realized with today's decision processes. In particular, we note a lack of general methodologies able to capture the effects on performance due to interdependencies between the logical and physical design choices. Moreover, physical DB design techniques usually produce an "optimal" set of access path structures in order to minimize the time spent in I/O operations. However, they do not provide insight into the behaviour of the system running the real workload; in other words no prediction is given about performance in terms of transaction throughputs and response times. On the other hand performance analyses can be conducted only by taking into account the resource contention also whose effects are not well considered at logical and physical design levels.

As investments in data base applications have become relevant to many enterprises and organizations, a motivated demand has grown for merging performance and cost-benefit analyses to provide practical guidance to the designer in making decisions that are consistent with management's view of DBMS operation. Again systematic methodologies are still not well developed; in fact, by contrast with the problem of performance analysis, the design problem is to compute the system parameters directly from costs, benefits, performance and workload specifications (14, 15, 16). The ultimate objective of the data base designer is also to provide a system configuration which meets user requirements at the lowest overall cost. As systems and usage become more complex, the decision process becomes increasingly difficult due to the high number of factors to be considered.

In (17) Sevicik proposed a layered model intended to provide a framework on which to examine the interrelations between various design decisions, using an analytical model based, at the lowest level, on queuing network models. By analytical techniques, the workload description at one level and a set of design choices are transformed into the workload description at the next lower (more fully specified) level.

Queuing network models are widely used to analyze the performance of multiprogrammed computer systems, given the workload parameters and system configuration. Reference (18) is a tutorial for the best known applications and solutions of queuing network models. The aim of the paper is to discuss the most relevant problems in employing such models for evaluating DBMS performance. Attention is first focussed on general concepts of the logical and physical design phases; then we propose a model for evaluating the performance

----

* This work has been partially supported by the project DATAID of the Italian National Council of Research (C.N.R.)

of a System-R like DBMS (19), taking into accoun the effect produced by the contention for the buffer pool shared among concurrent users and managed by the research storage system. Finally, we briefly deal with the problem of deriving inputs to the queuing network model from the outputs of the logical and physical DB design decisions. According to this kind of approach, a relational data base evaluator is currently being developed at CIOC - C.N.R. Istituto di Elettronica of the University of Bologna.

2. Summarization of logical and physical data base design

This section briefly discusses the most important steps and decisions made at logical and physical levels, in order to explain how to relate the amount of information which flows through these levels to the particular data needed for the performance evaluation level. In the following we will use the term "transaction" to indicate either elementary statements (both non-procedural queries and updates) or groups of statements possibly imbedded in a host language (i.e. SQL statements in PL/I procedures).

The logical design of a relational DB consists of deriving a relational model from a conceptual scheme including among others:

a) characteristics of the relations, that is for each relation:
   - number of tuples
   - number of columns and type of attributes
   - cardinality for each column
b) characteristics of the workload, that is
   - the set of the most relevant transactions grouped in classes;
   - for each class a description involving the rate of occurrence, information about its distribution over the observation period and particular constraints on the response time;
   - for each class again an estimate of the selectivity factors of the selection predicates used in the transactions.

It is worth observing that both categorizing transactions into various classes and making realistic assumptions on the workload parameters are crucial operations and require detailed knowledge which is not usually available at the design level. The most common approach to this problem relies on simplifying assumptions, such that the values of each attribute are uniformly distributed over its domain and there is no correlation between attribute pairs (17).

Physical design involves the choice of record and file structures and their allocation on mass storage devices. The most recent relational DBMSs allow alternate access paths, such as indexes, link and hashing in order to improve performance for some transactions. Given a set of access structures, the selection of efficient access paths to the stored data is the task of a DBMS software module which is commonly named optimizer (2,5,20,21,25). The optimizer is a program able to deal with any DB. Consequently, even by making suitable choices as a rule, some of its criteria may not exactly match with workload characteristics. In fact present optimizers estimate costs by making general assumptions on correlations of attribute values which may be quite different from real statistics for a specific DB.

Clearly, because of update activity and memory constraints, it is not reasonable to create indexes on every column. Thus physical design must include the search for the optimal set of access paths taking into account both the above limitations and workload characteristics. This problem has to be solved not only on installation of the DB but also for subsequent tuning when the workload is expected to change. Recently methodologies and automated tools have been developed to help the DB administrator in making decisions. A general framework of the major steps to be followed in solving the index selection problem is drawn in Fig. 1. It should be noted that step 1 includes precise consideration of: 1) the physical implementations of access structures (i.e. indexes based on tuple identifiers or on page identifiers, B^+-tree, inverted lists, etc.); 2) the access methods, i.e. one index per relation to access tuples in executing a statement or intersection of tuple identifiers taken from more than one index, etc.; 3) strategies used in solving transactions, in particular join methods and index maintenance techniques; 4) optimizer algorithms and criteria. In other words it should never happen that the designer suggests an access path which would never be chosen by the optimizer.

step 0: specification of DB statistics, workload characteristics and constraints on storage.

step 1: expected cost estimate for each transaction class and for each possible access path.

step 2: access path effectiveness comparison in order to determine those which are obviously useless or providing minor benefits when others are present.

step 3: generation of an optimal set of access paths.

Figure 1: Methodology for index selection

In System-R a tool named DBDSGN is available which accepts all valid SQL statements in the workload and solves the combined problem as to the column on which each relation should be ordered (clustered index) and on which other columns, secondary indexes, (unclustered) should be created (10). A methodology for secondary index selection, primarily addressed to small/medium size DBMSs whose characteristics are more limited than those of System-R, has been developed at the University of Bologna within the Informatica (DATAID) project supported by the C.N.R. (12, 13, 22).

3. Performance evaluation

This section discusses the most relevant problems in building models for evaluating DBMS performance. To cope with performance evaluation of a complex system, we usually abstract its essential features into a configuration model. Any performance estimation must be undertaken in relation to a certain known workload, that is, we need to model the demand placed by the user on the various system resources.

In our opinion, performance prediction, as the last step of the DB design, can be conducted well by using a macro-model, that is a model which is able to provide sufficient information from the little and perhaps approximate knowledge which derives from the previous steps of the design. In other words, it seems more reasonable at this level to focus attention only on the effect produced by the resource contention, an aspect which has not been adequately considered at logical and physical design levels. To better explain this concept we rank in the following the major "rubs" raised by the desire to have a detailed model versus its credibility and cost of solution.

3.1 Considerations on DBMS architecture

For the sake of exposition we make an explicit assumption on the type of DBMS to which the analysis can be addressed. A functional architecture like System-R is hypothesized. Fig. 2 gives a functional view of System-R major interfaces and components (19, 23) on an IBM 370 under VM/CMS operating system.

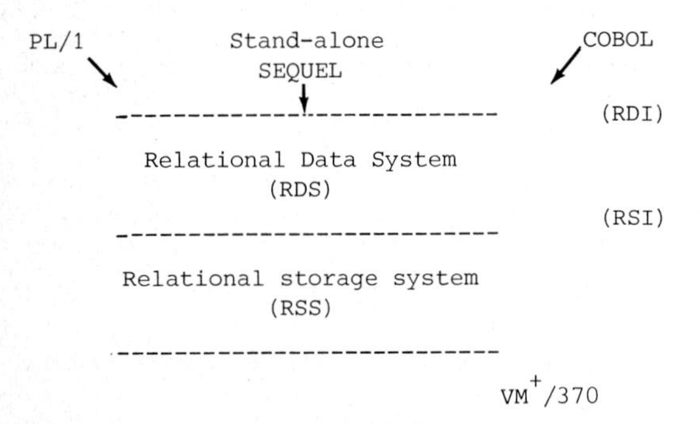

Figure 2: System-R interfaces and components.

The Relational Data System (RDS) implements the Relational Data Interface (RDI), which provides high-level data independent facilities for data retrieval, manipulation, definition and control. The major components of RDS are listed below:
- parsing
- authorization checking
- integrity constraints
- catalog
- access path selection (optimizer)
- code generation

One of the basic goals of System-R is to support two different categories of processing: (a) SQL stand-alone statements which are usually executed once in awhile and (b) canned programs which are executed hundreds of times. For this reason, the RDS module is split into two distinct functions:
- a precompiler to precompile host-language programs and install them as "canned programs" under System-R;
- an execution time system which controls the execution of canned programs and also provides execution of SQL statements for the ad hoc terminal user.

After precompilation and consequent compilation of a modified host language program, an access module, that is, a machine code ready to run on RSS, is generated. On the contrary an ad hoc statement is run-time optimized and its corresponding access module is dynamically generated. Now, from the point of view of workload characterization a first difference is to be pointed out: execution of a canned program does not include operations for parsing, optimization and code generation; in fact preoptimized statements are passed directly to the RSS component. In the following we will suppose that the workload is constituted only by canned programs and ad-hoc statements, in particular we will exclude the case of the possibility allowed by System-R to use the PREPARE statement inside a program which implies, also for canned programs, reinvoking the optimizer at run time (23).

The Relational Storage System (RSS) implements the Research Storage Interface (RSI), which provides simple, record-at-a-time primitives on base tables. Operators are also supported for data recovery, transaction management and data definition. Inside the RSS component a paged buffer acts as a cache memory preserving the most recently used pages. The major functions performed by RSS are listed below:
- space/device management
- index management
- link management
- concurrency control (locking)
- logging and recovery

RSS is responsible for execution of access modules produced by RDS. RSS uses a buffer pool, which can be either in real memory or a virtual address space and is shared across all concurrent users. The buffer pool is composed by pages each of 4096-bytes; the size of the buffer pool can be adjusted depending on applications. The advantage of a shared buffer in a data base environment results in decreasing the miss ratio (that is, fraction of times a page is referenced for the first time in a RSS call and the page is not found in the buffer) (24). The buffer pool is managed via a modified global LRU replacement technique in the sense that it is possible to fix for each query the minimum number of frames required by the query to run (no matter how efficiently). This ensures against their replacement when space is needed by every other concurrent user.

Buffer pool size is a primary design consideration for identifiying storage requirements in the System-R partition. Clearly the I/0 traffic for paging in the buffer pool is strictly dependent on the replacement policy used and the requests behaviour. To avoid some deficiencies of the LRU technique, a recent paper (25) has proposed a buffer management algorithm based on the hot set model to describe the database requests. The hot set model characterizes the buffer requirements of general queries on the basis of the looping properties of their evaluation plans and can be used to define a scheduling strategy which avoids global trashing. According to this policy, a query is eligible to run if a number of frames equal to its hot set is available in the buffer pool.

A remark is perhaps in order on VM/370 facilities and extensions made to support multiuser environment of System-R (26). In particular VM/370 has been modified to allow virtual machines to share read/write virtual memory and provide efficient communication among the various virtual machines Each of the latter can be viewed as a dedicated data-base machine which contains all codes and tables needed to execute all data management functions.

3.2 Configuration model

In this section we describe the model proposed for evaluation of System-R performances. Some simplifying assumptions are made in order to derive a configuration model which can be solved by means of approximate analytical techniques. Referring to Fig. 3, we hypothesize an interactive environment in which a set of user terminals submits stand-alone SQL statements and pre-optimized canned programs to the system. The overall structure of the model is that of a closed queuing network with several job classes, substantially the same as is used in (27) for modeling the VM/370 systems at scheduler level. We apply the decomposition approach, which consists of solving portions of the model in isolation ("offline") and gathering the results together ("online") to produce a solution of the whole model. We use an iteration algorithm in order to correct the inaccuracy in subnetwork representation. The overall model is analyzed in three stages. One chain of type Q and one of type C are considered. The Q chain groups classes of stand-alone statements; the C chain groups classes of canned programs. Differences between various typical DB operations are taken into account inside the lowest level submodel (CPU-I/0 subsystem) by means of an appropriate set of service classes. We distinguish between the type of operations involved by the submission of stand-alone statements and that of canned programs; in such a way contention for the RSS buffer pool is modeled.

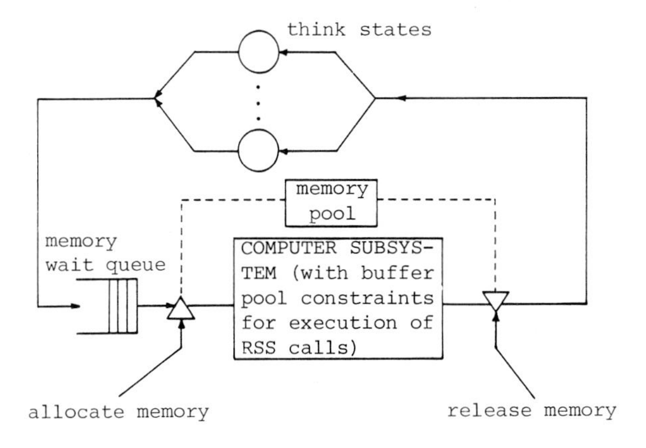

Figure 3: Configuration Model

References

[1] Astrahan, M.M. et al., System-R: a relational data base management system, IEEE Computer (May 1979).

[2] Astrahan, M.M., Kim, W. and Schkolnick, M., Evaluation of the System-R access path selection mechanism, Proc. IFIP Conf. Tokyo, Melbourne (October 1980).

[3] Stonebraker, M., Wong, E. and Kreps, P., The design and implementation of INGRES, ACM TODS, 1,3 (September 1976).

[4] Stonebraker, M., Retrospection on a database system, ACM TODS, 5,2 (June 1980).

[5] Bonfatti, F., Maio, D., Spadoni, M. and Tiberio, P., An indexing technique for relational data bases, Proc. IEEE COMPSAC Conf., Chicago (October 1980).

[6] Hawthorn, P. and Stonebraker, M., Performance analysis of a relational data base management system Proc. ACM SIGMOD Conf., Boston (May 1979).

[7] Yao, S.B., An attribute based model for data base access cost analysis, ACM Trans., 3,1 (March 1977).

[8] Yao, S.B., Optimization of query evaluation algorithms, ACM Trans. on Database Systems, 4,2 (June 1979).

[9] Schkolnick, M., A survey of physical data base methodology and techniques, Proc. Very Large Databse Conf., Berlin (September 1979).

[10] Schkolnick, M. and Tiberio, P., Considerations in developing a design tool for relational DBMS. Proc. IEEE COMPSAC Conf., Chicago (November 1979) and in Data Base Management in the 1980's, IEEE cat. n. EHO-181-8 (1981).

[11] Schkolnick, M. and Tiberio, P., A note on estimating the maintenance costs in rela-

tional database, IBM Res. Rep. (1981).

[12] Bonfatti, F., Maio, D. and Tiberio P., Selezione degli indici in una base di dati relazionale, Proc. AICA Conf., Pavia (September 1981).

[13] Bonfatti, F., Maio, D. and Tiberio P., Alcune considerazioni sul calcolo dei costi di esecuzione delle interrogazioni in una base di dati relazionale, Proc. AICA Conf., Pavia (September 1981).

[14] Trivedi, K.S. and Kinicki, R.E., A model for computer configuration design, IEEE Computer, 13,4 (April 1979) 47-54.

[15] Lazzarati, G. and Maio, D., Simulation tool to design distributed computer systems, Jrnl. of Modelling and Simulation, 1,2 (1981) 163-167.

[16] Bucci, G. and Maio, D., Merging performance cost-benefit analyses in computer system evaluation, IEEE Computer Magazine (September 1982).

[17] Sevcik, K.C., Data base system performance prediction using an analytical model, Proc. of IEEE Very Large Database (1981).

[18] ACM Computing Surveys, vol. 10, no. 3 (1980).

[19] Astrahan, M.M. et al., System-R, a relational approach to database management, ACM Trans. Database Syst. 1,2 (June 1976).

[20] Selinger, P.P. et al. Access path selection in a relational database system, ACM SIGMOD Conf., Boston (May 1979).

[21] Youssefi, K. and Wong, E., Query processing in a relational database management system, Proc. of Very Large Databases Conf., Rio de Janeiro (October 1979).

[22] Bonfatti F., Maio, D. and Tiberio P., A separability-based methodology for secondary index selection, to appear (North-Holland).

[23] Blasgen, M.W., et al., System-R: an architectural overview, IBM System Jrnl., 20,1 (1981).

[24] Blasgen, M.W., Eswaran, K.P., Storage access in relational data bases, IBM Systems Jrnl., 16 (1977).

[25] Sacco, G.M. and Schkolnick, M., Buffer management in relational database management systems, IBM Rep., RJ 3354 (December 1981).

[26] Gray, J.N. and Watson, V., A shared segment and inter-process communication facility for VM/370, IBM Res. Report, RJ. 1579.

[27] Bard, Y., The VM/370 performance predictor, ACM Computing Surveys, 10,3 (September 1980) 333-342.

Data for Science and Technology, P.S. Glaeser (ed.)
North-Holland Publishing Company

THE RELATIONAL INTERFACE TO THE CDS ISIS DATA BASE

J. Bańkowski, K. Biesaga, J. Dobosz, A. Figura, S. Romański,
J. Rybnik, M. Sulej, A. Wojewoda, E. Zabża-Tarka

Institute for Scientific, Technical and Economic Information
00-950 Warsaw, Al. Niepodległości 188, Poland

A relational interface to the information retrieval system CDS ISIS is presented as an auxiliary tool providing facilities not covered by the basic system. Due to the proposed interface users can obtain aggregated and statistical data characterizing the overall structure and content of objects described by original data base records. A relational data base is produced by extraction of data from the CDS ISIS data base. For various transformations and tabular printout descriptions the user language based on relational algebra is provided. The proposed interface is a useful tool in all domains where data bases containing data of several types (i.e. numerical data and textual data) are used. Particularly it is suitable to all users examining various data relations to obtain information implicitly defined by the data base content.

1. GENESIS

Most present-day information retrieval systems (IRS) usually supply information either as references (descriptions, abstracts) or factual data (source information). For this reason the main functions of typical IRSs concern:
1) retrieval of information in response to users' questions,
2) data base maintenance (i.e. data entry verification and updating),
3) edition of retrieved data, system catalogues, indexes, etc.

Furthermore, answers to users' questions are usually built up from data taken word by word from the data base. In some applications, however, a need for aggregated and statistical data characterizing the overall structure of stored information or cross-dependencies between descriptions of particular objects may appear. To derive such information, several facilities, not available in typical IRSs, may be needed, e.g.:
- computation of totals, averages and other user-defined values,
- interlacing of retrieval and computations,
- auxiliary input for user-defined classifiacation patterns,
- data restructuring,
- specialized printouts.

On the other hand, the non-programmer users very often encounter difficulties in understanding the data base structure, access methods and other implementation-dependent problems. This results in the need to formulate an implementation-independent and conceptually simple data model.

As a solution to this problem, implementation of a specialized interface to an information retrieval system is proposed.

One of the most natural ways of viewing data is in the form of simple tables. As the very convenient mathematical model of a table is a relation, the relational model of data (1), (2) has been recognized as the most suitable for this purpose. Transformations carried out on relations by means of relational algebra operators may be interpreted by non-programmer users in terms of operations acting on rows and columns of tables. This enables them to formulate a wide range of their application problems in an easy and transparent way.

It has been assumed that the proposed system should closely cooperate with IRS to complement its facilities rather than to replace them. Such an IRS, the CDS ISIS package, was chosen. In general, the CDS ISIS package covers only the basic features of IRS and does not allow the user to obtain more sophisticated information processing than sorting.

Therefore, according to user needs, for data base maintenance, bulletin and index editing or mass data entry and verification, CDS ISIS procedures are called upon, and for more complex computations and printouts, relational data manipulation programs are executed. In other words, when a simple list of documents (facts, descriptions) is needed, CDS ISIS procedures are used and when cross-dependencies between data base objects or some statistical information is desired, the relational interface procedures are activated. This approach enables us to avoid most of the data base maintenance problems and allows us to focus all efforts on data manipulation and aggregation processes.

2. THEORETICAL BACKGROUND

As mentioned above, the theoretical concept of the presented project is based on Codd's relational model of data (1),(2). The elementary unit of information accessible in the relational model of data is called a p-value (primary value) and is from the system's point of view an indivisible piece of information. P-values correspond to single values of ISIS record fields or subfields, whereas domains refer to sets of all possible values of fields or subfields in the data base records.

Associations between p-values are represented by Codd's relations. A relation is defined over a finite set of attributes, each one of them associated with a domain. Several attributes may be associated with the same domain. Attributes correspond to the names of ISIS fields or subfields and, in the simplest case, a relation may directly correspond to a file of records in the original data base.

Let $A = \{a_1, a_2, \ldots, a_n\}$ be a set of attributes and D_{a_i} be a domain associated with the attribute $ai \in A$. A set $T = \{P_{a_1}, P_{a_2}, \ldots, P_{a_n}\}$ of n p-values where $P_{a_i} \in D_{a_i}$ is called a tuple. Any finite set of tuples T is called a relation over A.

Such a definition of a relation shows the direct link between the record, tabular and relational approach to data representation. The correspondences between the mentioned data views are shown below:

record data view	tabular data view	relational data view
simple value	table element	p-value
set of all possible values of a field (subfield)	set of all possible values of a column	domain
field (subfield) name	column name	attribute
record occurrence	row	tuple
set of all occurrences of a record of given type	table	relation
IRS data base	collection of tables	schema

As both a tuple and a relation are sets, so in tabular representation of data, tables which differ in order of their rows or columns represent, in principle, the same information.

3. SYSTEM PRINCIPLES

The implemented version of the system consists of two separate parts:
- data extraction programs, extracting the data specified by the user from the CDS ISIS files into a separate relational data base,
- procedures operating on the relational data base.

According to the system implementation principles, there are two phases of system activity: generation of an appropriate relational data base and execution of user-defined application programs. The user controls these processes using Data Description Language and Data Manipulation Language statements, respectively (3), (4), (5).

In the presented system the first phase is achieved by a two-level sequential process. On the first level, a description of the relational model, called Schema, is created. It is formulated by the Data Base Administrator and consists of relation and attributes descriptions, and data extraction specifications. The transformation of CDS ISIS records, fields and subfields into relations and their attributes is performed according to the set of prespecified rules. On the second level, users define their own Subschemata, i.e. subsets of relations of the Schema. They may pass Schema relations to their Subschemata without any modifications or may restrict Schema relations to sets of attributes they are interested in. They may also redefine attribute descriptions and define and input auxiliary fixed-value relations, which may be used for data aggregation. On this level the knowledge of ISIS data base details is not necessary. For Schema and Subschemata descriptions a specialized, CDS ISIS-oriented Data Description Language (DDL) is used. According to specifications from the Schema and Subschemata, for each Subschema, a separate relational data base is generated.

This two-level approach provides valid transformation of the hierarchical CDS ISIS model of data into a relational one and separates the non-programmer user from implementation and access methods problems. Furthermore, as the user operates on his own data base and has no possibility to affect the source information or other user's data, the system is foolproof.

After the relational data base has been created, the user may process it by means of data manipulation programs written in the Relational Data Manipulation Language based on relational algebra. The RDML allows the user to perform various transformations of the Subschema relations creating new ones according to arising needs. New relations defined as the result of the execution

of the RDML statements may be one of the following types:
- temporary - accessible only in the scope of the data manipulation program,
- permanent - stored by specific RDML statement and accessible to the other data manipulation programs working on the same Subschema. Due to this feature data transmission between data manipulation programs is possible.

In this phase of data processing all of the proposed extensions of IRS functions are realized. Manipulation programs enable the user to obtain very sophisticated information representing either internal record data dependencies or links between particular records.

4. RDML STATEMENTS

There are three types of RDML statements (4):
- the assignment statement used to create new relations by means of relational operators,
- the auxiliary statement used for relation value and description maintenance (e.g. to store a relation on a disk between subsequent runs, to change the status of a relation, etc.)
- the print statement used to describe various forms of tabular printouts.

The assignment statement assigns a value of a relational expression being its right part to a relation specified as its left part. Relational expressions are built up from names of relations and attributes and relational algebra operators listed below:

+ union (set theory union) -
produces a relation which contains tuples of both source relations,

intersection (set theory intersection) -
produces a relation which contains identical tuples of both source relations,

- difference (set theory difference)-
produces a relation which contains those tuples of the first source relation, which have no identical counterparts in the second source relation,

: selection -
produces a relation which contains those tuples of the source relation, which satisfy criteria specified by a logical expression,

% projection -
produces a relation which contains only those attributes of the source relation, which were specified by the user,

* natural join -
produces a relation, which contains all attributes of the source relations and its tuples are generated as a Cartesian product of all groups of those tuples, which have identical values in common attributes,

$ glueing -
produces a relation which contains those attributes, which were specified by the user and new attributes, the values of which are obtained as the result of the computation of a specified function over attributes of the source relation.

The first six operators of the presented list are traditionally associated with relational algebra and in RDML their standard meaning is preserved.

Glueing is an operation of special importance. It allows the user to perform arithmetic operations on tuples and to obtain aggregated data and statistical information. A general form of glueing is:

A = B $ (X,Y)

where A,B are relation names, X is a list of copied attributes and Y is a list of definitions of new attributes, and each definition has the following form:
Z = F(E), where Z is a name of a new attribute, F is a function name and E is an expression (usually arithmetic). The execution of this operation follows the three-step algorithm:
1) for each tuple of relation B all E expressions are evaluated,
2) tuples, which have identical values of X attributes, are grouped together,
3) each group is reduced into a single tuple with common values of X attributes; Y attribute values are evaluated using functions F; the functions operate over the set of values obtained for each tuple by evaluating an appropriate expression in a given group.

Glueing is similar to projection except that values of omitted attributes may affect values of new attributes (in projection, values of omitted attributes are entirely dropped).

There are five standard F functions that may be used in the definition of new attributes:

AVG - arithmetic mean of evaluated values,
MAX - maximum evaluated value,
MIN - minimum evaluated value,
SUM - sum of evaluated values,
MPL - product of evaluated values.

The above list may be extended. In the next release of the system simple statistical functions will be implemented. Note that the COUNT function is implemented indirectly by the particular use of the SUM function, i.e. SUM (1). In this case, the value of the expression for each grouped tuple is 1, therefore, the value of the function is exactly equal to the number of tuples in each group.

From the non-programmer user's point of view all operations of relational algebra may be understood as simple operations on table elements, e.g.
- selecting those rows of the table which meet given criteria (selection, intersection, dif-

ference),
- selecting those columns of the table which contain information the user needs and dropping the ones of little importance (projection, glueing),
- joining source tables on rows (sum) or columns, (natural join) in order to obtain a new table.

The second statement of particular importance is the print statement. It allows the user to describe the external representation of a computed relation.

In general, a relation is printed in the form of a table. In the simplest case the table consists of all tuples of the relation outputted one by one and preceded by the relation and attribute names.

More complicated forms of output are also possible. The user may define a division of a relation on columns, subcolumns, rows and subrows. A value of an appropriate attribute is a header of each of those elements. Furthermore, the user may specify additional elements (i.e. additional column, subrow, etc.) for aggregated data, which are obtained as a function over all values in a given element. It enables us to produce totals, averages and other user-defined functions for each element of the table.

The format of the exteral representation of data (column widths, significant positions of numerics, etc.) is also controlled by the user.

5. AN EXAMPLE

Although the relational interface and, particularly the RDML, was not designed for scientific data processing purposes, the example given illustrates a RDML program processing the data obtained in a biological experiment.

The experiment described was carried out on isolated rabbit aortic strips. A lot of strips were tested for average experimental results. Each of the strips was exposed to the same series of drug doses. The strip response to the drug dose was compared with the response to the dose of a control substance. For every strip, a series of pairs of responses - the first one to the drug dose and the second one to the control substance dose - was obtained as the result of the experiment. The experiment was repeated for the series of different control substance concentrations.

Exemplary results of the experimental series presented in the following table were evaluated to obtain the dose-response curve. The strip response to the drug dose was obtained as a percentage of the control substance response and then the mean strip response was calculated.

STRIP No.	CONTROL DOSE CONCENTRATION	DRUG DOSE	CONTROL RESPONSE	DRUG RESPONSE
001	1.0	1.5	3.0	0.0
001	1.0	2.5	4.0	1.0
001	1.0	10.0	8.5	25.0
002	1.0	1.5	10.0	0.5
002	1.0	2.5	14.0	5.0
002	1.0	10.0	7.0	18.0
003	1.0	1.5	6.0	0.0
003	1.0	2.5	6.0	2.0
003	1.0	10.0	4.0	10.0
004	2.0	1.5	10.0	0.0
004	2.0	2.5	9.5	2.0
004	2.0	10.0	12.5	20.0
005	2.0	1.5	18.0	0.0
005	2.0	2.5	20.0	4.0
005	2.0	10.0	8.5	15.0
006	2.0	1.5	20.0	0.5
006	2.0	2.5	27.0	5.0
006	2.0	10.0	17.0	25.0

Let the Subschema, called EXPERIMENT, contain the relation RESULTS that refers to the table presented above. The following RDML program performs the described evaluation of experimental results and presents them in the form of a table.

```
INVOKE EXPERIMENT
   REL_A:= RESULTS $ (STRIP_NUMBER,
           CONTROL_DOSE_CONCENTRATION,
           DRUG_DOSE,N_RESPONSE:=SUM
           (DRUG_RESPONSE/CONTROL_
           RESPONSE*100));
   REL_B:= REL_A $(CONTROL_DOSE
           CONCENTRATION,DRUG_DOSE,
           RESPONSE:=AVG(N_RESPONSE));
   PRINT HEADER='MEAN AORTIC STRIP
                RESPONSE'
   ATTR RESPONSE→'MEAN RESPONSE'
   FROM REL REL_B
   COL=DRUG_DOSE_CONCENTRATION→
        'CONTROL DOSE CONCENTRATION'
END
```

MEAN AORTIC STRIP RESPONSE

CONTROL DOSE CONCENTRATION	DRUG DOSE		
	1.5	2.5	10.0
1.0	1.6	31.3	267.0
2.0	0.8	19.8	161.1

REFERENCES

[1] Codd, E.F., A relational model of data for large shared data banks. CACM 13(6), 377-387 (1970).

[2] Codd, E.F., Extending the data base relation-

al model to capture more meaning. ACM Trans. Database systems 4(4), 397-434(1979).

[3] Dobosz, J., Romanski, S., Szymanski, B., Zabza-Tarka, E., Relacionnyj dostup k bazam dannych informacionnoj sistiemy: issliedowania, razrabotka, primienienie. Nauczno-tiechniczieskaja informacija 2(4), (1982), in Russian.

[4] Dobosz, J., Symanski, B., Wojewoda, A., Jezyki uzytkownika w systemie relacyjnego dostepu do baz dannych pakietu CDS ISIS. Prace IPI PAN, no. 398, (1980), in Polish.

[5] Dobosz, J., Szymanski, B., An implementation of relational interface to an information retrieval system. Inform. Systems 6(3), 219-228 (1981).

Data for Science and Technology, P.S. Glaeser (ed.)
North-Holland Publishing Company

STATUS AND TRENDS OF NUMERIC DATA BANKS

John Rumble, Jr.

*Office of Standard Reference Data, National Bureau of Standards
Washington, DC 20234, U.S.A.*

This paper discusses the present-day status and trends of scientific numeric data banks. The main emphasis is on the user interfaces to data banks which provide the extra capability to computer-readable data banks that distinguishes them from the paper data banks.

1. INTRODUCTION

Sometime in the 1450s, Johann Gutenberg printed the Vulgate Bible, the first large book set in movable type. The impact that this particular book had upon the development of human society is probably small, but there is no doubt that the method of publishing it, namely, movable type, revolutionized the dissemination of human thought. In actual fact, the final product, a visual version of people's thought, was not different in any way from handwritten or carved block typeset manuscripts. But the method added orders of magnitude of possibilities to the number of copies, to the number of different publications, and to the exposure of the masses to these publications.

About 500 years later, computers started a second revolution in the dissemination of human thought. In many ways, the impact of computers will be far greater than that of the invention of movable type. Indeed, one cannot even conceive of what the next 500 years will bring. However, the past 30 years have already changed so much and the immediate future holds forth so many changes that everyone realizes the impact of computers will be tremendous.

With this broad perspective in the background, I want to examine in detail what I personally see as some of the present-day impacts of computers on the dissemination of numeric scientific data and how they might change in the future. This paper will not present a catalog of available data banks; those available in the United States and Western Europe are covered extremely well in at least two recent publications. (1) Nor will I cover textual data banks. The concern will be strictly on numeric data for scientific and technical purposes.

2. A BRIEF HISTORY

The common stages that occur in developing a data bank are given in Figure 1 and, in fact, these stages outline the historical development from the earliest computers to the present. Obviously, the first files were those created by a user for his own use; for example, a file of mass spectral peaks. Just as obviously, if the file was of some general use, his friends and colleagues wanted to access it. Usually this went smoothly, and little or no documentation was needed.

But if the file was really useful, persons outside the institution where the file was built became aware of its existence and wanted to use it. Thus, scientific data exchange began, though at a slower pace than software exchange. By the mid- to late 1960s, groups were exchanging several types of data worldwide on a regular basis--for instance, neutron cross-section data. These were the days of paper tape, boxes of cards, and unrecoverable tape-read errors. By then, also, the problems of exchange formats and documentation became evident.

In the early to mid-1970s, telecommunications became more widespread, and two important changes took place. First, people with a modem and terminal could make a telephone call and directly access a computer far away. Second, one computer could routinely access a remote computer in the same way. Public access networks and computer networking have since become commonplace. As this happened, two parallel developments began which today are what I see as the most important issues concerning numeric scientific data banks, namely, combining data banks to create data systems and user interfaces to data banks.

The vocabulary of computerized data collection has become so murky and complex that it is necessary to give my definitions. "Data bank" refers to a machine-readable data collection on one subject--for example, the Evaluated Neutron Data File (ENDF), the Cambridge Crystallographic File, the NIH-EPA-NBS-MSDC Mass Spectral Data Base. A data bank may have several "data elements" which can be related in many different ways, and a data bank may or may not have "software" associated with it. By "software" is meant the computer programs that search, manipulate, or otherwise use the data bank. A "data system" is a collection of data banks and associated software which are accessible to a user via one computer host.

By the term "user interface," I mean not only all the information necessary to make use of a data bank, such as query languages and other such information given to a user on a terminal, but also all the software that uses a data bank and is accessible to the user.

With these definitions in mind, I will now discuss two subjects: 1) combination of data banks to make data systems, and 2) user interfaces to data banks.

3. COMBINING DATA BANKS IN DATA SYSTEMS

The concept is simple: to provide a user with access to two or more data banks via one computer. One obvious advantage is that very often the end user needs several different types of data at one time. For example, an analytical chemist may wish to have the melting point, color, crystal structure, and C^{13} NMR spectrum for a substance. An automotive engineer may want the mechanical performance properties for several alternative alloys and plastics. Just as is presently the case with printed handbooks, data banks with this different information will be built by many different groups.

The motivation is thus simple, but the problems are complex and arise from both user desires and technical considerations. Over the past 10 years, many successful data systems have been built and used and some very important lessons learned about data systems.

The first problem comes from the fact that numeric scientific data almost always refers to an entity, a chemical substance, a material, a mineral, or a strain of bacteria. Unfortunately, in scientific and technical disciplines, the rules of nomenclature have been set only well after multiple common and technical names have been used. Consequently, to be effective, any data system which contains data on one substance in two or more data banks must insure a uniformity of nomenclature. Acetone cannot be CH_3COCH_3 in one and acetone in another. The use of unique registry numbers has progressed, but problems still exist.

The problem above leads directly to another problem--that of different data bank formats. Each individual data bank is usually built in a format tailored to the type of data, computer software and hardware requirements, and human whims. Two approaches have been taken to address this issue.

One approach is for the data system to define a format and for each individual data bank producer to reformat its data into the data system format. This approach has been used very successfully for bibliographic and other textual data; for example, the Lockheed Dialog System. The EXFOR System, used by the International Atomic Energy Agency (IAEA) and various national nuclear data centers, is an example of a successful use of this approach for numeric data. However, many numeric scientific data vary so much in their form that it is difficult to imagine generic data formats becoming prevalent.

The other approach has been used by the U.S. NIH/EPA Chemical Information System (CIS) and others and involves accepting each data bank in its own format and tailoring the data system software as necessary. While this makes life much easier for data bank producers, it can greatly add to the cost and time of the integration process.

Two other problems have started to come into the foreground with respect to integrating data banks into data systems, both of which have traditionally been areas scientists have stayed away from--economics and politics. (2) It is obvious that there is an initial investment that is needed to put a data bank into a data system. Who pays for this work is a very important issue. To date, very few data banks can recover that investment from user fees.

Another important point is the distribution of user royalties between the data bank producers and data system operators. This is especially vital in the area of engineering properties where traditionally various professional societies have been a major source of data and have used revenue from their data publications activities to help support other professional activities.

The political implications of combining data banks into data systems, especially where the country of origin of the two is different or where the system is offered internationally, are just now being addressed. Hopefully, in the long run, scientific numeric data will be recognized for what it is and will not be considered national property. One is less sanguine about the posture that will be taken on engineering and more applied data.

There have been some marvelous successes with respect to creating large useful data systems; much more activity is about to burst forth. Some of these efforts are described more fully in this volume and will be demonstrated at this CODATA Conference. Table I contains a partial list of these, and some will be mentioned in more detail in the next section.

4. USER INTERFACES TO DATA BANKS

It is in this area that data banks have made the greatest changes in the past few years and will provide new developments over the next 10 years. The basic reason for this is that the real attraction of computer dissemination of scientific numeric data, or almost all data for that matter, is not the table look-up uses that are associated with printed pages, but the extra capability which computers can add. (3) These are summarized in Figure 2. The first three developments on this list--frequent updating, convenient use, and easy extrapolation and interpolation--have long been recognized as advantages of computerized data banks. Now, after several years' experience with many data banks, some conclusions can be made regarding their realization.

4.1 Updating

Updating of data banks has turned out to be not as easy as once imagined. There is a lot of computer and people overhead required even for routine monthly updates. The flow of information, especially numeric data, is usually intermittent and not continuous so that a sufficient quantity of data to make updating worthwhile is available on a semiannual or annual, but not weekly, basis. Thus, updating scientific numeric data banks has turned out to be similar to yearly updates to a handbook. This is especially true if a great deal of evaluation of the data is done.

4.2 Convenience

The convenience of having data banks on computers accessible from a scientist's or engineer's office is only now being realized in a major way. One reason is that only in the past five years have computer terminals really been sufficiently inexpensive. The other reason is the user "friendly" or rather "unfriendly" software associated with data banks and data systems. The multiplicity of logon procedures, query languages, thesauri and other interface peculiarities has been exasperating for many. However, it is evident, especially to data systems operators, that the user friendly interfaces must become friendlier or competitors will attract the users. As that happens, the use of search specialists will diminish because the average scientist and engineer, <u>in the long run</u>, wants to look up the data himself.

4.3 Missing Values

As for extrapolation and interpolation, or better, filling in missing data, this need has only been marginally met by existing data banks. As data banks cover more applied areas, this problem will have to be better answered.

4.4 Nomenclature

The next group of features, nomenclature options and data subbanks, have had much interest in the last few years, and things have significantly changed. For chemical data banks, the multiplicity of nomenclatures and structure has been successfully addressed by several large data systems in the United States and Western Europe including the CIS, Chemical Abstracts, and DARC, among others. In the USSR and Japan, other groups have made similar progress. In other areas, such as materials and biological substances, only the first steps have been taken. A major problem is interpreting areas where the nomenclature refers to a substance which can legitimately exist in a range of compositions, such as an alloy, polymer, etc.

4.5 Data Subbanks

By specialized data subbanks is meant smaller data banks which have been set up from larger and more comprehensive data banks and data systems--for instance, selecting from the complete NIH-EPA-NBS-MSDC Mass Spectral Data Base all mass spectra for pesticides. Another example would be information for titanium binary alloys including phase stability data and associated thermodynamic and crystallographic data. Under this category would be included partial or complete data banks leased or bought from a data bank producer(s) and put on a minicomputer. Also included would be the creation of a private data bank on a commercial data system.

These approaches have been tried, and variations are continuing to appear. Several groups lease or sell scientific numeric data banks for internal use. The JCPDS-International Diffraction Data Centre distributes its powder diffraction file and the NBS Crystal Data ID file in this manner. The Office of Standard Reference Data of the National Bureau of Standards has also available files on thermodynamic and mass spectral data. Many others similarly exist. In another approach, F*A*C*T, Canada, has made private data bank facilities available for use with their high temperature thermodynamic data system of which several companies take advantage.

As yet, few if any scientific numeric data banks are available for use on home computers or small office/laboratory minicomputers. Already though, several types of analytical instruments are available with data banks located on minicomputers inside the instruments, e.g., mass spectra and powder diffraction. This arrangement allows for rapid analysis and identification of unknown substances.

The last group of user interfaces concerns linking software to data banks either by direct input to external modeling software, by using modeling software itself as a data bank, or by data systems. Several significant achievements have already been made in these areas; the coming decades will bring more.

4.6 Input to Other Software

While "table" look-up accounts for a substantial portion of data bank usage, there is an ever-expanding desire to use data banks in connection with computer software to model various phenomena and processes, such as fusion plasmas, the flow of feedstocks through a chemical plant, and materials selection in computer-aided design, plus thousands more. To date, most of this modeling software has been developed by the groups directly concerned, and interfacing data banks to these external software systems is of considerable interest. For example, the U.S. National Academy of Sciences has been studying this problem for CAD/CAM Systems.

4.7 Software as a Data Bank

But this is just one side of the coin. The other is the casting of a data bank in the form of modeling software. For instance, the Fluid Properties Data Center of NBS has created a data bank on various thermophysical properties of hydrocarbon mixtures such as those found in liquid natural gas (LNG). There are data on over 50 hydrocarbon compounds, and properties such as viscosity or density can be determined for a range of temperatures, pressures, and concentrations. This data bank is distributed as modeling software rather than massive tables. Other examples are the Facility for Analysis of Chemical Thermodynamics (F*A*C*T) or the ASPEN project. Both of these systems distribute or allow access to software modeling of various chemical engineering processes, which depend essentially on underlying data banks. In reality, the software-data bank package is a kind of super data bank which will come into greater importance as scientists and engineers take advantage of existing investments in modeling software, combine that with numeric data banks, and make the resulting package available to the public.

4.8 Data Systems

The last interface feature is data systems, which have been discussed throughout this paper. These large systems will ultimately provide all the features mentioned above, and, in fact, some, such as the NIH/EPA Chemical Information System, have many already. Today these systems are still in their infancy, but there will be an explosion of such systems in the next 10 years and beyond. The limit to their growth is most likely the rather small number of numeric scientific and technical data bases available. For instance, only 42 machine-readable scientific numeric data banks are listed in Cuadra's guide [1]. (See Table II)

5. FINALE

When the more traditional paper data sources are converted to data banks, when computer hardware becomes cheaper and more available, and when data systems expand their capabilities and interface with the user the way the user wants, then the real age of computer data will arrive. It is instructive to note that it was several hundred years after the invention of movable type before the regular distribution of scientific numeric data began. It is clear that scientists and engineers won't have to wait that long for computers to answer their needs.

FIGURE 1.

Stages of Data Bank Development

Single User File
Multi-Users, One Machine
Public Access Networking, Data Exchange
Combining with Other Data Banks
User Friendly Interface

FIGURE 2.

User Interface Features Which Can Be Supplied By Computers

1. Frequent Updating
2. Convenient Use
3. Easy Extrapolation and Interpolation
4. Nomenclature Options
5. Specialized Data Subbanks Integration into Instrumentation
6. Direct Input to External Modeling Software
7. Modeling Software
8. Data Systems

TABLE I.

Partial List of On-Line Data Systems Which Have Numeric Scientific Data

System	Country
Chemical Information System	U.S.
ChemShare Corp.	U.S.
GEISCO	U.S.
GIDSfT	Fed. Rep. Germany
INKA - Karlsruhe	Fed. Rep. Germany
F*A*C*T	Canada
Thermodata	France
University Computing Company	U.S.
United Information Service, Inc.	U.S.

Some On-Line Data Systems Which Have Indicated an Interest in Numeric Scientific Data

System	Country
BRS	U.S.
Chemical Abstract Services	U.S.
Dialog	U.S.
Japan Information Center of Science and Technology	Japan
SDC Search Service	U.S.
Telesystems - Questel	France

TABLE II.

42 Scientific Numeric Data Banks Now On-Line

Chemistry	19
Geology	2
Toxicology	4
Hydrology	3
Materials	4
Nuclear	2
Biomedical	1
Metrology	2
Chemical Engineering	5

REFERENCES

[1] Cuadra, R. N., Abels, D. M., and Wanger, J., Directory of Online Databases, Volume 3, Numbers 3 and 4 (1982). Available from Cuadra Associates, 2001 Wilshire Blvd., Suite 305, Santa Monica, CA 90403 U.S.A.

Directory of Online Information Resources (1981). Available from CSG Press, 11301 Rockville Pike, Kensington, MD 20895 U.S.A.

Williams, M. E., Lannen, L., and Robin, C. G., Computer-Readable Databases. A Directory and Data Sourcebook (1982). Available from Knowledge Industry Publications, Inc., 701 Westchester Ave., White Plains, NY 10604 U.S.A.

Hilsenrath, J., Summary of Online or Interactive Physico-Chemical Numerical Data Systems, National Bureau of Standards, Technical Note 1122 (1980). Available from the Superintendent of Documents, U.S. Government Printing Office, Washington, DC 10402 U.S.A.

[2] See Proceedings of the Third National Online Meeting, New York (1982). Available from Learned Information, P.O. Box 550, Marlton, NJ 08053 U.S.A.

Heller, S. R., Economics of Online Data Dissemination, Proceedings of 7th International CODATA Conference, Kyoto, Japan (1980). Pergamon Press.

[3] Hampel, V. E., "Fact Retrieval for the 1980's," UCRL Preprint 85749, Lawrence Livermore Laboratory, Livermore, CA 94550 U.S.A.

Data for Science and Technology, P.S. Glaeser (ed.)
North-Holland Publishing Company

ORGANIZATION OF DATA BANKS IN THE U.S.S.R.

V.V. Sytchev, Yu.S. Vishnyakov, L.V. Gurvich, A.D. Kozlov, N.G. Rambidi

Soviet National CODATA Committee
Academy of Sciences of the U.S.S.R.

Our report represents first of all a brief review and is intended to introduce the specialists attending this lecture to the basic directions of the research carried out in the U.S.S.R. on the creation of data banks concerning material properties and fundamental constants. We shall discuss a number of projects, research and organizational measures directed at increasing the labor efficiency of the scientist, engineer and designer, and illustrate typical examples showing how the components of an integrated system under development cover all aspects of present-day scientific user activities We shall also outline some of the modern approaches to solving the arising problems.

The automation of scientific research, as in other countries, is realized simultaneously in several directions. The scientist today is provided with a number of automatic systems differing in their functional purpose, types of information circulating in them and user languages. Let us briefly describe these classes of systems.

1. Data collection and primary processing systems

These systems are designed to collect experimental data through the use of special gauges and enable on-line gathering of data directly from the subjects of research, transforming them into computer readable formats by executing the simplest conversions. As a rule, such systems are based on micro-computers with the developed peripheral facilities.

2. Research systems

These systems are intended to make scientific and engineering computations as well as to answer traditional inquiries for direct scientific results such as the arrays of numerical values, values of constants, etc. which comprise the factual data banks, and permit the user access to the structured information in conversational mode.

3. Application software packages

These systems already have a long history of existence and, in some cases, are connected to factual data banks of the simplest structure. They represent sets of standard processing procedures frequently employed by a large number of users, for example, the programs of statistical processing, numerical methods and strength analyses. Written in various algorithmic languages, the components of these packages can be connected to the user's own program. The packages are distributed in the form of standard libraries of routines tailored to the corresponding operating systems.

4. Information retrieval and archive systems

The typical versions of these systems can be represented by systems storing large files of abstracts of scientific papers, books, reports and dissertations with additional structures of linguistic support - the key words, thesauri, author and subject indices. Such systems are of great importance, especially at the first stages of scientific work when information retrieval assists in orienting oneself correctly within the studied subject field and partially prevents duplication of work. However, one also has the occasion to work with systems of this type at the final stages - when registering the results for presentation to information centres.

5. Publishing systems

This is the newest but a very actively developing class of system. These systems enable automation of one of the most labor-consuming stages of any scientific work, that is, publishing the results in polygraphic forms - books, booklets, preprints and journals.

The five listed types of systems are incompatible not only in format and language but also in organization. The basic inconvenience arising in dealing with various systems lies in the disconnection of the working cycle and the separation of different stages and operations for the user. Serious research work is being conducted on mutual compatibility of the developed systems and systems being created, as well as on meeting the present-day requirements imposed upon them, such as validation, security, correctness, minimum redundancy and modularity. The hardware on which automatic systems are based in the U.S.S.R. is rather diverse. For the most part, the hardware is comprised of standard devices and computers manufactured in the member countries of the Council for Mutual Economic Assistance (EC, CM, BESM-6, VIDEOTON, etc.). Some systems include hardware produced in other countries. In practice, all the systems are able to be connected to the computer network. Local networks and networks using intercity communication lines are being created. The architectural designs used in some of the systems are greatly varied. It suffices to mention examples of all the types of data structures in practice known in the world: hierarchical, relational, network and mixed ones. It is obvious that efficient separation and integration of the data resources are not possible without special compatibilities providing the interface between individual systems. This requires the generation of communicative formats for the data presentation and special intra- and interdisciplinary language-mediators between all the network systems. At present, both theoretical research and experimental work are under way in this direction. A user should be able to use comprehensively all the resources available in the network as a continuous working process.

Let us illustrate the continuity of the user's working process cycle with the following example. In accordance with the specified program, the system performs an experiment and collects the needed data arrays which are available to the user in interactive mode as a data base. The user executes the required semantic tasks of data conversion and starts his own procedures. From the terminal keyboard or from other systems, entry and editing of data are made for accompanying a text. Then, using newly collected and previously stored data, arrangement of the final document is made and is then moved in polygraphic form at the needed moment to the photocomposing devices. We may consider that the central idea of an integrated process lies in the possibility of obtaining the user's activities' results in the form of a finished product available for consumption by a wide circle of network subscribers. Of considerable importance are the role and responsibility of specialists in the support and maintenance of their own activities' results. Processing, interpretation and analysis of the data validation are mainly in the hands of professionals in the subject fields. This releases a great part of the centralized network services' resources for boosting the user developments. Problems of integrating and differentiating the activity products' maintenance and support functions define the efficiency of personnel use and financial resources. This especially holds true for the organization of the system distribution, testing and extension, as well as for user training problems. The programming languages' history points to the fact that the creation of a common general-purpose language is problematic. A more realistic approach seems to be the decision to develop a certain internal language-mediator performing all-system functions. This does not prevent the user from building a data description and manipulation language most fully satisfying his individual needs and tastes. In this case, the system should be provided with a user local language - language-mediator interface.

In the U.S.S.R., such work is carried out as a component of the "ACADEMNET" (academic network) project. The local network is adjusted to the most efficient linguistic, mathematic and hardware support of the user's subject field under careful observation of the results exchange discipline with the integrated network. A simple analogue of this idea is the problem of arranging books with different formats on standard bookshelves. You are able, in accordance with your library stock, to assemble the needed configuration of racks from a generalized set of standard constructions. Of course, the conception stated above requires new methods of data organization, information retrieval strategy and data processing.

It should be noted that in the U.S.S.R. intensive development of the recursive data retrieval and allocation mechanisms are underway and can be outlined as follows. There are several entities to be defined by the values of K attributes. Description of the entity according to a specified condition is required. Allowance is made, as a retrieval condition, for monadic predicates "more", "less" and "equal" type as well as for their combinations with the use of conjunction, disjunction and negation signs. The method's idea lies in finding the rated recursive function realizing the representation of K, that is, the mapping of a K-dimensional cube on a one-dimensional cube, which, in turn, is mapped in a unique technique on the linearly ordered set of computer memory cells. The entity retrieval is reduced to computation of the appropriate recursive function. This method enables us to update the files.

Let us now present another direction realizing the integrated technique of supporting scientific and practical activities. It is based on the following main principles: data are presented in standard constructs; standard data constructs are also processed by standard procedures - filters in accordance with a uniform technique; data constructs and filters are distributed in homogeneous layers with the help of standard procedures of the layers' distribution, synchronization and claying; final product - the result of activities represents a collection of fragments in the corresponding layers. An example of the standard data construct is the "grape" representing a particular case of the tree in which branching is allowed at the bottom level only. This enables preservation of the semantic properties of hierachical and relational structures, solving the problem of correct processing of multi-valued correspondencies and transitive dependencies. In this case, wide use is made of the theory of abstract data types which is well known to the scientific world. The data processing is reduced to starting a series of standard filters with each of them performing one simple operation on an optional data fragment. Sequential execution of the series of filters is equivalent to the mathematic and symbolic conversions of arbitrary complexity. In some senses, the system for the user is analogous to a pocket calculator that carries out operations not on individual numbers but on large groups of data in one step. The concept of reversibility of performed operations becomes a principal problem that substantially affects the general methodology of storage and transmission of data through the network. The layers are shaped based on such criteria as homogeneity, validation, access category and format characteristics of the information. When following the standardized process, one and the same layer takes part in the generation of several products intended for various users. From this standpoint, the reference book entitled "Thermodynamic Properties of Individual Substances" of the IVTANTERMO data bank (to be discussed later) is represented by a set of the following main layers: optional data such as fundamental physical constants, atomic weights of elements, atomic masses and spins of isotope nuclei, etc.; constants needed for calculating the thermodynamic functions of substances in a wide temperature range (molecular constants for gases, characteristics of phase transitions and polymorphic transformations, values of thermal capacity and changes in enthalpy for the substances in condensed state), these data are based on critical analysis of all primary literature; primary experimental data on the equilibrium constants of chemical reactions and saturated vapor pressure of each substance; intermediate constants obtained as a result of processing and expert data analysis; thermochemical constants (the enthalpies of formation and sublimation, dissociation energy, ionization potential, etc.) taken for each substance based on expert analysis of the primary literature; calculated tables of the thermodynamic properties for a wide range of temperatures and their satellite data in the format adopted in the "Thermodynamic Properties of Individual Substances"; documentation pictures of predictive type; cross references connecting the text layers of various subject fields.

Let us now consider the general characteristics of some data banks in the U.S.S.R. today. Some of these banks represent departmental, and others - institutional, systems; a number of systems are concentrated in problem oriented information centres. The individual systems are based on various standard aids of system support. Among them we can distinguish as analogues to foreign systems (OKA, KAMA) and native ones INES, DIS, GTO, COMPAS and others. The all-state systems also exist. Data banks and automated information systems on physical constants, properties of substances and materials are organized in the U.S.S.R. within the framework of the State Service of Standard Reference Data. The following problems are considered when determining the scope of systems and data banks, their number, functions, tasks and modes of functioning: analysis of the needs for data in science, technology and production, establishment of banks according to scope to reflect the results of this analysis; certification of data bases in banks; automation of base formation, development of data tables and data dissemination, organization of a subscription service for consumers.

Systems and data banks similar in scope are combined into topically oriented subsystems of the GSSSD. At present, such systems combine data bases on properties of metals and alloys, wood and lumber, petroleum and petroleum products, polymers, solutions, food products, etc. Data bases may include an approximately identical nomenclature of substances and materials, but differ in the nomenclature of properties determined in a system of data banks along the parameters of state, etc. Organizational and functional uniformity of data centres within the GSSSD is assured by a complex of methodic documents facilitating a uniform technical policy for the establishment and development of the centres as well as for compatibility of information, programs and technology. At present, the GSSSD has developed the following documents: 1. Standard (draft). Automated information system on physical constants and properties of substances and materials; 2. GSSSD basic classifier on properties of substances and materials; 3. GSSSD recommendation. The compilation of a local classifier on properties of substances and materials; 4. GSSSD recommendation. Requirements for the contents, the compilation of a technical assignment for setting up an automated information system of the GSSSD data centre; 5. GSSSD recommendation. The formation of data bases on properties of substances and materials using regulatory technical documentation in automated information search systems; 6. GSSSD recommendation. Temporary procedure of the certification of data bases of the GSSSD automated data centres; 7. Classifier of properties of polymers; 9. Classifier of properties of steels and alloys.

Among diverse automated systems or data banks that provide the basis for the GSSSD data centres, the following can be given: automated data bank on thermophysical properties of technically important fluids at the Moscow Institute for Energetics and at the All-Union Research Centre of the GSSSD; automated all-union system of thermophysical documentation (AVESTA) with the data centre on thermophysical properties of hydrocarbons and petroleum products (Kiev); automated roentgenographical computer system for inorganic substances and materials (ARIS) of the Institute for Physical Metallurgy, the Ukranian Academy of Sciences; IVTANTERMO data bank at the Institute of High Temperatures in the U.S.S.R. Academy of Sciences; data bank of the Data Centre on Atomic and Molecular Constants.

Information on several systems and data banks is presented below. The Data Centre on Atomic and Molecular Constants functioning at the All-Union scientific research centre is engaged in studying the properties of surface and vacuum and a base of data on molecular constants has been created here. The fullest and most continuously replenished base is one of experimental data on geometrical constants of diatomic molecules and ions in main and electronically excited states and on the geometrical configuration of nuclei and internuclear distances of multiatomic molecules, as well as the base of data on molecular constants obtained by a nonempirical method of quantum mechanics for diatomic and simple multiatomic systems. Creating large data arrays is carried on in accordance with the following principles: specific sequence of the molecules' location in the common data array enabling it to be considered as a sequentially alternating series of similar compounds; maximum availability of information and convenience from the standpoint of numerical data retrieval by the user; specific form of writing the numerical information convenient for developing programs of data storage and accumulation with help of the computer.

In the Centre, two card catalogs have been developed as one of the methods of storing numerical information: one of them represents a base of experimental data on internuclear distances of diatomic molecules and on geometrical configuration of the nuclei of multiatomic molecules. The numerical data are stored on bibliographic cards with each of them containing the following information: for diatomic molecules - a chemical formula of the molecule, type of the electronic state (main or excited), energy corresponding to the electronic transition of the molecule from the main state to an excited one (for the excited states), internuclear distance, method of research and literature source. The second card catalog contains key information on all nonempirical calculations of diatomic molecules and simple multiatomic ones. The card catalog for diatomic molecules formed the basis for developing the information retrieval system created based on the U.S. Hewlett Packard 9640A computer. The information about molecular constants and methods of computation is written on magnetic disks and can be printed out in the form of a table. Each line of this table contains a chemical formula of the molecule, type of electronic state, full energy of the molecule, energy of excitation of the given electronic state, as well as the literature sources. Based on the data available at the Centre, the Centre specialists have worked out a number of reference tables under a common name of "Geometrical configuration of nuclei and internuclear distances of molecules and ions in the gas phase". Since 1978, these tables are submitted for certification as tables of standard reference data in the State Committee of Standards. The Centre experts have developed information material on the methodology to provide to organizations and specialists of member-countries of the Council for Mutual Economic Assistance concerned with reliable data on molecular constants. We received and answered more than 30 inquiries from the Council's member-countries. Later on, these tables of reference data on diatomic molecules certified in the U.S.S.R. as standard reference data were assumed to be a standard of the Council for Mutual Economic Assistance.

Another system is IVTANTERMO. The IVTANTERMO issues information on thermodynamic properties of substances in the form of a complete table for selection of individual temperatures or properties, as well as in the form of equations approximating these properties over the entire range of temperatures. The information can be received on various data carriers, including magnetic tape and in a format readable on the computer EC seria; experience has also been gained in working with IVTANTERMO through telephone communication lines. At the end of 1982, the data base with thermodynamic properties of 1100 substances will be delivered to the leading institutions of a number of departments as well as to the Siberian branch of the U.S.S.R. Academy of Sciences. In the future, these materials will be updated and expanded yearly.

In 1982, we will have finished the preparation and publication of a new, completely revised and almost four times larger edition of the "Thermodynamic Properties of Individual Substances" hand-book. In this edition, based on the critical analysis of 17 000 original works as well as on our own calculations and estimates, a selection has been made of the values of molecular, thermochemical and other constants needed for computing thermodynamic properties of substances in a wide range of temperatures. The themodynamic properties of 50 elements and more than 1100 of their compounds in crystalline, liquid and gaseous states are rated for temperatures ranging from 100 to 6000 K (for 230 gases - up to 10 000 or 20 000 K). Validation is characterized of all the rated thermodynamic functions taken for caluculating the constants. All the calculated properties and thermochemical constants represent a system of internally matched values.For the number of considered elements and their compounds, the validation of calculated properties, the range of temperatures and for many other characteristics, this edition surpasses all analogous reference books known in the literature. At present, computation has been made in IVTANTERMO of the thermodynamic properties of approximately 1200 substances - compounds of 52 elements. It is expected that in 1990 the number of considered elements and their compounds will reach 65 and 2000-2200 respectively. To realize these plans, development has been started on the magnitudes of key thermochemical values for 17 elements which have not yet been considered by the international working group on key values for thermodynamics. Simultaneously, development is underway of the program on experimental and theoretical research on molecular, thermodynamic and other constants required for calculating thermodynamic properties of new classes of substances as well as for improving some of the data already included in the IVTANTERMO system.

The above analysis shows that the existing data bases cover the entire spectrum of scientific activities. We have mentioned only some of them. As of today, the individual systems are at different levels of development. This is due to the change of generations and types of standard computers. It should be added that one of the serious problems lies in the necessity of automating data bases in which the information has been accumulated over tens and hundreds of years. This is typical for traditional branches of knowledge such as zoology, botany, medicine, etc. In some cases, we face the necessity of investing substantial sums for pre-computer processing of the collected data. The problem of information support of scientific activities in the U.S.S.R. is is a problem of state importance and it is being solved at an all-state level.

Data for Science and Technology, P.S. Glaeser (ed.)
North-Holland Publishing Company

MICROCOMPUTER SPATIAL DATA BASES FOR HUMAN SETTLEMENTS PLANNING

Jerry Coiner*, Ignacio Armillas* and Vincent Robinson**

*United Nations Centre for Human Settlements (Habitat), Nairobi, Kenya
**Remote Sensing and Spatial Analysis Laboratory, Hunter College, New York, U.S.A.

The growing availability of inexpensive data-processing technology permits the use of low-cost information systems to support human settlements policy-making, planning and service-delivery activities. Better information on land, housing, infrastructure, social services and population not only assists the policy-maker and planner in their tasks but can also increase government revenues and minimize waste in the delivery of services. The use of inexpensive methods of data capture and management is of particular importance in developing countries, since it facilitates the creation of better information flows in situations where extensive technological and financial investments are neither possible nor desired. Although recent advances in data-processing technology have greatly improved the suitability of such technology for application in developing countries, knowledge of these advances and their applications in the field of urban and regional development are not widespread. The Urban Data Management Software (UDMS) package, designed for use with microcomputers, is a general data-management framework intended to assist in the operational activities of urban and regional planning agencies. The data base creation, analysis, display, and mapping needs of a broad range of users such as physical, social, health, environmental and land-use planners, housing officials, resource managers, regional scientists, and other human settlements professionals can be met with this package.

1.0 INTRODUCTION

The United Nations Centre for Human Settlements (Habitat) sponsored the development of a data management software package for use on a microcomputer for two reasons. One was to facilitate the transfer of microcomputer technology to developing regions of the world, since the technology is considered one of the most appropriate technologies for application in urban and regional planning organizations. The other, perhaps more important, reason was to use the software to train human settlements planners in the use and potential of microcomputing for planning applications (1; 12).

Besides providing an interactive system at a very low cost, there are many other factors making a microcomputer-based system an attractive alternative to large centralized mainframes for developing countries. Factors contributing to the attractiveness of microcomputer geo-processing systems relate to the microcomputer environment, maintenance, and software characteristics associated with microcomputer systems.

2.0. THE MICROCOMPUTER ENVIRONMENT

One of the most important aspects of the microcomputer system is that it does not demand a sophisticated and costly support environment as do larger computer systems. With a microcomputer system, the user has direct and near total control over the entire computing system. In addition, the user is not subject to the many intervening levels of administration and control which are so typical of centralized mainframe institutions. The emphasis therefore shifts to the application of computing from the administration of data processing.

2.1. Maintenance

If an organization has several microcomputer systems, it is protected from downtime due to hardware failures. To illustrate, if all terminals of an organization are linked to a central main frame, when that mainframe goes down due to hardware or environmental control problems, all activities end. With microcomputers, work continues by shifting the load to other functioning machines.

The microcomputer system does not require as costly maintenance as do larger mini- and mainframe computer systems. It also does not require as frequent repair; costs of repair are by comparison relatively low. Often the system can be repaired by the user with some training and experience. This is because repairing a microcomputer often consists of replacing a plug-in module or board. In Bangladesh, for example, a complete microcomputer system can be purchased with the money spent on two months of maintenance of a mainframe system.

2.2. Software

In comparison with software costs associated with larger systems, software costs for microcomputer systems are much less. However, for the human settlements planner there is a serious shortage of relevant software for the microcomputer system (2). Thus, software development for geoprocessing/planning applications is lagging far behind the development of the capabilities of the hardware. The lack of software is one of the greatest limiting factors in the transfer of microcomputer technology for human settlements planning purposes. To help overcome this substantial obstacle, the United Nations Centre for Human Settlements (Habitat) provides the UDMS package source code to member governments free of charge (13).

3.0 DESIGN PARAMETERS

Because of the dual purposes of the UDMS software and the international focus of the software development programme, the design parameters emphasized transportability and expandability. The package had to have near global transportability, be capable of being easily expanded or customized, recognize the spatial nature of most planning data bases, and incorporate the relationship between the analytical model and the data base. "Transportability" here means that the software had to be able to be easily provided to the users and be capable of running on a wide variety of hardware with a minimum of configuration problems.

3.1. The Hardware

The minimum hardware requirements for the UDMS package are as follows:

(1) a 48K RAM Z80-based microcomputer;
(2) approximately 500K or more of disk storage and a minimum of two disk drives;
(3) a video (ASCII) terminal; and
(4) a standard line printer.

These hardware components are among the most common and widely distributed microcomputer system components (11). However, this choice of components limits the type and quality of graphic output products. What is lost in aesthetics is gained in transportability. Even a microcomputer such as an APPLE, which is not Z-80 based, has the capability of being easily transformed into a Z-80 machine. This characteristic is important, since the most widely-used operating system, Control Program/Microcomputer (CP/M), depends upon the Z-80 type microprocessor.

3.2. The Operating System

In order to make UDMS as transportable as possible, the operating system must be one of the most common. The CP/M disk operating system is available for over 250 different types of microcomputers (3). In addition, CP/M or its derivatives will probably be the predominant microcomputer operating system for at least several more years (6).

3.3. The Programming Language

Given the hardware configuration and the operating system, the choice of the high-level programming language was based on the following considerations. The language had to be commonly available, low-cost, and high-level, because the software would run on many different types of microcomputers. Another consideration was the utility it would have in the expansion and/or customization of the package.

For the above reasons and its low cost, UDMS is written in CBASIC-2. CBASIC-2 is the least interactive dialect of BASIC and is not a particularly fast BASIC. However, it is easily learned by those who may be involved in the elaboration of the UDMS package. Since UDMS makes extensive use of disk resident files, an important characteristic of CBASIC is that its file structures, both random and sequential, are easier to use than those of Microsoft BASIC (7). Furthermore, in stark contrast to Microsoft BASIC, it provides a royalty-free environment for software development. Thus, CBASIC is one of the most prevalent languages for microcomputer applications programming (5). This has important implications regarding the current pool of experienced microcomputer programmers. Therefore, there is likely to be a much larger pool of personnel experienced in the development of CBASIC software for applications on microcomputer systems.

4.0 SYSTEM DESIGN

Due to the nature of the intended users, i.e. human settlements planners, UDMS is designed primarily for a polygon-based geographic information system. In addition to area data elements, point and network data also can be stored, manipulated, analysed, and displayed by various programmes contained in the package (12).

The current version of the UDMS package (1.03) is composed of 30 separate programmes, listed and described in Table 1.

TABLE 1

PROGRAMMES CONTAINED IN THE UDMS PACKAGE

Programme Name	Description
MAIN	Presents the menu allowing selection of tasks to be performed.
MAINB	Presents menu for the spatial analysis tasks.
CHECK	Determines if polygons are formed correctly.
CORD	Forms the co-ordinate file directory.
NETDIST	Determines the distance between each link in a network and creates a disk file of distances.
SCALE	Scales co-ordinates for input to mapping routine.
SORT	Sorts line segments for mapping purposes.
BDRY	Creates the map of polygon boundaries.

TABLE 1 (continued)

PROGRAMMES CONTAINED IN THE UDMS PACKAGE

Programme Name	Description
OVERLAY	Overlays point/network data onto boundary map.
MAP	Assigns variable values to the sorted line segments.
MAPOUT	Prints choropleth map on line printer/console.
CIRCLE	Performs point in circle search.
PSEARCH	Performs point in polygon search.
INTERSTN	Performs polygon intersection detection.
GRID	Performs a grid onto polygon outlay.
CLOCK	Transforms counterclockwise polygon coding to clockwise coding.
VARSTAT	Calculates and prints descriptive statistics.
REGEOMET	Calculates centroids, areas, and perimeters.
DENSITY	Calculates density distributions of data base variables
VARREGR	Performs simple linear regression.
PLOT	Produces plot of regression and gravity model results.
REGRAVIT	Performs a simple gravity model analysis.
LOC1	Solves the optimum single facility location problem.
LOCM	Solves the optimum multiple facility location/allocation problem
NETLOC1 NETLOC2 NETLOC3	Solves the optimum multiple facility location/allocation problem on a network.
SPATH1 SPATH2 SPATH3	Finds the shortest path through a network, Saves them on disk, and provides the input for the NETLOCK programmes.

The intermediate code (similar to object code) of the programme files requires 80K bytes of storage on the diskette. In addition, the CBASIC-2 interpreter, CRUN2, requires 18K bytes of disk storage space. Thus, when transient files are included, over 120K bytes of storage area is the minimum space required on the run disk. Although over 120K is required, this package can run on a microcomputer with 48K or more of RAM. The programmes are designed so that each programme performs a specific task, then is eliminated when the system "chains" to the next specified programme. The "chaining" operation consists of the programme resident in the computer's memory being replaced by the programme specified by the programme being replaced. Chaining proceeds quickly since most of the programmes are of less than 4K in code size. This modular programming design enables easy expansion and customization. In the future, more elaborate versions of UDMS can be of such size that they may require the larger, mass storage devices now becoming more available.

The set of tasks performed by UDMS address all the areas of spatial analysis suggested by Salomonsson (10). However, the package at this time emphasizes the spatial display and location/allocation functions.

UDMS consists of modules which perform the tasks outlined in Table 2. It is a menu-driven package which leads the user through each of the tasks.

TABLE 2

TASKS PERFORMED BY UDMS MODULES

I. DATA BASE FORMATION
   A. Co-ordinate Directory Check & False Region Creation
   B. Polygon Definition Check
   C. Network Link Distance File Creation.

II. MAPPING
   A. Polygon Boundary Map
   B. Overlay of Point/Network Data onto Boundary Map
   C. Thematic Maps of Variable Data

III. SPATIAL SEARCHES AND TRANSFORMATIONS
   A. Point in Polygon Search
   B. Point in a Circle Search
   C. Grid Overlay onto a Polygon

IV. SPATIAL DATA DESCRIPTION AND ANALYSIS
   A. Descriptive Statistics of Variables
   B. Geometrical Description of Polygons
   C. Simple Linear Regression
   D. Simple Gravity Model Analysis
   E. Optimal Facility Location/Allocation
      1. On a Plane
      2. On a Network
   F. Shortest Path Through a Network Algorithm

Figure 1 is a flow diagram of the UDMS package with regard to the tasks outlined in Table 2. Critical to the application of the UDMS package is the set of data base files which are formed using both CP/M ED (Editor) functions and programmes contained in UDMS (14).

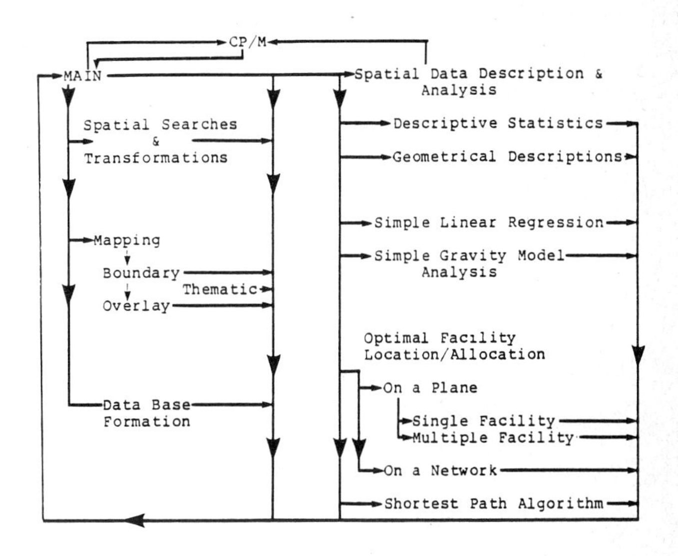

FIGURE 1

PROGRAMME FLOW THROUGH THE UDMS PACKAGE TASKS

5.0 DATA BASE DESIGN

The UDMS package is designed to operate using a set of data base files which have standard CBASIC-2 CP/M sequential formats. The data base is primarily a polygon-oriented geoprocessing system with capabilities to handle network and point data (Table 3). Design of the data base structures assumes that little or no sophisticated input technology is available. Hence, input of the data base elements is not dependent upon any particular input technology, e.g. co-ordinate digitizer.

TABLE 3

DATA BASE FILES USED BY THE UDMS PACKAGE

POLYGON DATA BASE

Co-ordinate Directory File
Region Definition File
Variable Data File

NETWORK DATA BASE

Node Co-ordinate Directory File
Link Definition File

Network Distances File

POINT DATA BASE

Point Co-ordinate File

SPECIAL USE FILES

Interaction or Interregional Flows File (for use in simple gravity model analysis)

Node Constraint File (for optimal facility location/allocation problem on a network)

Place-Weight File (for the optimal facility location/allocation problem on a network)

Source Node File (for optimal facility location/allocation on a network)

Node Subset File (for the shortest path through a network algorithm)

5.1. The Polygon Data Base

In general, polygon structures are preferred whenever spatial data storage and representation are important. There are many methods for encoding polygon structures. For the UDMS package, the polygon boundary is viewed as being composed of a series of line segments. The boundary of each polygonal region is recorded as a string of digital pairs for each vertex in an X/Y co-ordinate system. Assumptions underlying the processing of polygons are:

(1) the polygon is coded in a counterclockwise fashion;
(2) the polygon is a closed polygon; and
(3) the starting point of the polygon coding sequence is the highest and "rightmost" vertex.

These assumptions are primarily for development of maps. The logic employed in the mapping software is presented in more detail in MacDougall's (8) work on computer programming for spatial problems. Thus, the mapping software assumes that the map image is rectangular (or square). This is a constraint imposed by the output device, a line printer. Although the mapping routines are based on the work of MacDougall (8), the data base structures are based primarily upon the discussions presented by Baxter (3).

There are two files necessary for the storage of the polygon structures. One is the co-ordinate directory file and the other is the region definition file (Table 3). The co-ordinate directory file contains information concerning the total number (N) of co-ordinate pairs in the file and the largest value which any one point label number takes on. Furthermore, there are N records, each of which contains the point label number and the corresponding X/Y co-ordinates. It is assumed that these numbers are all integers. The last record of the file contains scale information.

The other file of the polygon data base is the region definition file. The first field of each region's definition is the region identification number, the second field contains the total number of points needed to define the polygon, and the following fields contain the point label numbers which define the polygon. These point label numbers permit the definition of the polygon by looking up the X/Y co-ordinates of the appropriate point in the co-ordinate directory. Thus, the order of the point labels in the region definition file must meet the assumptions of counterclock-wise coding, closure, and upper rightmost starting point location. Due to the assumption that the map image is rectangular, a "false" region needs to be included. A "false" region is a rectangular polygon that encloses all other polygons making up the map. It is co-equal to the neat line in conventional cartography. UDMS assumes that the last region in the region definition file is the "false" region.

The variable data file contains the socio-economic, demographic, land-use, and other data which describe each region or polygon. The first record contains the total number of variables and the total number of regions. Labels describing each variable are contained in the next set of records. Finally, the data values reside in the remaining records of the file.

5.2. The Network Data Base

Networks are considered to be composed of links which meet at nodes. Links therefore can be represented by line segments as degenerate polygons. By recording the location of each node as a point, it is easy to store the links as line segments. The method of storage reflects both the character of the network and the use to be made of it. The UDMS package assumes that the network is a transportation network, a common network encountered in human settlements planning. Given the characteristics of a transportation network, the present version of UDMS makes two primary uses of the network data base. The two uses are: (1) finding the shortest paths through the network and (2) optimally locating infrastructure through the use of the network location/allocation programmes.

To form the network data base, three files are developed. Of the three network data base files, the user must create only two. The first file is the node co-ordinate directory file, which uses the same form as the co-ordinate directory file discussed above. The second network file is the link definition file. This file is very similar to the region definition file discussed above; however, it is not the same in form. The difference results from allowing for one-way or two-way movement along a line segment. Once these two network files have been properly created, the UDMS package will form the network distance file and save it on disk in the proper format for use with the shortest path algorithm.

5.3. The Point Data Base

This data base is used to store the location of infrastructure entities, such as schools and hospitals, which can be represented by point locations. The single file necessary for this data base is the point co-ordinate file. The point co-ordinate file consists of N records, with each record describing the location of a particular point, using as elements of each record the point label number and its X/Y co-ordinates. At present, the utility of such files is limited to the location/allocation on a plane problem and in the mapping overlay for display purposes. In addition, the files are used by the spatial searching programmes.

6.0. OTHER CONSIDERATIONS

This paper presented the rationale and design of a microcomputer-based geoprocessing software package. For readers wishing more detail concerning the UDMS package, they are referred to the User's Manual (14). Most parameters concerning design of the package relate to the problems of technology transfer. Although the system of hardware necessary to support UDMS can be assembled for less than U.S.$10 000 (F.O.B. New York), there are problems other than cost which affect the transfer of microcomputer technology for geo-processing applications.

Salomonsson (10) has identified two problems, among others, of particular relevance to technology transfer. First, there is the need to make the potential producers and users in the country familiar with the appropriate techniques. Second, there is a need to secure software support for whatever is transferred. This second issue is of great importance when considering the transfer of microcomputer technology.

The UDMS package was developed largely in the context of an educational/demonstration tool which could be of use in an appropriate operational environment. In terms of educating local, regional, and national mid-level planners in the concepts, techniques, and technology of urban data management, this software package has proven to be a success. Much of the success has been attributed to the characteristics of the microcomputer system (1). UDMS has also proven itself as an operational system, e.g. in Sri Lanka and Colombia. The package is less than two years old and already a great deal of interest has been generated by workshops held in South America and Asia. Therefore, by next year, the system will be established on an operational basis in at least ten developed and developing countries.

There remain several human obstacles to technology transfer. A familiar obstacle to the technology is the general bias against computers. With microcomputers, thereis an additional dimension to the problem. The commercial marketing of microcomputing has been so heavily oriented towards games and other entertainment that it is difficult for some to accept the notion of such "small" computers having the capability of solving "real" problems. To some extent this can be overcome with the workshops being conducted under the auspices of the United Nations and the host governments. However, little headway will be made on a large scale until such systems begin to enter the operational environments of developing countries. At this point, the need for software support will become intense.

Of all the obstacles to the adoption of microcomputer geoprocessing technology, the most important is the general lack of software. This contributes to a vicious circle inhibiting the development of microcomputer geoprocessing systems. The lack of sophisticated and generally applicable software reinforces the perception that the hardware is inadequate for the job. It is hoped that the UDMS package contributes to a change in attitude toward the capabilities of microcomputers. Continued advances will make the micro-mainframe a reality in the very near future (9). It remains to be seen if the field of geoprocessing will be ready to exploit such technology in a manner which allows its transfer to applications in the developing regions of the world.

REFERENCES

(1) Armillas, Ignacio; Coiner, Jerry C.; and Robinson, Vincent R. Final Report: Data Management for Urban Environments Workshop, Bogota, Colombia, 23 March-3 April 1981. United Nations Centre for Human Settlements (Habitat), Nairobi, Kenya, April 1981.

(2) Auerbach Publishers. Computers in Local Government: Urban and Regional Planning Applications Software Directory, Pennsuken, New Jersey: Auerbach Publishers, 1980 & 1981 (supplement).

(3) Baxter, Richard S. Computer and Statistical Techniques for Planners, London: Methuen & Co., 1976.

(4) Digital Research Inc.. Infoworld, 16 March 1981: 14.

(5) Editorial. Infoworld, 18 August 1980: 8.

(6) Hagan, Thom. "Thom Hagan Picks Top Ten", Infoworld, 16 March 1981: 4.

(7) McDonnel, K. "CBASIC": "Two Views", Infoworld, 13 October 1981: 11.

(8) MacDougall, E. Bruce. Computer Programming for Spatial Problems, London: Edward Arnold, 1976.

(9) Robinson, Arthur L. "Micromain-frame is Newest Computer on a Chip", Science 212: 527-531.

(10) Salomonsson, Owe. "Geocoding for Developing Countries", Data For Development Newsletter, No. 13, January: 4-8.

(11) United Nations Centre for Human Settlements (Habitat). Urban Data Management Software Package (U.D.M.S.) - User's Manual (preliminary draft), Nairobi, Kenya: United Nations (Habitat), 1980.

(12) United Nations Centre for Human Settlements (Habitat): Science and Technology in Human Settlements, Nairobi, Kenya: United Nations (Habitat), 1979.

(13) United Nations Centre for Human Settlements (Habitat). "Urban Data Management Software (UDMS) Package" Nairobi, Kenya: United Nations (Habitat), 1981.

(14) United Nations Centre for Human Settlements (Habitat). User's Manual, U.D.M.S. Version .98 (CHS/PP.81-3), Nairobi, Kenya: UNCHS (Habitat), 1981.

ACKNOWLEDGEMENTS

We would like to acknowledge the support of the United Nations Centre for Human Settlements (Habitat) which made the development of the Urban Data Management Software (UDMS) package possible. All statements in this paper are the authors' and do not in any way represent the view, opinion, or policy of the United Nations. In addition, we would like to note the support of the Hunter College Remote Sensing and Spatial Analysis Laboratory in New York City.

******

Data for Science and Technology, P.S. Glaeser (ed.)
North-Holland Publishing Company

TECHNOLOGICAL EVOLUTION OF ENVIRONMENTAL DATA ACQUISITION SYSTEMS

*G. Bucci, **G. Neri, and ***F. Baldassarri

**ISIS, University of Florence, Italy*
***Istituto d'Automatica, University of Bologna, Italy*
****SIAP, Bologna, Italy*

We describe the evolution of a class of data acquisition systems, as a consequence of both microprocessor technology growth and upgrading user requirements. These systems range from simple autonomous peripheral stations to complex central stations operating in real-time. The former are implemented around the RCA Cosmac, a C-MOS microprocessor, while the latter make use of the 8086, a powerful 16-bit microprocessor. Motivations are described for the adoption of the 8086 in substitution of the previous CPU, which was microprogrammed and in-house fabricated. More specifically, we discuss the problems associated with software production and maintenance. In this context, it is shown that consistent benefits have been achieved by switching from assembly to high level language programming. The paper critically reviews the evolution of the past decade.

1. INTRODUCTION

This paper discusses the evolution of SIAP data acquisition and processing systems, occurring in the last decade as a consequence of microprocessor technology growth. The systems here described are used in weather forecasting, air traffic control and assistance, control of seismic events, agriculture and the like. Measured data include rainfall, wind speed, wind direction, temperature, water level, visibility, etc.

In its most general configuration an environmental data acquisition system consists of a central station and a number of peripheral ones that act as measurement points. A peripheral station is composed of one or more sensors for measuring physical quantities, appropriate transducers and an electronic section to amplify, convert and encode measured data. Peripheral stations may be located either in the close vicinity of the central station or remote from it. In the latter case, data is transmitted via phone lines, telegraphic channels or via radio transmission. Basically, a data acquisition system must perform all or part of the following functions:

a) data collection and recording on some mass storage device, such as magnetic tapes, cassettes, floppy disks, etc.;

b) data processing, which may include simple computation of some significant parameter or tracking algorithms;

c) presentation of relevant information on output devices, such as typewriters, alphanumeric displays and plotters;

d) alarm signalling when some incorrect operating condition is detected;

e) execution of special functions required by the operator, such as repetition of previous operations, peripheral checking, scale adjustment, etc.;

f) execution of special procedures such as flood forecast.

Data recording can be as simple as putting data in a sequential file, or managing a (sort of) data base.

In many applications, systems must only do a small fraction of the above functions. In addition, certain applications require that peripheral stations work autonomously, performing measurements and storing data on devices like mini-cassettes.

To cope with contrasting requirements SIAP has arrived at establishing two lines of products. The first line is currently based on the 8086, and is intended for medium/large central stations. The second line of products uses an 8-bit CPU, and is intended to cover the range from single autonomous data measurement points, to small/medium stations. This second line is made in two versions: the SM3200 and the SM3800. From the hardware point of view they differ only in that SM3200 uses an RCA 1802 CMOS microprocessor, while SM3800 employs an 8088.

This paper will review the developments of the past ten years which have led to the current system architectures. Of course, the major impacting factor has been the rapid growth of microprocessor technology, but also the expansion of user requirements (due largely to that growth) has been the forcing factor towards major project decisions, design upgrades and architecture changes. It will be shown that software production and maintenance is the key activity in one such context. Most of the design choices have actually been made with the objective of minimizing the cost associated with the entire software life-cycle.

2. THE USE OF A MINICOMPUTER

The use of a digital computer, as the heart of environmental data acquisition and processing systems, began in the early seventies. The first system designed had to collect data from 21 pluviometers and 3 level-meters, scattered through the basin of an Italian river, and to perform flood forecasting in the real-time (1). Flood forecasting was carried out in accordance with a straightforward mathematical model of the river's behaviour.

Pluviometers and level meters were equipped with the appropriate transducers and (hardwired) electronics for radio transmitting the actual measured values. Processing requirements associated with flood forecasting imposed the design of a minicomputer-based central station in replacement of the previous hardwired data collection systems. The choice fell on an HP2100, a minicomputer very successful at that time. Collected data were stored on magnetic tapes and presented through two typewriters. An additional plotting unit served to diagram measured levels against forecasts. The system is still in operation.

Looking in retrospect, from the current level of technology, the most noticeable aspect of this system is its low (core) memory capacity, which was limited to 8K words (16-bit). The system software operated in the real-time, supporting parallel operations among i/o devices and computation. Task coordination was obtained in the less expensive manner through a sort of "busy form of waiting". The system software was absolutely stand-alone and written in assembly language in order to keep program size small. Use of the 2100 minicomputer was continued for some years, though its cost discouraged employment in low end systems which were still hardwired.

3. FROM MINICOMPUTERS TO MICROPROCESSORS

The reasons for developing microprocessor based stations were the following:

1. Hardwired stations required long design and development times. They had no flexibility to accommodate future unanticipated changes, nor had they processing capabilities. Keeping alive this production line required an adequate level of investment. It was felt that this line had been eliminated with microprocessor technology, with consistent benefits for the manufacturer in terms of both internal organization and production costs. However, experience showed that, even with microprocessors, two production lines were still required to cope with end-users' needs.

2. Minicomputers gave rise to fairly high costs, and often led to overdimensioned stations. That is, although minicomputer sellers offer a certain range of choice, it was almost impossible to find modules tailored for specific requirements. In addition, minicomputers come with parts that are useless for many applications. The front panel, for instance, is often too fine for the non-skilled final user. It was judged that microprocessors had allowed more flexibility and therefore a better match between application requirements and system capabilities.

3. Associated with minicomputer usage were certain problems of operation and maintenance. Actually, it is unlikely that engineers and technicians working for a manufacturer will ever have a complete and detailed knowledge of the minicomputer used. Therefore, both end user and manufacturer had to rely on the minicomputer seller's organization as far as maintenance and repairs were concerned. In certain circumstances this gave rise to misunderstandings and inconsistencies, and led to overhead and service inefficiency. It was felt that the new technology gave the manufacturer complete knowledge of the system produced, so he might provide maintenance service for the entire system.

A thorough discussion of the motivations for entering the game of microprocessor technology is in (2).

The objectives of the project of the central station have been the following:

a) to replace the minicomputer previously used;

b) to increase system availability, from the point of view of both Mean Time Between Failures (MTBF) and Mean Time to Repair (MTTR);

c) to enhance the 'easy to use' characteristics of the system mainly for those users not acquainted with digital computers;

d) to reduce the overall system cost.

The resulting system was called MP76. It used a 16-bit microprocessor: the National Semiconductor IMP-16C. The choice of IMP-16C was essentially dictated by the requirement of compatibility with the previous system. But the market situation had an important role as well (3).

Mechanical and electrical standards were those imposed by the IMP-16C CPU card. The system was equipped with a C-MOS real-time clock board, with buffered power supply, to maintain and update timing even in case of power failure. (A C-MOS real-time clock is included in any system built since MP76.) System software was structured as a set of cooperating sequential processes. The project entailed the design of a real-time software kernel (4). Programming was entirely done in assembly language for the lack of any acceptable compiler.

During the implementation of MP76 it became clear that low-end applications required a smaller system. It was evaluated that, if carefully designed, one such system could become the basis of any peripheral device. The design of SM3200 was therefore started, based on RCA's COSMAC (a C-MOS microprocessor). Both mechanical and electrical standards resulted from design choices. Every system card is in-house manufactured. Use of the 1802 microprocessor made it possible to manufacture a standard peripheral station which may be arranged for several different sensors and storage devices. SM stations are multi-role devices and can operate: (1) as a simple peripheral with few attached sensors; (2) as a node within a hydrometeorological network; and (3) as a small autonomous central station (this last case applies better to SM3800 described later).

A thorough investigation of the motivations that led to the design of SM systems is in (5).

4. A MICROPROGRAMMED CPU

After some years of production of MP76, the firm was faced with certain applications which required more processing power than that possessed by MP76. These applications were essentially the same as before, but, as a consequence of evolving user requirements, demanded more sophisticated, time-consuming mathematical computations. At about the same time rumors arose about a possible IMP-16C discontinuation. We evaluated that emulating the IMP-16C with bipolar bit-slice microprocessors would be the most cost-effective solution.

The new CPU, called MP80, was a great improvement with respect to the IMP-16C. It was 6 times faster, had a better I/O system and better arithmetic. A most noticeable feature of MP80 was the microprogrammed software kernel. As a result, the hardware itself implemented a set of cooperating sequential processes, and synchronization primitives were simply executed as machine instructions. A detailed description of MP80 is in (6).

MP80 was quite an improvement over the MP76 and was used in several implementations. MP80 had the great feature of providing, as machine instructions, a set of well-understood and well-tested kernel primitives. This allowed easy implementation of real-time systems, safe process interaction and no risk of kernel corruption as a consequence of program bugs.

However, MP80 was not the definite solution to central station implementation. Actually, programming was still done in assembler, and the cost associated with software production and maintenance began to rise beyond acceptable limits. This depended on the fact that central stations had to be manufactured for applications that were only partially repetitive: every delivered system had some relevant differences with respect to the others, thus requiring several specific programs. This was primarily due to growth of users' requirements, which, in turn, was mostly due to the pressure of technology evolution and microprocessor diffusion.

The other reason (though not so important) that eventually led to stopping the MP80 was associated with certain problems encountered in manufacturing a complex, dense card as MP80 CPU.

5. PASSING TO THE 8086

Before deciding to stop implementing new systems on the MP80, we considered the possibility of overcoming the problem of software production through the development of a compiler for a PASCAL-like language. The compiler had to be designed taking into account MP80 real-time software architecture, that is, we wanted to use synchronization primitives in a straightforward manner and to keep compatibility with assembler-written programs. Obviously, the compiler was not required to possess all the features of a commercial, general purpose compiler. We rather preferred a compiler tailored to our needs.

A preliminary design was then started, in order to quantify the investments that were needed. (Development of a compiler had forced development of some other software components such as loaders, etc.)

It became apparent that the level of investment for developing a compiler was so high that we began considering switching to a commercial 16-bit microprocessor of the new generation. Using a commercial product was considered beneficial in terms of:

a) conversion time much shorter than that for in-house compiler development;

b) greater reliability of commercial compilers with respect to an in-house compiler;

c) large choice of support software, system software and application software, available from the seller and from the community of microprocessor users;

d) reduced risk of obsolescence.

We had clear that all the above advantages could be achieved only by selecting a microprocessor whose producer was strongly committed to it for the future. It was evaluated that Intel 8086 best met our requirements.

It is worth mentioning that 8086 has a powerful real-time operating system called iRMX 86. Real-time applications are developed on top of it. Programming is done exclusively in high level language. So far, we have been using only PL/M 86. We are evaluating employment of FORTRAN, mostly for input/output formating reasons. We think that 8086 central stations have a long way ahead.

A beneficial consequence of adopting the 8086 was the subsequent abailability of the 8088, i.e. the 8-bit version of 8086. This microprocessor has great processing capabilities for an 8-bit machine. In addition, it is perfectly software compatible with the 8086. This has motivated the design of a new CPU card for SM stations. This card is obviously compatible with the bus architecture of SM3200, but adds to these systems the greater processing capacity of 8088. The new peripheral stations are called SM3800. They find their application in stations of intermediate complexity. Obviously SM3800s are programmed in PL/M 86. A detailed description of SM3800 hardware/software architecture is in (7).

5. A CRITICAL VIEW

We learned several lessons from these almost 10 years of development. The following is a discussion of the most relevant issues.

Software costs. In medium to large applications, unless the number of identical systems sold is very large, the costs associated with software development are, by far, the largest fraction of overall costs. In this situation, rather than trying to optimize hardware size and utilization, any effort must be attempted to reduce both costs and times associated with software production.

Languages. Program productivity (measured as the average number of debugged instructions produced per day) seems to be almost independent of the programming language, no matter if it is an HLL or an Assembler. The consequence is: at the stage of microprocessor selection, give preference to that for which there is at least a reliable compiler.

Development systems. A good development system has the potential of easing the programming task, thus increasing programmers' productivity. Interactive editors and sophisticated emulators greatly reduce program development and debugging times. Debugging tools must receive the proper credit, actually microprocessor software is still in its infancy and, very often, software products (including compilers) contain some undiscovered bugs. The system programmer must be prepared to fix this kind of malfunctioning, that can determine consistent waste of time. This can only be done with appropriate tools.

To buy or to build. The alternative refers to hardware and must be evaluated in any specific context. It strongly depends on both the number of systems that will be manufactured and the industrial organization of the manufacturer. Currently we buy the hardware of central stations and build that of SM stations.

Availability and support. The success of a project may strongly depend on the future availability of the components used. This refers primarily to the microprocessor. It is essential to estimate the time a given product will survive, if it will be replaced with a compatible one and whether or not it will be supported for the future. These are elements of risk that may impact a project much more than the architecture of a given microprocessor.

Development techniques. Software production requires appropriate development methodologies and programmers' organization. Structured programming and top-down program development cannot be avoided. Program implementation may (at least in our experience) be done as a mixture of bottom-up and top-down. Chief programmers team can be the appropriate organization for the first projects. As long as programmers become expert they pretend a more creative job than that implied by this organization.

6. CONCLUSION

Selecting a microprocessor has a certain degree of risk and can strongly impact the future. Preference must be given to those devices which, because of their support and level of technology, guarantee that the hardware/software system they implement will grow (possibly through compatible systems and devices) rather than become obsolete.

REFERENCES

[1] Bucci, G. and Ghilarducci, G., A system for forecasting Ombrone river floods, Rivista Italiana di Geofisica, vol. XXIII (1974), n. 5/6, pp. 313-316.

[2] Ghilarducci, G. and Guidelli Guidi, G., Impiego dei microprocessori nei sistemi di acquisizione dati idrometeorologici, Proc. XXIV Congresso per l'Elettronica, Roma, March 1977, pp. 147-155.

[3] Bucci, G. and Neri, G., Using microprocessors in hydrometeorological data acquisition and processing systems, in Microprocessor Applications, International Survey of Practice & Experience (Infotech International, Maidenhead, U.K., 1979), pp. 267-278.

[4] Bucci, G. and Neri, G., Hardware and software structure of a microprocessor-based system, in Niemi, A. (ed.), A Link Between Science and Application of Automatic Control (Pergamon Press, N.Y., 1979), pp. 691-696.

[5] Baldassarri, F. and Ghilarducci, G., Applicazione di microprocessore C-MOS in apparati di acquisizione dati ambientali ad alimentazione autonoma, BIAS '78, Milano, Nov. 1978.

[6] Bucci, G., Neri, G. and Baldassarri, F., MP80: a microprogrammed CPU with a microcoded operating system kernel, Computer, Oct. 1981, pp. 81-90.

[7] Neri, G., Bucci, G. and Baldassarri, F., An 8088 based system for meteorological data acquisition and processing, to be published in ISMM Journal.

Data for Science and Technology, P.S. Glaeser (ed.)
North-Holland Publishing Company

THE CHESAPEAKE BAY INFORMATION SYSTEMS (BIF) FOR ENVIRONMENTAL DATA STORAGE AND ACCESS

Angel Bailey and Rita Colwell

University of Maryland Sea Grant College
College Park, Maryland, U.S.A.

This paper addresses current technology and services becoming available on modern data bases. The overall design, partial implementation, and a list of services connected with the Chesapeake Bay Information Facility (BIF) are detailed. The BIF is a combination data base designed to contain source, numeric, and bibliographic data. A new way of accessing external data bases has been developed and a compendium of presently available environmental data bases constructed.

I. INTRODUCTION

Information retrieval is concerned with the structure, analysis, organization, storage, searching, and dissemination of information. An information retrieval system operates on the one hand with a stored collection of information, and on the other with a user population desiring to obtain access to the stored items. Thus, it is designed to extract from the files those items most nearly corresponding to specific user needs as reflected in requests.

In recent years, the information retrieval area has been of concern to an increasing number of people interested in environmental and biological technology, in part because of the continued outpouring of potentially useful information. These new informational uses have significantly changed the ways and styles in which policy can actually be made. The need has especially increased for utilizing available information for regional decision making. This need has led to a recent development of a number of environmental and biological data bases with a potential regional application.

However, efforts to meet the need for information in regional decision making, unfortunately, have not been organized or coordinated in the Chesapeake Bay region. In fact, establishing whether data required for a given purpose exist, and whether it can be accessed has been a difficult task. This is partly due to a lack of an effective promotion effort by the institutions involved concerning the location or availability/accessibility of all the data and information developed and being developed within the Chesapeake Bay region. Also, the problem is that public officials and citizens do not know how to request scientific information (especially regarding environmental management problems). In addition, even if received, that information or data poses a problem because it is not translated into layman's terms. These systems are technically oriented and difficult for an untrained person to use. This may explain why a large segment of users in the daily mainstream of Chesapeake Bay affairs fails to take advantage of these sources.

Clearly, environmental planning and management require timely access to data from one or more sources. A mechanism by which the source(s) is identified and the means of interfacing the data in a user friendly fashion must be established. In this paper, we address the background, the objectives, and the overall design of a user friendly integrated data base system which characterizes a modern view of such a system and the factors relevant to its operation.

II. BACKGROUND

The Chesapeake Bay is the largest and most productive of the 850 estuaries of the United States. The Bay is heavily used and the intensity of use is expected to increase. It presently sustains a commercial and recreational fishery with an annual value exceeding one hundred million dollars. It also provides the region with recreation, transport, and secondary products. More than almost any major body of water in the United States, the Bay has drawn the attention of scientists, regulatory agencies, commercial users, shipping and energy interests, and many of the nearly nine million residents of the Bay area.

There are more than one hundred agency organizations and more than thirty research institutions (Sea Grant 1980) concerned with the study and environmental management of the Bay. From federal government to state and local agencies, universities and colleges large and small, all manner of data and information have been accumulated about every aspect of the Chesapeake Bay. Environmental scientists, and policymakers need to access this information to answer environmental questions. To solve the complex problems resulting from interaction of such forces as population, economic growth, land and resource utilization, and environmental pollution, integrated approaches to the use of scientific data are needed. Data can support planning to address these problems, but it must be organized in a framework which spans the spectrum from basic data collection to regional policy development. Many Bay officials and citizen organizations, however, have a difficult time finding the technical information they need concerning environmental issues of interests. The problem has been one of not knowing where to go, whom to ask, or how to request what is needed.

Computer approaches for requesting information

offer a means of creating a framework. The use of data from many sources, integrated by modeling and display techniques, connects basic data to sets of planning problems. Recently, a number of environmental data bases have been developed that could be of benefit to the Bay users. These data bases, created and maintained by agencies of the federal government, academic institutions, and commercial enterprises are often excellent compendiums and potentially valuable research and management tools. However, these systems are generally not well publicized, access is difficult, and they are created by and for data information specialists. In addition, commercial and governmental data bases often are designed for restricted access and cannot be accessed by an individual. See table 3.

With these problems in mind, the University of Maryland Sea Grant College is creating an information facility to supply an end user data base to serve the community surrounding the Chesapeake Bay area. We conducted a comprehensive survey to find existing information sources in the Bay area and elsewhere in the United States to avoid duplication of effort and to effectively utilize previously developed information retrieval systems.

In order to determine the needs for Bay related information and the extent to which these needs are currently being met, various users from the government, university, and private sectors were invited to a workshop sponsored by the National Marine Fisheries Service (NMFS) and coordinated by the University of Maryland Sea Grant College on the 25th and 26th of June 1981. The results of the workshop were analyzed and a set of goals and objectives were established upon which the Chesapeake Bay Information Facility (BIF) was finally designed. Implementation was begun specifically to satisfy the needs of Chesapeake Bay area users.

III. BIF SYSTEM OVERVIEW

This section provides an overview of the BIF system in terms of its design objectives, its system configuration, its functional system description, and progress to date. This represents a total design and a partial implementation of the bibliographic system and the system standard terminal. The final implementation of the BIF system is in progress.

1. Design Objectives

With information gathered from user workshops and surveys of existing data bases, a set of guidelines emerged which were translated into a specific set of objectives for the BIF system. The overall goal is: a system to collect, via computer, information from all sources throughout the Chesapeake Bay region, as well as elsewhere in the United States applicable to Chesapeake Bay information needs, and to make this information readily available to government agencies, scientists, businessmen, private companies, watermen, civic organizations, and citizens.

Specific objectives of the BIF system then became to: 1) provide a user friendly system that could be menu driven to user specific requirements; 2) provide promotional and user data, on the use and benefits of the system for major elements of the estuarine community; 3) provide communications access points throughout the user area to facilitate access; 4) support a terminal end user functionality that will provide the user with supplemental utility and ease considerably the routine access to the information system support structure and data; 5) extend the system to alternative data bases of interest via query translation and file transfer techniques by external communications to these external data bases; 6) provide advanced statistical, data management, and graphics capability to University researchers; and 7) allow modular expansion and an upgrade capability for future needs.

The scope of the data base will continue to be defined by areas of interest. Current indicated areas of interest are listed in table 1. Specific services or information, which can be provided, defined by user interest, are outlined in table 2.

Table 1. BIF User's Areas of Interest

Air & Water Pollution	Oyster Propagation
Aquatic & Trstrl. Ecol	Permit Coordination
Biological, Chem. Data	Physical Data (geology)
Chesapeake Bay Strategy	Port Development & Economics
Coastal & Estuarine	Regulatory Control
Energy	Seafood & Shellfish Resources
Estuarine Studies	Seafood Ind. & Mark
Fisheries & Aquacult.	Sewage Treatment & Wastewater
Fisheries Laws & Regs.	Toxicology
Fish & Game	Toxics
Fish & Wildlife	Wastewater Management
Hazardous Substances	Water Quality
Hydrographic Info.	Water Resource Policy
Hydrologic Information	Water Supply
Marine Science	Watershed Management
Methods Development	Wetlands
Mining	
Oceanography	
Oil Spills	

Table 2. BIF Service Objectives

Analytical services
Bay Management History
Bay Scenario services
Bibliographic services
Consultant/Specialist Referral services
Directory services
Education and Outreach services
Issue/Conflict Involvement services
Library Reporting services
Objective Analysis services
Referral services
Statistical Analysis services

2. System Configuration

The BIF system comprises an IBM 4341 operating under CMS/MVS. System storage is 1500 megabytes of hard disk storage of which 500 megabytes are reserved for the BIF data bases. Communications is via an IBM 3705 configured for 60 channels. Direct connect channels are configured for 9600 BPS RJE and several dial in channels are configured for both 300 and 1200 Baud. The BIF system, being developed at the University of Maryland Sea Grant College will overlay the IBM operating system and is described below in the BIF functional description section. The system standard terminal is composed of a CP/M based microcomputer with 64 kilobytes of RAM and over 200 kilobytes of floppy disc storage with either a 300 or 1200 Baud modem and an Epson MX-80 or MX-100 dot matrix printer with graphic capabilities. A SAS package is available on-line for user applications.

3. Functional System Description

The BIF system consists of three separate data bases and two support systems which interface the user to the base operating system of the computer and its various utility and application packages. All of the data bases (bibliographic and numeric) have an extended access system attached to allow direct and indirect access to allied and linked alternative data bases. Figure 1 provides an overall system diagram.

An important element when considering the functionality of BIF is the inclusion of a system standard terminal. The system standard terminal is an independently functional microcomputer with optional modems and a dot matrix printer. This configuration provides several advantages to the user and the system. Some I/O dependent chores, such as screen editing, can be offloaded from the system while the user can utilize the independent functionality to perform such confidential tasks as accounting and market projection. The end user functionality can be increased by the inclusion of a users group package on the main BIF system which would distribute standard terminal community oriented information and programs (by download) that could be taken by the community for offline use. In addition, the user would be in a position to economize his real and connect time by organizing his system access off line at his convenience and allowing the terminal to connect and transact the previously organized tasks at an optimum rate on line. The user could also receive large amounts of output data for later off-line analysis and selective printing. The user could then maintain a journal of his transactions on floppy disk for later reference or as an aid in forming future uses of the system. The system standard terminal is chosen by its economic price and portability. A standard terminal is preferable to a general mixture of commercially available microcomputers due to the necessity of assigning scarce resources for added functionality rather than accomodation of several different types of languages, formats, and protocols. The system standard terminal is an "extra" and the system supports a full range of "dumb" terminals commonly available on the market.

Features of the system standard terminal include several software packages. Systems software consists of a full screen text editor (Wordstar), a spread sheet analysis package (Supercalc), the CP/M operating system, and a BASIC interpreter and compiler. Sea Grant provides a custom communications program that allows the user to connect to the BIF via an optional modem, which may be 300 or 1200 BPS. The communications program supports full auto dial capabilities and connection with the BIF may be accomplished by simply hooking up the telephone and hitting the return key. The terminal with the proper modem is capable of dialing the telephone, repeatedly if the line is busy, signing on the system, transacting queries, formatting

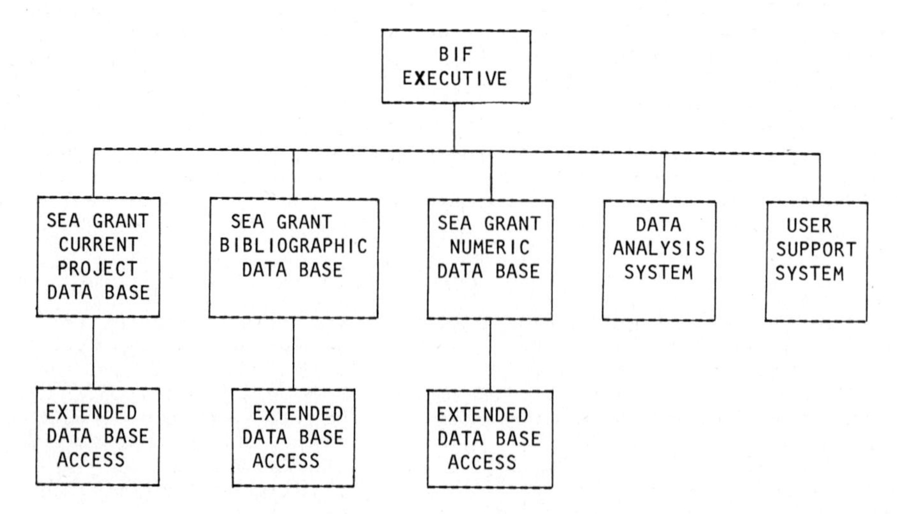

Figure 1. Bay Information Facility Functional Diagram

output, locally storing results, signing off the system, and disconnecting the phone. An audible alert will be made to the user if unexpected problems arise. The following paragraphs give a description of each subsystem.

3.1 Data Base Systems

There are three system data bases on the BIF. Each data base consists of a query help system, index files, a user permissions and validation module, a formats and forms translations module and an extended access module. See Figure 2 for a functional diagram of a data base subsystem.

The query help module provides optional menu and help assistance to the on-line user in the formulation of queries to the data base. The menu system allows the user to formulate a search by selecting the appropriate options from a table of options. The options include logical combinations of data elements (e.g., Author X but not Subject Y). This greatly simplifies a search by an infrequent or new user of the system. In addition the user may call for help at any time the computer is requesting input. The computer will respond with an explanation of the user's options at this point together with a number to select, which exercises that option.

Index files contain quick access points to the data base and are organized by the access elements of the data base, such as author or contract number. The compilation of index files allows a rapid response to structured queries. The index files contain data references to the actual file and are organized for efficient search (e.g., authors are arranged alphabetically). When a query is activated the system responds with the data if the number of finds is below a user settable level and with the number of finds if above that level. This allows the user to further reduce the finds by a tighter specification of the search to avoid printing out long lists of largely irrelevant data.

The user permissions module contains the extent to which a particular user can access the data base and what his permissions and restrictions are. If extended access is authorized, the user permissions will determine also the maximum limit of such access. The user permissions is a command file that tells the system what the scope and permitted access of each user is. In addition, the user permissions module will tell the system if the user is to receive automatic extended and/or additional cost services as a part of his session.

The formats and forms module provides formats and forms for interactive use and output of certain types of data. The forms provide a blank form to the screen with protections to allow input of data. The formats module contains preset common user formats for output of data which allows the user to structure output to preferred common element form when applicable.

The extended access module is a communications and translation module that allows the user to access external data bases. Two modes of operation are supported, direct and assisted. In the direct access mode the user is connected to the external data base, as if he were a registered user of that data base and directly connected with it. In the assisted access mode, the user will search the extended data base(s) as a part of his normal search on the system internal data base. His query is automatically formatted for the external data base and returned data are sized and reformatted to comply with internal system requirements and merged with system returned data automatically. This is accomplished for chosen systems by reformatting the user query in the external system format, posing the query to the external system, and merging the response from the external system in the BIF standard format. If the number of finds exceeds a specific number, the user is queried to see if he wishes to narrow the query or to accept the data as is. A specific duplicate entry program determines if the entries are duplicate of BIF data in the search and merges the data to the user output file tagged as to origin. A charge enquiry is then made to the external data base to allow the query to be charged to the user who initiated that query. These external queries are conducted over a privileged path to the external data base and look like RJE to that data base. For example, currently inaccessible data bases to be brought to the University, include the EPA Bay Program Data Base (Chesapeake Bay data and software developed over a five-year period), Maryland Department of Natural Resources Fisheries Data Bases (historical data obtained over 20 years), and Environmental Data and Information Service Data Bases (national marine data obtained in field study and remote sensing

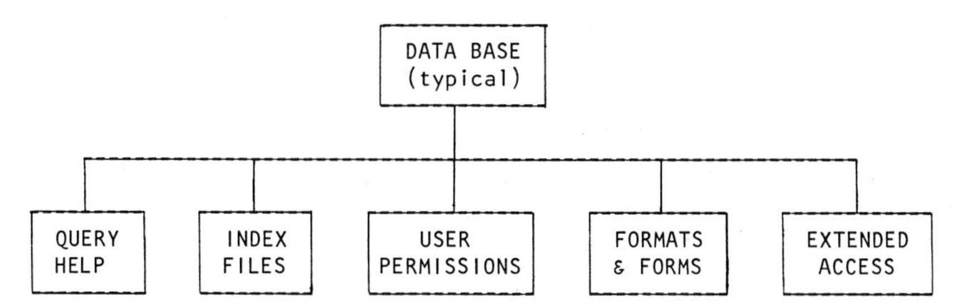

Figure 2. Typical Data Base Functional Diagram

programs). Access to data bases not linked to the BIF system will be handled by a referral file giving subjects of interest, contacts, and access methods. Table 3 gives a list of related data bases indicating those that could be linked with the BIF system.

3.1.1 Current Project Data Base Subsystem

The current project data base concerns active and near term projects under the control and direction of the University of Maryland Sea Grant College. This is a source data base and is updated periodically. It may be accessed by project, contract or grant, date, or by a full text search, as well as by principal researcher, project manager, office, and key words.

The extended current project data base is designed to be extended to other Sea Grant facilities nationwide to allow a total Sea Grant current project data base at some future date.

3.1.2 Bibliographic Data Base Subsystem

The largest data base is bibliographic and is built primarily upon the pre-existing FAMULUS data base. It provides access by author, date, subject, source, key words and a logical combination of these elements.

Extended access is provided via the bibliographic data base to similar bibliographic files with an auto merge function to format and delete duplicated data.

3.1.3. Numeric Data Base Subsystem

The third data base is numeric and contains coordinated value information concerning harvests, pollution indices, market information, salinity, water temperature, demographics, littoral area traffic, channel depths, hazards, navigation aids, currents, and tides. Wherever possible, the information is keyed by geographic location, for example, regional location.

Extended access is provided from the numeric data base to other numeric data bases with an auto merge function to eliminate duplications of data.

3.2 Data Analysis Subsystem

The data analysis subsystem (see Figure 3) is accessed, normally, after a user has obtained search results from one or more data bases. The data analysis subsystem components may be invoked in any sequence as necessary to process and format the user's data. This system contains several powerful tools for the manipulation, conversion, and reduction of data.

The cluster analysis module provides several clustering algorithms that may be used to group and classify collections of data and to identify common elements pertinent to an analysis or search. Two clustering packages are presently available. The first package consists of the NTSYS clustering analysis system developed at SUNY, Stonybrook. The second package was developed at Sea Grant specifically for the BIF system, and thus requires minimum translation to use. The Sea Grant clustering package includes the following algorithms:

1. Direct Joining Algorithm
2. Direct Splitting Algorithm
3. Distance and Amalgamation Algorithms
4. Ditto Algorithm
5. Drawing Trees Algorithm
6. Euclidean Distances Algorithms
7. Fisher Algorithm
8. Joiner-scaler Algorithm
9. Joining Algorithm
10. K-means Algorithm
11. Leader Algorithm
12. Minimum Mutation Algorithms
13. Mixture Algorithm
14. Profiles Algorithms
15. Quick Position Algorithms
16. Single-linkage Algorithm
17. Sparse Root Algorithm
18. Triads Algorithms
19. Variance Components Algorithm

The statistical analysis module interfaces to SAS and its powerful statistics and graphics capabilities. Graphics output from this module may be further processed by the graphic processor module.

The bacterial identification module (TAXAN 6) provides several techniques for computing simi-

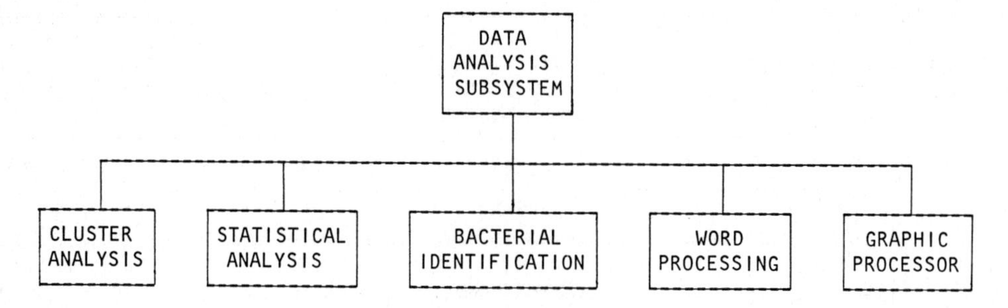

Figure 3. Data Analysis Subsystem Diagram

DATA BASE DESCRIPTION	ACCESSIBILITY	TYPE	COVERAGE	PERIOD
AFEE: Environmental Information	O/Spidel	R(B)	World	70-82
APTIC: Environmental Information	O/DIALOG, EPA	R(B)	U.S.A	66-78
Aquaculture: Aquatic Sciences	O/DIALOG	R(B)	World	70-82
Aqualine: Aquatic Sciences, Environment	O/DIALOG, ESA-IRS	R(B)	World	74-82
ASFA: Aquatic Sciences & Fisheries Abstracts	O/DIALOG, QL. sys. LTD.	R(B)	World	78-82
BIOSIS Review: Life Sciences Abstracts	O/DIALOG, SDC, CISTI,...	R(B)	World	76-82
BIOSTORET: Aquatic Biology Field & Lab. Data	B/Producer(EPA)	S(N)	U.S.A	73-82
CBI: Chesapeake Bay Inst. Oceanographic Data Bank	Contact Producer(CBI)	S(N)	C.Bay	49-82
CDI: Comprehensive Dissertation Index	O/DIALOG, SDC, BRS	R(B)	U.S.A	61-82
CDS: Air Pollution Compliance Status & Enforcement	Contact Producer(EPA)	S(N,B)	U.S.A	Weekly
CPI: Conference Papers Index.	O/DIALOG, SDC	R(B)	World	73-82
DMS: Environmental Data	O/Hydrocomp, Inc.	S(N)	U.S.A	*
Ecology & Environment: Chemical Aspects of Eco.& Env.	Cont.Prod.(Ch.Abs.Serv.)	R(B)	World	75-82
EDBD: Environmental Data Base Directory	O & B/NODC	Ref(B)	U.S.A	1850-
EDE: Environmental Information & Data	O/Da Centraler	S(N),R(B)	N.Am.	*
EIS: Environmental Information	O/BRS	R(B)	U.S.A	70-82
ENDEX: Environmental, Marine, and Coastal Resources	O & B/NODC	Ref(B)	World	*
Enviroline: Environment	O/BRS, DIALOG, SDC,...	R(B)	World	71-82
Environmental Bibliography	O/DIALOG	R(B)	U.S.A	73-82
ESIC: Inventory of Environmental Data Bases	O & B/Oak Ridge Lab.	MR(B)	U.S.A	*
FEDREG: Federal Register Abstracts	O/SDC	R(B)	U.S.A	77-82
Foundations Grants Index	O/DIALOG	S(B)	U.S.A	73-82
FWRS: Fish & Wildlife Reference Service	Cont.Prod.(Fsh&Wld.Ref.)	S(B)	U.S.A	55-82
Geoecology Project: Energy related assess.& plans	Cont.Prod./Oak Ridge Lab.	S(N)	U.S.A	65-82
GEOREF: Geological Literature	O/SDC	R(B)	World	1785-
INFOBANK: NY Times news & editorial matter	O/N.Y. Times	R(B)	World	69-82
INFORUM: Environmental Reports & Impact Statements	B/Atomic.Ind.Forum,inc.	S(B)	World	73-82
IRIS: Economics, Politics, Environment	O/DIALOG	R(B)	World	80-82
LSC: Life Sciences Collection	O/DIALOG	R(B)	World	*
MERRMS: Marine Env. Resources & Research Mng.	B/Contact Producer	R(B)	C.Bay	Curr.
MGA: Meteorological & Geoastrophysical Abstracts	O/DIALOG	R(B)	World	70-82
National Sea Grant, Marine Advisory Service	Contact Producer	S(B)	U.S.A	Curr.
Nature Conservatory Preserve: Natural Resources	O & B/N.Conservancy	S(N,B)	W.Hem	53-82
NAWDEX: National Water Data Exchange (Referral)	B/U.S. Geo. Survey	S(Ref)	U.S.A	Curr.
NRC: National Referral Center	O/Lib.of Cong.& DOE	S(Ref)	U.S.A	Curr.
NTIS: National Technical Information Service	O/DIALOG,SDC,BRS	R(B)	U.S.A	70-82
Oceanic Abstracts: Aquatic Science	O/BNDO,DIALOG,SDC	R(B)	World	64-82
OCEANIC: Oceanic & Man's Environment Abstracts	O/DIALOG,SDC	R(B)	World	64-82
OHM-TADS: Environment/Toxicology	O/ISC	S(N,B)	World	*
PIDS: Parameter Inventory Display System	O/NODC	S(N)	U.S.A	1893-
POLL: Pollution Abstracts	O/BRC,DIALOG,SDC,IRS	R(B)	World	70-82
ROSCOP: Oceanographic Observations Collected Inf.	O & B/NODC	S(N)	World	74-82
SAROAD: Aerometric Data	Contact Producer(EPA)	S(N)	U.S.A	57-82
SCISearch: Scientific Journals Abstracts	O & B/DIALOG	Ref(B)	World	74-82
SIS: County Level Data for Va.	Contact Producer (VPI)	S(N)	Va.	47-82
SSIE: Inf.on Proj.funded by Gov't & non-profit Inst.	O/DIALOG, BRS, SDC	R(B)	U.S.A	Curr.
State Natural Heritage Programs: Ident. of Elements	Contact State Source	R(B)	Local	Curr.
STORET: Water Quality Data	O/EPA	S(N)	U.S.A	1900-82
TOXLINE: Toxicology Stud.from Env.& Chemical Reac.	O/Nat.Lib.of Medicine	R(B)	World	71-82
UMD: Microbial Ecology Data	Cont.Prod.(Un.of Md.)	S(N)	C.Bay	72-82
VIMS: Hydrographic, Meteo., and Water Qu.Data	B/Va.Inst.of Marine Sc.	S(N)	C.Bay	*
WATERLIT: Environment	O/SDC	R(B)	World	76-82
WDROP: Environment	O/ISC	S(N),R(B)	U.S.A	70-82
Weather & Climate Records (Nat.Weather Service)	Contact Producer	S(N)	U.S.A	*
WRA: Water Resources Abstracts	B/Off.of Water Res.(DOE)	S(N,B)	World	68-82

O/ = On-line access R = Reference B = Bibliographic data base M = Multiple data base
B/ = Batch access S = Source N = Numeric data base Ref = Referral

Table 3. Existing Environmental, Biological, and Related Data Bases (Extended access data bases such as DIALOG, DOE/RECON, ORBIT IV, etc. are not considered here)

larities between strains and strain data sets.

Word processing may be invoked to ease formatting of data for use by the other packages and to format output. The word processing package provides screen oriented editing features to the user.

The graphics module provides translation capabilities as well as scaling and rotation routines to format graphics as needed for specific devices. This module in particular will provide device data translation to the system standard terminal. The graphics module will provide graphic and histogram translation from the other analysis packages, and also will provide background maps of the Chesapeake Bay area and environs for the purpose of correlating geographically tagged data. The output for some of this data is suitable for a color graphics presentation if the user is so equipped.

3.3 User Support Subsystem

The user support subsystem provides those elements necessary to form a user friendly system. Figure 4 illustrates the components of the User Support Subsystem. These elements provide help in the form of manuals, queries, warnings and calendar bulletins, and, if the user has the system standard terminal, several specific help elements geared to extended use of that system. This subsystem includes four modules, shown in figure 4.

The operating services and statistics module provides system standard menus, help messages, accounting, external query translation, system statistics, and the necessary interchange to accept user donated input. The user donation input element provides a feedback path that makes it easy and convenient for the user to provide corrections and to donate new material to the data base. Each element in the file is reviewed by an appropriate review committee and converted to the applicable data base or deleted as necessary. Additions to the data base are rewarded with system credits. The extended query translator provides the user with an interactive method of forming a query on external data bases with different query fomats and methods. These queries may be retained by the user in his file or, if he has a system standard terminal, on his terminal for later use on line with the data base in question. Each user may operate in direct command mode or in a menu driven mode. Either mode is connected to a help module which allows a user to examine his options at any point in an interchange. System statistics are available showing various conditions of the system such as number of users active, inactive memory, number of jobs running, etc.

The user reference aids module provides instructions, manuals directory information, and use directions for the other system modules and services. This allows a user to obtain specific instructions for local printout, or whole manuals if desired. Thus a user need not search for manuals or make special trips to solve problems of procedure or to expand his knowledge of the system.

The bulletins and warnings module provides system and other warnings of interest to the community, including weather advisories and navigation hazards. These bulletins and warnings will appear when a user signs on with a notation of

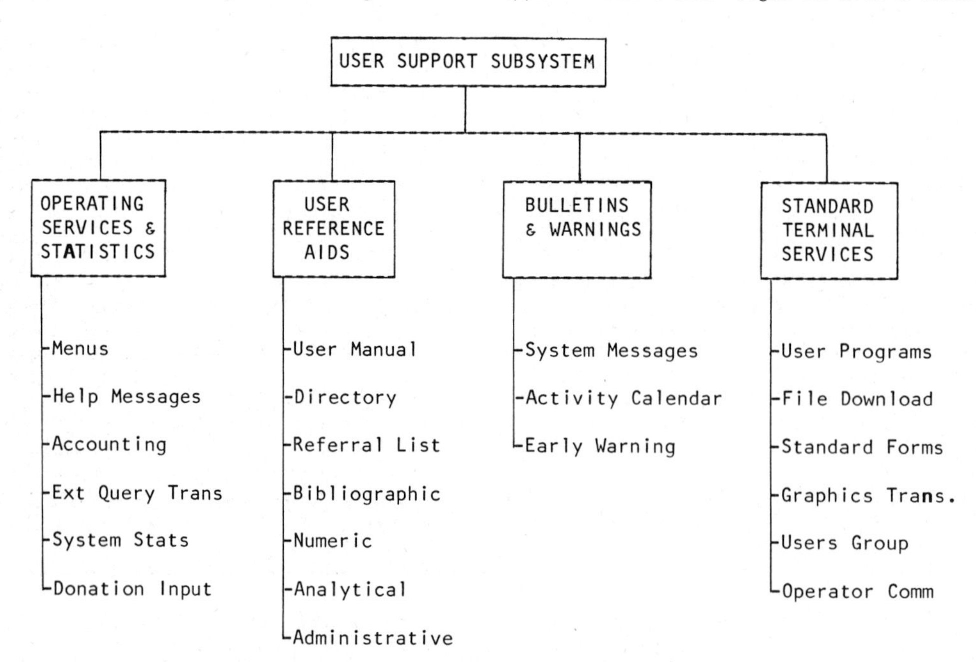

Figure 4. User Support Subsystem Functional Diagram

type allowing a user to access the actual warning or bulletin if he is concerned. Emergency bulletins such as severe weather warnings will be sent to on-line users, interrupting whatever they are doing unless the user specifically commands his terminal system to refuse such warnings. An activity calendar is included which outlines events by date of interest to the community. Each major element in the Bulletins and Warnings area can be configured to automatic warning of the user with audible tones if the terminal is so equipped. The user may input to the activity calendar via the operating services and statistics module.

The standard terminal services module provides user group type services to allow the exchange of user programs and BIF generated application programs to each user by file transfer and download systems. Programs of interest to users will include navigation computations, accounting packages, market analysis programs, and inventory control programs. Certain standard forms that are of use to the user will also be contained on the module system for those users who have the system standard printer with graphics. These will include environmental reporting forms and permit applications. Graphics output from the data analysis subsystem can be translated to system standard format and sent to the user via this module. In addition any user can communicate on-line with the system operator via this module.

4. Progress to Date

The implementation of the BIF consists of many steps from the initial concept, as the first phase, to the on-line system. The necessary steps and a synopsis of progress to date is outlined below.

BIF Data Base Acquisition - The initial bibliographic data base is an upgrade of the FAMULUS data base, available off-line on the University of Maryland's computer facility. This upgrade and the query system design and access have been completed and are ready to interface with CMS/MVS. Sea Grant has gathered a considerable amount of economic, industrial, environmental, and biological data concerning the use of Chesapeake Bay resources. Some of these data are collected through literature searches of publications from scientific, government, and private organizations, and other data are the result of a continuous research effort conducted by the Sea Grant College. These data are currently under adaptation to data base format. The source data base of Sea Grant Projects will be made after the initial system is operating. The data from this data base will consist of the status reports on file for all current projects.

BIF Data Analysis Subsystem - All data analysis modules are developed and/or procured. These modules are awaiting system installation for high level test.

System Implementation - The IBM system is procured and has been installed. Communication services are scheduled to be implemented as needed. The user system module will be the first implemented, other than the data base itself. The standard system terminal and the necessary software have been developed and are awaiting the installation of the system so that testing may begin. Graphics and color graphics user applications are presently under development by Sea Grant utilizing a RAMTEK color graphics hardware and software system.

Brochures explaining the availability and access of the system are currently in process and will be released coincident with system availability.

IV. FUTURE DEVELOPMENT - LONG TERM

Work is currently underway to determine the most efficient way of accessing the tremendous volume of information that deals with the Chesapeake Bay. Extensive computerized bibliographic and numeric data banks have been developed. However, ready access to, and use of, these data are hampered by the absence of a coordinated network of computer systems. Such a network is now in the development stage. Computer facilities will allow Sea Grant to ensure the participation of the University of Maryland in this network and will gain access to the information available, an important "gold-mine" of data to be applied to Bay research projects and management problems.

Through cooperative agreements and projects with the Maryland Department of Natural Resources, Sea Grant is continuing to develop extensive data bases dealing with Chesapeake Bay oysters, the use and effects of chlorine, and the migratory finfish of the Chesapeake region. To be useful for fishery managers, government officials and industry, these data bases must be easily accessed and periodically updated. The computer will also permit the expansion of these projects to include the use of the large number of federal data bases which are currently underutilized or unavailable in Maryland. What will be accomplished is the use of existing information sources, with specially designed software systems, to eliminate duplication of effort and to allow access to important information on environmental matters.

A data base to identify bacteria collected from the Bay has been developed and the approach is readily computerizable. The methodology has been developed by Sea Grant investigators and other University researchers. The research products will be shared with others, especially with state and county managers who have a need for such information. As additional information is added and mechanisms for identification improve, the data base will be updated and expanded. Such a data base, just one example of many potential numeric data bases, provides a tremendous potential for both researchers and public health officials. Resource economic data concerning the Chesapeake Bay will be developed. Currently the time lag in reporting data on catch statistics, prices, and related information ranges from a week for reporting on wholesale market data to a year for information on other markets and sections of the industry. With an on-line data base, the reporting time will be shortened dramatically, thus making the data base an important tool for resource managers, industry, and academicians in timely decision making. A shortened time lag will permit resource managers to evaluate short-run trends in fisheries which may need immediate attention. Industry will benefit from access to current information for input in planning, both short term and long range. Scientists will also benefit, since they will have access to current data obtained with significantly less cost and effort than at present. The development and use of this data base will be through a cooperative effort between Sea Grant and the Department of Agricultural and Resource Economics.

The expanded use of the system standard terminal in Chesapeake Bay research is an added area of interest. Through development of this area, the Bay research community will be made aware of current trends in the computer industry and their advantages. The University of Maryland will also obtain valuable feedback from the user community on how BIF will be utilized, given specialized support in an information industry. The microcomputer is expected to enhance the use and dissemination of the data base, and also is expected to generate a host of secondary on and off-line procedures. The on-line users group is designed to act as a clearing house for new ideas in this area and Maryland Sea Grant will generate programs for localized use wherever a suitable need is identified. In the future, Sea Grant will expand the system standard computer concept to include other capablities such as color graphics, and to extend the system to other microcomputers.

Long range plans call for the development of a distributed processing system, over an integrated host-to-host computer network, to allow expanded access and utility within the BIF. This will allow BIF to benefit from parallel research data bases elsewhere in the region, and across the country, and will also contribute significantly to the study of marine sciences.

V. ACKNOWLEDGEMENTS

We are grateful for the contributions of Ms. Mildred Pacl, Mr. Richard Jarman, and Mr. David Swartz who were instrumental in the formulation and implementation of the University of Maryland Sea Grant computer facility.

VI. REFERENCES

Sea Grant Report. (1980). University of Maryland Sea Grant Program.

Tucker, C. and Huber E. (1980). Inventory of Sources of Computerized Ecological Information. Oak Ridge National Laboratory, Environmental Sciences Division, Pub. no. 1561.

Guide to NOAA's Computerized Information Retrieval Services. (1979). U.S. Department of Commerce (NOAA/EDIS).

Data for Science and Technology, P.S. Glaeser (ed.)
North-Holland Publishing Company

DATA BANK ASTRA FOR THE INVESTIGATION OF CHEMICAL EQUILIBRIUM

B.G. Trusov, S.A. Badrak, I.M. Barishevakaya,
V.F. Perchik, V.P. Turov

Moscow Higher Technical School, Moscow, U.S.S.R.
Special Design and Technological Bureau
of the Information Systems, Kiev, U.S.S.R.

A data bank for investigating chemical and phase equilibria in multi-component thermodynamic systems has been developed. The data bank consists of an information subsystem for storing and processing thermodynamic properties of individual substances and also a subsystem for calculating equilibrium parameters of multi-component and multi-phase mixtures. The data bank is a universal computer program with a base of thermodynamic properties for 2000 individual substances.

Thermodynamic calculation of chemical and phase equilibrium are at present widely used at initial stages of investigation of different high-temperature processes. Examples of such processes are combustion or heating in heat engine chambers, technological processes of plasma plating, production of very pure substances or ultradispersive powders. Operating conditions for synthesis of heat-resistant and hard substances and compounds cannot be chosen without a thermodynamic analysis. Equilibrium calculations also give a large body of information for choosing optimum correlation between initial substances in high temperature metallurgy, for complex plasma processing natural raw materials, coals, natural gas, etc. Lastly, physicists working in the fields of spectroscopy and low temperature plasma also need thermodynamic data.

In all these cases a thermodynamic system is chosen. The state of the system is characterized by the mass content of chemical elements and the values of two parameters (for example, P and T or p and I). The other characteristics of state and also chemical and phase composition are calculated from the condition of a minimum thermodynamic potential. The most important computation result is the possibility to determine the composition of the system, that is the content of individual substances - components formed as a result of the reactions of dissociation, ionization and also processes of melting and evaporation.

Initial prerequisites, on which the calculation of thermodynamic equilibrium is based, formalize the description of any real system to be investigated. No kinetic or diffusion phenomena can be taken into account when making thermodynamic calculations. Every system under consideration is artificially enclosed within boundaries excluding any mass or energy transfer.

Nevertheless, in some cases the equilibrium assumption can give a fairly acceptable model for describing the real state. At a temperature of from 1500 to 5000 K and a pressure of up to 10 MPa, the time of establishing internal equilibrium is, as a rule, less than the relaxation time for diffusion or hydrodynamic processes. This allows us to use the concept of local thermodynamic equilibrium. If any process is studied as a whole, the equilibrium parameters may be used as those criteria relative to which the efficiency can be calculated.

Besides, thermodynamic calculations can provide such information, which is very difficult to obtain experimentally, for example, determination of partial pressure or mass content of components unstable at normal conditions or with low concentration. The computation method is also more accurate for determining some macroscopic properties such as enthalpy, specific heat capacity, speed of sound.

It is this possibility to receive a large body of information, without using special data, about the phenomenon that makes thermodynamic calculation of phase and chemical equilibrium so attractive.

Effective application of thermodynamic methods is possible only when computers are used. And in this case two requirements must be met: 1) the algorithms and the programs must be universal and must ensure that the solution for arbitrary thermodynamic systems should be obtained with minimum initial information for each calculation; only the chemical elements content in the system and the conditions of its interactions with the surroundings should be given; 2) information about thermodynamic properties of individual substances comprising the equilibrium system should be stored in the computer memory as the data base from which only necessary information is taken for each calculation.

It is clear that only when these requirements are met is it possible to achieve the necessary community of computational-theoretical approach to high-temperature systems with chemical and phase transformations. In other words, for complex solution of the problem it is necessary to create an automatic information-computation system or a data bank. The data bank must contain universal computation algorithms, a unified base of mutually correlated thermodynamic properties of individual substances and a set of routines for management.

ASTRA - an automatic system for collecting thermodynamic data and calculation of the equilibrium state is such a data bank. The system is in operation in the Data Centre on Physical Properties of Materials of the State Service of Standard Information Data of the U.S.S.R.

The ASTRA data bank consists of a data base, a set of service programs controlling this and a set of subroutines for calculating complex equilibria in multi-component and multi-phase systems.

The data base is an informational array with a cell structure. A cell contains information about an individual substance; its name; the chemical formula with the marks of phase and electric state; thermochemical constants (enthalpy of formation, enthalpy of phase and polymorphous transformations); polynomial coefficients approximating variations of Gibbs free energy with temperature; brief textual information (up to 900 symbols) about the method of determining and the accuracy of thermodynamic and thermochemical properties with reference to the source of the data; and the criterion of data reliability. The size of this information cell is from 300 to 1200 bytes. Thus the capacity of magnetic disk memory for the data base for 10 000 individual substances would be 8-10 Mbytes. At present, the data base contains information about the properties of 2000 individid-

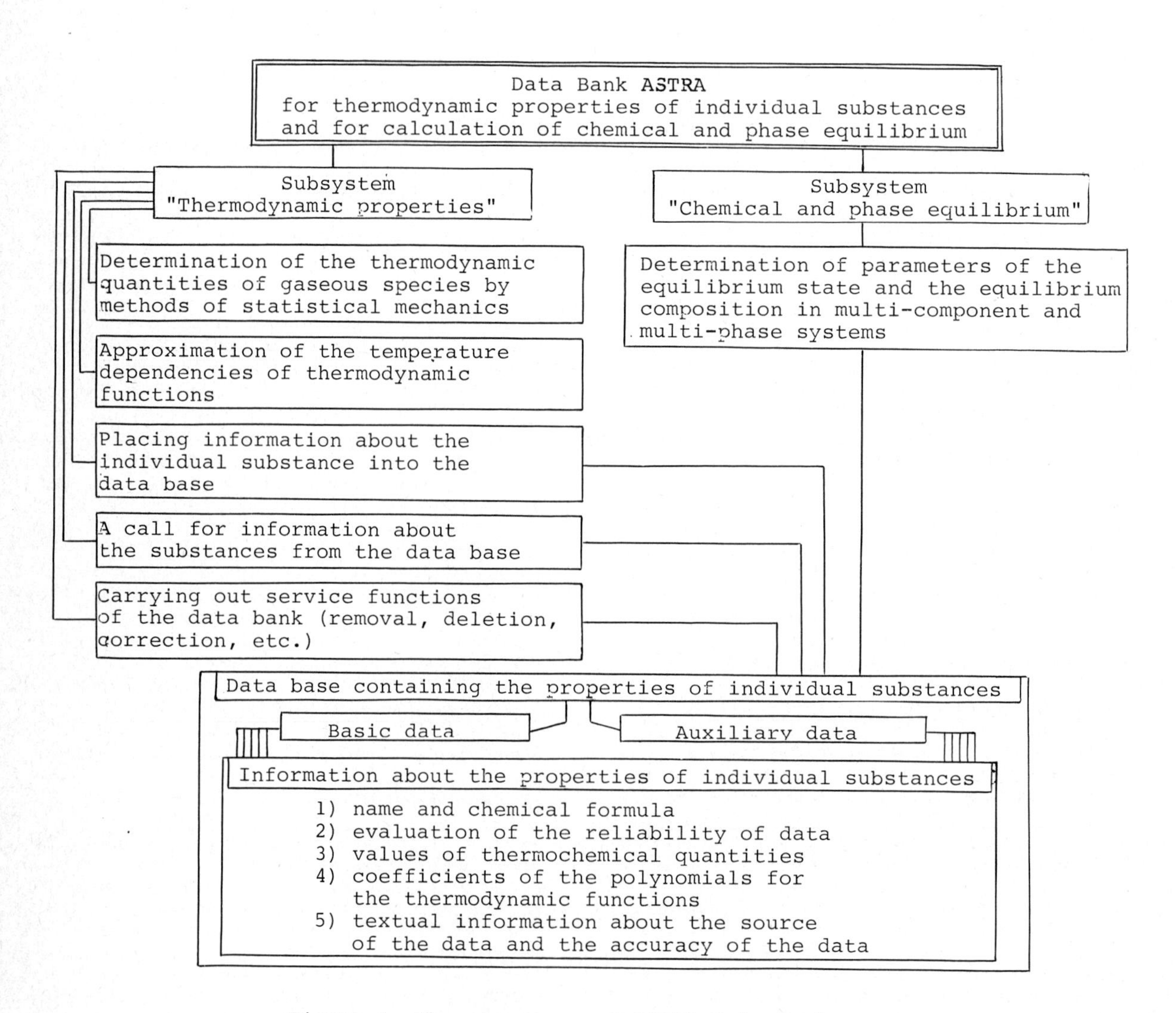

Figure 1. The structure of ASTRA Data Bank

ual substances, formed by 50 chemical elements.

The content of the data base is determined by the problems being solved. As the interest in thermodynamic research grows, the list of compounds in the data bank ASTRA steadily increases. Not less than 300 individual substances are added annually.

The chemical formula of an individual substance is taken as the key sign for search of properties of an individual substance. The formula is written in the conventional way because chemical elements are designated by means of symbols of the periodic table and the stoichiometric coefficients are given as integers. Special marks are used to indicate the phase state, electric charge, isomers and different crystalline modifications. The unique rules of formation of a chemical formula make it possible to use the formula as an identifier and as a compact form of writing data about the content and structure of the substance.

The thermodynamic properties of individual substances are stored in the data base as coefficients of approximating polynomials for reduced Gibbs free energy. As the approximation of $\Phi°(T)$ by means of the polynomial may be insufficiently accurate, the whole temperature range containing information about the properties of the substance may be divided into non-intersecting intervals with their own approximating polynomials. The polynomial with seven coefficients was chosen because it is widely used in Soviet technical literature. The properties of one substance may be described by means of one to five polynomials and their number depends on the nature of variations of $\Phi°$ with temperature. The accuracy of the approximation of reduced Gibbs free energy within 0.50 $J \cdot mol^{-1} \cdot K^{-1}$ is considered sufficient from the practical standpoint although the reference tables JANAF and the Institute of High Temperatures of the U.S.S.R. Academy of Science give the values of $\Phi°(T)$ with three significant digits after the point. The insignificant decrease of accuracy is compensated by the compactness of data representation. Using the well-known thermodynamic relationships the same approximation coefficients may be applied for calculating temperature dependence $Cp°$, $S°$, $H°(T)-H°(0)$.

Commentaries to the thermochemical constants and the thermodynamic properties taken are in a short textual form. This reference is not obligatory in the data base, but it is the only source of finding the origin of the data. It also points out the accuracy of calculation or approximation of the properties of an individual substance.

This information is indispensable when it is necessary to estimate the influence of inaccuracies in the properties of individual substances on the reliability of the calculation of equilibrium parameters or when the results obtained by means of different data bases are compared. The properties of many substances are determined only approximately and different authors give rather differing values. In these cases the ASTRA data bank makes it possible to store several sets of information. For this purpose the data base is divided into two - the basic data and the auxiliary data. The first is for storing the most reliable data and is the main source of thermodynamic data for carrying out the calculations of equilibrium parameters of multi-components systems. The auxiliary base may store the rest of the data for reference or for use in special cases.

The data base is controlled by means of a set of programs, containing 27 modules with a total capacity of about 3200 operators. They ensure carrying out the necessary operations during inquiry of the information data and during the placing of the information into the data base. Although the main purpose of the data base is to provide data for predicting chemical and phase equilibria, service programs may carry out the following independent operations; 1) placing into the data base thermodynamic properties of a substance as well as its thermochemical constants, name and textual information; 2) a call by inquiry of information about an individual substance from the basic fund or auxiliary fund; 3) a call by inquiry of the data base table of contents; 4) deletion of some thermodynamic and reference information from the data base; 5) calculation of thermodynamic properties of individual gaseous substances using molecular and spectroscopic data; 6) approximation of the thermodynamic functions by means of the polynomials with given accuracy.

At first glance, the problem of calculating thermodynamic properties of individual substances is not included in the scope of the problems for which the ASTRA data bank was created. But actually this function proved to be very important for determining equilibrium parameters. Reliable computation results for multi-component systems may be obtained only when two requirements are met: first, enough accuracy of known properties of individual substances: second, the completeness of the list of the components taken into account. That is why the addition of any individual substance which has not been analyzed before, even if its thermodynamic functions are calculated on the basis of approximate data, may result in greater accuracy of the general equilibrium parameters. The data obtained on the basis of estimations may be placed into the auxiliary fund of the data base and provided with appropriate textual information.

Formulation of definite tasks is carried out by means of a special input language, which includes a set of commands and parameters. Russian words are used as command codes. They are given in French quote and are recognized in the program by the first three letters of the key word. Thus the users of the data bank may introduce the command code either with cutting or in full form. There are twenty command codes. Their parameters may be: key words (directives), numerical constants, texts, chemical formulae of individual substances and symbols of chemical elements. The command codes arrangement is arbitrary. Special syntax checker blocks verify the correctness of the data record and the coding of the initial information and mutual compatibility of the command codes and the parameters.

As is seen from the above, the formulation of the task using the key principle instead of the positional principle of the data bank input language creates some freedom for the computer users. At the same time the unique character of the task is achieved only when all the necessary commands and parameters are introduced. A dialogue mode of communication has been developed for inexperienced users of the ASTRA data bank. The base of thermodynamic properties of individual substances (which was discussed earlier) is obviously the central part of the ASTRA data bank. But still storage, processing and even calculation of properties of simple substances are not its main tasks. Properties and composition of multi-component mixtures are of most interest for the main users of the data bank who are specialists in applied research (metallurgists, plasmachemists, spectroscopists). The characteristics of mixtures are not strored in the computer memory because of the infinite variety of thermodynamic systems of practical interest. They are determined by means of calculation.

The method of determining equilibrium parameters is based on the main principle of thermodynamics - achievement of the entropy maximum at constant internal energy and at zero value of the work performed at the system. Besides that, in case of a closed equilibrium system, the condition of conversation of chemical elements mass and total electric neutrality must be met.

Thus the problem of calculating equilibrium in multi component mixtures is reduced to the task of finding conditional entropy extremum observing the physically substantiated limitations on the region of the component concentration variations (numbers of moles).

The most general character of the problem leads to a non-linear system of equations in which the number of the unknown quantities exceeds the number of equations by two. These two parameters must be defined, to establish the conditions of interaction of the system considered with the surroundings. These parameters may be pressure and temperature, or specific volume and internal energy or any other two parameters of state.

The algorithm used in the data bank allows us: 1) to determine equilibrium parameters of the systems with arbitrary chemical composition, 2) to include in the number of expected components of equilibrium composition all individual substances from the basic and the auxiliary funds of the data base.

The solution based on the method used is carried out by means of successive approximations. At each step of the iteration procedure the reference system is linearized by the Newton method, is reduced to a set of linear relations and is resolved relative to the preceding approximation of the unknowns. Initial values from which the iteration procedure begins are produced by a program. To ensure successive convergence from an arbitrary initial approximation the iteration values are obtained at each step by correcting the values obtained at the preceding step. To increase the stability of the solution logarithmic variables are used.

The algorithm used makes it possible to consider heterogeneous systems containing not only pure condensed non-mixing phases but also ideal solutions. A solution model is the simplest and it can be realized using only the information from the data bank. The calculation carried out using this method allows us to determine the possibility of formation of solutions and the separate phases. In the case where the existence of the solution is possible, the concentration of its components is calculated. Thus the phase content of an arbitrary thermodynamic system as well as the chemical composition are calculated using only

```
Example 1.    <TASK> INFORM   <FUND OF DATA> BASIC
              <CHEM.FORMULA> CRO3, K*ZRO2   <DATA> T=1000,3500,50   <END>

Example 2.    <TASK> PROPS
              <REGIME> CONTROL
              <INPUT> TECHN.SYST.UNITS
              <DATA>DH=45000,OE=898.8,BE=.5286,OXE=6.5,AE=0.005,SIG=1,
                    E=0, 16584, G=10, 10,T=500,5000,50
              <CHEM.FORM.> CRO
              <TEXT> THE ENTHALPY OF FORMATION AND THE MOLECULAR CONSTANTS
                     FOR THE CALCULATION OF THERMODYNAMIC  FUNCTIONS WAS
                     TAKEN FROM [1]. METHOD OF MAYER AND GOEPPERT-MAYER
                     WAS USED.
              <REFERENCES> 1. JANAF THERMOCHEMICAL TABLES. 1975 SUPPL. - J.PHYS.
                           CHEM.REF.DATA, 1975, VOL.4, NO.1, PP.1-175.
              <NAME> CHROMIUM OXIDE
              <END>

Example 3.    <TASK> EQUILIBRIA   <INPUT>SI   <OUTPUT> SI, M
              <DATA> DATE= 5 OCT.1982, NUMBER= EXAMPLE, P=0.1,T=4000,3500,
                     3000,2750,2500, (10%C3H8),(80%N53.94 O14.48 AR 0.32),
                     (10% AL2O3)
              <END>
```

Figure 2. The examples of the sets of input information for ASTRA data bank (English version of key words).

those reference data for the individual substances which are stored in the data base.

The program using the universal method of calculating equilibrium consists of 17 modules and its capacity is 2400 operators. It is an integral part of the data bank and is called by using the control of sentences of the input language mentioned above.

The initial data must define the elementary composition of a thermodynamic system and the parameters determining the conditions of its equilibrium. A list of the individual substances defining the qualitative composition of the assumed condensed solutions is also included into the source data. All the other parameters and constants necessary for the computation process are given or determined in the program itself. This frees the users of the data bank from a large volume of preparatory work.

The ASTRA data bank has been used for some years on the computers of the Common System for simulating many technological and energy processes and has been shown as reliable and convenient in operation.

At present work is being carried out on developing the second version of the ASTRA data bank with the added possibility of solving the inverse problem of chemical equilibrium and of calculating the transport properties of multi-component mixtures.

Data for Science and Technology, P.S. Glaeser (ed.)
North-Holland Publishing Company

DATA BANK ON THERMOPHYSICAL PROPERTIES OF FLUIDS

V.V. Sytchev, A.D. Kozlov, G.A. Spiridonov

Soviet National CODATA Committee
Academy of Sciences of the U.S.S.R.

A thermophysical data bank on properties of fluids for cryogenic, heat power engineering and chemistry has been developed at the Moscow Power Engineering Institute and State Research Center of the U.S.S.R. Standard Reference Data Service. Users receive tables, diagrams and equations of thermophysical properties for any set of variables.

Today thousands of working substances and materials are used in various branches of science and technology which need data on various properties - thermophysical, electrophysical, dielectrical, nuclear, mechanical. There are also data needs on toxicity, thermostability, aggressiveness, etc., in solid, liquid, gaseous and plasma states. Information on properties of substances and materials is provided in the following three forms - tables, diagrams or equations. The set of properties, the range of parameters and the type of input variables are determined by the user.

In the area of thermophysical properties a user needs to know specific volume, density, compressibility, enthalpy, entropy, heat capacity, sound velocity, adiabatic index, fugacity, Joule-Thomson coefficient, thermal coefficients, phase transition heats, coefficients of viscosity, heat conductivity, autodiffusion, interdiffusion, Prandtl's number and many others. For pure substances the foregoing properties are needed in single and two-phase ranges along the equilibrium lines, along the extremum lines. Many data users are interested also in thermodynamic quantities' discontinuities by transition from single-phase to two-phase area of states.

The analytical means used for the determination of thermodynamic and transport properties is, in an overwhelming majority of cases, a system of equations which has density and temperature, or specific volume and temperature, as variables. In modern terminology such variables are qualified as the physical ones. However, in practice, problems of properties computation are often settled by various input variables such as pressure and temperature (P,T), pressure and enthalpy (P,H), pressure and entropy (P,S), pressure and degree of dryness (P,x), temperature and degree of dryness (T,x), entropy and degree of dryness (S,x), enthalpy and degree of dryness (H,x), and so on. The foregoing variables' combinations do not exhaust all the versions which one comes across. They are inherent mainly to cryogenic and power engineering computations. Other versions of input variables are possible in other branches of technology. Moreover users also want to have equations in variables which enable the straight computation of the needed thermodynamic and transport properties. This is important when the time required for properties' computation constitutes a substantial part of the work.

Thus by providing the user with thermophysical data we, in fact, deal with a great number of different versions of inquiries. Neither a monograph nor a handbook could serve as a source of such universal information.

The only way to solve the problem is to build a computer-aided information system which is able to generate data according to the user's inquiry. There seems to be two such opportunities.

In the first place a stationary system may be organized at the base of a highly productive computer with a developed system of terminals, which enables work in a conversational mode with time sharing. The necessary information on the properties of fluids is obtained by the users from the suitable terminals with the aid of standard program packs. By necessity the obtention of thermodynamic and transport data from any structural part of the foreground program may be organized.

It is also possible and it seems to be more realistic at present to organize information circulation with the aid of a magnetic carrier (tape or disk). The structure of the user's set consists of: tape cartridge or floppy disk, containing FORTRAN program for properties' calculation, user's instructions and microfiches with skeleton tables by node values of premises. The tape cartridge or floppy disk information circulation is oriented towards users who possess a minicomputer. In fact, information circulates via the "magnetic handbook", which satisfies users' demands and which is available to them from the moment the equations of thermodynamic and transport properties are obtained. Moreover, in this case "the magnetic handbook" on a 1 Mbit floppy disk will include the complete thermophysical information for 400-500 fluids, while information on one fluid consists of 200 numbers.

For many years at the State Research Center of the U.S.S.R. Standard Reference Data Service (VNITS GSSSD) and Thermodynamic Data Center at the Moscow Power Engineering Institute (MEI) research was carried out to create the thermophysical information-computing systems which could work both with developed terminal system high-productive computers and with minicomputers equipped with tape cartridges and floppy disks (diskettes). Several versions of the system have been developed to date. There are some differences between the versions determined by the peculiarity of a specific computer, but they are alike in their scientific methodology aspects. All the system's versions are destined to work both in a table calculation mode and in approximation mode. An application program pack is organized like the program modules in FORTRAN. The minimum necessary information on a substance is specified as a number file consisting of

$$\{b_{ij}\},\{\alpha_j\},\{\beta_j\},\{a_{Sj}\},\{a_{\lambda j}\},\{a^{\lambda}_{ij}\},\{a^{h}_{ij}\},\{a^{D}_{ij}\},$$

coefficients and an array of physical constants which are consistent with the substance. The sense of the coefficients' arrays is seen in the following experimentally grounded equation system which has the standard structure for all substances stored in the data bank.

$$Z = 1 + \sum_{i=1}^{r} \sum_{j=0}^{Si} b_{ij}\omega^i/\tau^j \qquad (1)$$

$$C^\circ_p/R = \sum_{j=0}^{m} \alpha_j\theta^j + \sum_{j=1}^{n} \beta_j\theta^{-j} \qquad (2)$$

$$\pi_S = \sum_{j=0}^{m} a_{Sj}\tau_S^j \qquad (3)$$

$$\pi_\lambda = \sum_{j=0}^{m} a_\lambda\tau_\lambda^j \qquad (4)$$

$$\lambda = \sum_{j=0}^{r} \sum_{j=0}^{Si} a^{(\lambda)}_{ij}\omega^i/\tau^j \qquad (5)$$

$$\eta = \sum_{i=0}^{r} \sum_{j=0}^{Si} a^{(h)}_{ij}\omega^i/\tau^j \qquad (6)$$

$$D = \sum_{i=0}^{r} \sum_{j=0}^{Si} a^{(D)}_{ij}\omega^i/\tau^j \qquad (7)$$

where (1) is the thermal equation of state, (2) the isobaric heat capacity in the ideal-gas state, (3) the saturation pressure, (4) the melting pressure, (5) the heat conductivity coefficient, (6) the viscosity coefficient, (7) the diffusion coefficient.

The system (1) - (7) of equations is first obtained by statistical processing of a large amount of experimental data. In accordance with the GSSSD program, the work is done by specialized data centers. Nowadays, jointly developed information-computing systems are giving reference data for water, nitrogen, oxygen, air, carbon dioxide, methane, ethane, ethylene, ammonia, hydrogen, helium-4, argon, krypton, xenon, neon, sulfur hexafluoride and some other technically important fluids. The software of the system enables us to add new program modules and new fluids data files to the system.

As for the systems operating in VNITS GSSSD the work on computing the thermophysical properties of gaseous and liquid mixtures from pure component data has been finished. The precision of the properties' calculation is not less than that of their experimental determination. The only restriction with the method is that the computations are made at temperatures above the triple point of the component with the highest boiling point. This restriction is characteristic of other computation methods also. The program for mixtures is organically chained with the programming support for the computation of the properties of pure substances.

The practice of thermophysical information-decision systems' operation has shown high efficiency. The systems developed on a minicomputer base and provided with tape cartridges and floppy disks have appeared to be the most convenient for the users. Such data carriers are a new type of thermophysical handbook. In fact, they may be easily conveyed, operatively "updated" and "disseminated".

REFERENCES

[1] Voronina, V.P., Kozlov, A.D., et al., The computer-aided system of standard reference data on the thermophysical properties of fluids. Survey information. Moscow, GSSSD, (1977).

[2] Grankina, L.N., Kasianov, Ju. I., Spiridonov, G.A., Some problems of thermophysical data banks' software for cryogenic substances. Thesis report. Leningrad (1981).

Data for Science and Technology, P.S. Glaeser (ed.)
North-Holland Publishing Company

EVALUATION OF THE EFFICIENCY OF COMPUTER-AIDED SPECTRA SEARCH SYSTEMS

Karl Schaarschmidt

Technical University of Dresden
German Democratic Republic

Information theory can be applied to evaluate the efficiency of retrieval systems aimed at the identification of substances by computer-aided comparison of spectra. Therefore, a significant number of search processes must be analyzed. The respective conditions and relations and some general results are presented.

In the literature much has been published on computer systems dealing with the storage and retrieval of molecular spectral data. Some systems were also presented at earlier CODATA conferences. (For recent review articles see [1...4] e.g.) It cannot be the purpose of this short communication to describe another system for infrared spectra, since we have been working successfully in Dresden for approximately five years [5]. The object of this paper is to present some ideas and experiences on the evaluation of such search systems.

It generally is the aim of spectra search systems to identify an unknown substance by comparing its spectral data with the data stored in a library. Only this object will be discussed here. For the evaluation of efficiency two points of view are considered.

The first one is that until now almost any search system, especially for infrared spectra, has not used the full information contents of the spectra, i.e. the complete spectral curve. They need some data compression. The result of such data compression may be a band table or a bit string or something else. It is obvious that such an approach is connected with some loss of information. Nevertheless, at present this cannot be avoided for reasons of storage capacity and computer speed, that is if you want to handle a sufficiently large data base. The question is: which is the most useful data compression system? Is it possible to compare different search systems in an objective way? These are questions which arise for anybody dealing with search systems.

The second point of view is that nowadays the acquisition of fully digitized spectra has become a common and useful technique, especially with the development of the FTS-technique and the coupling of common spectrometers with microcomputers. Only in some cases is direct comparison of such fully digitized spectra made. (For critical comparison of the applied algorithms cf. [6]) As already mentioned, data compression is generally needed, i.e. encoding. In former times encoding was done manually. The information not used became irreversibly lost. Now encoding can be done by computers, it can easily be repeated if new ideas on effective encoding are elaborated, cf. [7]. In this case, we also want to compare the different encoding instructions in an objective way.

For these purposes information theory should be applied. (General principles of information theory applied to analytical chemistry have been recently summarized [8]) In order to do this correctly the following conditions have to be met:

1) a sufficiently high number of test runs has to be performed for substances contained in the data file;
2) the test substances must be randomly selected, e.g. by an adapted random number generation program;
3) the spectra of the applied test substances must be of an origin different from that of the spectra stored in the data file (taking into account the influence of sample preparation and instrumental parameters).

If the data file contains N references, the probability $p(B_j)$ of reference j (j=1,2,3,...,N) to be searched for is given by:

$$p(B_j) = \frac{1}{N} \quad (1)$$

In the terms of information theory this is the probability of event B_j, and $H(\beta)$, the information contents or entropy (in bit) of the system is:

$$H(\beta) = -\sum_{i=1}^{N} p(B_j) \cdot \mathrm{ld}\, p(B_j) \quad (2)$$

(ld = logarithmus dualis)

This is the maximum information we can obtain from the identification system, provided that each test run will give us the correct answer without any ballast.

It is a well-known experience, however, that for various reasons (the spectroscopic ones will not be discussed here in detail) the correct answer is often found by the computer together with some other references, which obtained the same or an even better "figure of merit" by the applied algorithm. If for a test run N_i references in total are selected as possible "hits", the spectroscopist must visually compare the respective spectra in their original form. (Here the positive role of the ballast will not be discussed in more detail, this is considered in [4]). The result of a search run is referred to as event A_i (i=1,2,3,.. ..,N), if N_i references have been selected by the computer. Unfortunately, sometimes there are search runs with negative results: the correct reference is not found by the computer, it remains in the unselected part of the data file. In this case we set i≈N, in order to include such results into the totality of the search runs. With

$$p(A_i) = \frac{\text{number of events } A_i}{\text{total number of test runs}} \quad (3)$$

we get an (empirical) estimation of probabilities $p(A_i)$.

Applying the principles of information theory (cf. [9] e.g.) we consider the events A_i and B_j as possible results of two experiments α and β, respectively. α is the application of the computer. β is the unique identification of the correct reference. Both experiments reduce the degree of the previously existing uncertainty, i.e. they supply us with information. However α and β are not independent from each other. A special result A_i of α changes the probability distribution of $p(B_j)$ from the value given by (1), into $1/N_i$ for i references and zero for the remainder. These are conditional probabilities of B_j under the condition of realized experiment α with the special result A_i.

The totality of all test runs gives us the conditional entropy of β (entropy of β after realization of α) generally expressed as $H_\alpha(\beta)$. In the present case

$$H_\alpha(\beta) = \sum_{i=1}^{N} p(A_i) \cdot \text{ld } N_i \quad (4)$$

(for details of deduction cf. [10]). $H_\alpha(\beta)$ is the uncertainty with regard to β, which remains subsequent to the realization of α. It should be small, if not zero. The difference $H(\beta) - H_\alpha(\beta)$ is referred to as information gain $I(\alpha,\beta)$, which α presents with regard to β.

$$I(\alpha,\beta) = \text{ld } N - \sum_{i=1}^{N} p(A_i) \cdot \text{ld } N_i \quad (5)$$

all terms being expressed in bit. This is the quantity we need to evaluate the efficiency of a given search system.

An application of equation (5) to our search systems for infrared spectra has been described elsewhere [10]. Here some general principles only should be discussed.

1) Volume of data file: the influence of the quantity of references is directly contained in the term ld N.
2) Structure of data base: as pointed out in equation (2), $H(\beta)$ = ld N is only valid, if (1) is met. Mathematically, this can be postulated, for real search systems, however, this condition means: each reference must have the same chance to be searched for, i.e. to be a real analytical problem. This postulates a well balanced composition of the data base. Any reference which is never searched for will diminish the entropy of the system, as well as does any reference which is searched for too often.
3) Quality of data sets: spectra of poor quality and errors of encoding will cause the number of negative search results to increase. This is reflected in the second term of (5): it increases with high values of $p(A_i)$ for i=N. Another point of view is connected with the following.
4) Encoding instruction and algorithm of comparison; search systems can apply very strict criteria for the identity of data sets, but very soft ones, too. The former may be applied if the encoded spectra are of high quality, the latter, if the quality of the spectra is not so good, such as may be caused by instrumental conditions of older sources. If softer criteria are applied, the percentage of negative results will decrease, but ballast will increase, i.e. in (5) the $p(A_i)$ with higher values of i. It is obvious that finding a good algorithm (connected with the encoding instruction) is a problem of optimization which is mainly determined by the quality and structure of the data base.

Finally, it should be emphasized that theoretical considerations like these are completely in accordance with experiences which will be obtained more intuitively after having worked in the field of computer-aided search systems for a long time. In our case it was the analytical infrared spectroscopy which made us feel the need of the theoretical treatment of such problems. We think that these results may be generally applied to computer-aided search systems.

REFERENCES:

[1] Zupan, J., Anal. Chim. Acta 103 (1978) 273.

[2] Fisk, C.L., Milne, G.W.A. and Heller, S.R., J. Chromatogr. Sci. 17 (1979) 441.

[3] Hippe, Z. and Hippe, R., Appl. Spectroscopy Revs. 16 (1980) 135.

[4] Schaarschmidt, K, Z.Chem., to be published.

[5] Schaarschmidt, K., Z. Chem. 18 (1978) 337.

[6] Rasmussen, G.T. and Isenhour, T.L., Appl. Spectroscopy 33 (1979) 371.

[7] Buechi, R., Clerc, J.T., Jost, Ch., Koenitzer, H. and Wegemann, D., Anal. Chim. Acta 103 (1978) 21.

[8] Doerffel, K. and Eckschlager, K., Optimale Strategien in der Analytic (VEB Deutscher Verlag fuer Grundstoffindustrie, Leipzig, 1981).

[9] Jaglom, A.M. and Jaglom, I.M., Wahrscheinlichkeit und Information (VEB Deutscher Verlag der Wissenschaften, Berlin, 1967).

[10] Schaarschmidt, K., Anal. Chim. Acta 112 (1979) 385.

Data for Science and Technology, P.S. Glaeser (ed.)
North-Holland Publishing Company

DARC SYSTEM: TRANSFORMATIONAL RELATIONSHIPS AND HYPERSTRUCTURES IN CHEMISTRY

R. Picchiottino, G. Sicouri, J.E. Dubois

Institut de TOpologie et de DYnamique des Systèmes (ITODYS)
1 rue Guy de la Brosse, 75005 Paris, France

Aspects of the DARC-SYNOPSYS expert consulting system in the representation and use of knowledge for chemical synthesis are presented. Particular cases exemplify the structural description of chemical reactions, homogeneous with that of compounds, by the DARC-IGLOO method. Such an operator representation leads primarily to the controlled transition from specific data to more generic transformational relationships or productions. These in turn can be used as input to the PARIS chemical production system for synthesis planning. The resulting problem space, called a transformational hyperstructure, is discussed in terms of its nature, representation and confrontation to real cases.

1. INTRODUCTION

Recent achievements in the rational computer use of the DARC System [1] with data related to chemical compounds - e.g. structures, bibliographical references, spectra, thermodynamic or kinetic properties - provide the basis for an analogous work for reactions. In this case the difficulty of defining a reaction and of identifying the associated set of data leads to a more complex situation.

The perception of a chemical reaction proceeds gradually from qualitative descriptions - i.e. stoichiometry, medium, mechanism - to a more quantitative one: reactivity (Fig. 1). All of these can be expressed with three types of information: numerical, e.g. yield, rate, pressure, temperature; graphic structural, i.e. structures of all the known compounds occurring in the process; and textual, e.g. bibliographical references, comments. These data can be used first for simple retrieval and display of specific information, and then as a basis for computer-aided design. The proposed reaction descriptions - linear codes [2], computer routines [3], variation matrixes [4] - correspond to a compression of information directly influenced by a particular problem; e.g. synthesis planning, mechanism simulation, reaction nomenclature. Examples are

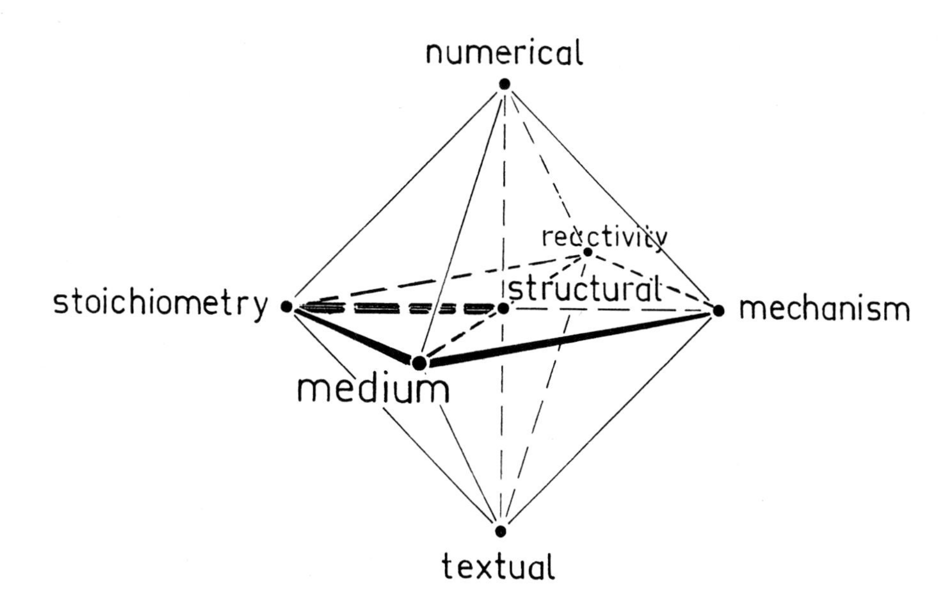

Figure 1: Reaction perception and information.

given of a more general language [5] leading primarily to the controlled determination of generic representations from specific ones and thereby making possible grouping and analogy searches. To this end, perception of a chemical reaction is restricted to the link between stoichiometry and structural information in Fig. 1. The use of such generic representations as production rules for chemical production systems is also discussed.

2. DESCRIPTION OF GENERIC TRANSFORMATIONAL RELATIONSHIPS

2.1 Compound oriented language

The KETO REACT specific reaction data base [6] contains a complete description (structural diagram, medium, yield, bibliographical references) of reactions leading to the synthesis of hypercrowded ketones [7]. The structural diagrams are described with block couples: a block is a set of chemical compound structures and the couple (Di, Ai) contains the blocks Di and Ai corresponding to the reactants and products of the reaction, respectively. The search associated with a chemical query such as "oxidation of an alcohol to a ketone" will be expressed in terms of compound substructures. The query language uses fuzzy structures where stars indicate that terminal carbon atoms can be followed by any kind of substituent [1]. The query just mentioned involves searching for the reactions (Di, Ai) in which *C-CHOH-C* is a substructure of Di and *C-CO-C* is a substructure of Ai. This kind of query, said to be by compound substructures, will be denoted by the couple: (*C-CHOH-C*, *C-CO-C*). The result of such a search of the KETO REACT data base will yield two main types of answers (Fig. 2). Firstly a population of specific relevant reactions which corresponds to the overall change indicated by the query. Secondly, a set of structural diagrams where (i) the change is not located on the same part of both structures and (ii) the changes in the diagram are greater than the change sought. The second type of answer constitutes the noise, i.e. answers inconsistent with the initial chemical query, basically due to a loss of localization of changes in the block couple description. So a more accurate description of structural diagrams is needed mainly to express changes common to a population of similar reactions.

2.2 Reaction oriented language: IGLOO method

The basic feature of the Invariant Graph and Localized Ordered Operators (IGLOO) method is to provide the reaction with a unique order [5]. Fig. 3 provides an example of determining the transformation structure (the Invariant and Localized Operators: ILO) modeling a reaction structural diagram. One of the major steps of the method [8] is to determine the Invariant modeled with the Greatest Common SubStructure (GCSS) to the structures on both sides of the arrow. The GCSS carries the order of the reaction and provides a quantitative localization of the variations on specified invariant sites. These sites are denoted by the -1 chromaticity which corresponds to a formal electron density defi-

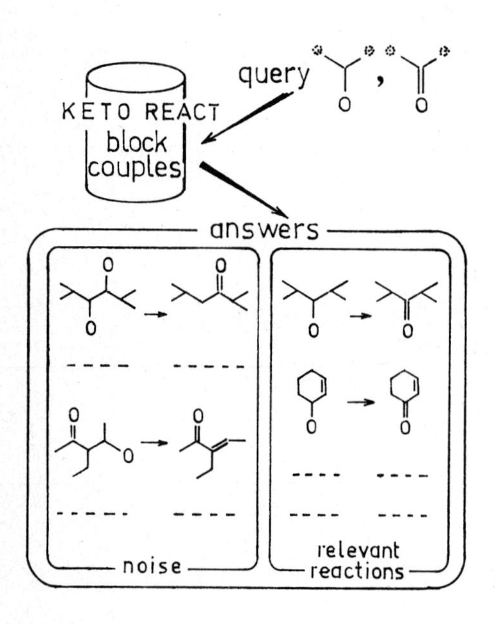

Figure 2: Example of a compound substructure search of KETO REACT.

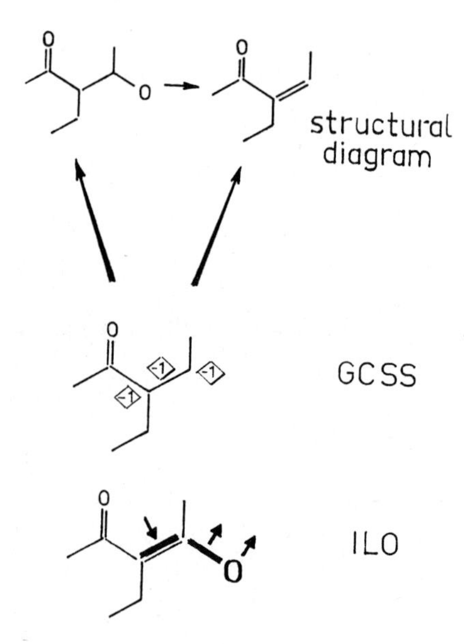

Figure 3: Transformation structure (Invariant and Localized Operators) associated to a reaction structural diagram (hydrogen atoms are implicit).

ciency of 1. The operations needed for rebuilding the reactants (ABlation Operator; ABO) and the products (ADjunction Operator: ADO) are indicated on these sites with outgoing and incoming arrows, respectively. By reformulating the problem of accessing KETO REACT with ILO descriptions of structural diagrams, it is clear that, with a relevant ILO substructure query, access is gained to the wanted chemical change (Fig. 4). In other cases the change is either absent from ILO or not limited to the desired change. It should be noted here that stars signify that terminal carbon atoms can be followed by any kind of invariant environment.

2.3 Scope of generic transformational relationships

The word reaction often designates a generic reaction representing a family of related-vicinal-alike reactions as opposed to a completely described specific reaction.

A generic transformational relationship corresponding to a homogeneous set of chemical reactions can be modeled by a Generic ILO structure. This GILO covers a set of Specific ILOs which are either determined from the data base or formally deduced from the GILO structure (Fig. 5). These SILOs correspond to organic chemical reactions that (i) can be carried out experimentally and are present or missing from the data base; or that (ii) have never been actually tested in the laboratory. Criteria leading to the restriction of large sets of SILOs with controlled predictive rules have been presented elesewhere [7]. The following example deals with the use of such GILO structures in a chemical production system for synthesis planning.

3. TRANSFORMATIONAL HYPERSTRUCTURES FOR SYNTHESIS PLANNING

3.1 PARIS chemical production system

Production systems developed in Artificial Intelligence are used in chemistry to solve problems such as elaborating synthesis design or simulating mass spectra [9]. A recently proposed formalism [10] describing Controlled Production System (CPS) distinguishes as input a control language, a state description, a production description and a search strategy which produce the problem space. One advantage of such CPS is that space determination, based on the control language, can be clearly separated from space generation, based on the search strategy. In chemistry, chemical structures (S) are the states, transformations of structures (TS) are the productions and organized populations of structures called hyperstructures (HS) are the problem spaces [11].

In the PARIS (Pertinent Analysis of Reaction by Igloo Simulation) CPS, the production data base

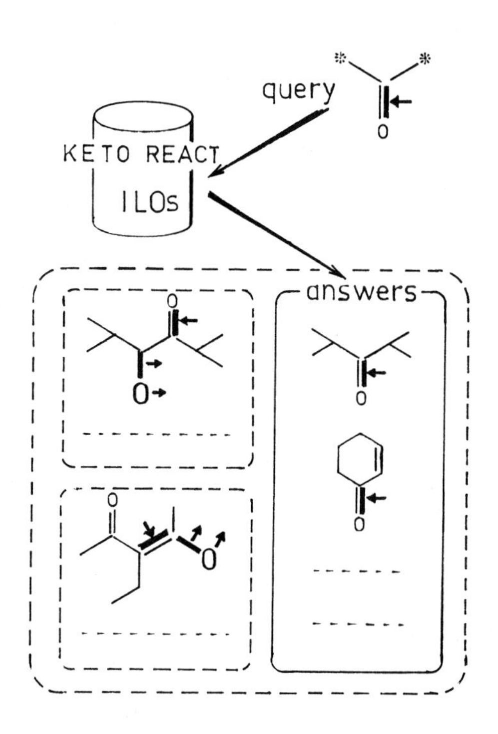

Figure 4: Example of an Invariant and Localized Operator (ILO) substructure search of KETO REACT.

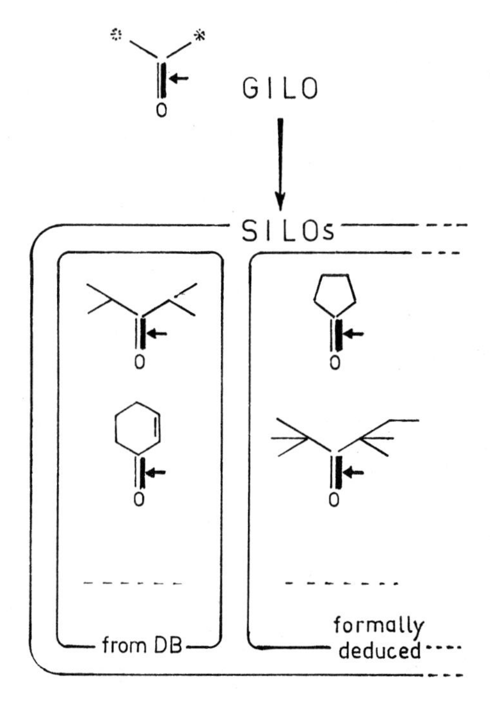

Figure 5: Scope of a generic transformational relationship modeled by a Generic ILO; corresponding set of Specific ILOs.

is composed of GILOs corresponding to usual types of organic synthesis reactions such as Diels-Alder, Michael,... The result of recursively applying such productions to a determined source state is a transformational hyperstructure problem space. Each GILO gives rise to two types of operator: a direct operator (in the chemical reaction direction) and an opposite operator (in the opposite direction). The control language limits the possible application sequences for these operators according to two criteria: type of operator deduced from GILO and application mode; hierarchy of reaction classes. The problem space is staked out by the goal condition (e.g. stop when compounds have less than seven carbon atoms) and the control language (e.g. opposite operator and no reaction class hierarchy). It is constructed by an overall breadth first search method [12] and without interaction. Using GILOs as productions and the information content of a transformational hyperstructure produced by the PARIS program in its 3.0 version [13] are dealt with hereafter.

3.2 Using a GILO as production

Each type of structural operator deduced from a GILO is specified by its application field and localized modifications. This corresponds to the precondition and action commonly used to define productions [10]. The precondition is that a specific structural block belongs to the population covered by a generic structural expression. The action is to carry out the indicated structural operations on the variation sites. As for the GILO 1 direct operator of Fig. 6, where R equals a hydrogen atom or a carbon atom indifferently substituted, the generic structure associated with the precondition is deduced from the GILO by applying the ABlation Operator (ABO) to the GCSS. The action is to break the bond between atoms 1 and 2 and to create a double bond between atoms 2 and 3. Reducing the production data base to the two GILOs of Fig. 6 yields a more pedagogical transformational hyperstructure for subsequent discussions.

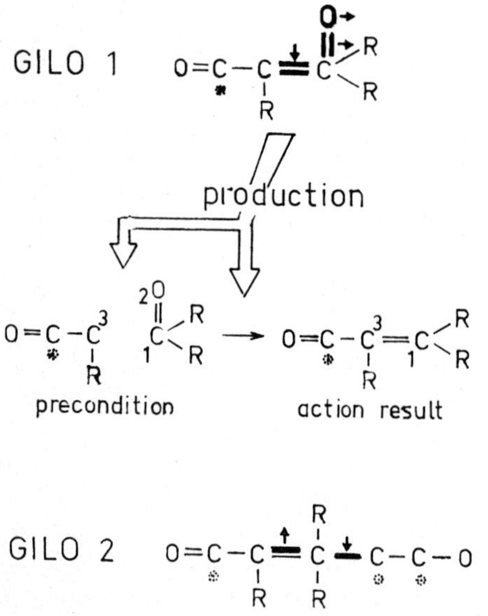

Figure 6: Example of using a GILO as production. (GILO variation sites have a different valency constraint than the usual one for chemical structures [8]).

3.3 Nature of the problem space

Representation problem

The transformational hyperstructure obtained as output to the PARIS program is a block hyperstructure [14]. It is defined by the couple HS=(P,B) of a population P of structures and a block relationship B over P. B is a binary relationship between structural blocks and the hyperstructure is represented as an oriented chromatic hypergraph. Representative graphs, in which structures and unitary transformations are specified by vertices, yield explicit displays.

Information content

The transformational hyperstructure representative graph produced by the PARIS program and corresponding to source structure 1 and using the two GILOs of Fig. 6 is displayed in Fig. 7. Two types of nodes are distinguished: one corresponding to the unique numbering of the structures (circled or not); and one corresponding to the number of the GILO opposite operator contained in the data base (diamond node). Dashed circled structures indicate that the goal condition (less than seven carbon atoms) is met for the specified structure. When two structures are identical, they have the same numbering and the search is continued only with the first one generated.

Choosing a realistic synthesis plan

The underlined pathway is the result of confronting a formally constructed synthesis plan and an experimentally proposed one [15]. On this pathway, a special notation indicates (i) that opposite operator 2 has two identical applications on structure 12 to produce structures 8 and 10 and (ii) that opposite operator 2 applied to structure 13 gives two identical structures numbered 10. For each structural block (3, 4+5, 6+7) the direct operator corresponding to the opposite operator used to generate these blocks has been applied. Such an enrichment of the problem space [16] can guide the choice of one path over others by topology criteria applied to the transformational hyperstructure. In this particular case, the path leading to 3 can be discarded because it produces nine competitive applications instead of three for the other two. In order to distinguish these last two, criteria based on structural reactivity are needed.

4. CONCLUSION

These examples suggest various applications of the IGLOO language for the description of chemical transformations. This language provides the basis for treating organic chemical reactions in the DARC-SYNOPSYS (SYNthesis OPtimization SYStem) expert consulting system. It provides the building of and access to specific reaction data bases that constitute the representation of knowledge, the first step in computer-aided design. Treating these data to extract analogies among them can yield the controlled definition of generic transformational relationships. These rules are used as input to chemical production systems, providing users with expert conclusions about synthesis planning.

5. REFERENCES

[1] Dubois, J.E., Approche informatique en chimie heterocyclique, 7th International Heterocyclic Chemistry Meeting, Marseille (1981); Attias, R., DARC substructure Search System: a new approach to Chemical Information, submitted for publication to the J. Chem. Comput. Sci.

[2] Hendrickson, J.B., Systematic synthesis design. 4. Numerical codification of construction reactions, J. Am. Chem. Soc. 97 (1975) 5784-5800; Littler, J.S., An approach to the linear representation of reaction mechanisms, J. Org. Chem. 44 (1979) 4657-4667.

[3] Corey, E.J., Wipke, W.T., Cramer, R.D., Howe, W.J., Computer assisted synthetic analysis for complex molecules. Methods and procedures for machine generation of synthetic intermediates. J. Am. Chem. Soc. 94 (1972) 440-459; Salatin, T.D., Jorgensen, W.L., Computer-assisted mechanic evaluation of organic reactions. 1. Overview. J. Org. Chem. 45 (1980) 2043-2051.

[4] Dugundji, J., Ugi, I., An algebraic model of constitutional chemistry as a basis for chemical computer programs, in Computers and Chemistry (Springer-Verlag, Berlin, 1973).

[5] Dubois, J.E., Structural organic thinking and computer assistance in synthesis and correlation, Isr. J. Chem. 14 (1975) 17-32; Dubois, J.E., Computer-assisted modeling of reactions and reactivity, Pure Appl. Chem. 53 (1981) (1313-1327).

[6] Mostaghimi, F., Système DARC: banque de données des cétones aliphatiques. Synthèse et propriétés physiques, Thesis, Paris 7 University (1978).

[7] Dubois, J.E., Panaye, A., Lion, C, Conception assistée par ordinateur. Notion de Domaine Structural Ordonné d'une Réaction (DSOR),

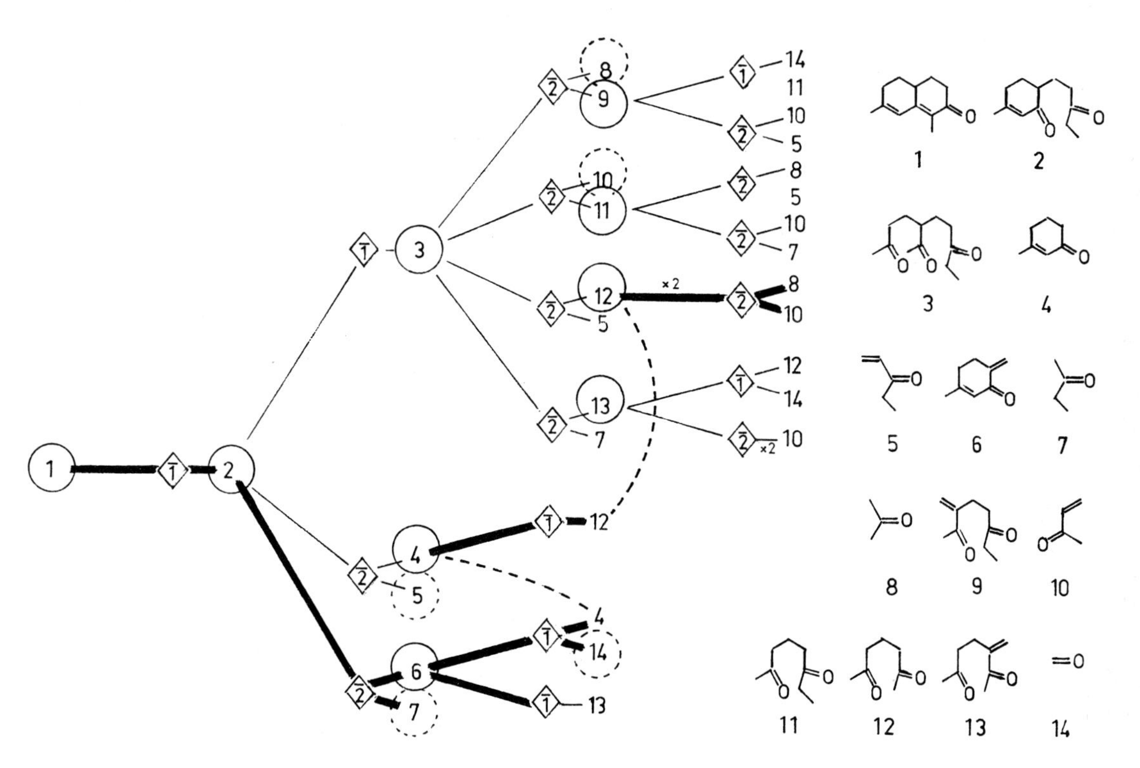

Figure 7: Example of a transformational hyperstructure representative graph generated by PARIS V3.0 and corresponding structures

Nouv. J. Chim. 5 (1981) 371-380; Dubois, J.E., Lion, C., Panaye, A., Evaluation des domaines de synthèse de cétones par alkylation et condensation magnésienne. Synthèse des cétones frontières hyper-encombrées, Nouv. J. Chim. 5 (1981) 381-391.

[8] Dubois, J.E., Panaye, A., Picchiottino, R., Sicouri, G., DARC System: structure of a reaction invariant, C. R. Acad. Sci. Paris Serie II (1982) 1081-1086.

[9] Computer Assisted Organic Synthesis, Edited by Wipke, W.T., Howe, W.J. (ACS Symposium Series 61, Washington D.C., 1977); Computer Assisted Structure Elucidation, Edited by Smith, D.H. (ACS Symposium Series 54, Washington D.C., 1977).

[10] Georseff, M.P., Procedural control in production systems, Artificial Intelligence, 18, (1982) 175-201.

[11] Dubois, J.E., Ordered chromatic graph and limited environment concept, in Balaban, A.T. (Ed.), The Chemical Applications of Graph Theory (Academic Press, London, 1976); Dubois, J.E., Laurent, D., Panaye, A., Sobel, Y., Système DARC: Concept d'hyperstructure formelle, C. R. Acad. Sci. Paris Serie C (1975) 687-690; Dubois, J.E., Laurent, D., Panaye, A., Sobel, Y., Système DARC: Hyperstructures formelles d'antériorité, C. R. Acad. Sci. Paris Serie C (1975) 851-854.

[12] Nilsson, N.J., Principles of Artificial Intelligence (Tioga, Palo Alto, 1980)..

[13] PARIS V3.0 Users Manual (1981)

[14] Dubois, J.E., Picchiottino, R., Sicouri, G., Sobel, Y., DARC System: block relationships and hyperstructures, C. R. Acad. Sci. Paris Serie I (1982) 251-256.

[15] Jacquier, R., Boyer, S., Recherches sur la base de Mannich de la methyl-1 cyclohexène-1 one-3, Bull. Soc. Chim. Fr. (1955) 8.

[16] Sicouri, G., Project DARC-SYNOPSYS: Application de la méthode IGLOO à la génération automatique de composées issues de réactions concurrentes, DEA, Paris 7 University (1978); Agarwal, K.K., Larsen, D.L., Gelernter, H.L., Application of chemical transforms in SYNCHEM 2, a computer program for organic synthesis route discovery, Comput. Chem. 2 (1978) 75-84.

6. LIST OF ABREVIATIONS

ABO: ABlation Operator

ADO: ADjunction Operator

DARC: Description Acquisition Retrieval Computer aided design

GCSS: Greatest Common SubStructure

GILO: Generic Invariant and Localized Operators

HS: HyperStructure

IGLOO: Invariant Graph and Localized Ordered Operators

ILO: Invariant and Localized Operators

PARIS: Pertinent Analysis of Reaction by Igloo Simulation

SILO: Specific Invariant and Localized Operators

SYNOPSYS: SYNthesis OPtimization SYStem

Data for Science and Technology, P.S. Glaeser (ed.)
North-Holland Publishing Company

CODING OF A CHEMICAL GRAPH

M. Uchino

Laboratory of Resources Utilisation, Tokyo Institute of Technology
4259 Nagatsuta-cho, Midoriku, Yokohama, 227 Japan

A theoretical approach to the unique coding of a chemical graph is described and a method to take in the concept of limited environments (J. E. Dubois) which is important in the structure-activity correlations, is presented. We define r-perfect environment of a vertex so as to provide an invariant property of the vertex which can be used to accelerate coding process of a chemical graph and the canonical code of the r-perfect environment itself can be used for the structure-activity correlation study. The present approach provides a unified mathematical method for unique coding, vertex classification, ring perception, computation of bond-atom symmetry, recognition of canonicity of code, and substructure perception.

INTRODUCTION

Chemical graphs can be used as carriers of chemical information such as chemical structures, reaction schemes, methods of experiments, etc. However, the machine processing of chemical graphs represents one of the so-called "Hard" problems. In this paper, we describe the problems of machine processing of a chemical graph as the problems of vertex numberings and present a unified mathematical approach where mathematical numbering rules are described in terms of "fitness" defined by a bond matrix, an atomic vector, invariant properties of vertices, and T-vector of a current numbering. In the method, the problem of duplicated numberings is solved by the use of "orbit-graph" and the limited environment of a vertex is used to provide the invariant property of great discrimination.

MATHEMATICAL FOUNDATION

Our approach to the problems of unique coding, vertex classification, computation of bond-atom symmetry, substructure perception, etc., is based on the following theorems of the "orbit-graph".

Theorem I. All representations of a chemical graph are produced without duplication by the numberings of nodes which obey its orbit-graph and the number of representations is given by $N!/(1 + D(v))(1 + D(v'))...(1 + D(v''))$, where $D(v)$ denotes the out-degree of node v in the orbit-graph.

Theorem II. Let $V_1, V_2, ..., V_i, ..., V_n$ be a sequence of nodes chosen in the construction of orbit-graph and construct, to each arrow V_iV_x in the orbit-graph, a permutation mapping V_i to V_x and fixing $V_1, V_2, ..., V_{i-1}$, and let L_i be the set of permutations thus constructed and a single identity permutation. Then, the bond-atom symmetry of chemical graph is given by the product $L_1L_2...L_n$.

Figure 1 illustrates the theorem II for cyclobutane. For vertex classification problem, we need only the union $L_1 \cup L_2 \cup ... \cup L_n$ and the computation of the product is not necessary.

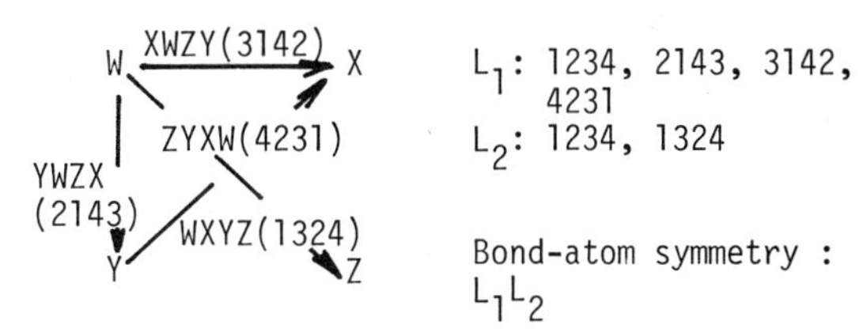

Figure 1. Computation of Bond-Atom Symmetry.

FITNESS AND T-VECTOR

Assume we have a rule f which assigns to each unnumbered node v a list f(i, v) depending on the previous assignment of the numbers 1, 2, . .., i - 1. We use the term pseudo-canonical numbering to denote a numbering obtained by choosing a node with the lexicographically maximum list for each i = 1, 2, .., n, and if the rule f satisfies the following conditions, each list f(i, v) will be called "fittness" of node v for the number i.

Condition 1. If the representations based on pseudo-canonical numberings are identical, then the sequences of the lists are identical.

Condition 2. If the sequences of the lists for pseudo-canonical numberings are identical, then the representations based on these numberings are identical.

The "fitness" can be introduced in various ways and provides mathematical description of numbering rules. The simplest method to give "fitness" uses the invariant property of node v and bond multiplicity, but more advanced method uses T-vector in addition to them. The T-vector of a numbering can be defined as follows. "T-vector of a numbering is an n-element vector (T(1), T(2), ..., T(n)) defined by T(i) = the smallest number among numbers assigned to nodes which are adjacent to nodes with numbers equal or greater than i." An example of "fitness" written in terms of T-vector, an invariant property s(v), and bond value m(v, v') is

$f(1, v) = s(v)$

$f(i, v) = (-T(i), m(v_{T(i)},v), s(v), m(v_{T(i)+1}, v), m(v_{T(i)+2}, v), \ldots, m(v_{i-1},v))$

$i = 2, 3, \ldots, n$

The pseudo-canonical numbering by this "fitness" proceeds from a center to the outer spheres and the components of the T-vector T(2), T(3), ..., T(n), provide a spanning tree of the graph (i.e., the edges $(v_{T(i)}, v_i)$(i = 2, 3, ..., n) define a spanning tree). The distance from v_1 to v_i is given by L(i) - 1 which can be easily determined by T-vector as follows. L(1) = 1, L(2) = L(T(2)) + 1, ..., L(i) = L(T(i)) + 1, ... The corresponding Fortran statements are

```
      L(1) = 1
      Do 1 I = 2, N
      IT = T(I)
  1   L(I) = L(IT) + 1
```

UNIQUE CODING

The canonical numbering can be determined by the following steps.
1. Determine an "orbit-graph" by the use of pseudo-canonical numberings.
2. Select one pseudo-canonical numbering from the pseudo-canonical numberings obeying the "orbit-graph", which provides the maximum sequence of the "fitness".

R-PERFECT ENVIRONMENT OF A NODE

R-perfect environment of a node can be determined by the following steps.
1. Assign the number 1 to a selected node.
2. Generate a pseudo-canonical numbering until the maximum k such that L(k) - 1 = R.
3. Determine the canonical numbering for the first k nodes, fixing the first node.

VERTEX CLASSIFICATION

1. Determine L_1, L_2, ..., by the use of pseudo-canonical numberings.
2. Make the nodes equivalent which can be mapped by a member of $L_1 \cup L_2 \cup \ldots \cup L_n$.

INVARIANT PROPERTY S(V) AND LOCAL NAME

The canonical name of r-perfect environment can be used to compute invariant property of great discrimination. First assign to each node the name of r-perfect environment or its lexicographic order. Thus, we have initial classification of vertices. Considering the name as node value, determine the new name of r-perfect environment and thus we have the second classification of vertices. We continue this process until no new class appears. By this method we can compute invariant property s(v) of great discrimination and the property s(v) thus determined can be used in the "fitness" for unique coding of whole structure. Figure 2 illustrates the process.

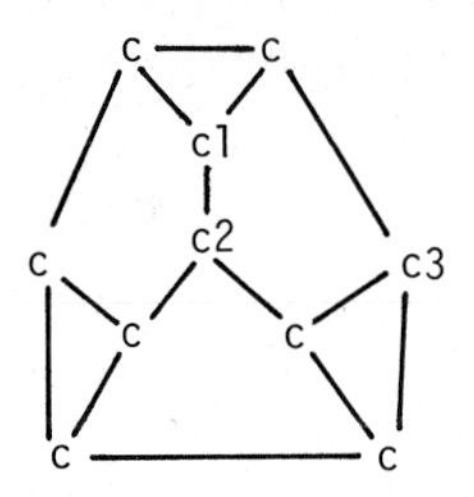

1st step: name of c1 = name of c3, name of c1 ≠ name of c2.

2nd step: to each node the name of 1-perfect environment is assigned and new names are determined. The name of c1 differs from the name of c3.

Figure 2. Vertex Discrimination by Local Name.

QSAR AND R-PERFECT ENVIRONMENTS

When the additivity rule of substituent constants fails in QSAR, r-perfect environments can be used for the correlation. In propylene dimerization catalyzed by nickel complexes, the content of dimethyl butenes and of 4-methyl pentene-1 in the product changes with the structure of the coordinated phosphine ligands and can be correlated with the 1-perfect environment of phosphorous atom. When the 1-perfect environments are characterised by the number of CH_2, CH_3, CH, @$(CH)_2$(where @ denotes aromatic carbon atom), dimethyl butanes are produced by CH and CH_2 in 1-perfect environments of phosphorous atom and 4-methyl pentene-1 is produced by CH and @$(CH)_2$. It is noted that $P(C_6H_5)(isopropyl)_2$ does not show intermediate property between $P(C_6H_5)_3$ and $P(isopropyl)_3$, and its property in dimerization is found to be identical with $P(p\text{-biphenyl})(C_6H_{11})_2$. Table I shows the result of multiple regression analysis for the content of 4-methyl pentene-1 .

Table I. 4-METHYL PENTENE-1 IN THE PRODUCT

R^2 = 0.933 R = 0.966 F Ratio 26.195

VARIANCE:

Source		DF
Due to regression	1368.	4
Residual	91.36	7
Total	1459.36	11

Variables	Coefficient	Error	T value
CH_2	1.0	1.672	0.598
CH_3	-2.394	1.632	1.467
CH	5.470	1.295	4.225*
@$(CH)_2$	15.061	2.648	15.061**
Y Intercept	7.0		

REFERENCES:

[1] Dubois, J. E., Viellard, H., Bull. Soc. Chim. Fr., No. 3(1968), pp. 900-919, ibid., No. 3(1971), pp. 839-843.
[2] Uchino, M., J. Chem. Inf. Comput. Sci., 20, 116 (1980); ibid., 20, 121 (1980); ibid., 20 124 (1980); ibid., 22, 201 (1982).

Data for Science and Technology, P.S. Glaeser (ed.)
North-Holland Publishing Company

GHS – A COMPUTER PROGRAM FOR THE EVALUATION OF THERMODYNAMIC PARAMETERS OF COMPLEX EQUILIBRIA

P.V. Krishna Rao, R. Sambasiva Rao and A. Satyanarayana

Department of Chemistry, Andhra University
Visakhapatnam, India

The computer program GHS written in FORTRAN IV calculates thermodynamic parameters for complex equilibria in solution phase. The algorithm is based on the assumption that the change in enthalpy is (i) a function of temperature and (ii) independent of temperature over a short range. In the temperature dependent method, calculations are carried out assuming ΔH to be a polynomial in T and logarithm of equilibrium constant as a parabolic function of temperature, the input parameters being the log equilibrium constants at different temperatures. Statistical tests are incorporated for the best fit of experimental data. The program was tested on an IBM 1130 computer with literature and several sets of experimental data in aquo-organic mixtures from this laboratory.

1. INTRODUCTION

A study of thermodynamics of complex equilibria throws light on understanding the nature of bonding in the complexes and in analysing the enthalpy and entropy changes in the stepwise formation of complexes. The thermodynamic parameters of interest, namely, ΔG, ΔH and ΔS are evaluated by (i) temperature variation method wherein the equilibrium constant is determined at different temperatures, (ii) calorimetric procedures in which the stability constant, and hence ΔG, is determined at a single temperature and the calorimetric measurements are made at the same temperature to obtain ΔH for the reaction and (iii) entropy titration wherein the integral heat change during the reaction is followed. Several equations relating the enthalpy changes in chemical equilibria to the temperature have been reported in the literature. These equations are based on the assumption that ΔH is either independent of temperature or dependent on temperature over a short range. We have chosen a few popular equations in an attempt to select a best mathematical model that satisfactorily explains the experimental results. These attempts have lead us to a new computer program GHS written in FORTRAN IV, which is described in this paper.

2. COMPUTATION OF THERMODYNAMIC PARAMETERS OF COMPLEX EQUILIBRIA

For the interaction of a metal ion and a ligand in aqueous or aquo-organic mixtures forming N complexes, the equilibrium for the formation of the jth complex may be represented as:

$$m(j)M+l(j)L+h(j)H = M_{m(j)}L_{l(j)}H_{h(j)} \qquad \ldots(1)$$

the overall formation constant is given by:

$$\beta_{m(j)l(j)h(j)} = \frac{[M_{m(j)}L_{l(j)}H_{h(j)}]_i}{FM_i^{m(j)} \times Fl_i^{l(j)} \times FH_i^{h(j)}} \qquad \ldots(2)$$

where FM_i, FL_i and FH_i are free concentrations of the metal ion, ligand and hydrogen ion respectively at the ith experimental point. Hereafter the equilibrium constant is abbreviated as "β" for simplicity.

From Table 1, it is evident that the equilibria representing proton-ligand complexes and metal-ligand complexes are specific cases of the general equilibrium represented by equation (1) and the program now developed is applicable to all these cases.

Two of the thermodynamic parameters, namely, the change in Gibb's free energy (ΔG), change in enthalpy (ΔH) or the change in entropy (ΔS) which are related to one another according to the equation

$$\Delta G = \Delta H - T\,\Delta S \qquad \ldots(3)$$

must be calculable for understanding the thermodynamic feasibility of complex formation. The

Table 1

m(j)	l(j)	h(j)	Equilibrium
0	1	⩾ 1	Acido-basic equilibria of ligand
0	1	< 0	Ligand hydrolysis
1	⩾ 1	0	Metal ligand complexes
⩾ 1	0	< 0	Metal hydroxide species
⩾ 1	⩾ 1	⩾ 1	Protonated, poly nuclear complexes

cumulative equilibrium constant "β" is related to the free energy change by the equation:

$$\Delta G = -RT \ln \beta \qquad ...(4)$$

A new algorithm for the computation of thermodynamic quantitites based on various literature methods wherein the change in enthalpy was considered as a function of temperature and independent of temperature is developed. A FORTRAN IV computer program GHS written using the above algorithm illustrated in the flow-chart (Fig. 1) handles equilibrium constant data at various temperatures and prints out the correlation coefficients along with ΔH, ΔS and ΔG for different regression lines. One can arrive at the best mathematical model from the analysis of these correlation coefficients and standard deviations of the data. A subroutine LLS82 written in our laboratory is used for linear regression analysis and the subroutine SOLE (solving of linear equations) is used for obtaining the coefficients of the simultaneous linear equations. Assuming the independence of ΔH with temperature, it follows from the Vant Hoff equation:

$$\frac{d \ln \beta}{dT} = \frac{1}{R} \times \frac{d(-\Delta G/T)}{dT} = \frac{\Delta H}{RT^2} \qquad ...(5)$$

$$\text{or } \ln \beta = \frac{-\Delta H}{RT} \qquad ...(6)$$

While equation (6) is applicable for gaseous reactions, for dilute solutions it is approximately valid. It is clear from the equation (6), that a plot of log β vs. 1/T yields a straight line of slope ($-\Delta H/2.303R$) from which ΔH can be calculated. Further, the entropy can also be determined from the slope of the plot of T log β vs. T using the relationship: $\Delta S = 2.303R$ (slope).

The Harned and Embree equation (1)

$$\log \beta - \log \beta_{ext} = -C(t - \theta)^2 \qquad ...(7)$$

being a generally applicable equation suggests the temperature dependence of log β (and hence ΔH). In this equation "C" is an empirical constant and has the value of $5 \times 10^{-5} deg^{-2}$, β is the value of the stability constant at t °C and β_{ext} is the value of the stability constant at θ °C. Rearranging the equation (7),

$$\log \beta + Ct^2 = \log \beta_{ext} - C\theta^2 + 2Ct\theta \qquad ...(8)$$

Thus a plot of $\log \beta + Ct^2$ vs. t (°C) gives a straight line of slope $2C\theta$ and an intercept of $\log \beta_{ext} - C\theta^2$ at t = 0 °C. Since "C" is known, θ and β_{ext} can be calculated. Incidentally, the intercept at any convenient value of "t" may be used for the evaluation of log β_{ext} since θ is immediately calculable from the slope.

From equations (5) and (7), it follows that

$$\Delta H = -2.303RT^2 \times 10^{-4}(t - \theta) \qquad ...(9)$$

Harned et al. (2) found, however, in the case of acetic acid dissociation ΔH varied approximately linearly with temperature. According to these authors, the results could best be represented assuming ΔH to be a non-linear (quadratic) function of temperature (cf. equation 10).

$$\Delta H = a + bT = cT^2 \qquad ...(10)$$

Combining equations (10) and (5) and on integration it follows

$$\log \beta = -\frac{a}{2.303R} \times \frac{1}{T} + \frac{b}{R} \times \log T + \frac{c}{2.303R} \times T + d \qquad ...(11)$$

$$\text{or } \log \beta = u/T + v \log T + wT + d \qquad ...(12)$$

where $u = -a/2.303R$; $v = b/R$; $w = c/2.303R$ and "d" is a constant of integration.

Since "d" is an empirical constant, the constants a, b, and c can be calculated using data at three different temperatures, by suitable transformation of equation (12). For data at more than three temperatures, least-squares analysis will yield the coefficients u, v, w and d. If β_1, β_2 and β_3 are the stability constants at temperatures T_1, T_2 and T_3 respectively, the matrix of the coefficients u, v and w can be represented as:

$$\begin{bmatrix} \frac{1}{T_1} - \frac{1}{T_2} & \log T_1 - \log T_2 & T_1 - T_2 \\ \frac{1}{T_2} \quad \frac{1}{T_3} & \log T_2 - \log T_3 & T_2 - T_3 \\ \frac{1}{T_1} - \frac{1}{T_3} & \log T_1 - \log T_3 & T_1 - T_3 \end{bmatrix} \times \begin{bmatrix} u \\ v \\ w \end{bmatrix} = \begin{bmatrix} \log \beta_1 - \log \beta_2 \\ \log \beta_2 - \log \beta_3 \\ \log \beta_1 - \log \beta_3 \end{bmatrix}$$

or

$$[A] \times \begin{bmatrix} u \\ v \\ w \end{bmatrix} = B \text{ or } \begin{bmatrix} u \\ v \\ w \end{bmatrix} = [A]^{-1}[B]$$

Using the subroutine SOLE, the values of u, v, w, and hence of a, b, c can be calculated. In some cases, it has been observed in this laboratory, ΔH values are subjected to errors when equation (10) was employed. However, the experimental data can be better represented if ΔH is assumed to be a cubic function of temperature, i.e.

$$\Delta H = a + bT + cT^2 + dT^3 \qquad ...(13)$$

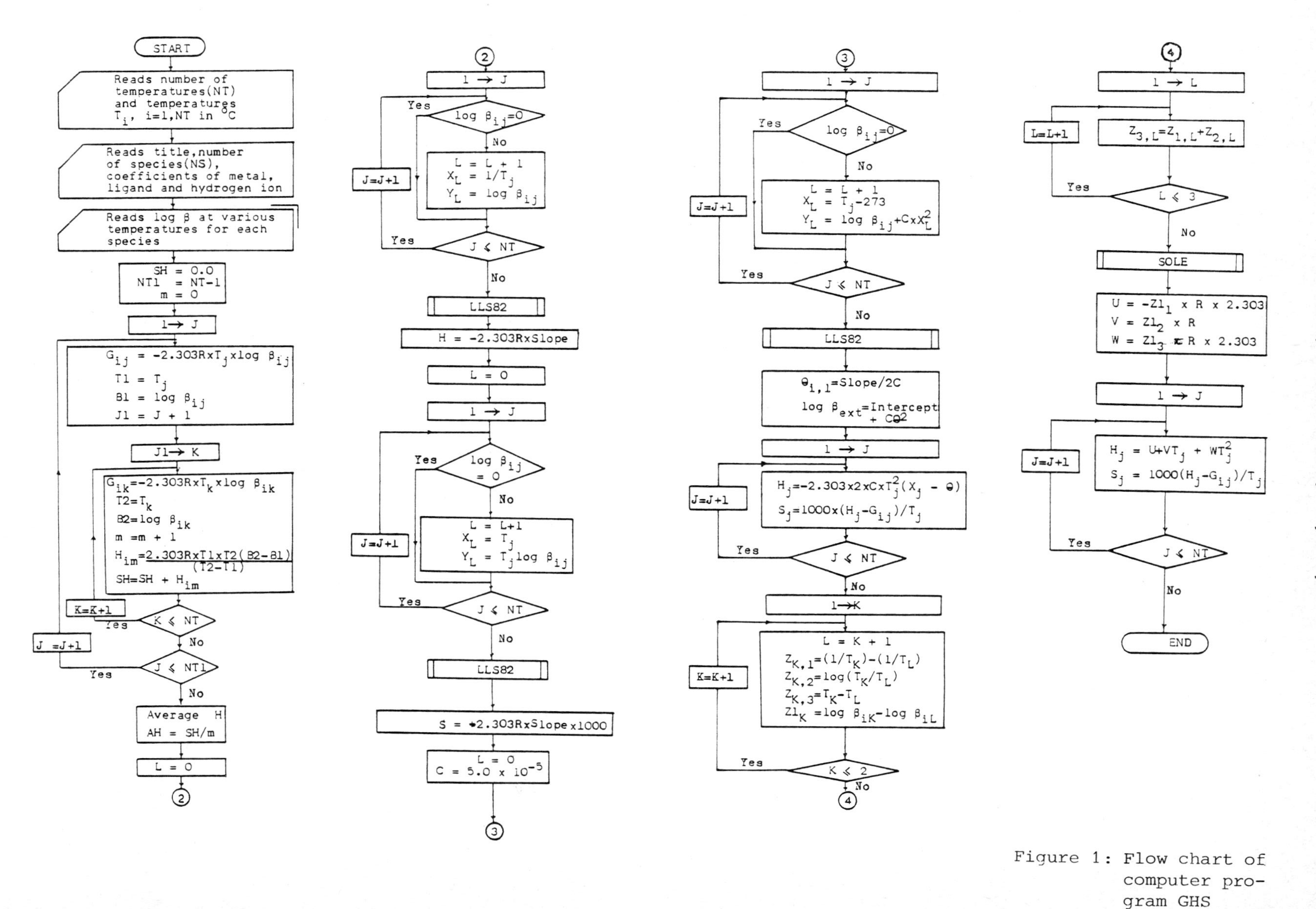

Figure 1: Flow chart of computer program GHS

Since theoretical considerations cannot predict the exact form of equation relating the equilibrium constant to the temperature, extensive efforts have been made in this direction. Of the various attempts to express the dependence of log β on T in an equation containing several terms, each being a different function of temperature, we have chosen a few popular equations which allow the data to be fitted in a final expression using linear regression analysis. From the correlation coefficients for various regression lines the best mathematical model representing the relationship between log β and T can be arrived at.

REFERENCES:

[1] Harned, H.S. and Embree, J. Am. Chem. Soc. 56 (1934) 1050.

[2] Harned, H.S. and Ehlers, R.W., J. Am. Chem. Soc. 55 (1933) 652.

Data for Science and Technology, P.S. Glaeser (ed.)
North-Holland Publishing Company

GASIFICATION AND LIQUEFACTION OF COAL AND DATA NEEDED FOR PLANT DESIGN AND OPERATION

Harald Jüntgen

Bergbau-Forschung GmbH
Essen, Federal Republic of Germany

Content

1. INTRODUCTION

Coal gasification and liquefaction are well-known techniques to convert coal into secondary energy, motor fuels and chemical raw products. The first industrial application of gasification of coal can be traced back to 1830. In this first period the primary material of gasification was coke, and the process of gasification using steam as gasifying agent took place in retorts, whose walls were heated to transfer the necessary heat into the reaction chamber. Since the heat transfer was limited the throughput of these units was only small. This process could not compete with the gas production in cokery-plants. In a second period of industrial application beginning in 1870 a new developed discontinuous process was introduced: In a first step coke was partly burnt with air and in this way heated up to about 1000 °C, in the following second step this hot coke was gasified with steam under consumption of reaction heat and therefore under decreasing of coke temperature. After cooling the coke below a limited temperature, the first step was repeated. Since the thirties of our century the third period of coal gasification began using modern continuous processes with steam and oxygen as gasifying agents. Since that time big industrial complexes of coal gasification are running worldwide producing NH_3, gasoline, town gas and chemicals.

The history of direct coal liquefaction began with the study of the basic chemical reactions of coal with hydrogen at high pressures by F. Bergius in the beginning of the century. The first industrial plant was operated at Leuna in 1927. In 1944 the annual gasoline capacity by direct coal liquefaction in Germany was about 4 million tons. Since the end of World War II no industrial liquefaction plant was in operation. At the present time some pilot plants of liquefaction are running in the Federal Republic of Germany and the USA to improve the process.

From historic remarks it follows that much knowledge on coal-gasification and liquefaction is available from its industrial application and that therefore coal and other data needed for plant design and operation are basically known too. However compared with coal combustion and coke manufacture from coal, the conversion technologies dealt with here are relatively recent with the consequence that the international classification system of coal is more directed to coal properties which are important in performing coal-combustion and coke manufacture. In the future it is expected that coal gasification and liquefaction will become more important and will be used industrially in more countries. Therefore, it is useful to discuss some fundamentals of these processes here and to derive the most important data which is necessary to design and to operate these plants.

In the following section a definition of coal and a short description of its origin and classification is given. After the definition of the process routes of gasification, direct and indirect liquefaction, the techniques of coal gasification and coal liquefaction are described. From there the most important data are derived.

2. DEFINITION, ORIGIN AND CLASSIFICATION OF COAL

Coal is a chemically and physically heterogeneous rock, mainly containing organic matter. This organic matter consists principally of carbon, hydrogen and lesser amounts of sulfur and nitrogen. Coal also contains ash forming inorganic components distributed as discrete particles of mineral matter throughout the coal substance.

Coals originated through the accumulation of plant debris that were later covered, compacted and changed into the organic rock we find today. The mechanism of this so-called "coalification" is, in the first stage, biochemical; in the second stage, chemical condensation reactions under splitting off of CH_4, CO_2 and H_2O. This transformation successively leads to peat, lignite, bituminous coal and anthracite. The progress in this coalification scale is called the rank of coal and is suitable for coal classification. Coal properties determining rank are listed in Table 1.

"RANK" AS PARAMETER OF THE DEGREE OF COALIFICATION:

RANK	INCREASING
CLASS OF COAL (SIMPLIFIED)	LIGNITE SUBBITUMINOUS BITUMINOUS ANTHRACITE

CLASSIFICATION OF "RANK":

COAL PROPERTY	RELATION TO RANK
MOISTURE OF COAL IN THE SEEM	DECREASING WITH RANK (SUITABLE FOR LOW RANK COALS ONLY)
CARBON-CONTENT	INCREASING WITH RANK
VOLATILE MATTER	DECREASING WITH RANK
REFLECTANCE	INCREASING WITH RANK

Table 1: Classification of Coal

3. DEFINITION AND SIGNIFICANCE OF GASIFICATION AND LIQUEFACTION OF COAL

Gasification of coal has the aim to produce gas from coal, gasification also means the process of complete conversion of the organic matter of coal into gases. Liquefaction of coal means the production of liquid fuels or other liquid products from coal. There are two process routes to liquefy coal: direct and indirect liquefaction. Indirect liquefaction provides the gasification to synthesis gas, containing mainly CO and H_2, and converts the synthesis gas by catalytic processes into liquids as methanol, hydrocarbons and other products. In direct liquefaction, coal as a coal-oil-slurry is hydrogenated at high pressure and low temperature to coal-oil and this is processed into fuels, gasoline or other products.

The main significance of coal conversion by gasification, direct and indirect liquefaction is given by the fact that, using these techniques, it is possible to convert coal into more valuable products as fuel gases, motor fuels and chemical products, which, at present, are being produced by oil refinery or by converting natural gas. In this way coal is a suitable replacement for mineral oil and natural gas. The product-producing-routes outgoing from coal gasification and liquefaction are complex and many products can be produced in different ways. Gasification and direct liquefaction especially compete in producing motor fuels. The optimisation of the processing of coal-gas and coal-oil is the subject of research and development.

The main product routes outgoing from the raw gas of coal gasification are given in Figure 2.

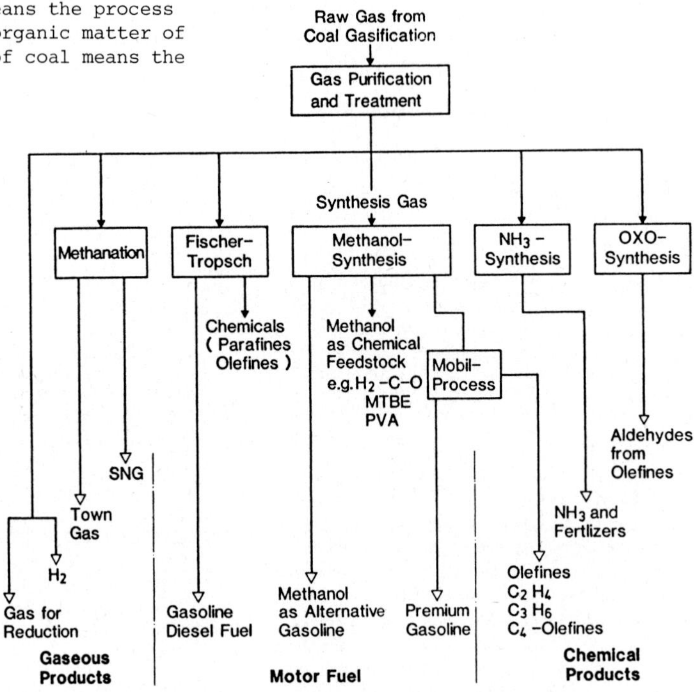

Figure 2: Gasification and indirect liquefaction: Production of fuel gases, gasoline and chemicals

A simple purification and treatment of the gas leads to hydrogen, reduction and low calorific combustion gas. By additional methanation town gas and substitute natural gas (SNG) can be produced. An important process route realized on a large industrial scale in South Africa is the Fischer-Tropsch-Synthesis leading to gasoline, diesel fuel and chemicals. Another process realized on a large industrial scale is the methanol-synthesis. Methanol is used as a solvent and is an important immediate product of the chemical industry. New utilizations are the production of lead free anti-knock agents such as MTBE or of acetic anhydrid and polyvinyl acetate. A future important utilization may be as an alternative motor fuel. Further processing of methanol by the Mobil-process leads to knock-proof gasoline and, under different reaction conditions, to chemical products such as olefines. This process is being developed. Synthesis gas also is the base of the NH_3-synthesis and the so-called "oxosynthese", where aldehydes are produced from olefines.

As can be seen in Figure 3, coal-oil from a hydrogenation plant leads to chemicals such as phenols, BTX-aromatics, olefines and higher nuclei aromatics, premium gasoline and diesel fuel. The first step of the processing of coal-oil is distillation, where light oil, middle oil and heavy oil are produced. The processing of these fractions include processes which are well known from the refinery of mineral oil. Since the composition of the coal-oil fractions differs from that of mineral oil fractions - they generally have a higher content of aromatic compounds - the process conditions have to be changed - e.g. higher temperature and pressure and longer residence time - or new and more effective catalysts have to be developed. The main processes applied here are removal of N, S and O, hydro-cracking and octane improvement. Additionally, the extraction of phenols and BTX-aromatics from light oil or higher nuclei aromatics from middle oil can be performed. The production of diesel fuel from middle oil requires an extensive hydrogenation of aromatic compounds. The same is true of pretreatment before steam cracking of light oil to produce olefines.

4. DESCRIPTION OF COAL GASIFICATION (1)

A simplified flow sheet of a coal-gasification-plant is shown in Figure 4. After preparation the coal is fed into the gasifier, the gas is purified and goes into the synthesis plant, from which the final product is obtained. A part of the coal is used in a steam generation plant, where steam and electricity are produced as gasifying agents and as energy sources for other plants. An oxygen generation plant is necessary, too. The oxygen is needed as a gasifying agent.

Before discussing the technical performance of coal gasification itself in more detail, the most important basic chemical reactions of coal gasification are briefly shown in Table 2. The first step is the pyrolysis of coal in which it is decomposed into char, mainly containing carbon and gaseous and liquid products. Since that pyrolysis takes place in the gasifier, all intermediate products - in parallel gasifiers - or only the solid carbon-product - in counter-

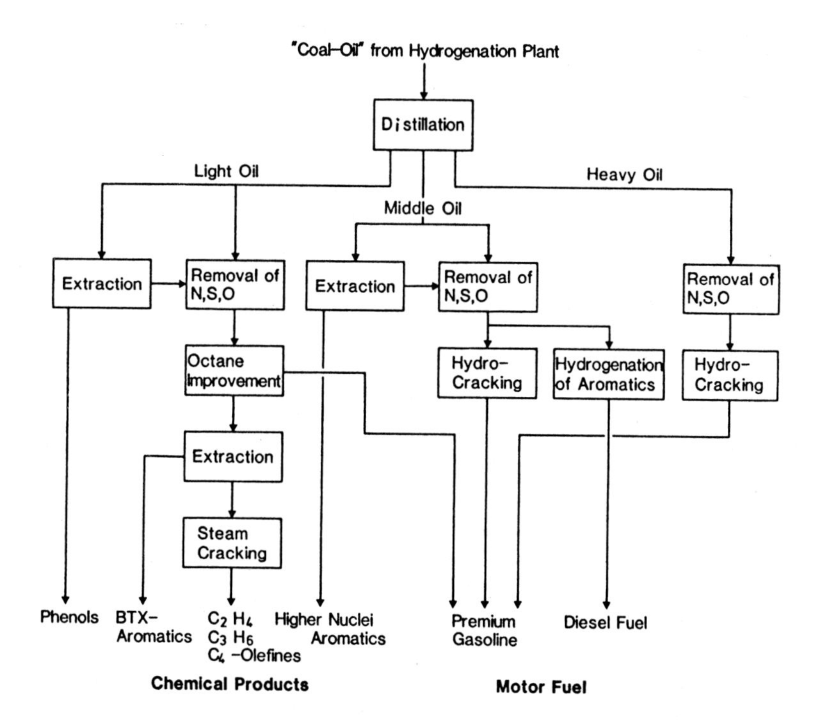

Figure 3: Direct Liquefaction: Production of fuel oil, gasoline and chemicals

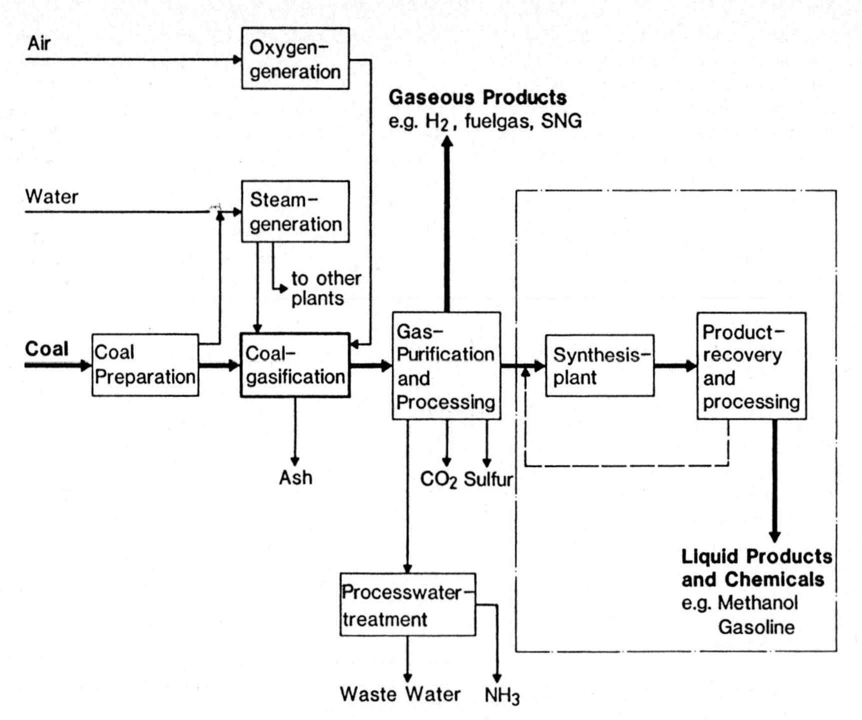

Figure 4: Simplified flow sheet of a plant for gasification and indirect liquefaction of coal

1. PYROLYSIS OF COAL

COAL → COKE (CARBON) + TAR + GASES (CH_4, C_2H_6, H_2, CO, CO_2)
(GENERAL COURSE OF REACTION)

$C_1H_xO_y \rightarrow (1-y)\ C + y\ CO + \frac{x}{2}\ H_2$ $\Delta H = +\ 17{,}4$ KJ/MOL

(SPECIAL COURSE OF REACTION, IF PYROLYSIS OCCURS PARALLEL TO HIGH TEMPERATURE GASIFICATION)

2. GASIFICATION REACTIONS OF CARBON

$C + H_2O \rightarrow CO + H_2$ $\Delta H = +\ 119$ KJ/MOL

$C + CO_2 \rightarrow 2\ CO$ $\Delta H = +\ 162$ KJ/MOL

$C + 2\ H_2 \rightarrow CH_4$ $\Delta H = -\ 87$ KJ/MOL

3. HOMOGENEOUS GAS-GAS-REACTIONS

$CO + H_2O \rightarrow H_2 + CO_2$ $\Delta H = -\ 42$ KJ/MOL

$CO + 3\ H_2 \rightarrow CH_4 + H_2O$ $\Delta H = -\ 206$ KJ/MOL

4. COMBUSTION REACTIONS OF CARBON (ONLY FOR AUTOTHERMAL GASIFICATION)

$C + \frac{1}{2}\ O_2 \rightarrow CO$ $\Delta H = -\ 123$ KJ/MOL

$C + O_2 \rightarrow CO_2$ $\Delta H = -\ 406$ KJ/MOL

Table 2: Basic chemical reactions of coal gasification

current gasifiers - react in consecutive gasification reactions. The main reaction of the formed carbon in the gasifier is the heterogeneous endothermic reaction with steam, the reactions with carbondioxide and hydrogen in general have only a minor significance. The consequence is that heat is consumed during the gasification reaction. This reaction heat can be provided in different ways as discussed later on. In the so-called "autothermal gasification" a part of the carbon is burnt by oxygen. These combustion reactions are exothermic. Autothermal gasification can be achieved by a suitable dosage of oxygen so that so much coal is burnt in the gasifier that the heat produced by combustion and other partly occurring exothermic reactions is equal to the heat consumed by the endothermic carbon-steam-reaction. Besides these heterogeneous reactions discussed, gas-gas-reaction between the products of gasification CO and H_2 with each other or with steam are also possible.

The two ways of heat transfer into the gasifier under technical conditions are represented in Table 3. The first possibility, as described before, provides heat generation in the gasifier

TERM	ALLOTHERMAL GASIFICATION	AUTOTHERMAL GASIFICATION
PRINCIPLE	HEAT GENERATION OUTSIDE OF THE GASIFICATION REACTOR HEAT TRANSFER INTO THE GASIFICATION REACTOR USING GASEOUS OR SOLID HEAT CARRIERS	HEAT GENERATION BY THE CHEMICAL REACTION OF A PART OF THE COAL WITH OXYGEN IN THE GASIFICATION REACTOR
SIMPLIFIED FLOW DIAGRAM	COAL ⇩; HEAT ⇨ / STEAM ⇨ [800-900°C, 1-40 bar] ⇨ GAS; ⇩ ASH	COAL ⇩; OXYGEN ⇨ / STEAM ⇨ [800-1800°C, 1-60 bar] ⇨ GAS; ⇩ ASH
STATUS OF APPLICATION	NOT YET REALIZED IN INDUSTRIAL PLANTS UNDER DEVELOPMENT USING HELIUM (950 °C) AS HEAT CARRIER FROM HIGH TEMPERATURE NUCLEAR REACTOR	REALIZED IN INDUSTRIAL PLANTS

Table 3: Two possibilities of synthesis gas generation by coal gasification

by partial combustion of coal with oxygen. This autothermal process is used in big industrial plants. Another possibility provides heat generation outside the gasifier - without use of expensive oxygen - and heat transfer into the gasifier using gaseous or solid heat carriers. A special version of this allothermal process is being developed. Here a hot helium stream with a temperature of 950 °C is produced in a high temperature nuclear reactor and flows through a heat exchanger immersed into the gasifier. In this way a part of the enthalpy of the helium can be transferred into the carrier.

In Figure 5 typical reactors of coal gasification are shown. In the moving bed reactor, coarse coal grains move slowly downwards, steam and oxygen flow in counter current. Because of the

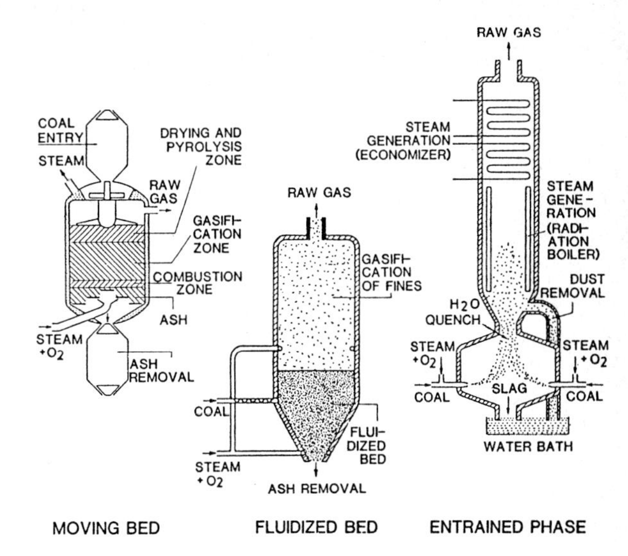

Figure 5: Typical reactor-types of coal gasification

different reaction rates of reactions with oxygen and steam, zones with different temperatures are formed: at the bottom the combustion zone with high temperatures, in the middle the gasification zone, and at the top, the pyrolysis and drying zone with low temperatures. In the fluidized bed reactor, grain sizes of solid particles and the flow rate of gasifying agents are adjusted in this way so that solid particles are in a fast movement in the gasifier without being carried over in the gas flow. Therefore, heat- and mass-transfer are very good and the gasifier shows a uniform temperature. This is limited in general to 1100 °C. For gasification at very high temperatures entrained phase reactors are used. Here the grain size of solids is small and the flow rate of gasifying agents extended.

Some design and operation parameters of gasification processes according to the three different contact patterns between solid and gas as described before are listed in Table 4. The systematic differences in temperature and grain size of coal are important. This leads to the fact that the ash removal is different for the reactor types, and that steam consumption and methane formation in the raw gas are decreasing and oxygen consumption is increasing in the direction of the moving bed, fluidized bed, entrained phase. The different grain size of coal causes the application of different coal feeding systems. It can be stated about the pressure of gasification that, in principle, all types can be operated at a pressure of 50 - 100 bar. For moving bed gasification, technical units operate at about 35 bar. For the other systems, which operate only at one bar under technical conditions, high pressure versions are being developed. Reactor size in industrially operating plants is according to the production of 17 000 - 50 000 m_n^3 of synthesis gas per hour.

	MOVING BED	FLUIDIZED BED	ENTRAINED PHASE
TEMPERATURE (°C)	PYROLYSIS ZONE <800 °C GASIFICATION ZONE 800-1100 °C COMBUSTION ZONE 1100-1300 °C	~ 1000	> 1300
PRESSURE (bar)	35 100 UNDER DEVELOPMENT	1 40 UNDER DEVELOPMENT	1 60 UNDER DEVELOPMENT
COAL SIZE (mm)	6-40	0-8	< 0,1
KIND OF COAL ENTRY	LOCK HOPPER SYSTEM	LOCK HOPPER SYSTEM SCREW FEEDER JET FEEDER (FOR CAKING COAL UNDER DEVELOPMENT)	COAL/WATER-SLURRY FEED SYSTEM LOCK HOPPER IN COMBINATION WITH PNEUMATIC INJECTION
KIND OF ASH REMOVAL	DRY LIQUID UNDER DEVELOPMENT	DRY	LIQUID
STEAM CONSUMPTION	DECREASING →		
OXYGEN CONSUMPTION	INCREASING →		
CH_4 CONTENT IN THE GAS	DECREASING →		
REACTOR SIZE FOR COMMERCIALIZED PROCESS (m_n^3 SYNTHESIS-GAS/h).	35-50 000 (LURGI)	17-20 000 (WINKLER)	20-50 000 (KOPPERS-TOTZEK)

Table 4: Design and operation parameters of Gasification Processes

5. DESCRIPTION OF COAL HYDROGENATION (2)(3)

The flow sheet of a plant of direct coal liquefaction (Figure 6) shows that this process is more complex than that of coal gasification. In contrast to coal gasification, hydrogenation is performed in the presence of an oil acting as a hydrogen donor. Therefore the finely ground coal must be dissolved in a recycle solvent first. A catalyst has to be added to this mixture, too. The coal-oil-slurry is fed together with hydrogen in the high pressure hydrogenation-reactor. The primary product-separation leads to recycle solvent, solid residue, gases from C_1 to C_4 and coal-oils. The solid residue is used for hydrogen production by gasification with steam and oxygen. The gaseous products have to be purified, the hydrogen is recycled and the hydrocarbons can be used as fuel gas with high calorific value. For processing of coal-oil the processes mentioned in chapter 2 are available.

As shown in Figure 7 the overall reaction of coal liquefaction takes place at high pressures between 150 and 300 bar and relatively low temperatures between 450 and 475 °C in the presence of catalysts, of molecular hydrogen and of hydrogen-donor-solvents. Under optimal conditions dependent on reactivity of coal-catalyst-system about 50 % liquid coal oils, 20 % C_1 to C_4 hydrocarbons and 30 % residue are received. A part of the residue can be recycled in the process as a hydrogen donor solvent, the rest can be used for other purposes, e.g. hydrogen-production or coke formation.

Figure 8 gives additional information on the reaction mechanism of coal hydrogenation. As is known after intensive investigations with different chemical and physical analytic methods, coal is a highly crosslinked polymer, consisting of stable aggregates connected by relatively weak crosslinks. The basic aggregates consist of some aromatic hydrocarbons of 2 to 3 nuclei

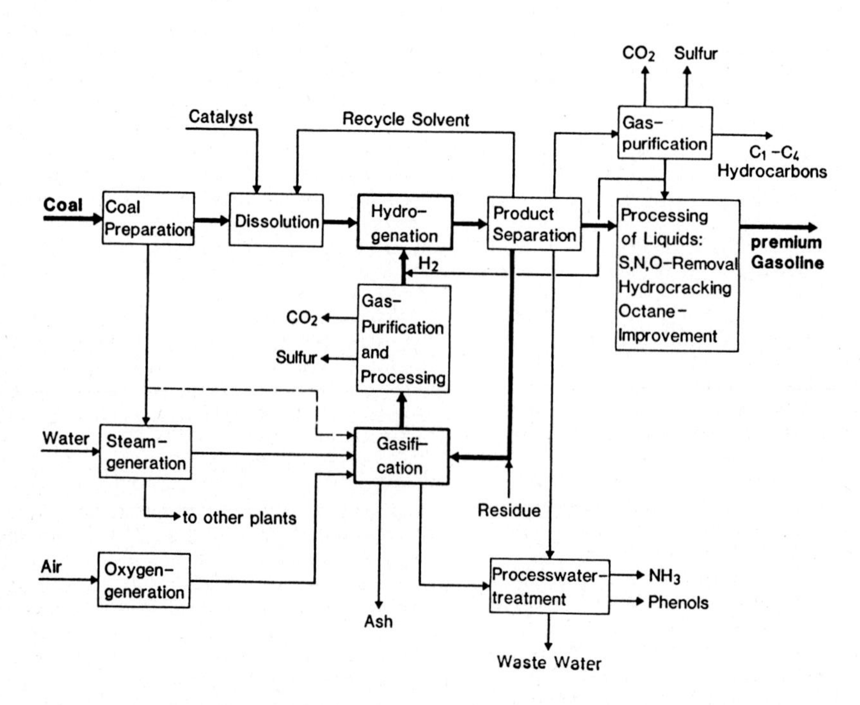

Figure 6: Simplified flow sheet of a plant for direct liquefaction of coal

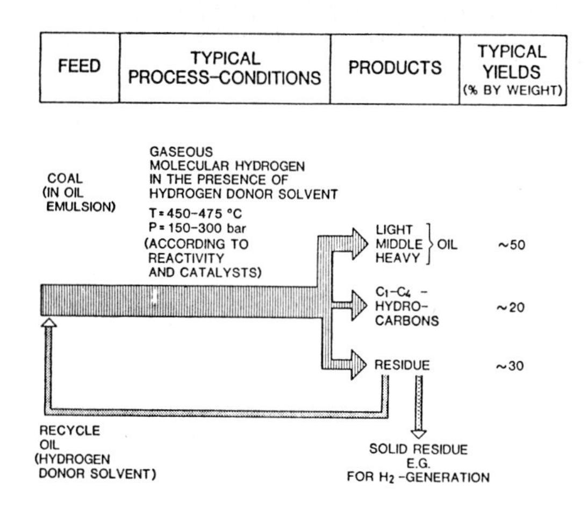

Figure 7: Overall reaction of coal hydrogenation

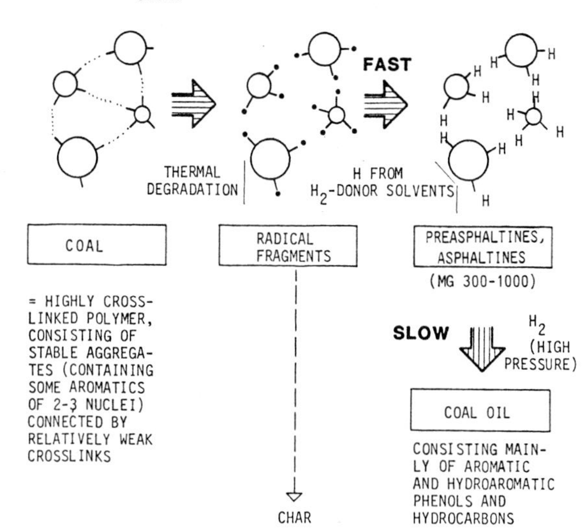

Figure 8: Reaction steps of direct hydrogenation of coal
(According to Whitehurst)

which are connected by aliphatic or other groups. The first reaction during thermal degradation is the breaking of the weak crosslinks between the aggregates of coal-structure and formation of radical compounds. In the presence of hydrogen donor solvents, they react in a fast reaction under take up of hydrogen, forming the so-called preasphaltenes and asphaltenes with molecular weights between 300 and 1000. These compounds react with molecular hydrogen in a slower consecutive reaction into coal oil. This is a mixture of numerous compounds, mainly aromatic and hydroaromatic hydrocarbons and phenols with different molecular weights, numbers of aromatic rings, portions of hydroaromatic structure, degree of substitution.

This reaction is performed in a long extended tube reactor standing upright and made from special steel (Figure 9). The mixture of coal-slurry and hydrogen is fed at the bottom of the

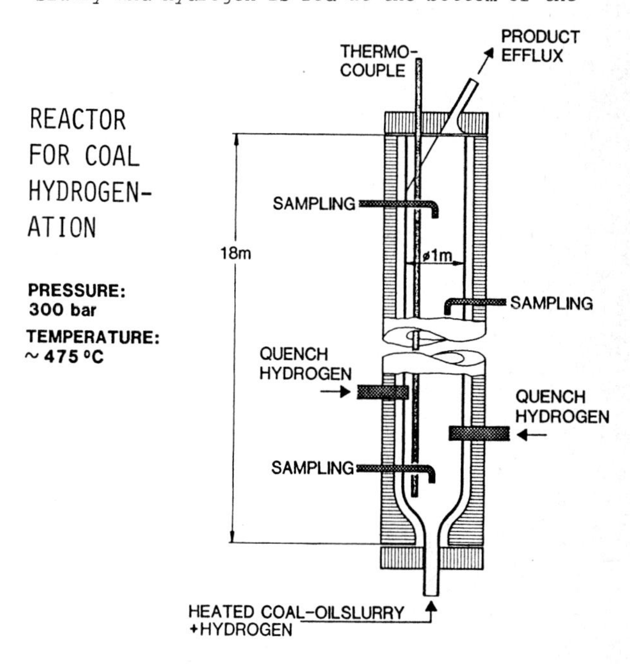

Figure 9: Reactor for coal hydrogenation

reactor, and the outlet for the reaction-products is at the top. Some inlets are provided for quenching the exothermic hydrogenation reaction with coal hydrogen. Three reactors were applied in a series. The size of each reactor used in the industrial plants of the forties was: internal diameter: 1 m and height: 18 m. The reactor used in the pilot plant of Ruhrkohle AG and Veba at Bottrop has the same order of magnitude. It is intended to scale up the reactor size for future industrial plants.

6. SOME REMARKS ON REACTOR DESIGN FOR INDUSTRIAL PLANT

Since most of coal and other data are needed for reactor design, it is necessary to make a few remarks on it. The first aim of reactor design is to select the most suitable reactor type for the performance of gasification or hydrogenation under given conditions with respect to the kind of coal used and the kind of products desired. The other aim is to determine reactor dimensions for a given throughput. For that much experience is available from running industrial or pilot plants.

The first step of design is the setting up of mass and heat balance on the process under consideration. In the mass balance, feed-coal and products are connected by the chemical reactions which take place during the process. For gasification they are described in Table 2, for hydrogenation a survey on overall reaction is given in Figure 7. In the heat balance enthalpies of feed coal and of products and the heat of reaction are also considered. In autothermal gasification the heat of reaction of overall reaction is zero since heat development of exothermic reactions and heat consumption of endothermic reactions are equal. In liquefaction there exists a great reaction heat due to the strongly exothermic reaction. Both mass- and energy-balance are linked by the rate of reaction which can be determined by kinetic experiments under the conditions of the technical process. Furthermore, both balances are influenced by the residence-time-distribution of gases and solids in the special reactor selected for technical performance of reaction.

7. DATA NEEDED FOR DESIGN AND PERFORMANCE OF COAL GASIFICATION AND HYDROGENATION

Of course it is not possible to specify all data needed for design and operation of coal gasification and hydrogenation plants in this short communication. Therefore the overview of the following section shall give a systematic classification of the kinds of data joined with some examples of available data. This collection considers the following topics:

- Standard coal data needed for the setting up of mass- and heat balances
- Kinetic data needed for reactor design
- Thermodynamic data needed for the setting up of heat balances
- Coal data needed for chemical engineering
- Data needed in connection with environmental protection

7.1 Standard coal data needed for the setting up of mass- and heat-balance

The standard coal data, their relationship to coal rank and other parameters are given in Table 5. Most data, with the exception of ash content, mineral matter content, S- and N-content and specific heat, can be related to the volatiles, which themselves are a parameter concerning the coal rank. In the next figures some examples are given for bituminous coals from the Ruhr-district according to Ruhrkohle-Handbuch. Correspondant relationships exist for other coal regions, too. Figures 10, 11 and 12 give the C-, H- and O-content of coals as a function of the volatiles (4). In Figure 13, porosites and true densities are plotted against the volatiles (5). From these two quantities the apparent density can be calculated. Figure 14 shows the relationship between net calorific value and volatiles (4).

DATA	NEEDED FOR	RELATIONSHIP TO COAL RANK	OTHER CORRELATIONS
MOISTURE CONTENT	MASS-, HEAT-BALANCE	ONLY FOR LOW RANK COALS	-
ASH CONTENT	MASS-, HEAT-BALANCE	-	-
MINERAL MATTER CONTENT	MASS-, HEAT-BALANCE	-	RELATIONSHIP TO ASH CONTENT
ELEMENTAL COMPOSITION OF COAL-SUBSTANCE:			
C-CONTENT H-CONTENT O-CONTENT	MASS-BALANCE	RELATIONSHIP TO VOLATILES	
S-, N-CONTENT		NO RELATIONSHIP	
TRUE AND APPARENT DENSITY	MASS-BALANCE	RELATIONSHIP TO VOLATILES	
SPECIFIC HEAT OF COAL	HEAT-BALANCE		RELATIONSHIP TO H/C AND O/C
SPECIFIC HEAT OF CHAR	HEAT-BALANCE		
CALORIFIC VALUE	HEAT-BALANCE	RELATIONSHIP TO VOLATILES	RELATIONSHIP TO C-,H-,S-CONTENT

Table 5: Standard coal data needed for the setting up of mass- and heat balance and correlation to coal rank

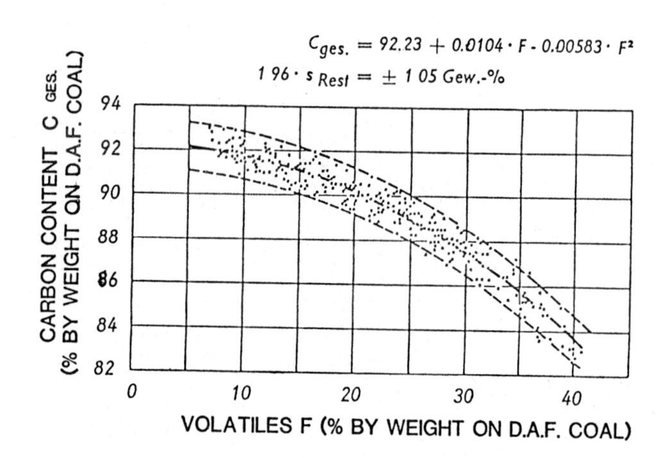

Figure 10: Carbon content dependent on volatiles

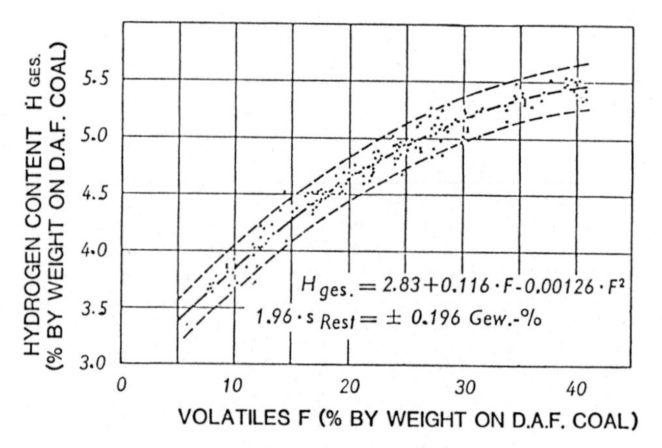

Figure 11: Hydrogen content dependent on volatiles

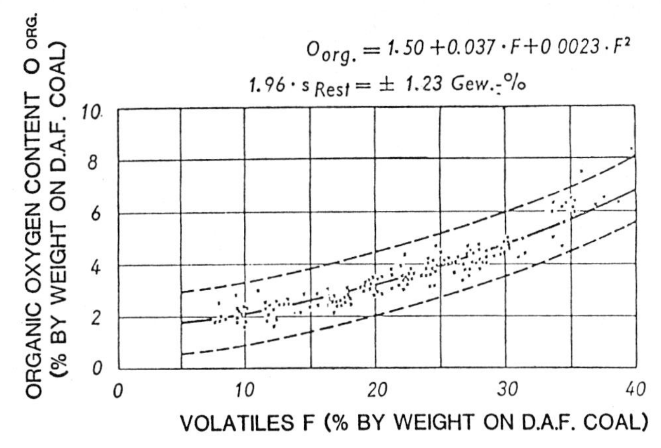

Figure 12: Organic oxygen content dependent on volatiles

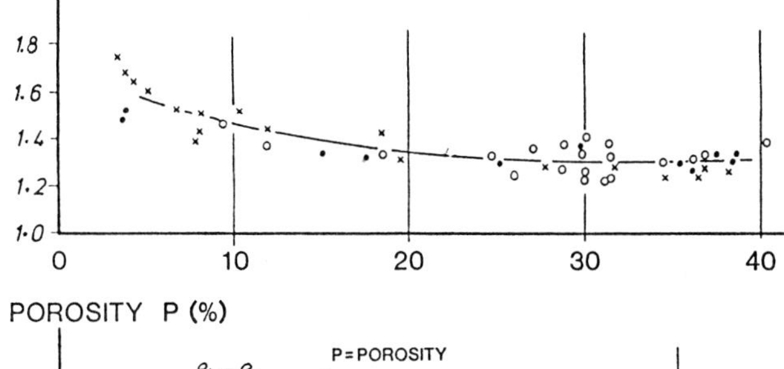

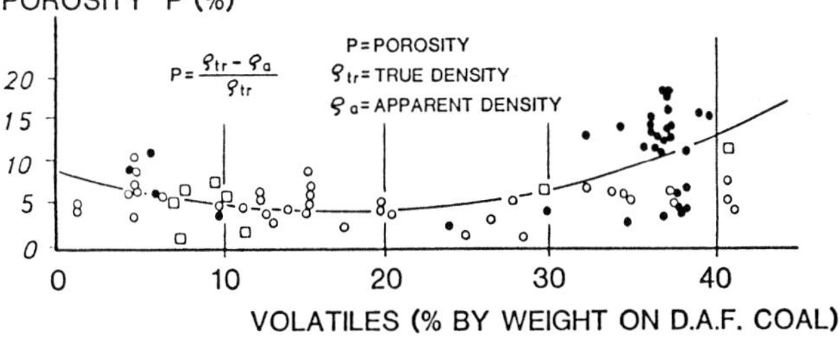

Figure 13: True density and porosity dependent on volatiles

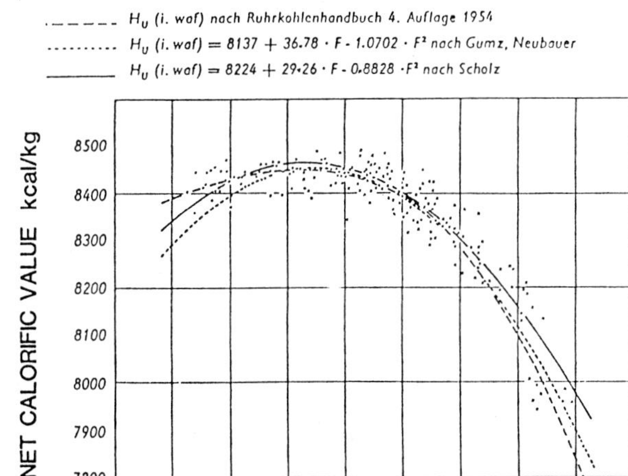

Figure 14: Net calorific value dependent on volatiles

7.2 Kinetic Data

The finding of suitable kinetic data for reactor design is very difficult. Kinetic data have to consider all possible relationships to process conditions such as temperature, pressure, composition of gaseous products, properties of coal used, and, therefore, they have to be determined under process conditions which in practice can vary in wide limits. Figure 15 gives a more detailed description of the kind of kinetic data

DATA	NEEDED FOR	COAL (CHAR) PROPERTIES INFLUENCING KINETIC DATA
"CHEMICAL" REACTIVITY OF COAL AND/OR CHAR AGAINST STEAM H_2/STEAM-MIXTURES UNDER PROCESS CONDITIONS	DESIGN OF GASIFICATION IN MOVING OR FLUIDIZED BED (TEMPERATURE < 1100 °C)	(RANK) PORE STRUCTURE CATALYTIC ACTIVE ASH CONSTITUENTS (ALKALI-, ALKALINE-EARTH-METALS)
KINETIC OF COAL PYROLYSIS UNDER PROCESS CONDITIONS	DESIGN OF PYROLYSIS ZONE OF MOVING BED GASIFICATION	RANK
REACTIVITY OF COAL IN THE PRESENCE OF HYDROGEN DONOR SOLVENTS AGAINST GASEOUS HYDROGEN	DESIGN OF HYDROGENATION	RANK CONTENT OF EXINIT AND INERTINIT CATALYTIC ACTIVE ASH CONSTITUENTS(FeS_2)

Figure 15: Kinetic data needed for reactor design

needed for the modelling of different reactions for gasification in moving and fluidized bed, for pyrolysis and hydrogenation. The coal properties influencing the kinetic data are also discussed briefly. As an example, Figure 16 shows reactivities of coal derived chars against temperature with the rank of coal and the pyrolysis conditions as parameters. The measurements are performed in a differential reactor with steam as the gasifying agent at 40 bar (6).

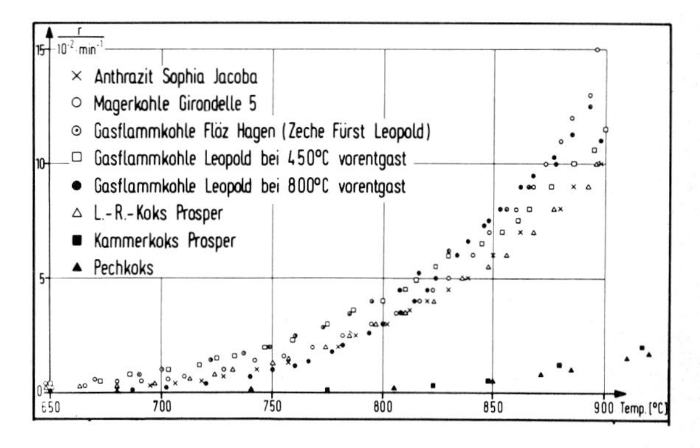

Figure 16: Chemical reactivity against H_2O (40 bar) depending on coal rank and pyrolysis conditions

It can be seen that the reactivities are only scarcely influenced by rank in the range of bituminous coals investigated here. Reactivities of lignites are plotted in Figure 17, measured under the same conditions (7). Here

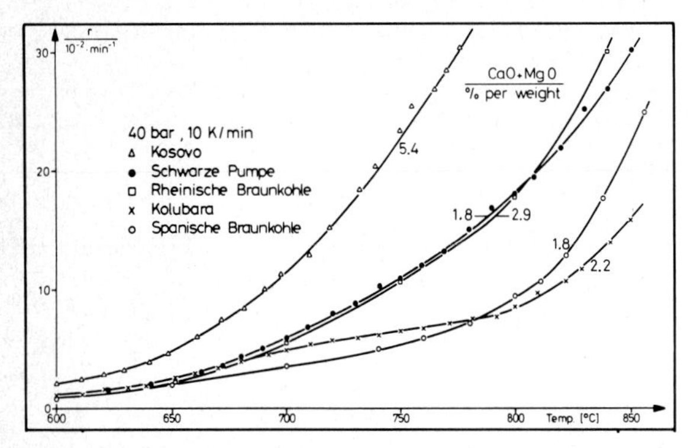

Figure 17: Chemical reactivity of lignites against H_2O (40 bar) depending on CaO+MgO

the content on catalytic active Ca-compounds is an important parameter. Reactivities measured under the conditions of coal hydrogenation against hydrogen in the presence of hydrogen donor solvent at 300 bar and 465 - 475 °C using an integral tube reactor are shown in Figure 18 dependent on coal reflectance which itself is a function of volatiles (8). The reactivities shown in Figures 16 and 17 are important for the design of moving bed or fluidized bed reactors for gasification, those shown in Figure 18 can basically be used for design of hydrogenation.

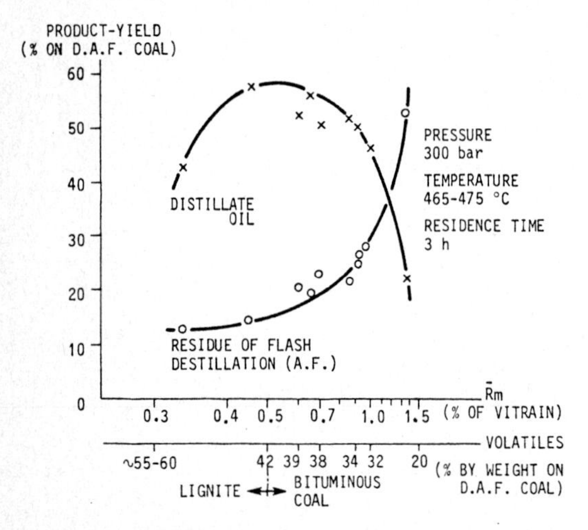

Figure 18: Reactivity of coals against H_2 in the presence of H donor solvents

7.3 Further data

Thermodynamic data needed for heat balance and reactor design are listed in Table 6. They can be found in the well-known standard tables. For the design of gasifiers at very high temperatures it can be assumed that thermodynamic equilibrium is achieved during the residence time of gases and solids in the gasifier. Therefore special kinetic data are not necessary.

DATA		NEEDED FOR
SPECIFIC HEAT FORMATION ENTHALPY HEAT OF VAPORIZATION	FOR H_2, CO, CO_2, CH_4 H_2O, HYDROCARBONS	HEAT-BALANCE
HEAT OF REACTION FOR ALL BASIC REACTIONS OF GASIFICATION AND HYDROGENATION		HEAT-BALANCE
FREE ENTHALPY OF REACTION ENTROPY OF REACTION EQUILIBRIUM CONSTANT	FOR BASIC REACTIONS OF GASIFICATION	DESIGN OF GASIFICATION AT HIGH TEMPERATURES

Table 6: Thermodynamic data needed for heat-balance and reactor design

Some data needed for chemical engineering of conversion plants are listed in Table 7. The grindability of coal is important for the design of gasification plants in the entrained phase and of hydrogenation plants because in these

DATA	NEEDED FOR	CORRELATION TO OTHER COAL PROPERTIES
GRINDABILITY (E.G. HARDGROVE TEST)	GRINDING OF COAL	VOLATILES
FREE SWELLING OR AGGLOMERATION	INJECTION OF COAL INTO THE REACTION VESSEL	RANK
FUSIBILITY OF COAL ASH E.G. -INITIAL DEFORMATION TEMPERATURE -SOFTENING TEMPERATURE -FLUID TEMPERATURE DURING THERMAL TREATMENT UNDER SPECIAL RATE OF TEMPERATURE INCREASE AND ATMOSPHERE	ASH REMOVAL (DRY OR LIQUID) FROM THE REACTION VESSEL	COMPOSITION OF MINERAL MATTER ($SiO_2/Al_2O_3/Fe_2O_3/CaO/MgO/TiO_2/Na_2O + K_2O/SO_3$ etc.)
CORROSION OF COAL ASH	OPERATION OF HIGH TEMPERATURE GASIFIER	COMPOSITION OF MINERAL MATTER

Table 7: Further coal data needed for chemical engineering

processes finely ground coal is needed. As an example the Hardgrove-index for coals from the Ruhr-district is plotted in Figure 19 against the volatiles (4). The knowledge of free swelling or agglomeration is necessary for gasification in moving bed- and fluidized bed-reactors. Coals with high agglomeration lead to the plugging of the reactor near the coal entry. The behavior of ash during thermal treatment is important for all gasification processes. For liquid removal the gasification temperature must exceed the softening temperature and vice versa for dry removal. Data of corrosion behavior of ash are important for high-temperature-gasification-processes.

In Table 8 some data are given which are needed

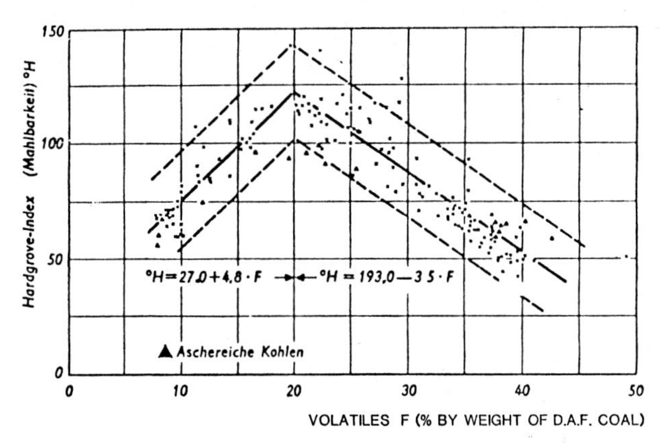

Figure 19: Hardgrove-index dependent on volatiles

S
N } IN COAL
"TRACE ELEMENTS"

BEHAVIOUR OF TRACE ELEMENTS DURING PYROLYSIS, GASIFICATION, HYDROGENATION

PROPERTIES OF TRACE ELEMENTS
(E.G. SOLUBILITY, TOXIC BEHAVIOUR)

TRACE ELEMENTS IN PROCESS WATER

TRACE ELEMENTS IN ASH OR SLAG

Table 8: Some data needed in connection with environmental protection

in connection with environmental protection. As for trace elements, their behavior during gasification, their concentrations in process water and their toxic behavior, there is only little information available.

8. CONCLUSION

The coal conversion techniques for gasification and liquefaction are very important for the replacement of oil in the future. They are able to convert coal into valuable secondary energy, motor fuel and chemical intermediates. Gasification is applied on a large industrial scale worldwide to produce NH_3, gasoline and fuel gas. A new generation of plants is in development. Hydrogeneration is practiced in some pilot plants with the aim of developing improved processes. Data for the design and operation of gasification and hydrogenation plants are known and in principle available. There is a difference in that data for standard coal and thermodynamic data for setting up mass- and heat-balances are generally available, while process specific data for reactor design and chemical engineering have to be measured under special conditions.

Selected Literature:

[1] H. Jüntgen u. K.H. van Heek: Kohlevergasung - Grundlagen und technische Anwendung, Thiemig Verlag München (1981)

[2] Kohleverflüssigung, Umschau Verlag, Frankfurt/Main (1978)

[3] D.D. Whitehurst, T.O. Mitchell, M. Farcasiu: Coal Liquefaction Academic Press New York, London,Toronto, Sydney, San Francisco (1980)

[4] Ruhrkohlen-Handbuch, 5. erweiterte Auflage, Verlag Glückauf, Essen (1969)

[5] Winnacker-Küchler: Chemische Technologie I, S. 285-289, Carl Hanser Verlag, München, Wien (1981)

[6] Wanzl, W., van Heek, K.H., Jüntgen, H.: Einfluß von Temperatur und Druck auf den Reaktionsablauf bei der Wasserdampfvergasung verschiedener Kohlen, Erdöl u. Kohle, Erdgas, Petrochemie, Compendium 78/79, S. 1023-1045

[7] van Heek, K.H., Jüntgen, H.: Zur katalytischen Vergasung von Kohle. Haus der Technik Vortragsveröffentlichungen Nr. 453 S. 53-59 (1982)

[8] Strobel, B., Friedrich, F., Kölling, G., Liphard, K.G., Romey, I.: Kohlehydrierung bei der Bergbau-Forschung, Erdöl u. Kohle, Erdgas, Petrochemie, Bd. 34, Heft 9, Sept. 1981, S. 391-396

Data for Science and Technology, P.S. Glaeser (ed.)
North-Holland Publishing Company

NUCLEAR DATA FOR NUCLEAR SCIENCE AND TECHNOLOGY

J.J. Schmidt

*International Atomic Energy Agency, Nuclear Data Section
Vienna, Austria*

Nuclear data form the numerical input information used by scientists and engineers for the solution of nuclear problems. Nuclear experts designing nuclear fission and fusion reactors need large computer libraries of neutron nuclear cross sections, whereas sciences employing nuclear methods need mostly comprehensive up-to-date handbooks and computer files with nuclear structure and radioactive decay data. 40 years of an intensive scientific effort have so far been devoted to identifying the nuclear data requirements and to measuring, computing and evaluating the required data. Mainly under the coordination by the Nuclear Data Section of the IAEA, supported by international networks of nuclear data centres, an efficient international system of compilation, exchange and dissemination of nuclear data has been developed over the past two decades.

1. INTRODUCTION

Let me start with a brief overview of the development which nuclear sciences and technology have taken over the past 40 years.

In 1942, by putting the first graphite moderated natural uranium reactor together, Enrico Fermi demonstrated the principal feasibility of a self- sustaining nuclear fission chain reaction and thus of nuclear fission power. In 1957, when the International Atomic Energy Agency was founded, two countries operated three nuclear power reactors with a total capacity of 105 MWe. In 1982, the year of the 25th anniversary of the IAEA, 277 nuclear power reactors totalling 157 500 MWe are in operation in 24 countries and produce about 9 % of the world's electricity. Conservative forecasts predict that the latter figure will rise to 18 % in 1990 and to 23 % in 2000 [1]. Nuclear power has thus become a significant reliable and economic source of energy. However, only in very few countries so far has a complete nuclear fuel cycle been realized.

Nuclear fusion, another field of nuclear technology presents one of the major scientific and technological challenges of our time. In spite of about 30 years of intensive efforts, the physical and technological feasibility of nuclear fusion as an alternate long-term source of energy has not yet been fully demonstrated, and commercial fusion reactors, if at all, will probably not become available before the second quarter of the next century [2].

In the field of nuclear and other applied sciences, our time witnesses a widespread use of nuclear radiations and radioisotopes [3-14].

The development of nuclear science and technology was enabled and accompanied by intensive basic and applied nuclear physics research. The development of nuclear fission power was made possible by the detection of the nuclear fission process accompanied by the emission of more neutrons than those initiating fission. This opened the possibility of controlled continuous fission chain reactions, the basis of every nuclear fission reactor. The presently most promising possibility for the realization of nuclear fusion energy relies on the fusion reaction between deuterons and tritons, $d+t = n+\alpha$, liberating neutrons which can be used to regenerate burnt tritium through reactions with Li isotopes. These two examples illustrate the fundamental importance of nuclear processes for the development of nuclear science and technology. All such processes are investigated in nuclear physics research, and their quantitative measure is briefly called "nuclear data". The following five basic types of nuclear data can be discerned:

(1) Nuclear constants such as nuclear masses and isotopic abundances;

(2) Nuclear structure data comprising nuclear ground and excited states and their quantum properties,

(3) Nuclear decay data comprising total and partial decay half lives of decaying ground and excited states of nuclei and decay branching ratios, energies and intensities of emitted α, β, γ and neutron radiations, neutron energy spectra from spontaneous fission of actinide isotopes etc.;

(4) Nuclear reaction data comprising the practically most important neutron nuclear reaction data, photonuclear and charged particle (p, d, t, ^{3}He, α) and light and heavy ion reaction data;

(5) Elementary particle data

In the context of this lecture we will only be concerned with nuclear data of types (1) to (4).

2. NUCLEAR DATA FOR NUCLEAR TECHNOLOGY

2.1. Nuclear fission reactor technology

2.1.1. Reactor design

Neutron nuclear reaction cross sections are the most important input parameters to nuclear fission reactor design.

Reactor neutrons originating from nuclear fission cover energies between 0 and 15 MeV. Reactors are roughly grouped into the two classes of fast and thermal reactors depending upon the degree of neutron energy moderation and the energy ranges where the bulk of the neutron nuclear reactions occurs.

In thermal reactors where the bulk of the neutron nuclear reactions takes place at energies below a few eV, only neutron fission, capture and elastic scattering are important, whereas in fast reactors with neutrons covering a much larger span from eV to MeV energies additional threshold reactions such as (n,p), (n,α) and (n,2n) plus resolved and statistical resonance parameters of the reactor constituents and energy and angular distributions of secondary emitted particles have to be considered.

Reactor design requires the theoretical prediction of the following more important parameters:

- space-dependent neutron energy spectrum and neutron reaction rates;
- effective neutron multiplication factor K_{eff} and its change as a function of reactor operating time;
- thermal and electric reactor power;
- nuclear fuel composition and enrichment;
- critical mass and reactor size;
- effect of absorber rods on K_{eff};
- reactor kinetic and dynamic properties, including nuclear safety coefficients;
- nuclear fuel burnup, discharge and reloading;
- neutron and γ-ray shielding.

Neutron cross sections enter as input parameters into the equations which are used for reactor design calculations. The materials to be considered depend on the specific type of reactor. The most important are:

fissile isotopes: ^{235}U, ^{239}Pu, ^{241}Pu
fertile isotopes: ^{238}U, ^{240}Pu
structural materials: Cr, Fe, Ni, Zr and others
cooling materials: H_2O, D_2O, He, CO_2, Na
moderating materials: H_2O, D_2O, Be, C (graphite)
control rod materials: B, Ta
oxide fuel: O ($U - PuO_2$)
carbide fuel: C (U - PuC)

^{238}U and ^{240}Pu are called fertile materials because through neutron capture (and in the case of ^{238}U two subsequent β-decays) they are transformed to the fissile materials ^{239}Pu and ^{240}Pu respectively. In thermal reactors less fissile material is generated than burnt, whereas fast plutonium-fuelled "breeder" reactors generate more plutonium than is burnt. This breeding of plutonium as well as attainability of higher fuel burnup are the attractive features of fast reactors, enabling a much higher utilization of nuclear fuel than in thermal reactors.

Nuclear reactor design theory and associated computer codes have been refined to a degree, that most reactor design parameters could be predicted to accuracies required by reactor utilities if the input nuclear data would be accurately known. For thermal reactors the required nuclear data basis is nowadays much more accurately established than for fast reactors. For fast reactors the following table shows, that for KeV - MeV neutron energies the target accuracies for the most important nuclear data have not yet been achieved [15-17].

Element or Isotope	Quantities	Accuracy (± %) requested	achieved
^{235}U	σ_f	1	3
^{252}Cf	$\bar{\nu}$	0.25	$\gtrsim$ 0.5
^{239}Pu	σ_f	1-3	5-10
^{239}Pu	σ_γ	5-10	10-20
^{238}U	σ_γ	2-3	5-10
^{238}U	$\sigma_{n'}$	5	10-20
Fe	σ_γ	5-10	10-20

As a consequence two of the most important fast reactor parameters can still not be calculated to the accuracies required by reactor designers.

Reactor parameter	Accuracy (± %) requested	achieved
K_{eff}	0.5	0.5-1.0
breeding ratio = $\frac{\text{Pu generated}}{\text{Pu spent}}$	2	5

For inaccurate data penalties have to be paid and extra reactor design margins must be foreseen to account for these inaccuracies, such as a larger reactor core or higher fuel enrichment than nominal. For the French fast SUPER-PHENIX reactor nuclear data uncertainties necessitated a higher fuel investment of 220 kg Pu [17].

2.1.2. Nuclear reactor safety

2.1.2.1. Reactor operation and safety coefficients

Reactors must be designed so that safe operation is ensured throughout their life time. This means in particular that in such incidents which cause a temperature increase of the fuel, reactor design must ensure, that K_{eff} drops below 1 and that the reactor is shut down, i.e. that the temperature derivation of K_{eff}

$$\frac{\partial K_{eff}}{\partial T} < 0$$

There are several such temperature coefficients of reactivity available in every reactor, which act as inherent safety mechanisms. Below we discuss only one of the more important mechanisms.

Neutron fission and capture reaction rates in nuclear fuel resonances increase with increasing temperature. In analogy to the well known similar phenomenon in acoustics, this effect is called Doppler effect, and its quantitative measure Doppler temperature coefficient. In a U-Pu fuelled fast reactor increase of neutron capture rates in ^{238}U and ^{239}Pu leads to a decrease of K_{eff}, but increase of fission rates in ^{239}Pu to an increase of K_{eff}. Thus ^{238}U must be admixed to ^{239}Pu to such an amount that a negative overall Doppler coefficient can be guaranteed. Typical ratios of U to Pu in fast reactors are around 6; the fuel of the SUPERPHENIX reactor consists of 85 % ^{238}U and 15 % ^{239}Pu [17]. For reliable Doppler coefficient calculations the parameters of hundreds of resolved and statistical resonances of nuclear fuel isotopes have to be accurately known, from eV to KeV neutron energies. Present accuracies of Doppler coefficients are about ± 20 % and have not yet met the reactor design targets of ± 10-15 %.

2.1.2.2. Radiation damage and neutron dosimetry

The damage caused by neutron irradiation of reactor structural materials such as reactor support structures and the reactor pressure vessel is one of the factors limiting reactor life time. Radiation damage consists of embrittlement caused by the atomic displacements connected with every neutron nuclear reaction, build-up of different elements/ isotopes and associated radioactivities through nuclear transmutations, and of swelling, blistering and surface breakups through (n,α) reactions and consequent He-build-up. The phenomenology of radiation damage to structural materials, e.g. stainless steel, as a function of neutron energy is summarized in reference [18].

The estimation of nuclear transmutations, radioactivities and He-build-up presupposes a knowledge of all (n, charged particle) cross sections of all structural materials. For the computation of radiation damage caused by atomic displacements the following two types of information are needed:

(i) cross sections for all neutron reactions with all structural material isotopes over the whole reactor neutron energy range;

(ii) damage functions which describe for each reaction its multiple atom displacement effect.

The data accuracy required for predictions of atomic displacement damage is typically about ± 5 % [16]; the accuracy achieved is, however, only about ± 20 %. Radiation damage thus represents still a wide field for material as well as nuclear data research.

2.1.2.3. Reactor shielding

Neutrons and γ-rays generated inside a reactor have to be shielded to radiation levels outside the reactor below tolerance doses. One discerns a thermal shield consisting typically of Pb or Fe for shielding of reactor γ-rays and a bulk shield of heavy concrete with Ba, Ca, Si, O and Fe as major constituents for the shielding of neutrons and γ-rays.

Refined Monte Carlo and discrete ordinates calculational methods have been developed over the past several years, replacing older rather crude methods. For their successful application they require a knowledge of all neutron and photonuclear cross sections of shielding materials over the whole energy range of reactor neutrons and γ-rays.

2.1.3. Nuclear fuel cycle

During the life time of a reactor the nuclear fuel has to be periodically discharged, stored, transported, reprocessed, refabricated and reloaded to the reactor, while the nuclear waste has to be safely disposed [19]. These are the more important stages of the nuclear fuel cycle.

From the nuclear data point of view the most important feature of the nuclear fuel cycle is the build-up of fission products, through neutron fission processes with fissile and fertile (above threshold) nuclei, and of secondary actinides through neutron capture in the main fissile and fertile actinides and subsequent α-and β-decays. Fission products and secondary actinides constitute the most important nuclear waste materials.

About 50 out of a total of 800 fission products and 30 out of a total of 200 secondary actinides are important, because of their long half lives and/or high yields and/or high capture and fission cross sections.

The nuclear data needed for the estimation of the production and the radiation hazard of fission products and actinides are as follows:

inside the reactor:	all neutron cross sections and yields for fission products and actinides for neutron energies from 0 - 15 MeV;
fuel cycle outside the reactor:	α-, β-, γ- and spontaneous fission decay characteristics, half lives, radiation energies and intensities for fission products and actinides respectively; (γ,n) and (α,n) reactions particularly in low-Z materials.

Most important are fission and capture reactions for which $\pm$ 10-20 % accuracies are required, but only $\pm$ 20-30 % achieved so far.

2.2. Nuclear fusion reactor development

Basic to all present nuclear fusion devices based on the magnetic confinement concept, such as TOKOMAKS, mirror machines and others, is the T(d,n) reaction, which produces 14 MeV neutrons and 3.5 MeV α-particles. In the most usual TOKOMAK devices the toroidal (d,t) plasma area is separated from the tritium-breeding lithium-blanket by a first wall of stainless steel. Concrete shielding protects the radiations generated in the device from the outside.

Fusion reactor physics calculations include the following important characteristics:

- space- and energy dependent neutron flux density;
- multiplication of neutrons in the first wall;
- neutron activation and induced radioactivities;
- structural material effects on neutron economy;
- radiation damage to structural materials; and
- tritium breeding in the reactor blanket.

The materials entering in such calculations are the following:

Fuel:	D, T
Tritium breeding materials:	^{6}Li- and ^{7}Li-compounds
Coolant:	e.g. FLiBe (F, Li and Be)
First wall materials:	SS constituents Cr, Fe and Ni, but also Ti, V, Mo, Nb, W
Other structural materials:	C (Carbide), Al
Neutron multipliers:	Be, Pb
Magnet conducting materials:	Cu, but also Be, Al, Sn
Reflector and moderator materials:	Be, B, C, O
Shielding materials:	B, C, O, Si, Fe, Ba.

For all of these materials and their stable isotopes all neutron cross sections are needed from KeV energies to about 16 MeV [20]. The tritium-breeding reactions ^{6}Li(n,t)α and ^{7}Li(n,n't)α need to be known to particularly high accuracy. Since the neutron spectra are much harder than in fast fission reactors and are centered in the MeV energy range, there is a higher emphasis on energy and angular distributions of secondary emitted particles and on partial reactions with thresholds in the MeV range such as (n,p), (n,np), (n,α), (n,nα) etc. Even lower-yield reactions such as (n,d), (n,t) and (n,He^3) are important through the radioactivities they induce.

3. NUCLEAR DATA FOR NUCLEAR SCIENCES

Nuclear sciences cover fields as different as medical nuclear dosimetry, diagnostics and therapy; nuclear chemistry; geosciences; archeology; environmental research; hydrology; forensic sciences; astrophysics and, of course, nuclear physics. In comparison to the rather compact nuclear data requirements for nuclear technology, the needs for nuclear science applications are more wide-spread and diverse, but nonetheless voluminous through the large variety of such applications. Apart from the field of nuclear physics, which uses its own products for its progress, the nuclear data requirements can best be grouped according to the main nuclear methods commonly employed in these fields [21].

Nuclear activation analysis with neutrons, charged particles and photons is mostly used for material spectrometry ranging from measurements of the full composition of complex materials to the measurement of traces in materials. For such analyses energy-dependent nuclear activation cross sections for neutrons, charged particles and photons, as well as half lives of the radioactive product nuclei and the energies and intensities of the measured radiations have to be known. Examples are the exploration of mineral resources [6, 7] and the investigation of non-radioactive environmental pollutants accumulated in human hair [22] by neutron activation analysis.

Tracer techniques require a knowledge of the half lives and characteristic decay properties of the radioisotopes used as tracers. Typical applications of these techniques are the use of tritium and other radioisotopes in surface water investigations of nuclear hydrology [10] and the use of iodine radioisotopes in medical diagnostics.

Radioisotopes are nowadays used for many applications in applied science and industry. To optimize the production rate and purity of a desired radioisotope one has to select among all reactions which lead to this isotope with an additional account for competing reactions leading to other unwanted isotopes. An important example is the production of pure

^{238}Pu whose α-radiation is used as a heat source for thermoelectric conversion in a heart pace maker [23].

In nuclear irradiations the cross sections for the interactions of the incident nuclear particles with the nuclei of the irradiated specimen have to be known as well as the energy and angular distributions of emitted particles and radiations resulting from these interactions.

An important example is the use of neutron irradiation in cancer therapy [5, 24]. It has been found recently that oxygen-deficient tumor cells are more sensitive to charged particles than to the electrons produced in classical x-ray therapy. Such charged particles can be produced via nuclear reactions with high energy neutrons generated in cyclotrons. Typical neutron-producing reactions are Be(p,n); Be(d,n); Li(p,n); Li(d,n) in which protons and deuterons with energies varying from 20 to 100 MeV produce neutrons whose energy spectra range from several MeV to about 60 MeV. These neutron spectra are now fairly well known so as to allow an optimum selection of p and d energies for optimizing the therapy.

For the main tissue elements H, C, N, O, P and Ca neutron cross sections for the reactions (n,n), (n,n'), (n,Xn), (n,p), (n,α) (n,Xn Yp) etc. and energy and angular distributions of secondary emitted particles for neutron energies up to about 60 MeV are needed, to measure and compute the effects of the irradiation on the afflicted tumor and surrounding healthy tissue.

Above 20 MeV neutron energy only for hydrogen the required neutron cross sections are reasonably well known, for all other tissue materials they have still to be investigated.

4. ASSESSMENT AND TRANSFER OF NUCLEAR DATA NEEDS FROM DATA USERS TO DATA PRODUCERS

The large demand for accurate nuclear data for the development of nuclear science and technology has led to the establishment of a world-wide co-operative effort among nuclear data centres, nuclear research laboratories and nuclear data users over the past 30 years. The aim of this effort is to initiate, support and coordinate the systematic generation and dissemination of the required data. The first step in this effort is to assess the nuclear data needs in the various fields of applications, to inform the nuclear physicists about these needs and to stimulate them to generate the required data. The assessment and publicizing of nuclear data needs proceeds mainly through scientific meetings between data users and data producers, publication of special reviews of data needs and status, periodic publication of a world request list for nuclear data, and co-ordination measures by nuclear data committees.

4.1. Scientific meetings

Scientific meetings in the field of nuclear data range from topical meetings to symposia and conferences.

IAEA held two international conferences in Paris in 1966 [25] and in Helsinki in 1970 [26], and a symposium in 1973 [27]. More recent international conferences were held at Harwell, UK, in 1978 [4], Knoxville, USA, in 1979 [3], Kiev, USSR, in 1980 [28] and at Antwerp, Belgium, in 1982 [29]. All of these conferences and symposia have helped in a global way to acquaint the nuclear physics community with the nuclear data requirements of the nuclear science and technology community and to review the progress in the generation of required data and associated techniques.

Specialist meetings between data users and producers in specific application or data fields have become another powerful tool for interdisciplinary communication for the purpose of satisfying nuclear data needs. The general structure of such meetings is to confront the needs for data with their status and to arrive at concrete conclusions and recommendations for further work needed. Many such meetings were held on national levels, e.g. in the USA, and under the auspices of the OECD Nuclear Energy Agency (NEA) and of IAEA.

As a typical example we quote the series of specialist meetings on fission product nuclear data held by the IAEA Nuclear Data Section in Bologna, Italy, in 1973 [30], Petten, Netherlands, in 1977 [31], and Vienna, Austria, in 1979 [32]. The results of these meetings were an improved coordination of the hitherto dispersed efforts [33] and a free international release and exchange of all existing fission product nuclear data.

4.2. Nuclear data review articles

Over the past 25 years numerous nuclear data review articles have been written and widely disseminated to the communities concerned. Their subject is usually a comprehensive survey and comparison of available experimental and theoretical data including a careful discussion of existing discrepancies due to systematic errors, followed by outlines of further clarifications and work needed. Many of these review articles provide in addition weighted averages or average curves of the analyzed data as a function of energy with point-wise confidence levels. As one example for many I may mention the review and critical evaluation of the 2200 m/sec neutron cross sections and related fission parameters for the major fissile nuclides ^{233}U, ^{235}U, ^{239}Pu and ^{241}Pu, which constitute the basic set of nuclear data required for thermal reactor design calculations [34].

4.3. WRENDA: World Request List for Nuclear Data

The most compact and efficient means of publicizing nuclear data needs to nuclear physicists has turned out to be WRENDA, a computer- based world-wide request list for nuclear data, compiled and published biannually by the IAEA Nuclear Data Section from contributions of 15-20 countries with major nuclear science and technology programmes. This list covers nuclear data requirements for the development of fission and fusion reactor research and technology and of safeguards analytical techniques, is widely disseminated in the nuclear physics community and is used as a guide for programme planning by experimentalists, evaluators and scientific programme managers. Input to this list is provided by national nuclear data committees through the channel of four neutron nuclear data centres (Brookhaven National Laboratory/ USA; NEA Data Bank/Saclay, France; Obninsk/USSR; and IAEA Nuclear Data Section). The most recent issue, WRENDA 81/82 [35], contains 1067 specific requests for fission reactors, 501 requests for fusion reactors and 106 requests for safeguards, where each request specifies the quantity to be measured or evaluated, energy range, accuracy, priority, and the origin and purpose of the request.

4.4. International nuclear data committees

International nuclear data committees such as the International Nuclear Data Committee (INDC) of IAEA and the Nuclear Energy Agency Nuclear Data Committee (NEANDC) of OECD/NEA have been instrumental in organizing and coordinating nuclear data efforts all over the world. INDC represents the nuclear data efforts of all IAEA Member States and acts as the advisory body to IAEA in all matters concerning nuclear data, NEANDC fulfills a similar role for all OECD Member States. While NEANDC is more a technical committee with nuclear data for fission reactors as primary concern, INDC is more a policy committee, dealing with all nuclear data for all applications and, more recently, placing a strongly increasing emphasis on the transfer of nuclear data technology to developing countries [36, 37].

5. NUCLEAR DATA GENERATION

5.1. Experiment

The transfer of nuclear data needs to nuclear physicists as described above has first catalyzed tremendous developments and refinements of nuclear data measurement techniques and instrumentation over the past 25 years. With neutron sources such as Van de Graaff and Tandem accelerators, cyclotrons, electron linear accelerators, neutron generators, and research reactors, total and partial nuclear, particularly neutron reaction cross sections and secondary particle energy and angular distributions have systematically been measured for the major portion of elements and isotopes over the whole range of atomic weights. The parameters of thousands of neutron resonances, up to a few hundred resonances per individual isotope, for eV and KeV neutron energies have been determined as well as the many diverse neutron reactions and their product distributions at higher neutron energies up to about 20 MeV. The total volume of nuclear data generated over the past 30 years amounts to about ten million data points.

The large neutron source strengths which have been developed over the past years usually result in small statistical errors. On the other hand the large background accompanying all measurements with neutrons causes large systematic errors and discrepancies between different experimental results which in many cases are very difficult to disclose and eliminate. This is a major concern of nuclear experimentalists today and the major reason, why the nuclear data basis of nuclear technology is still not fully satisfactorily established today.

5.2. Theory

Many nuclear data needed for fission and fusion reactor design, such as neutron reaction data for many radioactive fission product and actinide nuclides, as well as the diversity of nuclear interactions at higher MeV energies important for radiation damage and biomedical applications, are very difficult to measure. Nuclear models have been developed over the past 10 years to the extent, that with appropriate parameterisation the required limited accuracies for some of those "unmeasurable" data can approximately be met [38]. Parallel to these theoretical developments powerful computer codes have been developed which allow very detailed calculations of nuclear reaction cross sections. For comprehensive reviews of the state of the art the reader is referred to references [39, 40].

6. TRANSFER OF GENERATED NUCLEAR DATA TO USERS

Concurrently with the wealth of nuclear data produced the problem arose of the most efficient transfer of these data to their users. This general problem entailed the following partial problems:

(i) organization of centralized data compilation in universally acceptable computer-compatible formats;

(ii) organization of data dissemination in user-oriented computer-compatible formats and media; and

(iii) compilation and publication of nuclear data handbooks.

Applications in nuclear technology need predominantly nuclear, particularly neutron reaction data in the form of comprehensive computerized libraries of experimental and evaluated data ((i) and (ii)), whereas nuclear and non-nuclear sciences employing nuclear methods need mostly comprehensive up-to-date handbooks and computer files with nuclear structure and radioactive decay, and nuclear activation reaction data. In the following we deal essentially with the developments, functions and co-operation of nuclear data centres in meeting the need for efficient transfer of nuclear data from producers to users.

6.1. Neutron data compilation, history

Systematic nuclear data compilation efforts were started in the fifties first at Brookhaven National Laboratory in the USA [41], led to the first publication of neutron data handbooks [42, 43] with world-wide use, and were consolidated in 1977 to the US National Nuclear Data Centre (NNDC). In the beginning of the sixties the increasing data production no longer allowed the Brookhaven Data Centre to compile nuclear data from all the world and in 1964 three other neutron data centres were founded [44]:

- the Neutron Data Compilation Centre (NDCC) of OECD/NEA, located at Saclay near Paris, which forms today part of the NEA Data Bank (NEA DB);
- the Centre po Yadernym Dannym (CJD) of the Institute of Power and Energetics, Obninsk, USSR; and
- the IAEA Nuclear Data Section in Vienna, Austria.

6.2. EXFOR

Very soon after their foundation these four data centres started co-operation in the development of an international computer-based exchange system for experimental neutron cross sections, which is now known under the acronym EXFOR (EXchange FORmat). In 1970, a formal agreement was reached between the four neutron data centres which obliged all partners to a systematic collection of all neutron nuclear data from their service areas and to a regular magnetic tape exchange of the compiled data with the other centres; it also defined the responsibilities of each participating centre for the maintenance of the system [36]. According to a geographical subdivision of responsibilities, each centre was assigned a service area as follows:

NNCSC (today NNDC)	-	USA and Canada
NEA/NDCC (today NEA DB)	-	West European countries and Japan
CJD	-	USSR
IAEA/NDS	-	All other countries comprizing (mostly developing) countries in Asia (except Japan), Africa, Latin America, Eastern Europe, Australia and New Zealand.

The CJD Centre in Obninsk solved the difficult compatibility problem with Soviet computers and thus initiated the first computer-based numerical data exchange between Eastern and Western countries.

The development of the EXFOR system is discussed and coordinated at meetings of the four neutron data centre network held at regular intervals.

The basic features and principles of EXFOR can be summarized as follows [45, 46]:

(i) EXFOR is considered a data publication medium supplementary to conventional publications.

(ii) EXFOR is not a bibliographic system, but contains numerical data with cross references to pertinent publications.

(iii) EXFOR data are currently being updated, according to author's later revisions of his original data.

(iv) The numerical data contained in EXFOR are supplemented by explanatory text describing e.g. the facilities with which the data were measured and the various error components of the data. This information is particularly useful for an evaluator and is often not contained in publications.

(v) EXFOR is flexible enough to include all kinds of data and at the same time sufficiently structured for computer processing of the data. It is optimized for international data exchange suitable for a large variety of computers.

6.3. Compilation of charged particle and photonuclear data

During the seventies the need was growing for the compilation of charged particle and photonuclear data for purposes such as use of activation techniques, radioisotope production, neutron source properties, and nuclear fusion. This necessitated a generalization of EXFOR to charged particle and photonuclear data, i.e. all nuclear reaction data except heavy ion and elementary particle interaction data, and led to an expansion of the Four Neutron Data Centre Network to a Nuclear Reaction Data Centres (NRDC) Network which added five more data centres and study groups to the neutron data centre network.

The most important products which have so far resulted from the NRDC network co-operation are handbooks of integral charged paricle nuclear reaction data compiled by the Karlsruhe Charged Particle Group and published by FIZ [47] and the Bibliography of Integral Charged Particle Nuclear Data issued by the NNDC [48].

6.4. CINDA

The card index file CINDA (= Computer Index to Neutron Data), developed in the late fifties by Professor H. Goldstein and coworkers in the USA, gained world-wide recognition and developed to the primary international reference source for all neutron nuclear data. CINDA is a computerized compilation of experiment- oriented references in the form of index lines and is being fed and kept up-to-date by the four neutron data centres; it serves also as the index to the EXFOR file; the master file is being maintained by the NEA DB.

The input to CINDA is the result of a systematic scanning of neutron physics literature published all over the world. The content and format of CINDA is exactly tailored to the needs of nuclear scientists; it contains comprehensive, up-to-date information on all measurements, evaluations and theoretical computations of neutron nuclear data [36].

CINDA is published annually on behalf of the four neutron data centres by the IAEA [49]. The present content of the CINDA master file comprises more than 150 000 entries.

6.5. Nuclear structure and decay data

With the growth of applications of nuclear techniques and isotopes and the increasing importance of the nuclear fuel cycle, nuclear waste management and environmental safety during the seventies, the need increased for systematic and up-to-date compilations of nuclear structure and decay data (NSDD) [27]. Starting from the original co-operation between the USA efforts, coordinated initially by the Oak Ridge Nuclear Data Group and later by the NNDC, and the efforts at the Rijksuniversiteit at Utrecht, Netherlands, an international NSDD network was developed, in co-operation between NNDC and the IAEA Nuclear Data Section, which comprises now 19 data centres and evaluation groups in 11 countries and three international organizations. Through a topical subdivision of work by isobaric mass chains assigned to each participating group this network aims at a complete and continuous evaluation of all nuclear structure and decay data on a 4-5 year cycle, the continuous publication of these evaluations and their dissemination to the scientific community. Coordination meetings between the participating data centres are held every two years [50].

The evaluated mass-chain data resulting from this international effort are published in Nuclear Physics A [51] and Nuclear Data Sheets [52] and incorporated into an internationally available computer-based Evaluated Nuclear Structure Data File (ENSDF); they comprise the currently recommended "best values" of all nuclear structure and decay data.

6.6. Evaluated nuclear data

One of the main groups of customers of the nuclear data centres are evaluation physicists. Neutron data evaluators are mostly associated with reactor research centres and neutron physics projects, nuclear structure and decay data evaluators usually with nuclear physics research. The main task of evaluators consists in the comparison, critical assessment and selection of experimental data and associated statistical and systematic errors, followed by the derivation of self-consistent sets of preferred values by appropriate averaging procedures; nuclear models and systematics are used to fill gaps and remove inconsistencies in the available experimental data.

Extensive efforts have been devoted over the past 25 years to the build-up of comprehensive computer files of evaluated nuclear data. The content of these libraries serves as input to a large variety of computer code systems used in nuclear science and technology. Reviews of the history of the neutron data evaluation efforts and of the current state of the art in nuclear data evaluation can be found in references [44] and [53] respectively. Most of the evaluated data libraries are being made available to the data centres for exchange and dissemination to their respective customers.

6.7. Nuclear data dissemination

Nuclear data services are performed according to geographical areas. The bulk of the dissemination responsibilities rests with the four nuclear data centres at Brookhaven, Saclay, Obninsk and Vienna. I shall illustrate the dissemination functions with the example of the IAEA Nuclear Data Section [36, 37]. Since the exchange between the data centres proceeds on a costfree basis, IAEA/NDS renders costfree services, except for priced IAEA publications, upon request to customers in its service area, consisting in the supply of numerical data in computer medium, nuclear data documents and nuclear data handling codes.

IAEA/NDS maintains and disseminates currently more than 50 different experimental and evaluated computerised nuclear data libraries of up-to-date status. Retrievals are provided from individual sections to the full content of these libraries according to the requestors' specifications. Guidance is given to the requestors on computer-, physics- and application-related aspects of the data, and usually background documentation on the characteristics and use of the data is provided together with the data.

IAEA/NDS disseminates a variety of documents, comprising general compilations, handbooks, research reports and report series.

The most important report series is the INDC series of reports originating from IAEA Member States as well as from the Nuclear Data Section and the INDC Secretariat. INDC reports document experimental and theoretical nuclear data research and comprise also nuclear data progress reports from individual countries, the minutes of INDC meetings, summaries of the conclusions and recommendations as well as full proceedings of IAEA/NDS meetings. More than 1200 reports have so far been issued, complemented by more than 100 reports containing English translations of reports originally written in Russian.

A Nuclear Data Newsletter with a request return card, disseminated two to three times per year to about 2000 customers of the centre, has turned out to be a very useful advertising medium for the centre's services. It lists new data libraries, documents and codes received by NDS, meetings planned and other information useful for the centre's customers.

NDS issues furthermore a Nuclear Data Reference Report Series which documents in detail the actual formats and contents of the nuclear data libraries maintained by the centre.

Proceedings of larger specialists meetings and training courses are issued as edited IAEA Technical Reports or unedited IAEA Technical Documents; examples for this type of publications are references [30, 31, 39, 40].

Finally IAEA/NDS publishes on sale comprehensive bibliographic data indices such as CINDA and numerical data handbooks. A bestseller in the latter category was a Handbook on Nuclear Activation Cross Sections published by the IAEA in 1974 [54].

With the increasing number of disseminated nuclear data libraries the importance of providing also computer codes for the handling (merging, sorting, checking, correcting, plotting, averaging etc.) of these data is growing steadily. This represents also an area where an international data centre can fulfill an important coordination and verification function for the benefit of its customers. As an example I mention a project which IAEA/NDS very recently started concerned with the verification of nuclear cross section processing codes [55].

References

[1] Laue, H.J., International Atomic Energy Agency Bulletin, Supplement 1982, p.10.

[2] "Facts about Energy", brochure published by the IAEA, March 1982, p. 29.

[3] Clark, J.C., International Conference on Nuclear Cross Sections for Technology, Knoxville, U.S.A., October 1979, Proceedings, p. 458.

[4] Stöcklin, G., Qaim, S., International Conference on Neutron Physics and Nuclear Data for Reactors, AERE Harwell, United Kingdom, September 1978, Procceedings, p. 667.

[5] Broerse, J.J., reference 3, p. 440.

[6] Michaelis, W., reference 3, p. 615.

[7] Senftle, F.E., reference 3, p. 604.

[8] Quisenberry, K.S., reference 3, p. 599.

[9] Sanders, L.G., reference 4, p. 642.

[10] Payne, B.R., International Atomic Energy Agency Bulletin, Vol. 24, No. 3, p. 9, (1982).

[11] IAEA/FAO Symposium on the Sterile Insect Technique and Radiation in Insect Control, Neuherberg, 29 June - 3 July 1981, Proceedings published as STI/PUB/595, (1982).

[12] Libby, W.F., "Radiocarbon dating", University of Chicago Press, (1955).

[13] "Geochronology: Radiometric dating of rocks and minerals", Academic Press, (1973).

[14] Browne, J.C., reference 3, p. 627.

[15] Rowlands, J.L., in Minutes of 12th Meeting of the International Nuclear Data Committee (INDC), report INDC-37/U, p. 43, (1982).

[16] Rowlands, J.L., reference 4, p. 7.

[17] Hammer, Ph., reference 3, p. 6.

[18] Smith, D.L., reference 3, p. 285.

[19] Küsters, H., et al., reference 4, p. 518.

[20] Jarvis, O.N., reference 4, p. 1073.

[21] Schmidt, J.J., Proceedings of the Course on Nuclear Theory for Applications, International Centre for Theoretical Physics, January/February 1978, IAEA-SMR-43, p. 1, (1980).

[22] Ryabukhin, Yu.S., J. Radioanal. Chem. 60 (1), p. 7, (1980).

[23] Berger, R., et al.; IAEA Symposium on Nuclear Data in Science and Technology, Paris, March 1973, Proceedings Vol. I, p. 329.

[24] Cross, W.G., reference 4, p. 648.

[25] IAEA International Conference on Nuclear Data for Reactors, Paris, France, October 1966, Proceedings published in two volumes, (1967).

[26] IAEA Second International Conference on Nuclear Data for Reactors, Helsinki, Finland, June 1970, Proceedings published in two volumes, (1970).

[27] IAEA Symposium on Nuclear Data in Science and Technology, Paris, France, March 1973, Proceedings published in two volumes, (1973).

[28] 5th All-Union Conference on Neutron Physics, Kiev, USSR, September 1980, Proceedings published in 5 volumes, (1981).

[29] International Conference on Nuclear Data for Science and Technology, organized by the Central Bureau for Nuclear Measurements of the Joint Research Centre of the Commission of the European Communities, Antwerp, Belgium, September 1982, Proceedings to be published.

[30] IAEA Panel on Fission Product Nuclear Data, Bologna, Italy, November 1973, Proceedings published as IAEA Technical Report IAEA-169, Volumes I - III, (1974).

[31] Second IAEA Advisory Group Meeting on Fission Product Nuclear Data, Petten, Netherlands, September 1977, Proceedings published as IAEA Technical Report IAEA-213, Volumes I and II, (1978)

[32] IAEA Consultants Meeting on Delayed Neutron Properties, Vienna, Austria, March 1979, Proceedings published as report INDC(NDS)-107, (1979).

[33] Progress in Fission Product Nuclear Data, collected by M. Lammer, No. 8, published as report INDC(NDS)-130, (1982).

[34] Lemmel, H.D., International Conference on Nuclear Cross Sections and Technology, National Bureau of Standards, Washington, D.C., U.S.A., Special Publication 425, Vol. 1, p. 286, (1975).

[35] WRENDA 81/82, World Request List for Nuclear Data, edited by N. Dayday, published as report INDC(SEC)-78, (1981).

[36] Lorenz, A., Fifth International CODATA Conference, Boulder, Colorado, U.S.A., June 1976, Proceedings published by Pergamon Press, p. 209 f, (1977).

[37] Schmidt, J.J., Lorenz, A., IAEA Bulletin, Vol. 22, No. 2, p. 65 f, (1980).

[38] Young, P.G., et al., reference 2, p. 639.

[39] Course on Nuclear Theory for Applications, International Centre for Theoretical Physics, January/February 1978, lectures published as report IAEA-SMR-43, (1980).

[40] Interregional Advanced Training Course on Applications of Nuclear Theory to Nuclear Data Calculations for Reactor Design, International Centre for Theoretical Physics, January/February 1980, lectures published as report IAEA-SMR-68/I, (1981).

[41] Pearlstein, S., Nuclear News, November 1970, p. 73.

[42] Mughabghab, S.F., Garber, D.I., BNL-325, 3rd ed., Vol. I, Resonance Parameters, 1973.

Garber, D.I., Kinsey, R.R., BNL-325, 3rd ed., Vol. II, Neutron Cross Section Curves, (1976).

[43] Garber, D.I., et al., BNL-400, Vol. I, Z = 1 - 20; Vol. II, Z = 21 - 94, (1970).

[44] Schmidt, J.J., CODATA Newsletter No. 3, (1969).

[45] Calamand, A., Lemmel, H.D., Short Guide to EXFOR, IAEA-NDS-1, Rev. 3, (1981).

[46] Attree, P.M., Report on Generalized Data Management Systems and Scientific Information, published by the OECD/Nuclear Energy Agency, p. 268, (1978).

[47] Münzel, H., et al., Karlsruhe Charged Particle Reaction Data Compilation, published by FIZ Karlsruhe, Physik Daten No. 15-1, 15-2 and 15-Index, (1979).

[48] Burrows, T.W., Dempsey, P., The Bibliography of Integral Charged Particle Nuclear Data, Fourth (Archival) Edition, BNL-NCS-50640, parts 1 and 2, (1980).

[49] CINDA - An Index to the Literature on Microscopic Neutron Data, published by IAEA, Vienna; CINDA - A (1935-1976), 1979; CINDA - 82 (1977-1982), (1982).

[50] Summary Report of IAEA Advisory Group Meeting on Nuclear Structure and Decay Data, Zeist, Netherlands, May 1982, edited by A. Lorenz, report INDC(NDS)-133, (1982).

[51] Numerous nuclear structure data evaluations for $1 \leqslant A \leqslant 44$ published in Nuclear Physics A, North Holland Publishing Company, Amsterdam, Netherlands.

[52] Nuclear Data Sheets, produced by the National Nuclear Data Center for the International Network for Nuclear Structure Data Evaluation, Editor-in-Chief: M.J. Martin, Editor: J.K. Tuli, Academic Press, New York, USA.

[53] Workshop on Nuclear Data Evaluation Methods and Procedures, Brookhaven National Laboratory, USA, September 1980, Proceedings published as report BNL-NCS-51363, (1981).

[54] Handbook on Nuclear Activation Cross Sections, IAEA Technical Report Series No. 156, STI/DOC/10/156, IAEA, Vienna, (1974).

[55] Cullen, D.E., et al., report INDC(NDS)-134, (1982).

Data for Science and Technology, P.S. Glaeser (ed.)
North-Holland Publishing Company

DATA NEEDS, SOURCE AND METHODOLOGIES FOR WATER QUALITY PLANNING AND CONTROL

Jørgen F. Simonsen, Ph.D., and P. Schjødtz Hansen, M.Sc. Chem. Eng.

Water Quality Institute (VKI)
11, Agern Allé, DK-2970 Hørsholm, Denmark

By definition "water quality" is relative to the use of the water (drinking, bathing purposes, fishing, etc.). Accordingly, data needs for water quality planning and control depend on the specific use of the water. Once the use of the water is determined (e.g. by political means), the water quality standards and data needs can be defined by technicians (engineers, biologists, etc.). Methodologies for data treatment and application for water quality planning include computerized data processing (statistical evaluation), or, more adequately, mathematical modelling. Mathematical models are very useful in systematic data processing.

1. INTRODUCTION

The necessity of water quality planning and control is becoming more urgent in the modern society as the need for clean water is growing. Water quality planning should involve an integrated planning of the use of the limited water resource. Thus water quality is not merely a chemical and biological problem but rather an interdisciplinary problem, involving the whole hydrological cycle. The main elements in water quality planning are precipitation and evaporation, groundwater resources, run-off, quantity and quality of surface water, sewage treatment, water supply and land use. The water quality planning is strongly dependent on the wanted use of the water. Often the water use is defined by political means, and thereafter the necessary chemical and biological water quality is defined by the technicians. Having determined the water quality, standards for sewage treatment, rates of water catchments, etc. can be settled. In the following adequate methods for planning of *surface water quality* are mentioned, regarding data needs, data sources and methodologies.

2. WATER QUALITY PLANNING

Regarding surface waters, i.e. rivers, lakes and seas, there is a close relationship between water quality and sewage treatment and discharge. Therefore water quality planning and planning of sewage disposal should be co-ordinated. But also other factors should be considered, see Figure 1.

Quantity and quality data needed for planning of surface water quality are shown in Figure 1 as bold-faced arrows. The amount of data represented by bold-faced arrows in the figure may be considerably large, but absolutely necessary in order to obtain a rational basis for the planning. The information about these hydrological processes should often be rather detailed, i.e. seasonal, and in some cases diurnal variations are to be known.

Besides these hydrological data water quality data are needed, as mentioned below.

2.1 Water Utilization

In order to define the necessary water quality of a surface water the wanted utilization of the water should be decided. Examples of water utilizations are shown in Table 1. Of course, other kinds of utilizations than those shown in the table are possible.

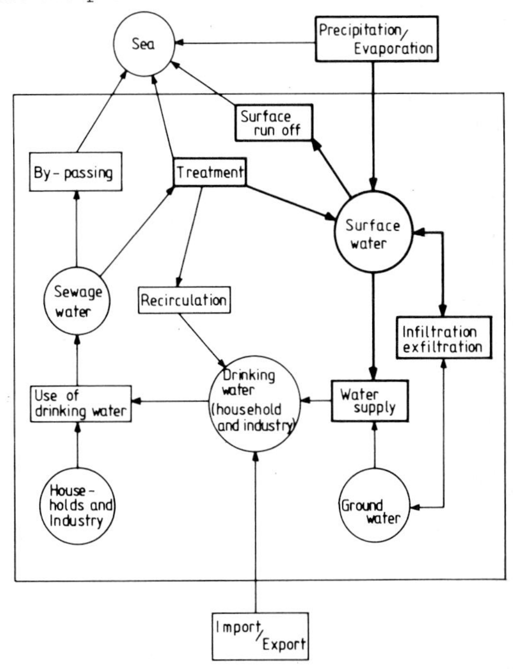

Figure 1: Diagram for water quality planning. Boxes represent 'processes' and circles represent 'states'. Bold-faced boxes and arrows show processes, which should be known in detail for water quality planning purposes.

The principle of defining the water quality interest and objective seems much more adequate than for instance a statement of 'the best possible water quality'. The possibilities of the utilization of the waters depend on the present use of the land in the catchment area, the present pollution, the technical and economic possibilities of sewage treatment. Thus a political and economic evaluation should be the first step in the process of planning the water quality. Often an iterative process is necessary, because the political goals and the economic possibilities will not be able to satisfy the interests connected with the water utilization. This planning process may be tedious and difficult, nevertheless it is the prime step in rational water planning.

Table 1: Examples of water quality interests and objectives.

Interest	Objectives
Environmental research	'Natural state'
Fishing	Hatching area for salmon and trout
	Living area for salmon and trout
	Living area for carp, eels, etc.
Bathing	Water suited for bathing purposes
Recreation	Aesthetically satisfactory
Water supply	Drinking water for households
	Watering of cattle
	Irrigation

2.2 Water Quality Parameters

Depending on the objectives of the water utilization, technicians (i.e. engineers, biologists, etc.) will be able to define the relevant water quality parameter and settle water quality standards.

Thus there are two main problems, namely which parameters are relevant and which standards are necessary. For the solution of these problems much literature is available. A total list of all relevant substances may be very extensive. In most cases, however, only a few parameters are needed. In Tables 2 - 4, the most relevant parameters, referring to the water quality objectives in Table 1, are shown. Regarding the interest 'environmental research' and the objective 'natural state', the list of relevant parameters might be very long and should in every case be specially designed for the special case.

The discussion of the relevant standard values is considered being outside the scope of the paper and thus omitted.

Table 2: Chemical parameters, relevant to fishing water. The maximum permitable values (concentrations) depend on the kind of fish (salmon, trout, carp, eels, etc.).

Parameter
Temperature
Dissolved oxygen
pH
Phenols
Hydrocarbons
Ammonia
Ammonium
Chlorine
Zinc
Ferro } components Ferri }
Suspended solids
BOD_5
Total phosphorus
Nitrite
Copper (dissolved)

Table 3: Some important water quality parameters for bathing water.

Parameter	
Microbiology	
Bacteria	(E.coli, Streptococci, Salmonella sp.)
Vira	(Enterovirus)
Physical/chemical	
pH	
Colour	
Hydrocarbons	
Detergents	
Phenols	
Transparency	
Dissolved oxygen	

Table 4: Important water quality parameters for water to be used for drinking purposes.

Parameter	Parameter
Colour	Fluoride
Odour	Chloride
Suspended solids	Nitrate
Temperature	Nitrite
pH	Sulphate
Conductivity	BOD_5
Ammonium	COD
Arsen	Hydrocarbons
Barium	Pesticides
Lead	Detergents
Cadmium	Phenols
Dissolved iron	Kjeldahl-nitrogen
Copper	Dissolved oxygen
Chrome	Coliforms
Mercury	E.coli
Selen	Streptococci
Zinc	Salmonella
Cyanide	

3. DATA NEEDS

The data need for water quality planning is often huge involving hydrological as well as water quality data. Furthermore, due to seasonal and diurnal variations, the necessary amount of data is extended to daily values during a period of at least one year. Besides the time variation of the data, at least two other factors are important when deciding the frequency of the measurements, namely the hydraulic retention time of the river, lake or sea and the rate of the chemical and biological processes, which are going on and are affecting the water quality.

In order to establish a sufficient knowledge of the water quality in a surface water a measuring program like the one scheduled in Table 5 should be carried out. In order to design the investigation plan the water quality part of Table 5 should be combined with the lists of parameters shown in Tables 2 - 4.

Table 5: Examples of data needs for water quality studies. Lakes I means lakes with relatively short retention time (less than 1 month). Lakes II means lakes with longer retention time. Regarding the measuring frequency ÷ means 'not relevant', H means 'hourly' (in principle), D means 'daily', W means 'weekly', M means monthly and O means 'once during the period of investigation'.

	Frequency			
	Rivers	Lakes I	Lakes II	Sea
Hydrology				
Precipitation	÷	D	D	M
Evaporation	÷	D	D	M
Surface run-off	H	W	M	M
Sewage discharge	O	W	M	÷
Exfiltration	O	W	M	÷
Water catchment for supply	H	W	M	÷
Water quality in				
Precipitation	÷	M	M	M
Discharge sewage	H	W	W	W
Exfiltration	O	W	O	÷
The water itself	H	W	M	M

For lakes and coastal areas the list of parameters should be extended with parameters necessary for an understanding of the eutrophication process. These data are the various nitrogen and phosphorus components, the biological data describing the algae, the zooplankton and the fish (biomass, and for algae concentration of chlorophyll, etc.).

As mentioned above an adequate period of data collection would be at least 1 year. For rivers, however, it is usually sufficient with an investigation period of a few days, i.e. a little more than the hydraulic retention time of the reach to be considered. The investigations of rivers, on the other hand, should be carried out in the 'critical season', which means in periods with high temperatures and/or low water flow.

It can be seen that the amounts of data outlined above may be very extensive. Two points should be noted in connection with that 1) the costs of the data collection may be very high and 2) electronic data processing is needed. Regarding the first point, one should note that some hydrological data are available from other investigations or running meteorological programs. Regarding the remaining part of the data, care should be taken during the planning of the investigation to minimize the number of gauging stations and frequency of the measurements.

4. DATA SOURCES

4.1 Planning of a Water Quality Study

A water quality study may be planned using the methods described above. The first point in the process will contain a total description of the data needed for the study. The second phase will consist of a preliminary study of the sources of pollution in the area and an investigation of the number and type of data collection institutions in the area. The latter may be universities and different authorities. Having an overall view of data 'generating' and data collecting activities the investigation program can be designed.

4.2 Studies and Investigations

The implementation and running of the sampling program are often the most critical part of the whole process. The transfer of data to the computer should be done as fast as possible and the data processing carried out concurrently in order to make adjustments (if necessary) to the gauging program.

In some gauging stations spot sampling is sufficient, but most commonly automatic continuous sampling is needed. Regarding the hydrological and meteorological measurements and physical/chemical measurements the 'in situ' method is used and data can be directly transferred to the data base. Most chemical parameters, however, are determined in a laboratory involving some delay. The quality control of the data produced by chemical laboratories should, however, be carried out with the aid of the computer.

One point should be considered if several institutions are involved in the investigation program and the chemical analyses. To ensure the consistency and homogeneity of the results an inter-calibration is necessary. That means that all institutions involved should once or twice during the program period make measurements and analyses on identical samples and (if necessary) adjust the methods.

4.3 Remote Sensing

As mentioned, data collecting may be tedious and expensive. Therefore new and unusual methods are appearing. One of these untraditional methods, which seems promising, is remote sensing from airplanes and satellites. So far the method is only applicable for the measument of chlorophyll in coastal areas, oceans and, in some cases, in large lakes. The aim is to minimize the gauging stations and sampling frequency by use of the information collected from airborne vehicles. By making a few measurements of chlorophyll from a boat and correlating the result with pictures taken by e.g. a satellite, it is possible to get an overall exact information of the chlorophyll concentration in much bigger areas.

This method involves data processing of a large amount of data as the information delivered by the satellites is enormous. Nevertheless the method has proved applicable in both base line studies and water quality control.

An example of the applicability of the method is shown in Figures 2 and 3. Figure 2 shows a map of a Danish fjord, the Roskilde Fjord west of Copenhagen, with 10 in situ gauging stations. The number of gauging stations is relatively large and will later on be reduced. The sampling, however, was part of a test of the method and therefore the sampling program was extended.

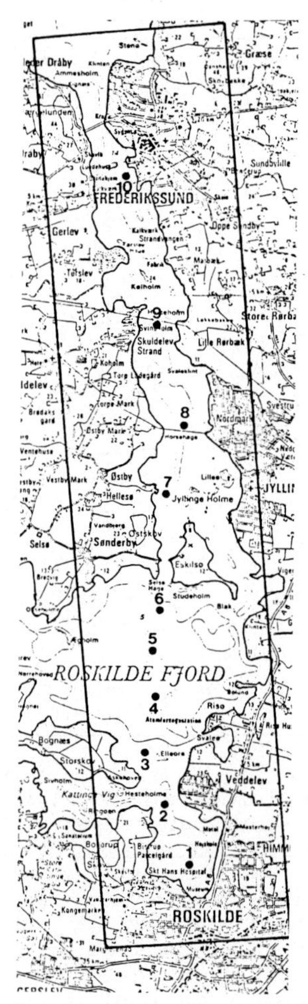

Figure 2: Remote sensing (from satellite) of a Danish fjord. The area covered by the satellite is shown as well as the in situ gauging stations.

Figure 3 shows the correlation between the in situ measurements and the radiance measured by the satellite.

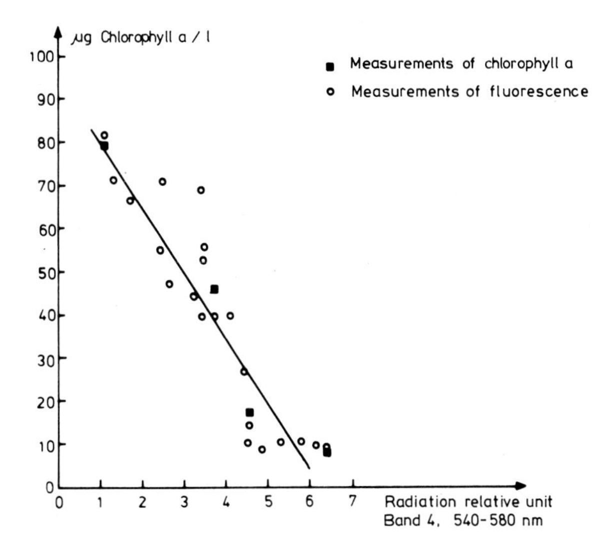

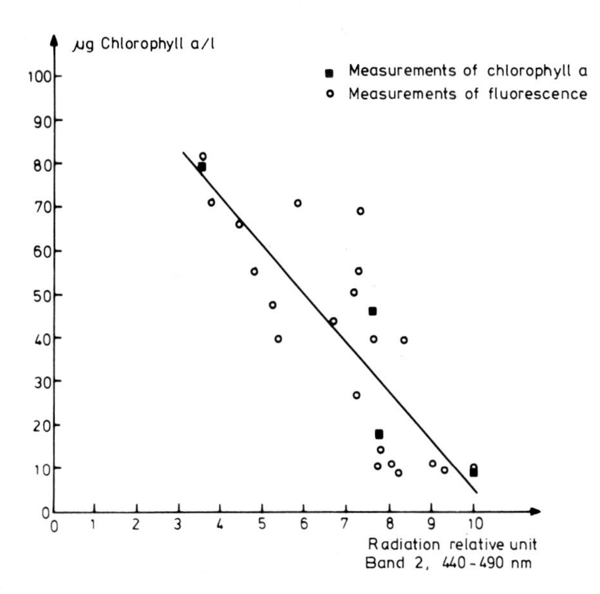

Figure 3: Correlation between in situ measurements of chlorophyll a and radiation measured from a satellite.

On the basis of the measurements, a colour map of the eutrophication of the fjord can be produced (resolution of satellite pictures 80 x 80 m or 900 x 900 m).

5. METHODOLOGIES

Having collected the relevant and sufficient amount of data either by means of a traditional base line study or by means of more sophisticated methods the hard task remains of evaluation and application of the data in the planning process or in the control of the water quality.

5.1 Data Processing

In most cases the amounts of data are so large that the aid of an electronic computer is needed. During the study phase the data should be transferred to a database so that in the evaluation phase the data processing can be carried out easily.

The purposes with the data processing by the computer may be

1) check of the validity of the collected data
2) presentation (e.g. plots) of the data
3) statistical evaluation, e.g.
   - calculation of means standard deviation maximum/minimum values
   - cluster analysis of the gauging stations
   - correlation analyses
   - trend analyses
   - time series analyses.
4) preparation of the use of data in mathematical modelling.

Regarding the latter point one of the most applicable tools in water quality planning is mathematical models.

5.2 Mathematical Models

The use of mathematical modelling in the water quality planning has several purposes. The main purpose, however, is to be able to predict the expected water quality under various external conditions, e.g. sewage treatment, water catchment, etc. Models can be divided into types, e.g.

1) 'qualified guess'
2) empirical/statistical models
3) dynamic/causal models
4) stochastic models.

The complexity growing from 1) to 4).

Mathematical modelling applies for rivers, lakes, coastal areas and open sea, but the mathematical description is different for the different types of waters.

5.2.1 River Modelling

One of the major topics in water quality of rivers is the concentration of dissolved oxygen. In order to support fish life a certain minimum concentration of dissolved oxygen should be maintained. The conditions in rivers are complicated due to extended diurnal variations of the water quality parameters. Therefore a river model should be a short term model (a few days) with a detailed description of the short term variations.

A diagram of the most important oxygen flows and transformations in rivers is seen in Figure 4. The most important state variables are oxygen and BOD concentration.

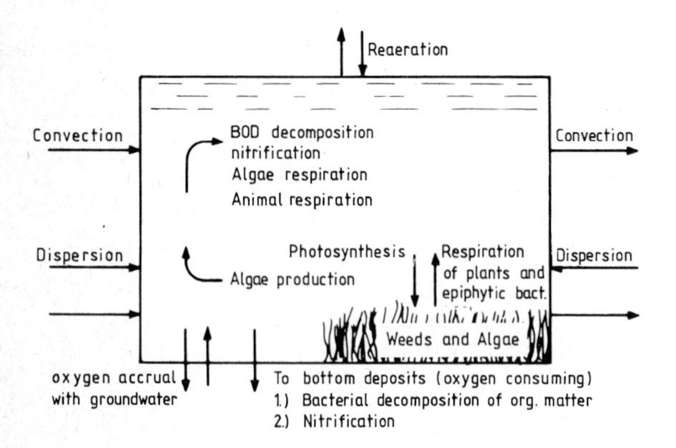

Figure 4: Oxygen flows and processes affecting the oxygen concentration in rivers.

In the following *an example* of the composition of a river oxygen model is given.

The forcing functions of the model are

- hydrology and hydraulics
- convection
- dispersion
- seepage of water and BOD, etc.
- equal distributed effluents
- sedimentation
- suspension
- exchange with the air
- biological processes
- chemical processes
- temperature
- incoming solar radiation
- discharge of oxygen consuming substances.

Among the biological processes are BOD decomposition, photosynthesis, respiration of plants and epiphytic bacteria, and nitrification.

Mathematical equations describing the reaction kinetics of the different processes linked together by material balance equation form a simulation model. The river is divided into sections and all important tributaries as well as waste water discharges are sampled, e.g. every second hour in a 24 hour period. Also within each section samples are taken in order to be able to calibrate the model. The main parameters to be estimated are:

$k_1(20)$	reaction constant for decomposition	day^{-1}
P_m	maximum oxygen production	mg O_2/l/h
R(20)	respiration	mg O_2/l/h
k_3	sedimentation constant of degradable organic matter	day^{-1}
D_B	benthic oxygen consumption	mg O_2/l/h
k_2	reaeration constant	day^{-1}

By the field measurement the physical parameters of the river are measured. These include the flow, width, slope, average velocity and dispersion coefficient in several stations in the river. A river divided into sections is seen in Figure 5.

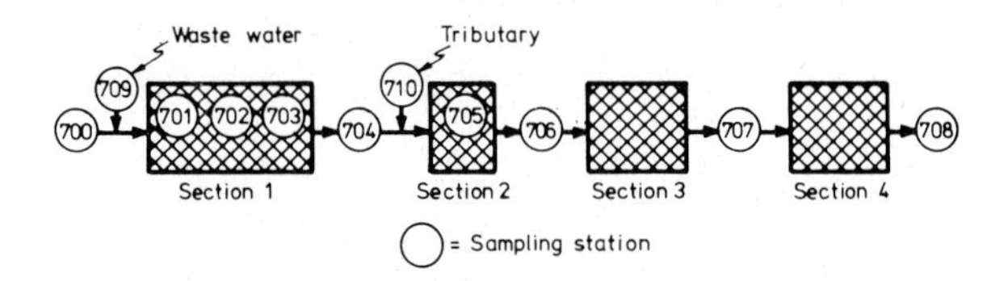

Figure 5: Sectioning of a river for calculation and sampling.

When the parameters are estimated, it is possible to calculate the effect of various discharges and the necessary treatment in order to achieve certain quality criteria for the given river.

5.2.2 Lake Models and Models for Coastal Areas

The most important problem in lakes and coastal areas is the growth of algae (eutrophication). Therefore the most applicable lake models are dealing with that problem. Opposite to river modelling, lake and fjord modelling is long term modelling describing the situation during at least one year.

Internationally the work on lake models have been concentrated on two model types. One of these types is the input-output oriented models, which based on measurements of loadings of a lake with phosphorus and nitrogen calculate some lake parameters (e.g. retention coefficient) on a statistical/empirical basis. The values of these parameters are then used for calculating consequences of changed loadings on concentrations of the substances in view. The classical models of this type are the models of Vollenweider (1969) and the later extended versions by Dillon and Rigler (1974). Generally the models are limited in their scope by several factors.

Firstly, the models are stationary, i.e. they cannot take yearly variations of loadings and parameter values into account. Secondly, in many cases the models do not describe the biological variables, which are of main interest, and what is very important, do not take into account the influence of biological processes on the model parameters.

Thirdly, because of the lack of description of biological processes important interactions between several phytoplankton limiting factors, light, phosphorus, nitrogen are not taken into account.

The advantage of some of the simplified versions of these input-output models is that a 'rule of thumb' calculation based on these models can be carried out in a few hours.

The other type of models, which has been developed internationally is the eutrophication models, which is dynamic, i.e. includes time dependent variations and takes important biological/chemical processes into account. Pioneers of these developments are Chen (1970), Di Toro et al. (1971), and Scavia et al. (1976). The first two authors have used their models in a management context. whereas Scavia et al. primarily have developed a research oriented model version.

In the Nordic countries work on ecological models was initiated at the Askö Laboratory in Stockholm (Sjøberg et. al. 1972). The scientists of this laboratory emphasized modelling of the Baltic ecosystems, however, the problems of these ecosystems were approached by the same basic methods as eutrophication models for lakes. In Sweden a eutrophication model for lake Norrviken has been developed by I. Ahlgren (1974). This model has been used for describing the consequences of a removal of waste water discharge from lake Norrviken and especially the role of the sediments in the transition period. The model developed by Ahlgren is one of the few existing models, which takes into account the yearly succession of the three groups of phytoplankton species (diatoms, green, and bluegreen).

In 1976 work on lake modelling was initiated in Finland by the National Water Board. The lake in view is Lake Paijänne and the EPA-ECO model, which has been used for lake investigation in the U.S.A., is applied on the lake.

In general the main emphasis on eutrophication modelling for lakes among the Nordic countries has been placed in Denmark starting in 1970. During the last decade the models have been accepted by planners and politicians as being most applicable in the process of making decisions.

The work on research-oriented lake models has primarily been catalysed by the Joint Lake Research Project (JLRP) in which several basic and applied research institutions are co-operating to gain understanding of lake eutrophication problems and possibly to model a number of complex aspects of the eutrophying processes. The work on this project started in 1972 and a number of publications have been the result (Joint Lake Research Project Reports 1974-1976, Gargas 1976, Jørgensen et al. 1976.). The work is continuing, and among others the emphasis is now on the general description of vertical gradients of phytoplankton biomass, nutrients, and dissolved oxygen.

In the same period work on management-oriented models has been carried out, and the research and management work have been mutually inspiring. The work on management models was implemented in 1970 on Haderslev Dam (VKI 1972). The aim of these initially very simplified models was to use the information obtained from a survey of Haderslev Dam to calculate the consequences of a proposed waste water treatment on concentrations of phosphorus and nitrogen.

In the years from 1973 to 1976 major surveys of approximately 20 Danish lakes were carried out, and in these surveys have been included comprehensive measurements of chemical and biological state variables and processes and of loadings with nitrogen and phosphorus and hydraulic properties.

The results of these surveys have been used by N. Nyholm to establish a eutrophication model for a shallow lake, i.e. a lake where complete vertical mixing can be assumed. A comprehensive documentation of this model is given in a VKI report (1978). Generally this model has been successful in simulating the measured values of state variables and processes for the lakes in view, and it is important to note that a reasonable agreement between simulated and observed values has been obtained by changing only very few (3-4) parameters within narrow intervals from lake to lake. However, there are limitations in the applicability of the model. Firstly, the simulation uses time averages of loadings and light intensities over weeks, i.e. short time fluctuations cannot be reproduced. Secondly, the model is limited in its possibilities for reproducing the diatoms in early spring. Thirdly, the model generally gives better results for lakes with a low hydraulic retention time.

The state variables, processes and forcing functions for the eutrophication model of a shallow lake is shown in Figure 6.

The state variables are phytoplankton-biomass, phytoplankton-nitrogen, phytoplankton-phosphorus, dissolved inorganic phosphorus, dissolved inorganic nitrogen, 'detritus' nitrogen, 'detritus' phosphorus.

The recycling of matter is described by the processes: phytoplankton growth, uptake of nitrogen and phosphorus in phytoplankton biomass, decimation of phytoplankton, sedimentation of phytoplankton and detritus, mineralization of detritus, and release of nutrients from sediments to phosphorus.

A comprehensive presentation of process kinetics is given in VKI 1978a. As important factors we can mention that the phytoplankton growth rate is related to intracellular concentrations of nutrients. Furthermore the release of nutrients from sediments is a function of the gross sedimented amount of nutrients. This implies that the model can describe a new state of equilibrium but not the length and course of the transient period.

The forcing functions for the model are light intensity of surface, water temperature, loadings with nitrogen and phosphorus, and water flow. The forcing functions are smoothed by appropriate averaging.

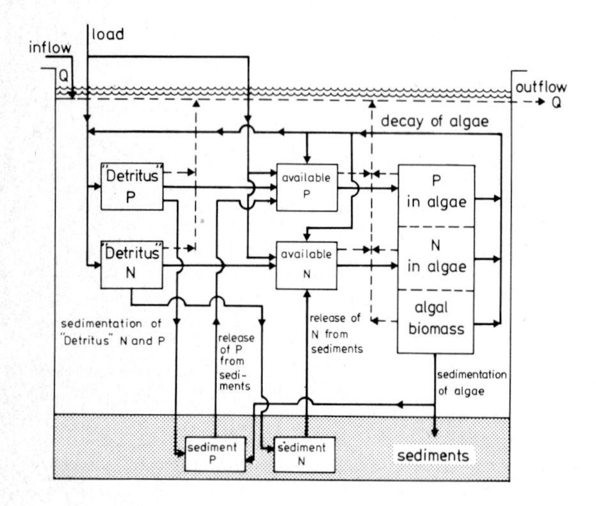

Figure 6: The structure of the eutrophication model for a shallow lake.

In many cases the simple hydraulic description (one totally mixed box) will not be sufficient. In such a case the lake description is a little more complicated (several connected boxes).

6. CONCLUDING REMARKS

In water quality planning a strategy involving water quality objectives and permissible loading seems more adequate than a strategy of best technical means.

The need of data for water quality planning is strongly dependent on the utilization of the water and the objectives ruled by this utilization.

Huge amounts of data are often collected. Data regarding hydrology, climatology and sewage discharge should be combined with water quality data in order to achieve a rational water quality planning.

The aid of digital computers are necessary due to the large amounts of data.

Modern and sophisticated methods should be applied in order to satisfy the growing need for data.

Useful tools in the water quality planning are mathematical models. Digital computers have made it possible to evaluate complicated models for practical use.

7. REFERENCES

Ahlgren, I. (1974): A model for the turnover of of phosphorus and the production of phytoplankton in Lake Norrviken. (In Swedish). Limnologiske Institutionen, Uppsala.

Chen, C.W. (1970): Concepts and utilities of ecological model. J. of San. Eng., Div. ASCE 96, SA5, 1085-1097.

Commission of European Communities, Environment and Consumer Protection Service (1975): Evaluation of Community Water Management Programmes. Final Report.

Dahl-Madsen, K.I. (1972): Operation analysis and water quality. (In Danish). VAND 1, 1-7.

Dahl-Madsen, K.I. & E. Gargas (1974): A preliminary eutrophication model of shallow fjords. IAWPR conference, Paris.

Dahl-Madsen, K.I. & J. Simonsen (1974): River Simulation. Report from the Water Quality Institute and Department of Sanitary Engineering, Techn. Univ. of Denmark, (in Danish).

Dillon, P.J. & F.H. Rigler (1974): A test of a simple nutrient budget model, predicting the phosphorus concentration in lake water. J. Fish.Res.Can., 1771-78.

Di Toro, D.M., D.J. O'Connor & R.V. Thomann (1971): A Dynamic model of the phytoplankton in the Sacramento-San Joaquin Delta. Advances in Chemistry Series, No. 100. Am. Chem. Soc.

Dresnack, R. & W.E. Dobbins (1968): Numerical Analysis of BOD and DO Profiles. J.San.Eng. ASCE. 94, No. SA5. Proc. Paper 6139, 789-807.

Gargas, E. (1976): A three box eutrophication model of a mesotrophic Danish lake. Rep. Water Quality Institute, Copenhagen, 1, 1-47.

Joint Lake Research Project (1974) Final Report 1973, Copenhagen, 104 pp. (in Danish).
- (1975) Final Report 1974. Copenhagen, 180 pp. (in Danish)
- (1976) Final Report 1975. Copenhagen, 112 pp. (in Danish).

Jørgensen, S.E., L. Kamp-Nielsen & O. Jacobsen (1976): An eutrophication model for a lake. Ecol Modelling 2, 147-165.

Ministry of the Environment, the Danish, Agency of Environmental Protection (1974): Waste Water Treatment 1972/82. Publ. No. 1974/1 (in Danish).

Nyholm, N. (1978): A simulation model for phyto-plankton Growth and nutrient cycling in eutrophic, shallow lakes. Ecol. Modelling 4 (1978), 279-310.

Scavia, D. & R.A. Park (1976): Documentation of selected constructs and parameter values in the aquatic model Cleaner. Ecol. Modelling, 2, 33-59.

Sjøberg, S., F. Wulff & P. Wåhlstrøm (1972): The use of computer simulations for systems ecological studies in the Baltic. AMBIO, 1, 217.

VKI (Water Quality Institute) (1972): Haderslev Dam 1971-72. Report to the municipality of Haderslev. Vandkvalitetsinstituttet, ATV, (in Danish).

VKI (1973): Roskilde-Isefjord. Report from Vandkvalitetsinstituttet, ATV (Water Quality Institute) (in Danish).

VKI (1974): Sørup Sø, Hvidkilde Sø, Nielstrup Sø, Ollerup Sø, Brændegård Sø, Nørre Sø, Arreskov Sø, 1974-74. Report to the municipality of Funen. Vandkvalitetsinstituttet, ATV, (in Danish).

VKI (1975): Gudenå-investigation 1973-1975. Report to the Committee of the Gudenå investigation. Vandkvalitetsinstituttet, ATV, (in Danish).

VKI (1978a): A program system for lake models, LAVSOE and BOXSOE. Vandkvalitetsinstituttet, ATV, (in Danish).

VKI (1978b): Gyrstinge Sø and Tystrup-Bavelse Sø. The effect of an increased water catchment in the Suså area. Report to the Environmental Protection Agency. Vandkvalitetsinstituttet, ATV, (in Danish).

VKI (1982): Remote sensing of coastal areas. Vandkvalitetsinstituttet, ATV, (in Danish).

Vollenweider, R.A. (1969): Möglichkeiten und Grenzen elementarer Modelle der Stoffbilanz von Seen. Arch. Hydrobiol. 66, 1-36.

Data for Science and Technology, P.S. Glaeser (ed.)
North-Holland Publishing Company

THERMOPHYSICAL PROPERTY DATA FOR ORGANIC COMPOUNDS OF INDUSTRIAL IMPORTANCE - A NEW DATA PROJECT

Michael J. Hiza, Bernard Le Neindre, Lucien Sobel,
Calvin F. Spencer *(Chairman)* and Andrzej M. Szafrański

CODATA Task Group on Data for the Chemical Industry

1. RATIONALE

It is becoming increasingly important for the chemical engineer, materials scientist, and the chemical, mechanical and electrical research worker to have access to reliable physical and thermodyanmic property data. Although handbooks contain material properties, equation coefficients, and other basic engineering and scientific data, the specific information that is needed always seems hard to locate, or out-of-date, or difficult to interpret.

In the days when chemical engineering was less sophisticated, the collection and tabulation of engineering and scientific data was fairly straightforward. Today the task is no longer the same. For many of the nearly 13 000 chemicals produced in about 3000 Western European companies in commercial quantities (more than 2000 kg/yr or valued at more than US $10 000) (1) most salient property data are missing. Furthermore, the data scattered throughout the literature, when finally compiled, prove often to be so inconsistent as to allow no results to be derived from then with confidence.

Conservative selection of design data leads to "overdesign", e.g. of separation equipment like a distillation column or other unit, while "optimistic" data jeopardize safety and reliability (2). In addition, the spread of crude data is often so great that nobody can tell whether the selected data are optimistic or conservative.

The above statements may be amplified by the following representative quote.

Professor Richard Montgomery Stephenson of the University of Connecticut, Editor of "Absorption/ Distillation/Extraction - A Newsletter" (April 1982) said:

"For several years, I have been a member of the Absorption, Distillation and Extraction Committee for the American Institute of Chemical Engineers. During this time I have had considerable contact with industry and have been impressed by the following three facts:

(1) the chemical and petroleum industries are still lacking much basic thermodyanmic and physical property data needed for process engineering calculations;

(2) some of this data is known to be available in specialized reports put out by universities and research institutes throughout the world.

(3) more active cooperation between American industry and these researchers could help to make this information available and could steer research programs into areas of greater interest to industry."

The need for amassed and evaluated thermophysical and thermodynamic data is steadily growing in view of:

(a) changes in cost and availability of raw materials, mainly crude oil and coal, which affect the feasibility and profitability of existing commercial processes;

(b) rises in process energy, plant safety, product transportation and like costs and also in the cost associated with the compliance with environmental control laws;

(c) growing demand for, and an increasing significance attached to computer modeling studies on the feasibility of new and modernization of old processes.

To sum up, the need for accurate, consistent and universally acceptable thermophysical and thermodynamic data and for data estimation routines is felt accutely by industry, government and universities alike throughout the world.

The response of the CODATA Task Group on Data for the Chemical Industry is to propose an effort of producing accurate, internally coherent thermophysical and thermodynamic property data bases in user-accessible, computer-compatible and internationally acceptable form.

The resulting information should be applicable to various situations covering industrial mixture separation processes, chemical process design, chemical fate modeling, experimental design, plant operation and like problems.

To reinforce this Rationale, let us cite the NRC CODAN Report on the "National Needs for Critically Evaluated Physical and Chemical Data":

"The benefits to the nation of having compilations of reliable data readily available are substantial. Such compilations save time for engineers and scientists in research and development....designs can be more precise, tolerances reduced, and R&D

options narrowed. The resulting savings can amount each year from one to several thousand times the cost of evaluation."

2. PRIMARY GOALS

The present Data Project has the following goals in view:

(1) To compile, critically evaluate, process and publish data on technologically salient properties of individual organic compounds that represent industrially important constituents of process streams.

The following physical constants are included in the present plans:

molecular weight
specific gravity
normal boiling point
critical properties
acentric factor
enthalpy of combustion
enthalpy of formation
Gibbs' energy of formation
and temperature-dependent property data from the melting point to the boiling point up to a reasonably high pressure:

liquid and vapor density (saturated)
liquid and vapor viscosity
liquid and vapor thermal conductivity
saturated vapor pressure
latent heat of vaporization
ideal gas heat capacity
liquid heat capacity
surface tension

(2) To provide property data preferably in correlation equation forms, e.g. as functions of state variables such as temperature of pressure, covering broad validity ranges and easy to input into any kind of electronic machine - from programmable micros to large computers.

(3) To provide property data not only for major process stream components including much-studied chemicals like benzene, cylohexane, etc., but also for so-called maverick compounds, e.g. cumene hydroperoxide, cyclopentadine and many others, which are of interest in some design areas but for which very few of the basic physical properties are known.

(4) To provide routines developed for use on handheld calculators (e.g. Texas Instruments TI-59 or Hewlett-Packard HP-41CV) applicable to:

(a) evaluation-from-formula calculations to decode property values, e.g. saturated vapor pressures, P^S, from complex literature correlation forms like Chebyshev polynomials, Frost-Kalkwarf and other $P^S = f(T)$ equations that require recursive or iterative solutions;

(b) predicting of property data by group contribution methods, e.g. Rihani-Doraiswamy for ideal-gas heat capacities, Lydersen for critical constants, etc.;

(c) curve fitting of newly measured or reported data to selected equations, e.g. P^S vs. T data to the Antoine equation, by using standard or weighted least-squares techniques;

(d) linear and nonlinear regression routines involving various functions as well as other useful programs applicable in research and engineering data handling practice.

All of these should include step-by-step instructions and worked-out examples.

(5) To provide concise chapters treating errors, effect of errors involved in input data on estimated property values and calculated design data, regression analysis, weighted least squares techniques, evaluation of coefficients of correlation, etc.

(6) To precede each numerical data section with a preview chapter describing the relevant property, the constraints to be met by, and the relative merits of the considered property correlation functions, serving both as the reminder of basic property definitions and relations and as a guide through the matrix of existing correlation equations (although not to the extent of a Reid, Prausnitz, Sherwood (3)) discussed in terms of the intrinsic thermodynamic and/or mathematical constraints and their fulfillment.

3. MODUS OPERANDI

Self-consistent P^S (or T), saturated liquid density, specific saturated vapor volume, specific enthalpies and entropies of saturated liquid and saturated vapor, specific enthalpy of vaporization, specific heat capacity of saturated liquid, and also v, z(=pv/RT), c_p, c_v, h and s data for the superheated vapor can be evaluated from a single correlation function, viz., the Helmholtz energy, by the use of standard thermodynamic relationships.

This method was applied by Keenan and co-workers to produce steam tables and also by the ESDU teams to prepare the tables and charts for thermodynamic properties of benzene (4), toluene, and isopropanol.

For the present Task Group team this method is entirely impracticable. It is too tedious to follow and requires too many observed data points* to be applicable to more than a very few

---

* For benzene, 332 critically selected experimental p-v-T or v-T-h data points were used to adjust a total of 50 least-squares constants in the Helmholtz energy function (4).

out of the hundreds of compounds of potential interest.

A considerable range of chemicals can be successfully covered with a known level of confidence by resorting to correlation equations applicable to T- and/or P-dependent property data.

For the purposes of illustration below are listed several property correlation equations by the British Institution of Chemical Engineers (adapted from CIE Newsletter No. 2, 15 December, 1977):

I. Cubic in temperature

$$Z = \sum_{i=0}^{3} a_i t^i$$

II. Extended Antoine

$$\log_{10} Z = a_1 + a_2 t + a_3/(t + a_4)$$

III. Function of reduced temperature

$$Z = a_o + \sum_{i=1}^{7} a_i (1 - T/B)^{i/3}$$

IV. Chebyshev polynomial

$$T\log_{10} Z = (a_o/2) + \sum_{i=1}^{i} a_i E_i(x)$$

$$x = (2T - T_{max} - T_{min})/(T_{max} - T_{min})$$

V. Wilhoit-Zwolinkski

$$Z = a_1 + a_2 y(1 + (y-1)(a_3 + \sum_{i=4}^{7} a_i y^{i-3}))$$

$$y = \frac{T/B}{1 + T/B} \qquad B - \text{constant}$$

VI. Vapor enthalpy equation

$$Z = a_o + a_1 x^{-1} + a_2 \ln x + \sum_{i=3}^{7} a_i x^{i-2}$$

$$x = (1 + T/B)^{-1}$$

VII. Liquid heat capacity equation

$$Z = \sum_{i=0}^{4} a_i t^i + a_5/(a_6 - t)$$

VIII. Liquid enthalpy equation

$$Z = \sum_{i=0}^{5} a_i t^i + a_6 \ln(a_7 - T)$$

Properties Z cross-refer against equation types as follows:

Liquid density: III; viscosity: II; conductivity: I, III; specific heat: I, VII; enthalpy: VIII; surface tension: I; cubical expansion: I. Vapor pressure: II, IV; viscosity: I; conductivity: I; specific heat: V; enthalpy: VI. Latent heat: III.

A more comprehensive approach to empirical and semiempirical equations suitable for correlation and/or estimation of thermophysical and thermodynamic data is presented elsewhere (5). Such equations used in conjunction with reported experimental data may help us establish internally coherent property data for selected chemicals including maverick compounds.

3.1 Critical vs. Uncritical Compilations

In view of recent critical compilations like

D.R. Stull, E.F. Westrum, G.C. Sinke, The Chemical Thermodynamics of Organic Compounds, Wiley, N.Y., 1969;

J.D. Cox, G. Pilcher, Thermochemistry of Organic & Organometallic Compounds, Acad. Press, London, N.Y. 1970;

J.B. Pedley, J. Rylance, Sussex-NPL Computer Analyzed Thermochemical Data: Organic & Organometallic Compounds, Sussex Univ., 1977;

R.C. Wilhoit, B.J. Zwolinski, Physical and Thermodynamic Properties of Aliphatic Alcohols, J. Phys. Chem. Ref. Data, vol. 2, Suppl. 1, 1973;

A.P. Kudchadker et al., Key Chemicals Data Books;

etc., the present Task Group regards the critical approach to be desirable especially for compounds whose data status will enable the aforementioned compilations to be matched in data assessment level or when some results can even be copied provided the compound's data status has remained unchanged.

Maverick compounds will be difficult to treat critically. A possible treatment can involve comparisons of estimated with observed (if any) data and consistency tests (cf. Wilhoit and Zwolinski, op. cit.) intended to check the internal coherence of data evaluated by generalized property estimation methods.

For maverick compounds, the present data project can offer

(a) property data estimation routines
(b) reported primary data points, if any, and annotated bibliography
(c) evaluator's comments.

The following example illustrates roughly what can be done for maverick compounds.

CUMENE HYDROPEROXIDE, C9H12O2

TEMPERATURE-DEPENDENT PROPERTIES OF COMPOUND

TEMP	DENSITY LIQUID	DENSITY VAPOUR	ENTHALPY LIQUID	ENTHALPY VAPOUR	HEAT OF VAPORIZ.	SPECIFIC HEAT LIQUID	SPECIFIC HEAT VAPOUR	VISCOSITY LIQUID	VISCOSITY VAPOUR	THERMAL CONDUCT. LIQUID	THERMAL CONDUCT. VAPOUR	SURFACE TENSION	VAPOUR PRESSURE
DEG C	G/ML	G/L	CAL/MOL	CAL/MOL	CAL/MOL	CAL/MOL/DEG C		CP	E-6P	KCAL/M/HR/DEG C		DYNES/CM	MM HG
20	1.061	.0002	1160.	18098.	16937.	59.6	43.9	1.569	.612E+02	.078	.672E-02	29.34	.0
25	1.057	.0002	1460.	18319.	16859.	60.4	44.6	1.455	.622E+02	.079	.693E-02	28.92	.0
30	1.053	.0004	1764.	18544.	16780.	61.1	45.2	1.353	.633E+02	.079	.714E-02	28.50	.0
35	1.049	.0006	2072.	18772.	16700.	61.9	45.9	1.261	.643E+02	.080	.736E-02	28.08	.1
40	1.045	.0009	2383.	19003.	16620.	62.7	46.5	1.178	.654E+02	.081	.758E-02	27.66	.1
45	1.041	.0014	2698.	19237.	16539.	63.4	47.2	1.102	.664E+02	.081	.780E-02	27.25	.2
50	1.037	.0020	3017.	19474.	16457.	64.2	47.8	1.033	.674E+02	.082	.802E-02	26.83	.3
55	1.033	.0030	3340.	19715.	16375.	65.0	48.4	.971	.685E+02	.082	.825E-02	26.42	.4
60	1.029	.0042	3667.	19959.	16292.	65.7	49.1	.914	.695E+02	.083	.848E-02	26.00	.6
65	1.025	.0060	3997.	20206.	16209.	66.5	49.7	.861	.706E+02	.083	.871E-02	25.59	.8
70	1.020	.0084	4331.	20456.	16124.	67.2	50.3	.813	.716E+02	.084	.894E-02	25.18	1.2
75	1.016	.0117	4669.	20709.	16039.	68.0	50.9	.769	.727E+02	.084	.918E-02	24.77	1.7
80	1.012	.0160	5011.	20965.	15954.	68.7	51.5	.729	.738E+02	.085	.942E-02	24.36	2.3
85	1.007	.0218	5357.	21224.	15867.	69.5	52.1	.691	.748E+02	.085	.966E-02	23.95	3.2
90	1.003	.0292	5706.	21486.	15780.	70.2	52.7	.656	.759E+02	.085	.991E-02	23.54	4.3
95	.998	.0389	6059.	21751.	15692.	71.0	53.3	.624	.769E+02	.086	.102E-01	23.14	5.8
100	.994	.0513	6416.	22019.	15603.	71.7	53.9	.594	.780E+02	.086	.104E-01	22.73	7.8
105	.989	.0670	6776.	22290.	15514.	72.5	54.5	.567	.791E+02	.087	.107E-01	22.33	10.3
110	.984	.0868	7140.	22564.	15423.	73.2	55.1	.541	.801E+02	.087	.109E-01	21.93	13.5
115	.980	.1117	7508.	22840.	15332.	74.0	55.6	.517	.812E+02	.087	.112E-01	21.52	17.6
120	.975	.1425	7880.	23120.	15240.	74.7	56.2	.494	.823E+02	.088	.114E-01	21.12	22.7
125	.970	.1805	8255.	23402.	15147.	75.5	56.8	.473	.833E+02	.088	.117E-01	20.73	29.1
130	.965	.2271	8635.	23688.	15053.	76.2	57.3	.454	.844E+02	.088	.119E-01	20.33	37.0
135	.961	.2838	9017.	23976.	14958.	76.9	57.9	.435	.855E+02	.088	.122E-01	19.93	46.8
140	.956	.3524	9404.	24266.	14862.	77.7	58.4	.418	.865E+02	.089	.125E-01	19.54	58.7
145	.951	.4349	9794.	24560.	14765.	78.4	59.0	.402	.876E+02	.089	.127E-01	19.14	73.2
150	.945	.5336	10188.	24856.	14668.	79.2	59.5	.386	.887E+02	.089	.130E-01	18.75	90.6

MAVERICK: Cumene hydroperoxide

Primary data points (6)

d_4^{18}, g cm^{-3}	1.062 - 1.064				
p^s, mmHg	1	3	7	8	15.5
t, °C	74	88	97.5	100.5	116.5

Viscosity, cP	0.174	0.105	0.068
t, °C	20	30	40

Heat of vaporization (Knudsen, 10-60 °C)
(cf. Cox, Pilcher), Kcal mol^{-1} 16.70 ± 0.04

The evaluator may comment on the reliability (method, error, coherence of data within homologous series, etc.) and acceptability of primary data. If found acceptable, the data may serve for comparison with the estimated values (cf. accompanying printout (7)) to help deduce the final conclusions , e.g.

"The density and enthalpy of vaporization estimated at respectively 18 °C and av. t = (10+60)/2 = 35 °C are in excellent agreement with the respective experimental values. Density, heat capacity, enthalpy, heat of vaporization, saturated vapor pressure and presumably also thermal conductivity of liquid cumene hydroperoxide are well estimated by the generalized methods applied (7). Experimental vapor pressures as well as observed and estimated viscosity data are doubtful."

The example serves only to stress some aspects of the intended treatment of mavericks and leaves numerous features entirely untouched.

4. RETROSPECT vs. PROSPECT

Several ongoing data projects and major compilations are acknowledged; these will provide experience and help to identify the gaps.

The Thermodynamic Research Center, Texas A&M (K.R. Hall, R.C. Wilhoit, B.J. Zwolinski et al.) has been operating for 40 Years as a service organization to collect and critically evaluate thermodynamic data for organic compounds. F.D. Rossini instituted the Center in 1942 at NBS, Washington, D.C., as API Research Project 44. The project has covered about 5000 compounds.

The NBS Chemical Thermodynamic Data Base (D. Garvin, V.B. Parker, D.D. Wagman et al.) has recently completed a substantial critical evaluation of chemical thermodynamic measurements on inorganic and C1 to C2 organic compounds. The values selected for some 12 000 substances, based on a collection of 250 000 measurements, are being both published and distributed via the Chemical Information System. During this work, data banks of several types have been designed and are now being constructed: extracted unevaluated data plus bilbiography, evaluated measurements (catalogs of reactions), and selected chemical thermodynamic properties of individual substances.

NBS Monographs by R.D. Goodwin, W.M. Haynes, et al. represent a series dedicated to thermophysical properties of hydrocarbons and other fluids.

A continuous effort is being made to add new species to the JANAF Thermochemical Tables (M.W. Chase, P.A. Andreozzi, J.R. Downey, R.A. McDonald, E.A. Valenzuela et al.) and to revise existing outdated tables. Prime interest is centered on the increasing use of computers in data retrieval, generation and updating of bibliographic files, maintenance of data files and accessiblity of the adopted thermochemical values. Additional information covering heats of atomization, crystal structures, and more detailed transition data is being included. Calculational techniques are being re-evaluated and upgraded or extended, where necessary.

The Thermodynamic Research Laboratory at Washington University, St. Louis, has just released an extensive data bank containing thermodynamic data on pure compounds. The data bank was the result of combing through 25 000 literature documents and collecting, processing and critically evaluating the data and organizing them in a computer file. The data system is an ongoing project that is far from complete. The released data bank covers 383 hydrocarbons in the C3 to C9 range. Entries for these include pure compound constants such as melting point, boiling point and critical properties; vapor pressure, liquid density and second virial coefficients as a function of temperatures; and equation-of-state data.

The IUPAC-sponsored International Thermodynamic Tables of the Fluid State (S. Angus et al.) include, among others, volumes on methane, ethylene en propylene.

The DIPPR Project was extensively presented by Professor D. Zudkevitch at the IUPAC 6th Conference on Thermodynamics, Merseburg, GDR (26-29 Aug. 1980).

The EROICA Data Base for the Fundamental Physical Properties of Organic Compounds has been reported by Yukio Yoneda (8a).

Other data bases that deserve mention:

ORGANIZATION	BANK NAME	COMPOUNDS COVERED	PURE-COMPOUND PROPERTIES
Inst. of Chem. Engineers	PPDS	ca. 800	31
ICI Mond. Div.	DATABANK	500	40
Monsanto	FLOWTRAN	180 m	10
Solvay & Cie.	CBM	200	29
DSM	TISDATA	120	25
Engineering Sciences Data Unit	ESDU	931	thermal cond.
		760 m	p^s, crit.
		113	viscosity

m - mixtures included

Together with those mentioned before, the major critical and uncritical compilations of physical and thermodynamic property data on pure *organic* compounds published over 1960-1980 include more than 20 volumes (9-28).

In the present data project correlation equations will be the preferred form used to describe properties; adjusted coefficients will be tabulated with equation validity ranges and possibly also confidence intervals duly specified. Single values like heat of combustion will be listed in tables. Primary data will be collated particularly for maverick compounds. Each section containing equation coefficients and numerical property values will be preceded by an introductory or a preview chapter.

Flexible presentation formats will be sought to make passing from one correlation equation to another possible with no distortion in table shape and size.

5. COMPOUND OPTIONS

Just which, and how many, compounds may be labeled as industrially important? Dr. Schönberg's Category 1 chemicals (8b) (characterized by US production and/or 1979 price levels of more than 10^6 lb/yr and less than US $1 per lb, resp.) number about 800. If these chemicals are major process components, the chemical engineer will also be interested to know property data of associated mavericks to design suitable process equipment.

The following examples are illustrative of the maverick compound option.

MAJOR COMPONENT	PHYSICAL PROPERTY DATA REQUIRED FOR
p-Cumylphenol, 98.9% pure	m-Cumylphenol
	2,4-Diphenyl-4-methyl-1- and -2-pentenes
	1,1,3-Trimethyl-3-phenylindane
Phenol	Acetone
	Cumene
	α-Methylstyrene
	Methylstyrene dimer
	Mesityl oxide; isooxide
	Cumylphenol
	2-Methylbenzofuran
	Phorone
	Acetophenone
	Dimethylphenylcarbinol
Cyclohexanol	Butylcyclohexane
	Pentylcyclohexane
	Pentylcyclohexene
	Butyloxycyclohexane
	Pentyloxycyclohexane
	Methylcyclohexanones
	1-Cyclohexen-3-one
	1-Cyclohexen-4-one
	Cyclohexyl formate, acetate
	Bicyclohexyl
	2-Cyclo-1-hexenol
	n-Pentyl ether
	Butylbenzene
	2-Methylcyclopentanone

Of the 14 minor constituents indicated in the cyclohexanol rectification case only ten were available in conventional handbooks, merely m.p., b.p. and liquid density.

Another compound option is carbo chemicals. At the present moment about 100 compounds have been recovered from classical high-temperature carbonization coal tars in commercial quantities. Coal liquids produced by traditional or modern coal conversion processes (29) consist primarily of aromatic components with structures ranging from one to seven or eight condensed rings per molecule. These aromatic structures can be associated with one or more fully saturated napthenic rings, ring systems containing heteroatoms, phenolic functional groups, and side chains. Data on pure compounds are needed, especially on polynuclear aromatics and S-, O- and N-containing compounds (30).

6. PROSPECTIVE USERS

Those intended to benefit from the data project include primarily

(a) chemical engineers, process equipment designers and chemical technologists: their interest is focused mainly on

   (i) feedstock components
   (ii) intermediates
   (iii) products
   (iv) undesirable impurities

   In terms of the types of physical property data required, the needs of this group of users cover a rather broad spectrum of data and are almost always desired in the form of functions of state variables (T or P).

(b) research workers: their areas of interest comprise the whole of chemistry and thus the number of chemicals and data types of interest appears to be unlimited.

7. ASSESSMENT

A 2% error in critical temperature can cause a 10% error in liquid specific heat, a 2% error in latent heat, a 2% error in liquid density, and a 10% error in liquid thermal conductivity (assuming all other parameters to be correct).

Estimation techniques are many and varied; so are correlation equations. For example, more than twenty empirical and semiempirical equations have been suggested to describe the latent heat of vaporization as a function of temperature. For the functions to be used to correlate this property, mathematical rather than thermodynamic constraints can be formulated which are helpful in screening incorrect forms.

Preview chapters planned for the properties to be included in this data project are intended to cover such questions.

8. HANDHELD CALCULATOR ROUTINES

Programmable micros are expected to be playing an increasing part in the work of practicing chemical engineers, materials scientists and researchers. The most recent products like ESDUpacs and COMpacs are aimed at these users.

Commercial modules contain regression routines that do not meet chemical engineers' needs, and the area appears to be open for initiatives to supply regression packages for programmable minis that meet real needs of research workers and chemical engineering professionals.

Monographs (31, 32) have recently appeared dedicated to program sets intended for the solution of e.g. chemical engineering type design, construction and operation type problems; reference (31) contains programming needed for both TI and HP type handheld calculators.

This data project will provide also for encoding various group-contribution methods of property data estimation.

9. FINAL PRODUCT

The final product of the data project is expected to serve the individual user as well as provide input to data banks. In the forthcoming decade the latter will certainly grow in view of the need to handle an expanding volume of data. A conventional dissemination route involving a printed, hopefully computer produced, loose-leaf data book presentation, is anticipated. The aim is a relatively inexpensive publication with supplemental issues published on a continuing though irregular basis.

10. REFERENCES

[1] Directory of chemical producers - Western Europe, 5th ed., SRI International, 1982.

[2] Zudkevitch, D., Hydrocarbon Proc., 54, 97 (Mar. 1975).

[3] Reid, R.C., Prausnitz, J.M., Sherwood, T.K., The properties of gases and liquids, 3rd ed., McGraw-Hill, N.Y., 1977.

[4] ESDU Item No. 73009.

[5] Szafranski, A.M., Cholinski, J., Wyrzykowska-Stankiewicz, D., A new data project - computer-aided calculation of consistent physicochemical property data for maverick organic compounds, 8th CODATA Conference, 4-7 Oct. 1982, Jachranka (Poland).

[6] Kruzhalov, B.D., Golovachenko, B.I., Coproduction of phenol and acetone, (in Russian), Goskhimizdat, Moscow, 1963, and references therein.

[7] Cholinski, J., Szafranski, A.M., Wyrzykowska-Stankiewicz, D., Computer calculation of physicochemical property data for maverick organic compounds, CHISA 1975, Prague, Czechoslovakia.

[8] Glaeser, Ph. S., Ed., Data for science and technology, Proc. 7th Intern. CODATA Conf. (Kyoto, Japan, 8-11 Oct. 1980), Pergamon Press, Oxford, 1981; (a) p. 254-257; (b) 447-450.

[9] Wilhoit, R.C., Zwolinski, B.J., Handbook of vapor pressures and heats of vaporization of hydrocarbons and related compounds, College Station, Texas, TRC, Texas A&M Univ., 1971.

[10] Boublik, T., Fried, V., Hala, E., The vapor pressure of pure substances, Elsevier, N.Y., 1973.

[11] Ohe, Shuzo, Computer aided book of vapor pressure, Data Book Publ. Co., Tokyo, 1976.

[12] Ambrose, D., Vapor-liquid critical properties, NPL Report No. 107, NPL, Teddington, Feb. 1980.

[13] Tatevskii, V.M., Ed., Physicochemical properties of individual hydrocarbons, Gostoptekhizdat, Moscow, 1960.

[14] Karapet'yants, M.K., Karapet'yants, M.L., Handbook of thermodynamic constants of inorganic and organic compounds, Humphrey Publ., Ann Arbor, Michigan, 1970.

[15] Glushko, V.P., Medvedev, V.A., Eds., Thermal constants of substances (in Russian), USSR Acad. Sci., Moscow, vol. IV, 1970.

[16] Vargaftik, N.B., Handbook of thermodynamic properties of gases and liquids (in Russian), 2nd rev. ed., Izdat. Nauka, Moscow, 1972.

[17] Raseev, S., Istrati, S., Physical constants of hydrocarbons and related compounds (in Roumanian), Edit. Tehn., Bucharest, 1964.

[18] Gallant, R., Physical properties of hydrocarbons, Gulf Publ. Company, Houston, 1970.

[19] Riddick, J.A., Bunger, W.B., Organic solvents - physical properties and methods of purification, 3rd ed., Techniques of Chemistry, vol. II, Wiley-Interscience, N.Y., 1970.

[20] Gas Encylopaedia, Div. Sci. L'Air Liquide, Elsevier, Amsterdam, 1976.

[21] Jamieson, D.T., Irving, J.B., Tudhope, L.S., Liquid thermal conductivity - a data survey to 1973. NEL, Her Majesty's Stationery Office, Edinburgh, 1975.

[22] Vargaftik, N.B., Filippov, L.P., Tarzimanov,

A.A., Totskii, E.E., Thermal conductivity of polyatomic liquids and gases, State Committee on Standards, Moscow, 1981.

[23] Domalski, E.S., Selected values of heats of combustion and heats of formation of organic compounds containing the elements of C, H, N, O, P, and S. J. Phys. Chem. Ref. Data, 1, 221-272 (1972).

[24] Jasper, J.J., The surface tension of pure liquid compounds. Ibid., 1, 841-1014 (1972).

[25] Nelson, R.D., Lide, D.R., Maryott, A.A., Selected values of electric dipole moments for molecules in the gas phase. NSRDS-NBS 10, Washington, 1967.

[26] Osipov, O.A., Minkin, V.I., Garnovskii, A.D., Handbook of dipole moments (in Russian), Izdat. Vysshaya Shkola, Moscow, 1971.

[27] Dymond, J.S., Smith, E.B., The virial coefficients of gases - a critical compilation, Clarendon Press, Oxford, 1969: 2nd ed. 1980.

[28] Cholinski, J., Szafranski, A.M., Wyrzykowska-Stankiewicz, D., Second virial coefficients for pure organic compounds and binary organic mixtures - compilation, correlation and evaluation, PWN, Warsaw, in press.

[29] Szafranski, A.M., Needs for critically evaluated physicochemical and thermodynamic data for coal and petro chemicals. 8th Intern. CODATA Conf., 4-7 Oct. 1982, Jachranka, Poland.

[30] Lencka, M., Szafranski, A.M., Maczynski, A., Verified vapor pressure data. Vol. 1. Organic compounds containing nitrogen. PWN, Warsaw, in press.

[31] The Editors, Chemical Eng. Mag., Calculator programs for chemical engineers, McGraw-Hill, N.Y., 1981.

[32] S. Jaganath, Calculator programs for hydrocarbon processing industries, Vol. I, Gulf Publ. Company, Houston, 1980.

Data for Science and Technology, P.S. Glaeser (ed.)
North-Holland Publishing Company

SEQUENTIAL PARAMETER ESTIMATION FOR CONSTRUCTING PARAMETER TABLES OF GROUP CONTRIBUTION MODELS

Sándor Kemény, Gábor Chikány

Technical University of Budapest, Department of Chemical Engineering
1521 Budapest, Hungary

Parameter tables are usually built up gradually. The number of pair interaction parameters estimated in one step is limited so the preestimated and fixed parameters cannot be improved according to a new dataset. A sequential estimation method is proposed which allows to make use of all experimental information for estimation of every group interaction parameter without fixing any of them during the estimation of some new ones. This algorithm hopefully improves the abilities of group contribution models as parameter tables estimated this way contain a greater information volume.

Introduction

Parameter tables are built up gradually. A measured data set usually contains information on interaction of several groups, though it is not utilized. A typical way of constructing parameter tables is described by Fredenslund et al. [1] . In order to estimate the parameters for interaction between CCOH and CH_2CO groups from an alcohol - ketone data set the parameters for the CCOH, CH_2 and CH_2CO, CH_2 must be estimated previously from an alcohol - alkane and ketone - alkane data set, respectively. In the course of estimation the interaction parameters for the CCOH, CH_2CO pair the remaining parameters are not estimated simultaneously, they are fixed at their previously determined value.

A theoretically possible way for utilizing both the previous and a new piece of information on all pairs would be the simultaneous estimation of of all parameters from all data sets.

In practice there are two serious obstacles:

- the computation time and memory requirement would be enormous
- in order to include a newly published data set the whole procedure is to be repeated from the very beginning.

A sequential estimation method is proposed in this work which is equivalent to the above mentioned method but it requires reasonable computation time and core size. A similar procedure was proposed by Neau and Peneloux for different reasons [4] .

The method is developed in details for binary VLE data sets but it may be applied to any other type of measured data. The method is more fully explained in [2].

The assumptions are the same as for the usual application of maximum likelihood method to thermodynamic data [3] , namely:

a.) the model is adequate, i.e. it would describe the error-free measurement data

b.) the distribution of measurement errors is a normal distribution with zero expected value and σ^2 variance and the errors in different variables are independent of each other.

Let us consider two measured data sets (I and II), which are independent of each other. The objective function, irrespective of its inner structure may be decomposed into two independent parts:

$$\Psi(\underline{\Theta}) = \Phi_{I}(\underline{\Theta}) + \Phi_{II}(\underline{\Theta}) \qquad (1)$$

where $\underline{\Theta}$ is the parameter vector to be estimated.

The theoretically correct way of parameter estimation utilizing all the experimental information would be to minimize the $\Psi(\underline{\Theta})$ objective function with respect to the elements of the $\underline{\Theta}$ parameter vector.

Let us suppose that the first data set is an old one and the parameter estimation procedure has been executed on it, while the second data set is a new one. Thus instead of including the whole first data set it can be approximated by its second order Taylor polynomial:

$$\phi_I(\underline{\Theta}) \approx \phi_I(\underline{\Theta}^o) + \underline{q}^o \underline{\delta\Theta} + \frac{1}{2}\underline{\delta\Theta}^T \underline{\underline{H}}^o \underline{\delta\Theta} \qquad (2)$$

where

$\phi_I^o \equiv \phi_I(\underline{\Theta}^o)$ the value of the first objective function at its minimum

$\underline{q}^o \equiv \left(\frac{d\phi_I}{d\underline{\Theta}}\right)_{\underline{\Theta}^o}$ the gradient vector; in our case (no constraints between elements of $\underline{\Theta}$) it is a zero vector

$\underline{\underline{H}}^o \equiv \left(\frac{d^2\phi_I}{d\underline{\Theta}^2}\right)$ the Hessian, its negative inverse is equal to the covariance matrix of the approximately normal posterior density function of $\hat{\underline{\Theta}}$

$$\underline{\delta\Theta} \equiv \underline{\Theta} - \underline{\Theta}^o$$

This approximation is allowed if the $\hat{\underline{\Theta}}$ parameters estimated from the combined objective function [Eqn (1)] are sufficiently near to $\underline{\Theta}^o$.

If this assumption is proved to be invalid, only the original method is applicable, that is reducing the two data sets together. In the context of thermodynamics serious deviation of $\hat{\underline{\Theta}}$ estimated model parameter values shows either the inconsistency of data sets or the unsuitability of the model applied. In both cases the estimation of parameters is meaningless and one should reject either one of the data sets or the model.

In the next part of this paper examples are shown in order to demonstrate the applicability of the approximation expressed by Eqn (2).

Comparisons based on the UNIQUAC model

To illustrate the approximation let us take the example with two parameters (binary system with UNIQUAC [4]). Having done the reduction of a data set one obtains a 2x2 Hessian matrix. The curve which describes the parameter values giving the same level of the objective function is an ellipse [2]:

$$\varepsilon \equiv \phi_I(\underline{\Theta}) - \phi_I(\underline{\Theta}^o) = \frac{1}{2}\underline{\delta\Theta}^T \underline{\underline{H}}^o \underline{\delta\Theta} \qquad (3)$$

The axes of the ellipse are defined by the eigenvectors of the Hessian. The greater the eigenvector, the shorter the corresponding axis is:

$$a_i = \sqrt{2\,\varepsilon/\lambda_i} \qquad i=1,2 \qquad (4)$$

where

a_i the length of the i-th half axis

λ_i the i-th eigenvalue of the Hessian

Having known the axes, the ellipse may be drawn to any value of ε and can be compared with the shape of the surface of the objective function at the same level. The goodness of approximation expressed by Eqn (2) is illustrated in Fig. 1. on the example of system ethyl alcohol-benzene.

The parameter estimation was carried out for several binary systems using the two ways: treating the two data sets simultaneously and sequentially.
The results are shown in Table I.
It can be seen that the estimated parameters and the optimum values of the objective function found by simultaneous data reduction are approximated well by the sequential method.

In order to investigate the effect of goodness of a data set simulated using fixed UNIQUAC parameters, then random errors with prescribed variances and zero expected value were added to them. A simple but general case of systematic errors can be modelled if the true parameters of the two data sets for the same binary system are different.

The following cases were investigated: (see Fig 2.)

a.) the two parameter sets are identical (data sets A and B)

b.) the parameters of the second data set are different from those of the first one but they are lying in the valley of the objective function of the first data set (data sets A and C)

c.) the parameters of the second data set are lying outside of that valley (data sets A and D)

The results are shown in Table II.

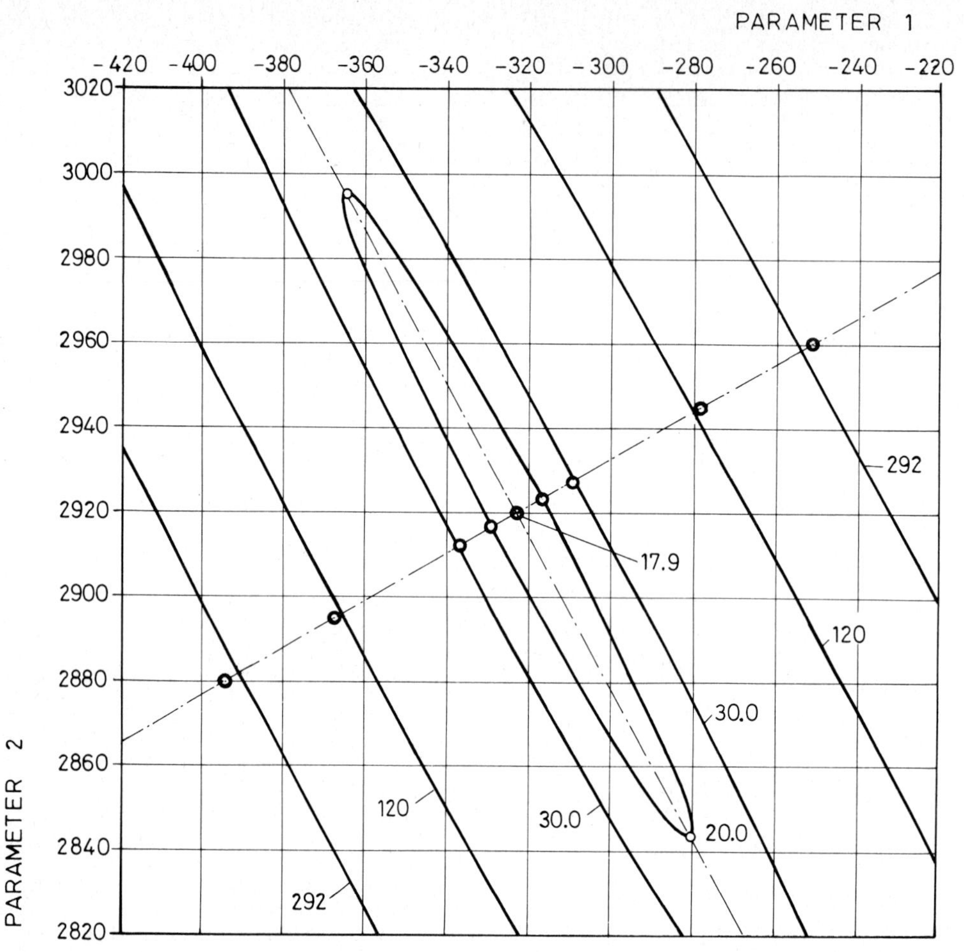

Figure 1: Contour line map of the objective function and the axis points of the approximating ellipses Data set: [5]

Table I: Parameter estimation by simultaneous and sequential way, results

ethanol-benzene			
data set	objective function value	parameters	
I (isobaric, 13 points [5])	17.9	-322.1	2918.2
II (isobaric, 9 points [6])	60.9	-401.2	3130.2
I+II (simultaneous)	99.2	-327.6	2945.7
I→II (sequential)	99.3	-326.4	2943.6
acetone-chloroform			
I (isotherm, 9 points [7])	18.6	-1139.2	641.1
II (isobaric, 9 points [8])	1.7	-1358.8	700.4
I+II (simultaneous)	28.6	-1383.6	725.2
I→II (sequential)	28.8	-1356.5	685.7

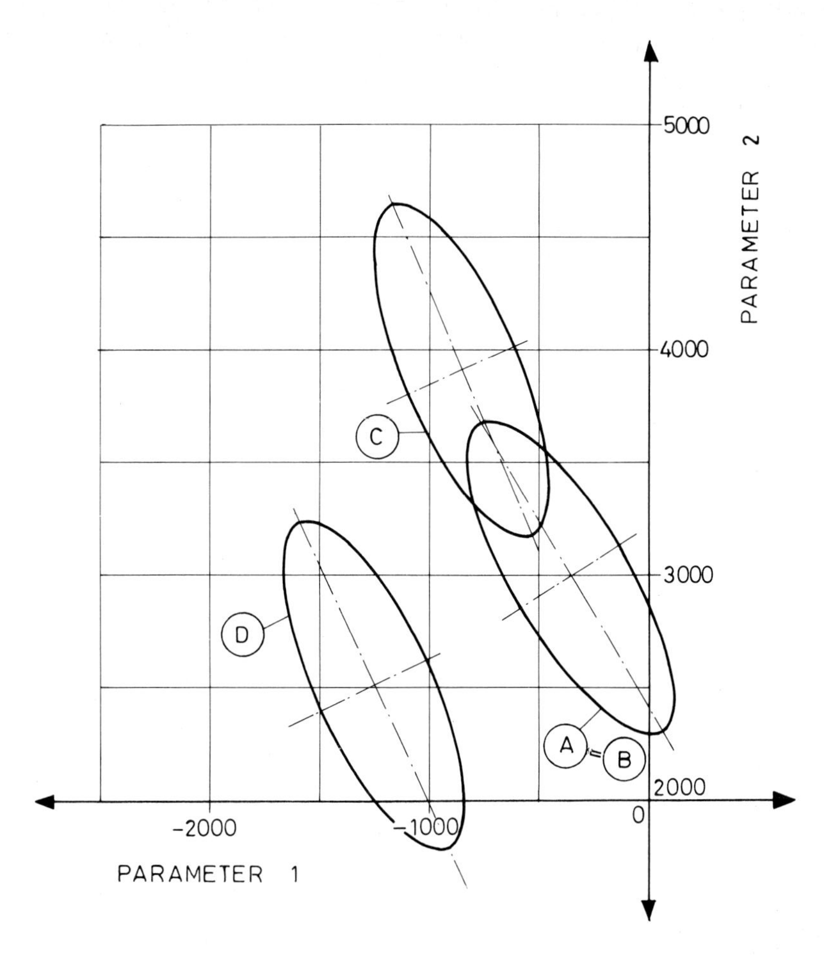

Figure 2: Positions of the parameter pairs generating the different data sets ('ethanol-benzene')

Table II: Parameter estimation from generated data sets, results

data set (the generating values of the parameters)	objective function value	parameters	
A (-350, 3000)	4.4	-337.0	2970.1
B (-350, 3000)	23.5	-445.3	3173.2
A+B (simultaneous)	32.8	-388.1	3064.3
A→B (sequential)	32.8	-388.4	3064.3
C (-850, 3910)	11.6	-829.2	3843.7
A+C (simultaneous)	190.2	-579.1	3366.6
A→C (sequential)	154.4	-656.5	3505.1
D (-1250, 2506)	4.9	-1243.9	2495.3
A+D (simultaneous)	17 800.0	-764.3	2845.1
A→D (sequential)	18 400.0	-788.4	2837.7

Application to group contribution models

A particular pair of groups may appear in systems of totally different components. Thus the parameter vectors of the old and the new data set may overlap partly or totally; in the typical case for a new data set there are parameters which have been estimated from a previous data set and there are new parameters.

However if the two data sets (old and new) do not contain common pairs of groups (characterized by common parameters) the elements of the parameter table can be filled in separately for the two data sets. In this case their Hessian matrices will not contain common elements, and those elements of the Hessian which would describe the correlation between parameters obtained from the two data sets will have zero value.

Including a new data set three types of pairs of groups may be distinguished:

a.) appearing only in the new data set

b.) appearing both in the old and new data set

c.) appearing only in the old data set

In the case of simultaneous parameter estimation (thus also for a well established sequential method) the parameters of type a are estimated without any difficulty from the new data set, while parameters of type b are refined in order to include the old and new piece of information for the particular pair of groups. However the parameters of type c also change because of the interdependency (correlation) of old (type c) and old-new (type b) parameters. The previously estimated values for the old (type c) parameters were determined so that they should minimize the objective function of the first data set together with the previously estimated values of the type b parameters. The change of the values of parameters of type b involves changes also in the old parameters (type c) . This way the refinement of the old-new parameters (type b) in order to include the new piece of information would require the use of too many variables. E.g. assuming 34 groups as in the UNIFAC model, with two parameters for each pair, the parameter table contains about 1000 elements. The reliable solution of a multivariable nonlinear optimization problem is hopeless even if the required core size and computation time are available.

The dimension of the problem being handled numerically can be reduced to the number of parameters occurring in the new data set if one makes use of the constraints and the approximate analytical expression of the interdependency of parameters.

In order to clarify the ideas let us take an example with two parameters only. The first data set (I) contains information on both parameters (Θ_1 and Θ_2), while only Θ_2 appears in the second data set (II).

Applying Eqn. (2)

$$\phi_1(\underline{\Theta}) = \phi_I^o + \frac{1}{2} H_{11}^o(\Theta_1 - \Theta_1^o)^2 + H_{12}^o(\Theta_1 - \Theta_1^o)(\Theta_2 - \Theta_2^o) + \frac{1}{2} H_{22}^o(\Theta_2 - \Theta_2^o)^2 \tag{5}$$

The first derivatives of the combined objective function Ψ must be equal to zero at the place of minimum:

$$\frac{\partial \Psi}{\partial \Theta_1} = \frac{\partial \phi_I}{\partial \Theta_1} = H_{11}^o(\Theta_1 - \Theta_1^o) + H_{12}^o(\Theta_2 - \Theta_2^o) = 0 \tag{6}$$

(the second data set does not contain information on Θ_1)

$$\frac{\partial \Psi}{\partial \Theta_2} = H_{12}^o(\Theta_1 - \Theta_1^o) + H_{22}^o(\Theta_2 - \Theta_2^o) + \frac{d\phi_{II}}{d\Theta_2} \tag{7}$$

The second derivatives (elements of the Hessian):

$$H_{11}^{new} = \frac{\partial^2 \Psi}{\partial \Theta_1^2} = \frac{\partial^2 \phi_I}{\partial \Theta_1^2} = H_{11}^o \tag{8}$$

$$H_{12}^{new} = \frac{\partial^2 \Psi}{\partial \Theta_1 \partial \Theta_2} = \frac{\partial^2 \phi_I}{\partial \Theta_1 \partial \Theta_2} = H_{12}^o \tag{9}$$

$$H_{22}^{new} = \frac{\partial^2 \Psi}{\partial \Theta_2^2} = H_{22}^o + \frac{d^2 \phi_{II}}{d\Theta_2^2} \tag{10}$$

This H_{22}^{new} would be obtained by seeking the minimum with respect to both Θ_1 and Θ_2.

If, however, the search routine is used with Θ_2 as the only variable, a $\Psi[\Theta_2, \Theta_1(\Theta_2)]$ function is minimized. The first derivative:

$$\frac{d\Psi}{d\Theta_2} = \left(\frac{\partial\Psi}{\partial\Theta_2}\right)_{\Theta_1} + \left(\frac{\partial\Psi}{\partial\Theta_1}\right)_{\Theta_2}\left(\frac{d\Theta_1}{d\Theta_2}\right) =$$

$$= \frac{d\Phi_{II}}{d\Theta_2} + \frac{\partial\Phi_I}{\partial\Theta_2} + \left[H_{11}^o(\Theta_1 - \Theta_1^o) + \right.$$

$$\left. + H_{12}^o(\Theta_2 - \Theta_2^o)\right]\left(\frac{d\Theta_1}{d\Theta_2}\right) \qquad (11)$$

The second derivative found by the search routine will be:

$$H_{22}^{num} = \frac{d^2\Psi}{d\Theta_2^2} = \frac{d^2\Phi_{II}}{d\Theta_2^2} + H_{22}^o + H_{12}^o\left(\frac{d\Theta_1}{d\Theta_2}\right) +$$

$$+\left[H_{11}^o(\Theta_1 - \Theta_1^o) + H_{12}^o(\Theta_2 - \Theta_2^o)\right]\left(\frac{d^2\Theta_1}{d\Theta_2^2}\right) \qquad (12)$$

From Eqn. (6)

$$\frac{d\Theta_1}{d\Theta_2} = -\frac{H_{12}^o}{H_{11}^o} \quad \text{and} \quad \frac{d^2\Theta_1}{d\Theta_2^2} = 0 \qquad (13)$$

Thus the following expression is obtained:

$$H_{22}^{num} = \frac{d^2\Phi_{II}}{d\Theta_2^2} + H_{22}^o - \frac{(H_{12}^o)^2}{H_{11}^o} \qquad (14)$$

From Eqns. (10) and (14) the right value of H_{22}^{new} is:

$$H_{22}^{new} = H_{22}^{num} + \frac{(H_{12}^o)^2}{H_{11}^o} \qquad (15)$$

In general case the partial derivatives of the combined objective function with respect to the parameters occurring only in the old data set (type c) can be expressed from the second order approximation of the objective function of the old data set. This way a set of linear equations may be established and solved. The solution gives the new optimal values of parameters of type c as a function of parameters appearing in the new data set.

So all the parameters are optimized by a numerical minimum search algorithm formally treating fewer variables. The estimated parameter vector contains information on the old and the new data set as well. Though the numerical search routine gives a matrix of second partial derivatives of the combined objective function, value of its elements must be corrected using the derivatives obtained from the previously treated (old) data set, as the search algorithm cannot recognize that the type c parameters are also changed with changing the old-new (type b) parameters.

The new second order approximation of the objective function of the old+new data sets may substitute the combined data sets for further building the parameter table.

References:

[1] Fredenslund, Aa., Gmehling, J. and Rasmussen, P.: Vapor-liquid equilibria using UNIFAC (Elsevier, Amsterdam-Oxford-New-York, 1977)

[2] Bard, Y.: Nonlinear Parameter Estimation (Academic Press, New York and London, 1974)

[3] Kemény, S., Manczinger, J. Skjold-Jørgensen, S. and Tóth, K.: Reduction of thermodynamic data by means of the multiresponse maximum likelihood principle, AIChE Journal 28 (1982) 20-30.

[4] Neau, E., and Peneloux, A.: The successive reduction of data subsets a new method for estimating parameters of thermodynamic models, Fluid Phase Equilibria, 8 (1982) 251-269

[5] Ellis, S.R.M. and Clark, M.B.: Chem. Age (India) 12, 377 (1961)

[6] Wehe, A.H. and Coates, J.: AIChE J. 1, 241 (1955)

[7] Kogan, L.U. and Geisenrot, I.V.: Zh. Prikl. Khim. 48, 2757 (1975)

[8] Kudryavtseva, L.S. and Susarev, M.P.: Zh.Prikl. Khim. 36, 1231 (1963)

Data for Science and Technology, P.S. Glaeser (ed.)
North-Holland Publishing Company

COMPUTER AIDED DATA APPROXIMATION AND PREDICTION FOR G^E AND H^E

Klaus Sühnel

Karl Marx University, Department of Chemistry
Leipzig, German Democratic Republic

Modified lattice models by Barker and Tompa are used for the prediction of vapor-liquid equilibria and enthalpies of mixing in binary and ternary nonelectrolyte mixtures. Applying the "predictive Barker method" we were able to calculate the total vapor pressures in mixtures of n-alkanes with different polar components (e.g. alcohols, ketones, amines, chloroform, benzene, alkylbenzenes, ethers) with three to five parameters.
In homologous series the "predictive Tompa method" requires two constant parameters and a third parameter depending linearly on the chain length of n-alkanes. All results were compared with those given by UNIFAC.
Information based on a study of extensive data banks shows that the results can fill gaps in such data banks.

1. INTRODUCTION

Thermodynamic mass data providing a basis for the calculation and modeling of mass separation processes is of extreme importance to chemical technology and to the chemical industry as a whole. The cost of experimental collection of such data is constantly increasing and the calculation of thermodynamic mixture characteristics from data obtained for pure substances also is not readily possible at the present time.
Constituting a compromise between experiments and exact theories are the model theories of which the mathematical expense could not be accommodated until the advent of high-speed computers. On the other hand, however, such model theories enable correlations between the necessary parameters to be found by taking the approximation of mixture data of individual systems as a base. By using these correlations, it is possible to subsequently predict the thermodynamic characteristics of unknown mixtures. Part of these model theories are the lattice models of Barker (2), Tompa (4), and Abrams and Prausnitz (5) that have been used for the prediction of Gibbs' and enthalpies of mixing in binary nonelectrolyte mixtures.

2. THEORETICAL FUNDAMENTALS

The use of model theories for purposes of prediction leads to predictive methods by which the excess behavior of individual mixture systems that have been very exactly studied experimentally are modeled on the basis of comprehensive data banks and attempts are made to recognize, predetermine, and augment the physical implications of the required parameters. The number of the necessary experimental data points is increasing and the requirements of computer engineering are becoming greater at the same rate as the number of these parameters increases. The important advantage of the lattice models used here resides in the fact that only two to six parameters are required for each homologous series.
All of the models employed are based upon lattice theory of Guggenheim (1), and the predictive use of the UNIQUAC model (5) represents the very widely used UNIFAC method of Fredenslund (6). The predictive method of Barker is based upon initial studies performed by Sweeney-Rose (8), Chao et al. (9) and Smirnova (10), retaining the dependence of the separation of molecules upon the molar volume, which has been proposed by Barker (2), and the coordination number, z = 4, as well as the possibility of using temperature-dependent interaction energies according to equation (1).

$$A^{\mu\nu} = W^{\mu\nu} + T \cdot (\partial A^{\mu\nu} / \partial T)_V \quad (1)$$

It also introduces uniform interaction parameters which are not, within a homologous series, dependent upon the chain length of the individual partners (Table 1).

component	$A_{ij}(W_{ij})$ J·mol	n-hexane r = 6	n-decane r = 10	n-tetradecane r = 14
benzene r = 4	π-H	- 3140(- 3140)	- 3140(- 3140)	- 3140(- 3140)
	π-S	- 2090(- 2090)	- 2090(- 2090)	- 2090(- 2090)
	H-S	42(42)	42(42)	42(42)
toluene r = 5	π-H	- 3265(- 2890)	- 3265(- 2890)	- 3265(- 2890)
	π-S	- 2090(- 1970)	- 2090(- 1970)	- 2090(- 1970)
	H-S	42(42)	42(42)	42(42)
ethylbenzene r = 6	π-H	- 3265(- 2890)	- 3265(- 2890)	- 3265(- 2890)
	π-S	- 2090(- 1970)	- 2090(- 1970)	- 2090(- 1970)
	H-S	42(42)	42(42)	42(42)
n-butylalcohol r = 5	O-H	-14200(-19700)	-14200(-19700)	-14200(-19700)
	O-S	- 840(- 2100)	- 840(- 2100)	- 840(- 2100)
	H-S	42(42)	42(42)	42(42)
n-octylalcohol r = 9	O-H	-14200(-19700)	-14200(-19700)	-14200(-19700)
	O-S	- 840(- 2100)	- 840(- 2100)	- 840(- 2100)
	H-S	42(42)	42(42)	42(42)

Table 1: Parameters of the predictive Barker method

Insofar as the classification of molecules into spherical and cylindrical species, the coordination number, and the calculation of interaction energies by means of a quasichemical approximation are concerned, the predictive method of Tompa (7) is based upon the original model (4). Computationally, the principal problem consists in solving fourth-degree equations that were solved using Newton's method. The predictive method of Tompa is characterized by the following assumptions for the example of predicting thermodynamic properties for mixtures of alkanes and a more polar second component, for the alkanes are the spherical species, the second components get the cylindrical form. Parameter "a" depends on this component, but "b" or "c" are considered as parameters for the disturbance of association. For this disturbance we assumed a linear dependence on the chain length of alkanes. Table 2 contains those parameters and also provides an overview of all homologous series that were studied using both of the predictive methods, and all of the results were additionally compared with those that were obtained on the basis of the UNIFAC method (6). The range of temperatures was between 0 and 120 °C, with temperature-dependent Barker parameters (3,15) or special Tompa parameters being occasionally required for certain temperature intervals.

3. RESULTS

The results calculated using the parameters in Table 1 and Table 2, respectively, are given in Fig. 1 for the benzene - alkane systems and in (12) for all aromatic - alkane mixtures. Because of the relatively small values of Gibbs' energies that were found experimentally for the systems with longer-length n-alkanes, the deviations in total vapor pressures are demonstrated in Table 3 too. The errors involved are calculated using Eq. 2.

$$\partial Z = \frac{1}{n}\sqrt{\left(\frac{Z_{calc} - Z_{exp}}{Z_{exp}}\right)^2}\cdot 100 \quad (2)$$

By using the ketone-alkane mixtures as an example, it was possible to show not only good agreement between experimental and calculated liquid-vapor equilibria but also after the publication of predicted heats of mixing (13) these values were confirmed experimentally by Benson (14).

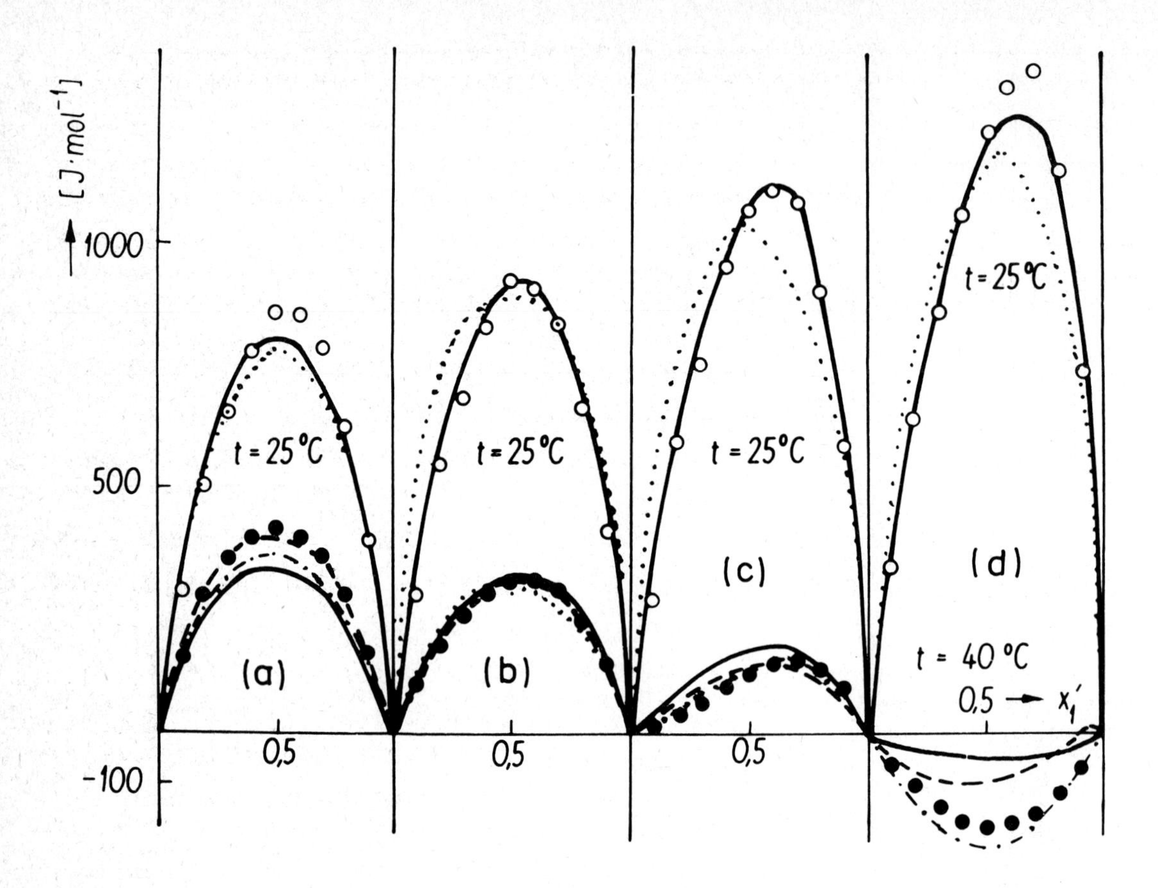

Figure 1:

Comparison of experimental and predicted G^E and H^E in benzene (1) - n-alkane (2) mixtures

- ● experimental G^E
- o experimental H^E
- —— predicted values using the predictive, temperature-independent Barker method
- --- predicted values using UNIFAC
- -·- predicted values using the predictive Tompa method basing on G^E
- ••• predicted values using the predictive Tompa method basing on H^E

(a) n-pentane (2)

(b) n-heptane (2)

(c) n-dodecane (2)

(d) n-heptadecane (2)

component		n-hexane	n-decane	n-tetradecane	algorithm
benzene	a	0,047	0,047	0,047	constant
	b	0,250	0,250	0,250	constant
	c_G	0,480	0,400	0,320	$0{,}60-0{,}02\cdot C_n$
	c_H	0,440	0,520	0,600	$0{,}32+0{,}02\cdot C_n$
toluene	a	0,300	0,300	0,300	constant
	b	0,800	0,800	0,800	constant
	c_G	0,750	0,670	0,590	$0{,}87-0{,}02\cdot C_n$
	c_H	0,730	0,810	0,890	$0{,}61+0{,}02\cdot C_n$
propanone	a	0,010	0,010	0,010	constant
	b	0,240	0,240	0,240	constant
	$c_G=c_H$	0,200	0,260	0,320	$0{,}11+0{,}015C_n$
n-butylalcohol	a	0,700	0,700	0,700	constant
	b	0,800	0,800	0,800	constant
	$c_G=c_H$	5,500	4,700	3,900	$6{,}70-0{,}2\cdot C_n$

Table 2: Parameters of the predictive Tompa method

The results shown in Figure 1 are typical of the homologous series studied, with all of the results obtained. References to the original literature are given in Table 3.

In general, it can be said that a common prediction of Gibbs' free energies and enthalpies of mixing is most readily possible with the use of the predictive Barker-method (7). Above longer-length alkanes ($C \geq 8$), the results for VLE also are far better than those which were obtained using the UNIFAC method or predictive Tompa method (7), respectively. On the other hand, the latter method, when used for shorter-length alkanes, enables both Gibbs' energies and enthalpies of mixing to be satisfactorily predicted using only four different parameters.

In concluding, the predictive methods of Barker and Tompa may, within the limits established here, be considered to be all-important contributions to attempts at optimizing the experimental complexity of determining both VLE and H^E of unknown nonelectrolyte mixtures as well as interesting possibilities of extending existing data banks, but prediction cannot replace special experimental investigations.

Table 3: Deviations of experimental and predicted values using different methods

(1)	C_2	T/K/	∂P_T	∂P_B	∂P_U	∂G^E_T	∂G^E_B	∂G^E_U	∂H^E_T	∂H^E_B /%/	Lit.
benzene	5- 8	293-348	1- 4	2- 5	3- 5	10-20	11-17	5-15	9-15	4- 6	(12)
benzene	9-20	298-353	3-10	2- 5	3- 6	15-50	15-35	15-30	15-25	5-10	(12)
toluene	5- 7	293-353	2	2	1- 4	6-30	6-30	5-20	8-30	4- 9	(12)
toluene	12-17	333-353	4	2- 6	3- 4	100	50	50	20-50	2-10	(12)
propanone	3- 7	233-338	4- 6	6	3- 6	8-20	8-11	7-12	5-15	4-12	(13)
propanone	9-16	298-333	6- 9	4- 6	3-10	13-50	9-25	6-30	15-20	10-15	(13)
butanone	6- 8	298-333	3- 5	2- 5	2- 5	10-16	10-15	6-11	7	7- 9	(13)
butanone	12-16	298-353	9-20	4- 6	9	50	20	20-25	8- 9	10-15	(13)
butylamine	5- 6	293-333	3- 6	1- 4	1- 2	10-18	16	8- 9	10-17	6-17	(11,15)
butylamine	7-16	298-338	11	3	3	50-80	16	9	8	5-12	(11,15)
octanol	5- 8	288-373	4- 5	2- 3	2- 4	8-11	9	7-10	7-40	15-40	(11)
octanol	9-12	288-393	5-35	5- 8	3- 5	40	8-30	15-20	5-40	7-40	(11)

4. REFERENCES

[1] Guggenheim, E.A., Mixtures (Clarendon Press, Oxford, 1952)

[2] Barker, J.A., Cooperative orientation effects in solutions. J. Chem. Phys. 20 (1952) 1526-1532.

[3] Barker, J.A., and Smith, F., Statistical thermodynamics of associated solutions. J. Chem.Phys. 22 (1954) 375-380.

[4] Tompa, H., Quasichemical treatment of mixtures of oriented molecules. J. Chem. Phys. 21 (1953) 250-258.

[5] Abrams, D.S., and Prausnitz, J.M. Statistical thermodynamics of liquid mixtures: A new expression for the excess Gibbs energy of partly or completly miscible systems. A.I.Ch.E. J. 21 (1975) 116-126.

[6] Fredenslund, Aa., Gmehling, J., and Rasmussen, P., Vapor-liquid equilibria using UNIFAC - a group contribution method (Elsevier Scientific Publ. Co., Amsterdam, Oxford, New York, 1977).

[7] Quitzsch, K., Messow, U., Pfestorf, R., and Sühnel, K. Eine kritische Analyse halbempirischer und molekulartheoretisch begründeter Modellkonzeptionen zur numerischen Bestimmung thermodynamischer Exzeßfunktionen
in: Bittrich, H.-J.: Modellierung von Phasengleichgewichten als Grundlage von Stofftrennprozessen, (Akademie-Verlag, Berlin, 1981);
Sühnel, K., Die Anwendung von Modelltheorien zur Approximation und Vorausberechnung von thermodynamischen Mischungseigenschaften (Dissertation zur Promotion B, Leipzig, 1982).

[8] Sweeney, R.F., and Rose, A., The prediction of vapor-liquid equilibria using a theory of liquid mixtures. A.I.Ch.E. J. 9 (1963) 390-393.

[9] Kuo, C.-M., Robinson jr., R.L., and Chao, K.Ch., Quasilattice theory and paraffin-alcohol mixtures. Ind. Eng. Chem. Fundamentals 9 (1970) 564-568.

[10] Smirnova, N.A., On the methods for calculation of activity coefficient Preprints 4th Intern. Congress CHISA'72, Prag, 1972, F.2.11.

[11] Sühnel, K., Höppner, F., Hofmann, G., and Salomon, M. Die Anwendung von Gittermodellen zur Berechnung der Exzeßeigenschaften flüssiger Mischsysteme. Part 2: Die Eignung des Modells von Barker für die Vorausberechnung der thermodynamischen Mischungseigenschaften von n-Alkanen mit polaren Komponenten. Chem. Techn. 29 (1977) 505-508.

[12] Sühnel, K., Wittig, M., and Müller S. dto. Part 4: Vergleich der mit den Modellen von Barker und Fredenslund (UNIFAC) gewonnenen Ergebnissen in binären Aromat-Alkan-Mischungen, ibid. 32(1980)471-474.

[13] Sühnel, K., Messow, U., and Salomon,M. Thermodynamische Untersuchungen an Lösungsmittel-Paraffin-Systemen. Part 10: Die Anwendung des Gittermodells von Barker zur Vorausberechnung der thermodynamischen Exzeßeigenschaften in Keton-Alkan-Systemen. Z. phys. Chemie (Leipzig) 260(1979)142-148.

[14] Kiyohara, O., Handa, Y.-P., and Benson, G.C. Thermodynamics properties of binary mixtures containing ketones. Part 3: Excess enthalpies of n-alkanes and some aliphatic ketones.
J. Chem. Thermodynamics 11 (1979) 453-460.

[15] Sühnel, K. Application of the quasichemical lattice theory by Barker to approximation and prediction of thermodynamic properties of mixtures of different polar components.
Polish J. Chem. 55 (1981) 871-878.

Data for Science and Technology, P.S. Glaeser (ed.)
North-Holland Publishing Company

THE USE OF CHARACTERISTIC DATA FOR PREDICTING PHASE EQUILIBRIUM BEHAVIOUR

Horst Schuberth

Martin-Luther-Universität Halle-Wittenberg
Chemistry Section, Applied Thermodynamics
Halle, German Democratic Republic

The knowledge of exact data which characterize the thermodynamic behaviour of a liquid mixture (e.g., data for solubility, pressure above saturated solutions, azeotropism, alyotropism, extreme point of activity coefficients) permits the prediction of interesting phase equilibria of solvent systems with electrolytic or nonelectrolytic components. With that, it is possible to gain important parameters for process technology which allow us to estimate details of the practicability of a distillation or extraction separation task. The importance of using data in this direction is demonstrated with some examples.

A well known method is to determine the parameters of a modelling equation for the description of thermodynamic behaviour (e.g. REDLICH-KISTER, WOHL, WILSON, NRTL, UNIQUAC) from characteristic phase equilibrium data of the liquid mixture. From this, it will be possible to make predictions about all the behaviour of this system simply by knowing these supporting data. The problem exists solely in examining the attainable accuracy of a prediction. Let me give some examples of this method by means of some modelling equations for calculation of phase equilibrium situations developed by our working-group.

Case A: The two-sided PORTER-equation for the dimensionless molar free excess-enthalpy of ternary liquid systems with sufficiently wide distance from critical solubilities:

$$\bar{Q}' \approx [A_{12} + B_{12}(\varphi_1' - \varphi_2')(1 - x_3')]\varphi_1'\varphi_2'(1 - x_3')^2 + \sum_{i=1}^{2} \Lambda_{\sigma(i)}\varphi_i'(1 - x_3')x_3' \qquad (1)$$

$$\text{with } \varphi_i' \equiv \frac{x_i'}{1 - x_3'} \text{ for } i = 1,2 \text{ and } \sigma = S, L$$

Here, the constants A_{12}, B_{12} can be taken from data collections on vapour-liquid phase equilibria, and the constants $A_{S(i)}$, $A_{L(i)}$ can be obtained from the tie-line condition.

$$\Delta[\ln x_k^{(\sigma)} + \ln f_k^{(\sigma)}] = 0 \qquad (2)$$

$$(k = 1,2,3;\ \sigma = S, L)$$

$$\text{with } \ln f_k' = \bar{Q}' - \sum_{i \neq k} x_i' \left(\frac{\partial \bar{Q}'}{\partial x_i'}\right)_{T, x_j'}$$

$$\text{and } \Delta[\phi^{(\sigma)}] \equiv \phi^{(L)} - \phi^{(S)}$$

For the border (i) = (1) resp. (2):

$$A_{S(i)} = \frac{(x_{i(i)}^{(L)})^2 \ln \tau_{i(i)} + (1 - x_{i(i)}^{(L)})^2 \ln \tau_{3(i)}}{(x_{i(i)}^{(S)})^2 (1 - x_{i(i)}^{(L)})^2 - (x_{i(i)}^{(L)})^2 (1 - x_{i(i)}^{(S)})^2}$$

$$A_{L(i)} = \frac{(x_{i(i)}^{(S)})^2 \ln \tau_{i(i)} + (1 - x_{i(i)}^{(S)})^2 \ln \tau_{3(i)}}{(x_{i(i)}^{(S)})^2 (1 - x_{i(i)}^{(L)})^2 - (x_{i(i)}^{(L)})^2 (1 - x_{i(i)}^{(S)})^2} \qquad (3)$$

$$\text{with } \tau_{i(i)} \equiv \frac{x_{i(i)}^{(S)}}{x_{i(i)}^{(L)}};\quad \tau_{3(i)} \equiv \frac{1 - x_{i(i)}^{(L)}}{1 - x_{i(i)}^{(S)}} \equiv \frac{x_{3(i)}^{(L)}}{x_{3(i)}^{(S)}}$$

$$(i = 1; 2)$$

see Figure 1. Thereby we can make assertions about the distillation and extraction behaviour of separating nonelectrolytic systems in cases of the above mentioned applicability of this equation.

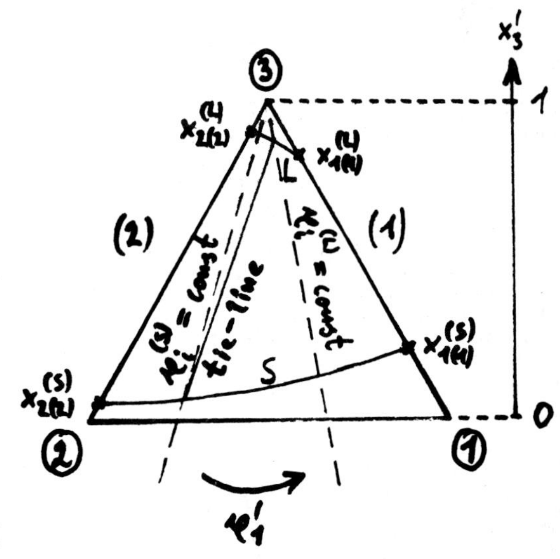

Figure 1: Ternary systems with liquid components with liquid components and liquid-liquid miscibility gap sufficiently wide from critical solubilities.

Table 1: Comparison between calculated and measured characteristic phase equilibrium data of the system n-heptane/dimethylformamide for 65 °C

Symbol	RMW-equation of FEIX	two-sided PORTER-equation of SCHUBERTH	experimental values of QUITZSCH
$x_1^{(2)}$	0.265	0.240	0.240
$x_1^{(1)}$	0.833	0.835	0.835
$\lg f'_{1(2)}$	0.922	0.986	?
$x''_{1\ HZ}$	0.894	0.890	0.893
p_{HZ}	34.22 kPa	33.81 kPa	34.04 kPa
$x''_{1\ Az}$	0.921	0.934	0.932
p_{Az}	34.48 kPa	34.14 kPa	34.26 kPa
Remarks:	use of $\ln f'_{1(2)}$ and $\bar{Q}'_{12(eq)}$ for calculation	use of $x_1^{(2)}$ and $x_1^{(1)}$ for calculation	p_{01} = 33.79 kPa p_{02} = 4.28 kPa

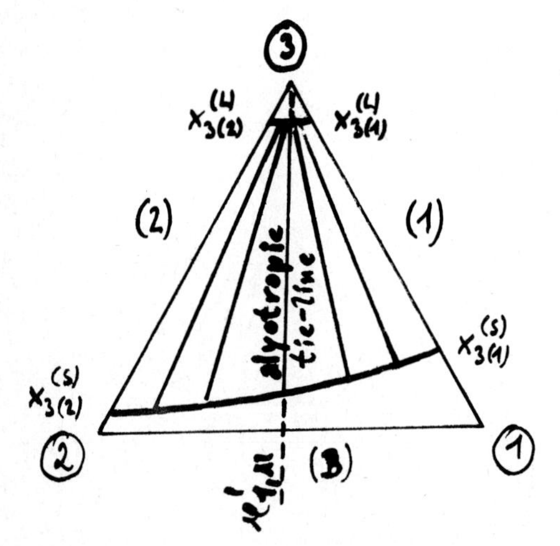

Figure 3: System furfural/ethyl-acetate/water (25 °C), schematically.
Exact values:
$x_{3(1)}^{(L)} = 0.984$; $x_{3(2)}^{(L)} = 0.984$
$x_{3(1)}^{(S)} = 0.222$; $x_{3(2)}^{(S)} = 0.141$
Integral excess-value for the basic system 1/2: ${}^{(10)}A_{12} \approx 0.1$

Calculation: ${}^{(10)}A_{S(1)} = 1.069$; ${}^{(10)}A_{S(2)} = 1.144$

${}^{(10)}A_{L(1)} = 1.797$; ${}^{(10)}A_{L(2)} = 1.810$

$\lg \beta_{(1)} = -0.080$; $\lg \beta_{(2)} = 0.087$

and in first approximation (see [3]) $\varphi'_{1,Al} = 0.498$

Experimental result in comparison: $\varphi'_{1,Al} = 0.46$

Example A1: For the calculation of heteroazeotropic and homoazeotropic data in the system n-heptane/dimethlyformamide at 65 °C, see the diagram Figure 2. Table 1 shows the results of our calculation in comparison with the results of calculations using the more pretentious equation of FEIX [1]. Starting data are: the vapour pressures of the pure components and the boundary solubilities on the borders of the miscibility gap. The application of the UNIQUAC-UNIFAC method (an incremental method) by using the data in the publication [2]

$r_{CH_3} = 0.9011 \; ; \; q_{CH_3} = 0.8480$

$r_{CH_2} = 0.6744 \; ; \; q_{CH_2} = 0.5400$

$r_{DMF} = 3.0856 \; ; \; q_{DMF} = 2.7360$

$a_{DMF/CH_3} = a_{DMF/CH_2} = 485.3 \; K^{-1}$

$a_{CH_3/DMF} = a_{CH_2/DMF} = -31.95 \; K^{-1}$

$P_{oHept} = 33.79 \text{ kPa} \; ; \; P_{oDMF} = 4.28 \text{ kPa}$

gives way to impossible results:

$p = 49.35$ kPa for $x_1' = 0.240$

$p = 37.99$ kPa for $x_1' = 0.835$

$p = 37.56$ kPa for $x_1' = 0.932$

Example A2: The estimation of the extractive significant quantities "capacity"

$$\ln \varkappa_i \equiv \Delta[\ln x_i^{(G)}] = -\Delta[\ln f_i^{(G)}] \qquad (4)$$

and "selectivity"

$$\ln\beta \equiv \Delta\left[\ln \frac{\varphi_1^{(G)}}{1-\varphi_1^{(G)}}\right] = \ln\frac{\varkappa_1}{\varkappa_2} = -\Delta\left[\ln\frac{f_1^{(G)}}{f_2^{(G)}}\right]$$
$$\approx \Delta\left[(A_{G(2)}-A_{G(1)})x_3^{(G)} - A_{12}(1-2\varphi_1^{(G)})(1-x_3^{(G)}) + B_{12}(1-6\varphi_1^{(G)}\{1-\varphi_1^{(G)}\})(1-x_3^{(G)})^2\right] \qquad (5)$$

and the ascertainment of the position of possible alyotropic tie-lines defined by

$$\ln\beta = 0 \rightarrow \frac{\ln\beta_{0,Al}}{x_{3,Al}^{(L)}-x_{3,Al}^{(S)}} \equiv S_{0,Al} = \frac{\Delta[(A_{G(2)}-A_{G(1)})x_{3,Al}^{(G)}]}{x_{3,Al}^{(L)}-x_{3,Al}^{(S)}}$$
$$= A_{12}(2\varphi_{1,Al}-1) + B_{12}[1-6\varphi_{1,Al}(1-\varphi_{1,Al})] \times [(1-x_{3,Al}^{(L)})+(1-x_{3,Al}^{(S)})] \qquad (6)$$

see Figure 3, especially demonstrated on the case of the system furfural/ethyl-acetate/water at 25 °C, see [3].

Example A3: Finally, let us make assertions about the changeability of the "selectivity power"

$$S \equiv \frac{\ln\beta}{x_3^{(L)}-x_3^{(S)}} \qquad (7)$$

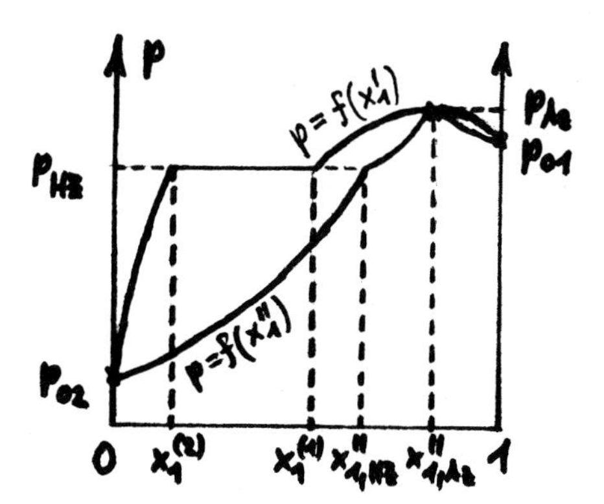

Figure 2: p-x_1-diagram of the system n-heptane/dimethylformamide (65 °C), schematically.

of some extractive agents for hydrocarbon systems with a thorough miscibility gap on the strength of literature data, only using boundary solubilities $x_{3\,(i)}^{(G)}$ in the binary systems (i). For this, we define an "average selectivity power" by

$$S_{(av)} \equiv \frac{1}{2}\sum_{i=1}^{2} S_{(i)} \equiv \frac{1}{2}\sum_{i=1}^{2}\frac{\ln\beta_{(i)}}{x_{3(i)}^{(L)}-x_{3(i)}^{(S)}}$$
$$\approx \frac{1}{2}\left(\sum_{i=1}^{2}\frac{\Delta[(A_{G(2)}-A_{G(1)})x_{3(i)}^{(G)}]}{x_{3(i)}^{(L)}-x_{3(i)}^{(S)}}\right) \qquad (8)$$
$$\text{for } B_{12}\approx 0$$

So we find, for example, the values shown in Table 2. It should be added that we could verify the accuracy of the calculated values of selectivity by means of gas-chromatography.

Table 2: Selectivity power of aniline, sulfolane, dimethylformamide and methylalcohol for some hydrocarbon systems and selected temperatures.

System	t/°C	${}^{(10)}S_{(av)}$
$C_6H_{11}CH_3/n\text{-}C_7H_{16}/C_6H_5NH_2$	25	0.312
$C_6H_{11}CH_3/n\text{-}C_6H_{14}/C_6H_5NH_2$	25	0.192
$C_5H_9CH_3/n\text{-}C_6H_{14}/C_6H_5NH_2$	25	0.272
$C_5H_9CH_3/n\text{-}C_6H_{14}/C_4H_8SO_2$	20	0.337
$C_5H_9CH_3/n\text{-}C_6H_{14}/C_4H_8SO_2$	25	0.329
$C_5H_9CH_3/n\text{-}C_6H_{14}/C_4H_8SO_2$	30	0.323

$C_5H_9CH_3/n\text{-}C_6H_{14}/C_4H_8SO_2$	40	0.330
$C_5H_9CH_3/n\text{-}C_6H_{14}/HCON(CH_3)_2$	20.0	0.150
$C_5H_9CH_3/n\text{-}C_6H_{14}/HCON(CH_3)_2$	25.0	0.139
$C_5H_9CH_3/n\text{-}C_6H_{14}/HCON(CH_3)_2$	29.3	0.128
$C_5H_9CH_3/n\text{-}C_6H_{14}/HCON(CH_3)_2$	38.7	0.100
$n\text{-}C_6H_{14}/C_6H_{11}CH_3/CH_3OH$	20	0.074
$n\text{-}C_6H_{14}/C_6H_{11}CH_3/CH_3OH$	25	0.079
$n\text{-}C_6H_{14}/C_6H_{11}CH_3/CH_3OH$	30	0.096

Case B: The "EF-Equation" for saliferous solvent-mixtures:

$$\hat{\bar{Q}} \equiv \sum_{l=1}^{L} \hat{x}'_l \ln \hat{f}'_l + \sum_{e=1}^{E} \hat{x}'_e \ln \hat{f}'_e = \sum_{l=1}^{L} \sum_{e=1}^{E} \hat{\bar{Q}}'_{le,Coul} + \sum_{i=1}^{K-1} \sum_{j=i+1}^{K} \hat{\bar{Q}}'_{ij} + \Delta \hat{\bar{Q}}'_{poly} \tag{9}$$

with

$$\hat{x}'_l \equiv \frac{x'_l}{1+\sum_e (\nu_e - 1) x'_e} \; ; \; \hat{x}'_e \equiv \frac{\nu_e x'_e}{1+\sum_e (\nu_e - 1) x'_e} \tag{10}$$

Premises: the salts are nonvolatile and "ionophorous"; degree of dissociation $\alpha_{De} \approx 1$ continuous; validity of the DEBYE-HÜCKEL-formula for highly diluted solutions is accepted. Then it is possible to formularize

$$\hat{\bar{Q}}'_{le,Coul} = \left[(E_{le} + F_{le} \hat{x}'_e)(\hat{x}'_e)^{\frac{1}{2}} \right] \hat{x}'_l \hat{x}'_e \tag{11a}$$

$$\hat{\bar{Q}}'_{ij} = \hat{\bar{Q}}'_{ij,NCoul}, \text{ for example} = \left[A_{ij} + B_{ij} (\hat{x}'_i - \hat{x}'_j) + C_{ij} (\hat{x}'_i - \hat{x}'_j)^2 + \cdots \right] \hat{x}'_i \hat{x}'_j \tag{11b}$$

$\Delta \hat{\bar{Q}}'_{poly}$ for example for ternary systems with solvents 1 and 2

$$= \left[(\Delta b + B_{13} + B_{23} - 2C_{13} - 2C_{23}) + (\Delta c_{21} + 4C_{13} + 3C_{23}) \hat{x}'_1 + (\Delta c_{12} + 3C_{13} + 4C_{23}) \hat{x}'_2 \right] \hat{x}'_1 \hat{x}'_2 \hat{x}'_3 \tag{11c}$$

with

$$E_{le} = \frac{2}{3} \cdot \frac{a_l}{\sqrt{M_l}} \cdot \frac{-\left| \prod_j z_{e_j} \right| \left(\frac{1}{2} \sum_j z_{e_j}^2 \nu_{e_j} \right)^{\frac{1}{2}}}{\left(\sum_j \nu_{e_j} \right)^{\frac{1}{2}}} \tag{12}$$

and

$$a_l = a_w \cdot \sqrt{\frac{\rho_{ol}}{\rho_{ow}} \cdot \frac{\varepsilon_{ow}^3}{\varepsilon_{ol}^3}} \; ; \tag{13}$$

$$a_w = 1{,}131 + 1{,}4052 \cdot 10^{-3}\, t/°C + 1{,}122 \cdot 10^{-5} (t/°C)^2$$

see [4]. With the help of characteristic equilibrium positions, the physical attributes of the pure components and the excess behaviour of the solvent mixtures assertions will be obtained

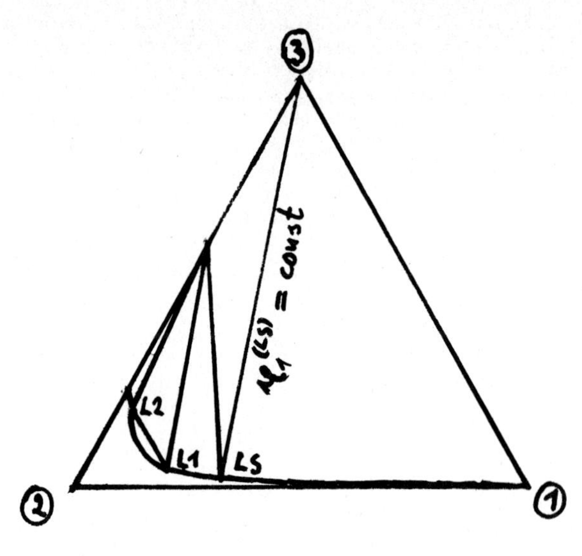

Figure 4: System methylalcohol/water/disodium-hydrogenphosphate (60 °C), schematically.

about the distillation and solubility properties of saliferous solutions.

Example B1: The prediction of the solubility, the vapour pressure and the salt-effect behaviour in the system methylalcohol/water/disodiumhydrogenphosphate at 60 °C. Starting data are (see Figure 4) the vapour pressure P_{oi}, densities ρ_{oi} and dielectric constants ε_{oi} of the pure solvent components i = 1 and 2; the vapour pressures of the saturated water solution $p_{H(2)}$ and the ternary saturated solution $p_H^{(LS)}$ at the point of the hydrate transformation $\mu_1^{(LS)}$, the excess behaviour of the basic system represented by A_{12} and B_{12}, the boundary solubilities on the borders of the miscibility gap ($\mu_1^{(L1)}$; $x_3^{(L1)}$) and ($\mu_1^{(L2)}$; $x_3^{(L2)}$) and in the binary systems $x_{3(1)}^{(L)}$; $x_{3(2)}^{(L)}$ and $x_{3(2)}^{(S)}$. If we can assume linear changeability of ln ($a_2^{(L)}$: $a_3^{(L)}$) with $\mu_1^{(L)}$ outside the miscibility gap up to the point of the hydrate transformation $\mu_1^{(LS)}$ and beyond this point ln $a_3^{(L)} = 1$ along the boundary line of solubility, it is possible to calculate all constants in (11). In the case of the above mentioned system, we get with the experimental values of BRUDER [5]

A_{12} = 0.6318 ; B_{12} = -0.1907

E_{13} = -0.3705 ; E_{23} = -0.1394

A_{13} = 11.0513 ; A_{23} = 2.4783

B_{13} = 2.3026 ; B_{23} = -0.7523

Δb = 3.6142

The comparison is demonstrated in Table 3. Especially the salt-effect:

$$\ln \frac{\alpha_{(L)}}{\alpha_{(B)}} = \ln \frac{x''_{1,H}\,(1 - x''_{1(B)})}{x''_{1(B)}\,(1 - x''_{1,H})} \tag{14}$$

can be calculated by using the equation

$$\ln \frac{\alpha_{(L)}}{\alpha_{(B)}} = \ln\left(\frac{f_1^{1(L)}}{f_2^{1(L)}}\right)_{x_3^{1(L)} \neq 0} - \ln\left(\frac{f_1^{1'}}{f_2^{1'}}\right)_{x_3^{1'} = 0} \tag{15}$$

$$= k_{1(\alpha)}\, x_3^{1(L)} + k_{1,5(\alpha)} (x_3^{1(L)})^{1,5} + k_{2(\alpha)} (x_3^{1(L)})^2$$

where $k_{n(\alpha)} = f(E, A, B \text{ and } \mu_1')$. The result is given in Figure 5; our calculations are in good conformity with the experimental statement (within the margin of error).

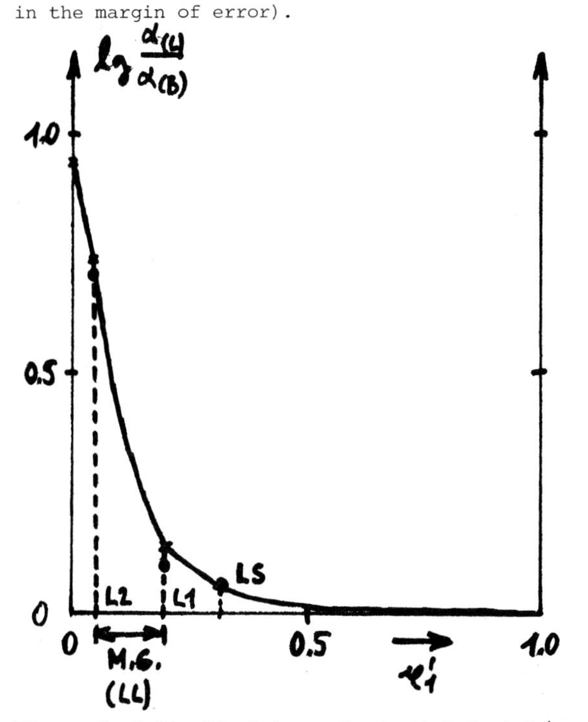

Figure 5: Salt effect in system methylalcohol/water/disodiumhydrogenphosphate (60 °C)
o experimentally measured
x calculated resp. predicted

Example B2: If the added solid is a nonelectrolytic component with negligible vapour pressure, so we have $E_{1e} = F_{1e} = 0$. For instance we can estimate the phase equilibrium behaviour of urea in the solvent mixture n-propanol/water by calculating the constants of (11) on the base of the experimental values of STIMMING [6] for 35 °C

$A_{13} = 3.4447$; $A_{23} = 2.2685$

$B_{13} = 0.2044$; $B_{23} = 1.5642$

$C_{13} = 0$; $C_{23} = 0.5387$

$A_{12} = 1.7503$; $\Delta b = 5.6632$

$B_{12} = -0.4979$; $\Delta c_{21} = -5.6047$

$C_{12} = 0.2313$; $\Delta c_{12} = -6.3304$

Starting data are: vapour pressures of the pure components, the azeotropic mixture and the saturated as well as half-saturated solutions, the vapour compositions on the azeotropic and equimolar points of the basic system, and also the boundary solubilities for some selected relative mole fractions. Supposition: $\lg a_3^{(L)} = 1$ along the boundary line of solubility. The comparison of the calculated and measured values of the vapour-liquid equilibrium together with the solubility equilibrium is demonstrated in (6) and especially for the solid-effect (14) in Figure 6; we see sufficient good conformity. In the same way we can predict for the system ethylalcohol/water the disappearance of the azeotropic point by adding urea up to the saturation; we had the opportunity to verify this [7].

If we analyse the methods used, we can state: for a good prediction of phase equilibrium properties the number of supporting data used shall not be too low, and a high accuracy of these supporting data must exist. This again de-

Table 3: Some predicted (and measured) values for the system methylalcohol/water/disodiumhydrogenphosphate (60 °C).

x_1'	$x_3^{(L)}$(pred)	$x_3^{(L)}$(meas)	p_H(pred)	p_H(meas)	$x''_{1,H}$(pred)	$x''_{1,H}$(meas)
0 000	0.2326	0.2326	16.1 kPa	16.1 kPa	0.000	0.000
0 025	0.203		39.7 kPa		0.611	
0.040	0.1866	0.1866	46.9 kPa	45.1 kPa	0.677	0.651
0.191	0.0380	0.0380	46.9 kPa	45.1 kPa	0.677	0.651
0.200	0.0348	0.0334				
0.300	0.0125	0.0140				
0.313	0.0119	0.0135	51.2 kPa	51.3 kPa	0.718	0.719
0.500	0.0026	0.0024	59.5 kPa	60.0 kPa	0.797	0.794
1.000	0.00016	0.00016	84.3 kPa	84.0 kPa	1.000	1.000

mands the delivery of reliable experimental data especially of liquid mixtures with significant properties (boundary solubilities, azeotropic mixtures, alyotropic tie-lines, and so on); as a rule these data are easily obtainable by experimental measurements.

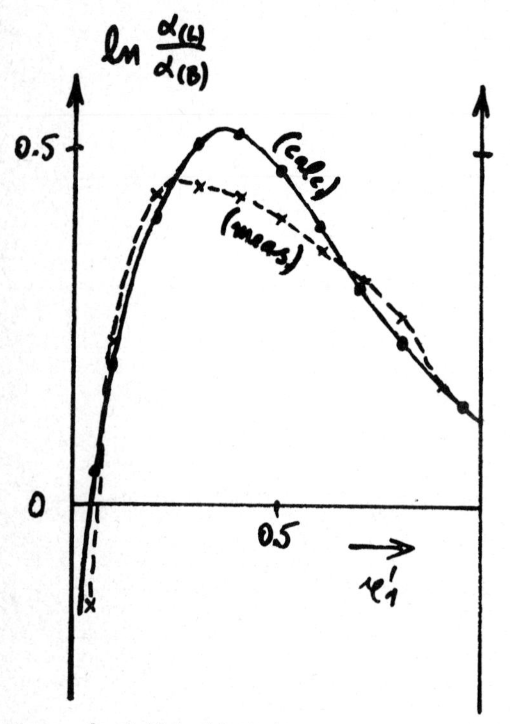

Figure 6: Solid effect in system n-propylalcohol/water/urea (35 °C);
o calculated resp. predicted
x experimentally measured

REFERENCES

[1] Feix, G., Bittrich, H.J. and Lempe, D. Coll. Czech. Chem. Comm. 45 (1980), S. 2035.

[2] Gmehling, J., Rasmussen, P. and Fredenslund, A., Phase Equilibria and Separation Processes: VL - Equilibria by UNIFAC Group Contribution. Revision and Extension II (Lyngby 1980).

[3] Schuberth, H., Chem. Techn. 27 (1975), S. 671.

[4] Schuberth, H., Z. phys. Chem., in press.

[5] Bruder, K., Vohland, P. and Schubert, H., Z. phys. Chem. 258 (1977), S. 721.

[6] Stimming, R.,and Schubert, H.,Z. phys. Chem. 253 (1982), S. 417.

[7] Schuberth, H., Z. phys. Chem. 261 (1980), S. 777.

Data for Science and Technology, P.S. Glaeser (ed.)
North-Holland Publishing Company

CALCULATION OF VAPOR-LIQUID EQUILIBRIUM BY EQUATION OF STATE IN THE ALCOHOL AND SEMI-INERT COMPOUND SYSTEMS[1]

Herbert Wenzel* and Adam Skrzecz**

**University of Erlangen-Nürnberg, Erlangen, Federal Republic of Germany*
***Institute of Physical Chemistry, Polish Academy of Science, Warsaw, Poland*

A van der Waals type equation of state was used to calculate vapor-liquid equilibrium in systems where one component, so-called semi-inert, is not self-associated, but is able to form solvated species with self associated species of alcohols. 54 binary systems involving altogether 144 data sets (isobar or isotherm) were considered. Only one single parameter was used for a binary system. The accuracy of representation was found to be comparable with that of the NRTL equation. The constant temperature dependence of this adjustable parameter was found for systems alcohol-ester and alcohol-ether.

The method of calculation of phase equilibria by equation of state (EQS) has been used by various authors. The different extensions of the van der Waals cubic equation are used very often. They may be written in general form

$$p = \frac{RT}{V-b} + \frac{a(T)}{V^2+ubV+wb^2}$$

where p is pressure, V is molar volume, T is temperature, and a and b are two adjustable parameters. When different values are assigned to the parameters u and w, various types of equation are obtained:

u	w	
0	0	van der Waals eq.
1	0	Redlich-Kwong-Soave eq.
2	-1	Peng-Robinson eq.
1+3ω	-3ω	Schmidt-Wenzel eq.

where ω is Pitzer's acentric factor. The parameters a and b for mixtures are obtained by following mixing rules:

$$a = \sum_i \sum_j x_i x_j (1-\theta_{ij})\ a_{ii} a_{jj}$$

$$b = \sum_i x_i b_i$$

$$\omega = \sum_i z_i \omega_i$$

where θ is the interaction parameter between species i and j, and Z_i a weighing factor defined by Goral as

$$z_i = \frac{x_i\sqrt{a_{ii}}}{\sum_j x_j \sqrt{a_j}}$$

The mixing rules above use only *one* binary interaction parameter θ_{ij}, which is determined by adjustment to binary experimental data sets.

EQS of the van der Waals type are applicable to fluids in which the individual molecules exist. An associated substance is a mixture of monomeric molecules and a number of components consisting of distinct associated species. Representing such a mixture as a single monomeric component leads to an incorrect representation of the binary system. One way of proceeding is to represent the associated substance monomerically and empirically to modify the mixing rules in some way that permits the VLE to be correctly represented. That leads to an increase in the number of adjustable parameters. An alternative approach is to represent the various components present in an associated substance by means of a chemical association model as proposed originally by Dolezalek in 1908.

In this work associated substances were also represented as a mixture of monomeric and associated components by means of an association model. The equilibrium between the various components of the model was defined under ideal gas condition by sets of chemical equilibrium parameters. Using these parameters and correcting them for conditions of nonideality by means of EQS the associating equilibrium can be calculated in both liquid and vapor phase. The temperature dependence of equilibrium constant K_{pi} is given by van't Hoff's Law:

$$\ln K_{pi} = \frac{-\Delta H_i}{RT} + \frac{\Delta S_i}{R}$$

where enthalpy and entropy are assumed to be independent of temperature. The equilibrium may be calculated from the additional constraint that the composition must sum to unity. In practice, many binary systems involve two associating substances. Such mixtures can be expected to solvate. There are also many substances which although not

[1] *Correspondence should be sent to A. Skrzecz, who has written the article.*

self-associated, can solvate with associated substances - the tertiary amines, esters and ethers being prime examples. We call them semi-inert.

The main problem in accounting for solvation is that the equilibrium constants for solvation and the EQS parameters of each solvated species are unknown. In order to avoid treating them as additional adjustable parameters, it was found possible to set all interaction parameters between solvating components to zero and to estimate the remaining parameters by a set of combining rules from the corresponding parameters of the pure substance. For this purpose, use was made of observation that for the components of a self-associated substance there exists an approximately linear relationship between the parameters $\sqrt{a_i}$ and u_i versus b_i as well as between the association parameters ΔS_i and ΔH_i versus the number of monomeric molecules contained in the associated species. Several rules were tried to establish the values of parameter b_{ij} for systems containing alcohols, and the following rules have been used:

$$b_{1-1} = b_{A-A} + b_B - b_A$$

$$b_{i-1} = b_{A-\ldots-A} + b_B \qquad i=1,2,3,4,\ldots$$

The indicator i-1 means that the solvated species consist of i molecules of alcohol and 1 molecule

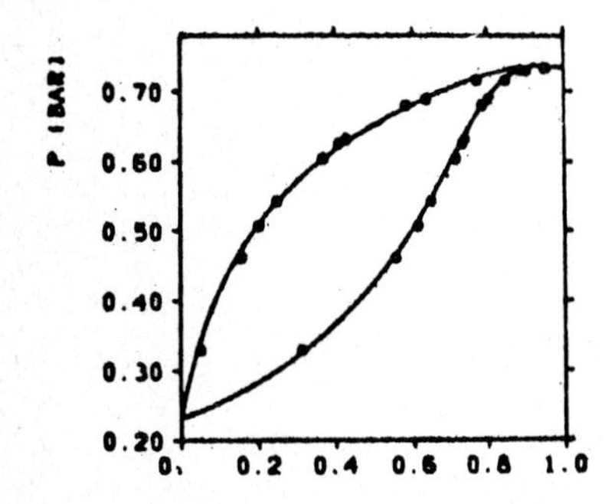

Figure 1: VLE in the system Ethanol-Ethyl formate at T = 328.2 K
Experimental data by Nagata et al. (1976). Model of cross-association (1,1)-(2,1)-(4,1)

of semi-inert compound. We assumed that the only adjustable parameter between solvated components will be parameter DS controlling the degree of solvation:

$$\Delta S_{i,solv} = \Delta S_{i,estimated} + DS$$

where ΔS_i,estimated was calculated by the rules described above. In this way, it was again possible to represent the binary system with the reasonable degree of accuracy by one adjustable parameter. We calculated all accessible systems with alcohols. The models of associations, presented in Table 1, were taken from previous works [1,2].

For example 1-propanol is a mixture of monomers,

Table 1

	Association model	No. of systems	No. of data sets
Methanol	1-2-...-13-14	13	45
Ethanol	1-2-4-8-12	11	46
1-Propanol	1-2-4-8	6	13
2-Propanol	1-2-4-8	6	11
1-Butanol	1-2-3-4	13	21
2-Methyl-1-propanol	1-2-3-4	1	2
2-Methyl-2-propanol	1-2-3-4	1	1
1-Pentanol	1-2-3-4	1	1
2-Methoxyethanol	1-2-3-4	1	3
2-Ethoxyethanol	1-2-3	1	1

dimers, tetramers and octamers; 2-ethoxyethanol - monomers, dimers and trimers. The columns on the right side of the table show the number of binary systems and the number of data sets which were investigated. 54 binary systems involving altogether 144 data sets (isobar and isotherm) were considered. Some results of the calculation are presented in graphical form in Figures 1 - 5. Circles denote the experimental points, and curves are the calculated values by EQS. The model of

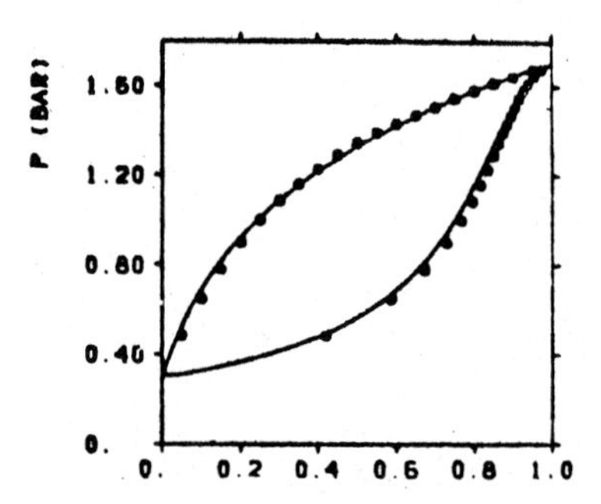

Figure 2: VLE in the system Ethanol - Diethyl ether at T = 323.2 K
Experimental data by Nagai et al. (1935). Model of cross-association (1,1)-(2,1)-(4,1)

cross association and the source of experimental data are written under the graph. The examples of numerical data for the 2-propanol binary systems only are presented in Table 2. The model of cross-association was assumed to be (1,1)-(2,1)-(4,1).

We calculated all binary systems of alcohols with esters and ethers, contained in the collections of Gmehling and Onken (3,4) and found, as is shown in the following examples, that an accuracy of representation by EQS with solvation taken into account is comparable with that of the NRTL equation.

We were interested in the prediction possibility of that method too. In spite of scatter of the various experimental data sets, we found that the adjustable parameter DS is a linear function of temperature (Fig. 6). Its slope is the

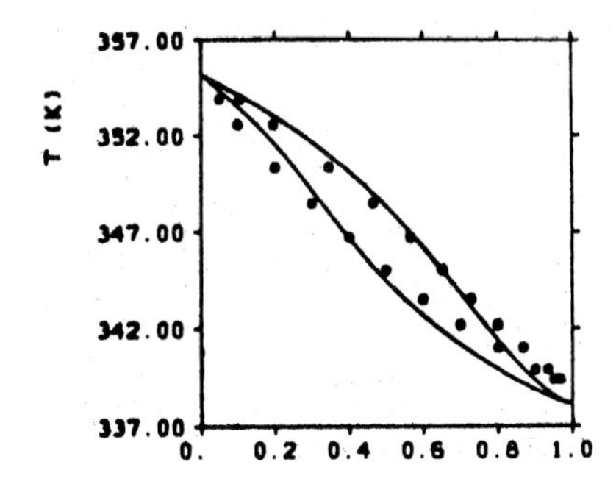

Figure 3: VLE in the system 2-propanol - Tetrahydrofuran at P = 101.3 kPa Experimental data by Shnitko et al. (1969). Model of cross-association (1,1)-(2,1)-(4,1)

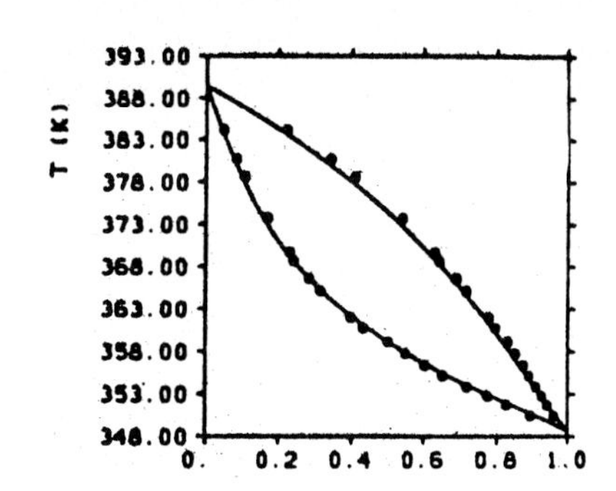

Figure 5: VLE in the system Butanol - Ethyl acetate at P = 97.3 kPa Experimental data by Mainkar et al. (1965). Model of cross-association (1,1)-(2,1).

Table 2

2-Propanol +		AVERAGE DEVIATIONS EQS $\Delta \bar{y}$ % mole fr.	NRTL $\Delta \bar{y}$ % mole fr.	EQS $\Delta \bar{P}$ kPa	NRTL $\Delta \bar{T}$ K	Authors
Ethyl formate	T=318.2	0.88	0.38	1.02	1.29	Nagata (1976)
Ethyl acetate	T=313.2	0.46	0.31	1.04	0.84	Murti (1958)
	T=328.2	0.36	0.40	1.22	1.44	Nagata (1975)
	T=333.2	0.65	0.48	1.20	1.25	Murti (1958)
	P=101.3	0.30	0.41	0.60	0.39	Nishi (1972)
	P=101.3	0.24	0.25	0.62	0.34	Murti (1958)
Diisopropyl ether	P=101.3	0.16	0.36	1.38	0.84	Miller (1940)
	P=101.3	0.28	0.49	0.83	0.56	Verhoye (1970)
Tetrahydrofuran	P=101.3	0.54	0.18	1.52	0.27	Shnitko (1969)
1,4-Dioxane	P=101.3	0.58	0.24	2.44	0.79	Choffe (1960)
4-Methyl-2-pentanone	P=101.3	0.17	0.17	0.63	0.86	Ballard (1953)

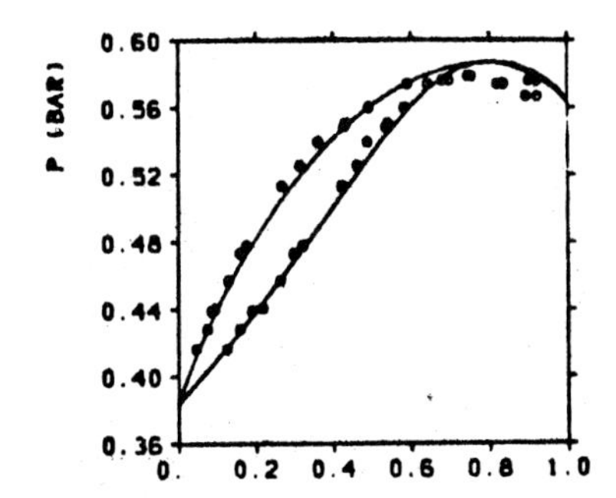

Figure 4: VLE in the system 2-propanol - Ethyl acetate at T = 333.2 K Experimental data by Murti et al. (1958). Model of cross-association (1,1)-(2,1)-(4,1)

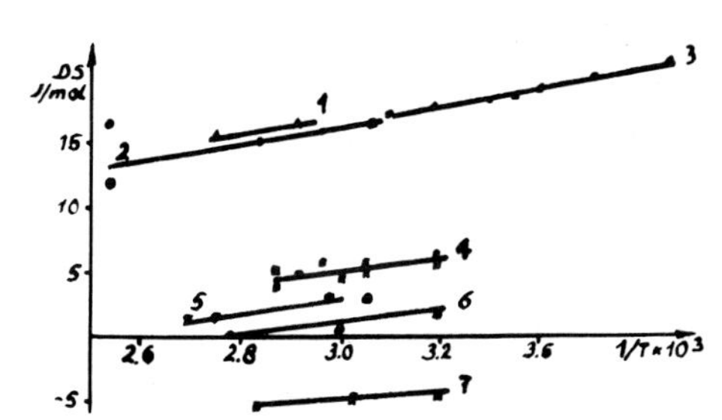

Figure 6: Correlation of adjustable parameter DS at various systems.
1 - 2-Methoxyethanol - Ethyl acetate
2 - Butanol - Butyl acetate
3 - Ethanol - Diethyl ether
4 - Ethanol - Ethyl acetate
5 - 1-Propanol - Propyl acetate
6 - 1-Propanol - Ethyl acetate
7 - 2-Propanol - Ethyl acetate

same for all the investigated ester and ether systems. It allows us to predict VLE at another temperature even if the system was measured only at one temperature.

REFERENCES:

[1] Baumgartner, M., Thesis, University Erlangen-Nurnberg, Erlangen 1981.

[2] Moorwood, R.A.S., Thesis, University Erlangen-Nurnberg, Erlangen 1982.

[3] Gmehling, J., Onken, U., Vapor-Liquid Equilibrium Collection, Organic Hydroxy compounds: Alcohols, Dechema, Chemistry Data Series vol. 1/2a, Frankfurt 1977.

[4] Gmehling, J., Onken, U., Arlt, W., Vapor-Liquid Equilibrium Data Collection, Organic Hydroxy Compounds: Alcohols and Phenols, Dechema, Chemistry Data Series vol. 1/2b, Frankfurt 1978.

Data for Science and Technology, P.S. Glaeser (ed.)
North-Holland Publishing Company

ESTIMATION OF CRITICALLY EVALUATED MODEL PARAMETERS FOR THE CALCULATION OF MULTICOMPONENT LIQUID-LIQUID EQUILIBRIA

Dieter Lempe, Monika Grassmann, Hartmut Krüger and Claudia Carl

Technical University 'Carl Schorlemmer' Leuna-Merseburg
DDR-4200 Merseburg, German Democratic Republic

To get model parameters for calculating LLE which are quantitatively and qualitatively suitable, the data of hydrocarbon - solvent systems were checked by using the following criteria
- independent experimental determination of the binodal curves and tie lines,
- correlation within homologous series,
- correlation by using empirical methods,
- investigation of the structural behaviour of the model equations, especially the stability conditions,
- comparison of the calculated thermodynamic functions with independent experimental data.

1. INTRODUCTION

To carry out the model calculations of processes and systems in chemical engineering large amounts of reliable data of mixtures in form of special data models are required. In recent years this has become more and more a central problem.

The yield of reliable model parameters for activity coefficients for calculating liquid-liquid equilibria (LLE) requires an extensive and careful examination of
- the experimental original data
- the used models with respect to the flexibility, extrapolation possibility, description of the stability behaviour, as well as
- the sets of model parameters with respect to the quantitatively satisfactory, qualitatively correct and thermodynamically consistent description of the systems and subsystems, the latter including the independent experimental determination of further suitable thermodynamical data.

Especially the qualitative aspects are very important, because the phase separation in the liquid state is caused only by the real mixing behaviour of the system.

The problems discussed here are related to systems hydrocarbon - N-methyl-ε-caprolactam - ethylene glycol [1]. But they are of general significance although in special cases the situation may be less complicated.

2. STATISTICAL AND SYSTEMATIC TEST OF THE EXPERIMENTAL LLE DATA

As an essential prerequisite for obtaining suitable sets of data the independence or the verification possibility of the determination methods for the boundaries and tie lines have to be warranted.

Therefore the direct application of the principle of the lever cannot be recommended. Likewise this is valid for the successive Othmer's cloud point titration because a remarkable error propagation occurs especially if it is difficult to disperse mechanically one of the phases in the other one. This was observed in the systems discussed here as examples.

Good results were yielded through the refractive index method. This method consists of measuring the refractive indexes along the boundary for constructing calibration curves and the following determination of the concentrations of the equilibrium phases which are obtained from defined heterogeneous mixtures at a fixed temperature through the measurement of their refractive indexes. The known total concentrations allow the exactness of the measured data to be proved. Not all connecting lines intersect these points as it would be necessary. Here experimental values were only used as a tie line, if the deviation was less than 0.3 mole per cent.

The reproducibility can be improved by repeat experiments. But in general more than two or three of those experiments are too expensive so that only a rough estimation of the random errors of the single measurements is possible.

To recognize the systematic errors is much more complicated. For this purpose it is necessary to use different correlation methods. In the case of LLE data in multicomponent systems there are e. g. the following test possibilities

- comparison of data obtained by different measurement principles
- looking at the progress of a sequence of the tie lines
- looking at the progress of the boundaries and tie lines in dependence on the temperature
- comparison of the LLE data within homologous series, here e. g. within hydrocarbon series in the solvent system NMC - EG.

The tests can be carried out by using

- the graphs in the triangle coordinate system
- the empirical correlation methods of Bachmann, Hand, Othmer etc. [2].
- characteristic single quantities like e. g. area fractions of heterogeneous regions in the Gibbs triangle, maximum concentrations of one of the components on the boundary, empirically yielded critical miscibility points, etc.

Such treatment was made use of with the ternary systems hydrocarbon - NMC - EG. Some examples are illustrated in figures 1 and 2.

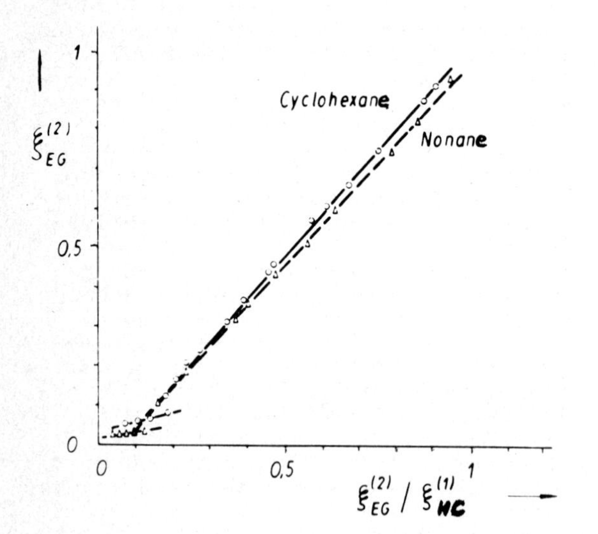

Figure 1 : Bachmann correlation in systems hydrocarbon - NMC - EG at 40 °C

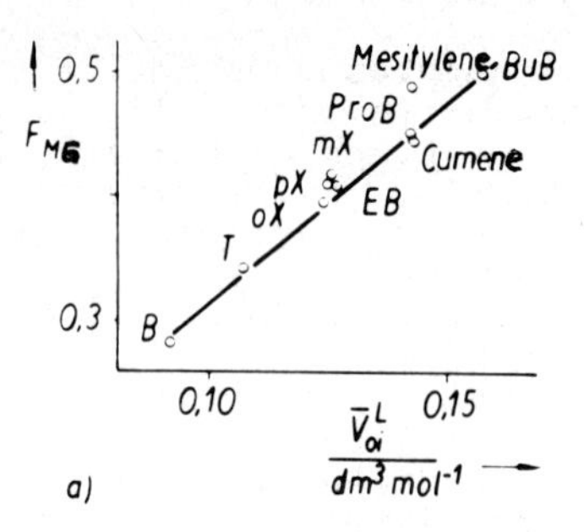

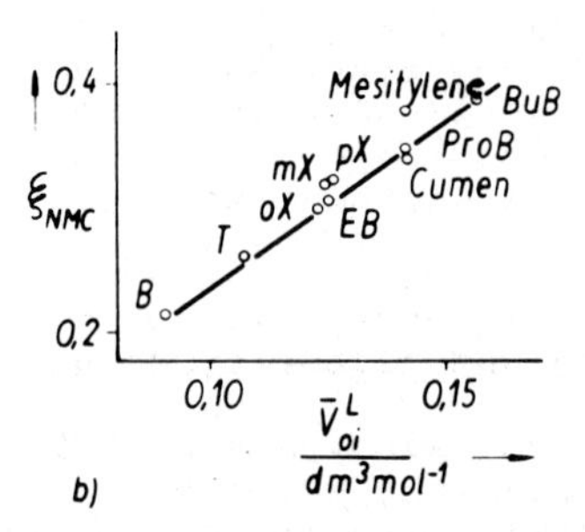

Figure 2 : Correlation of the area fractions of the heterogeneous region (a) and the maximum mass fractions of NMC (b) with the molar volumes of aromatic hydrocarbons in the systems HC - NMC - EG

Good correlations were yielded by using the simple Bachmann equation

$$x_{EG}^{(2)} = A \frac{x_{EG}^{(2)}}{x_{HC}^{(1)}} + B$$

1 HC-rich phase

2 EG-rich phase

also near the critical miscibility points.

It is shown, that corresponding experimental data were obtained. The use of the different test criteria results in the rejection of about 10 per cent of the experimental points.

At last it is necessary to state that the application of such methods do not guarantee the actual correctness of the LLE data. They only increase its probability.

3. STRUCTURAL INVESTIGATIONS OF THE MODEL EQUATIONS

Numerous examples show that most of the quantitative and qualitative defects in the description of LLE are caused by insufficiencies of the different model equations for the activity coefficients, although the parameters have been obtained from suitable experimental data. Such insufficiencies are

- the imperfect flexibility and the insufficient range of values
- the poor representation of the temperature dependence of the activity coefficients
- the problems with the transition to limiting values
- the wrong description of the stability behaviour of the binary systems (e. g. more than 2 zero points in the function GMII = $(\partial^2\Delta^M G/\partial x_1^2)_{T,p}$ in the case of simple miscibility gaps)
- occurrence of wrong plaits on the Gibbs free energy surfaces in the ternary and higher systems.

For the most widely used NRTL and UNIQUAC models the following statements can be made [3, 4, 5, 6].

- The NRTL equation is very flexible, but there are many problems with the description of the stability conditions (4 zero points in GMII in certain parameter regions). A description of LLE is possible only with positive G^E.
- The UNIQUAC equation is not so flexible and gives only 2 zero points in GMII. Phase separation can be described also with negative G^E (in dependence on the pure component parameters r_i and q_i; figure 3).

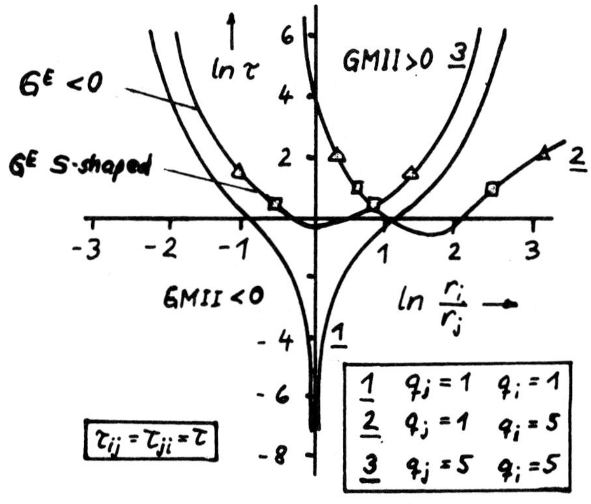

Figure 3 : Stable/metastable and unstable regions described by the UNIQUAC equation

- With both equations the calculation of wrong plaits on the Gibbs energy surface of the three- and more-component systems cannot be excluded ad hoc (figure 4; NRTL parameters from [7]), even if the description of the binary systems is good (UNIQUAC equ. [8]).

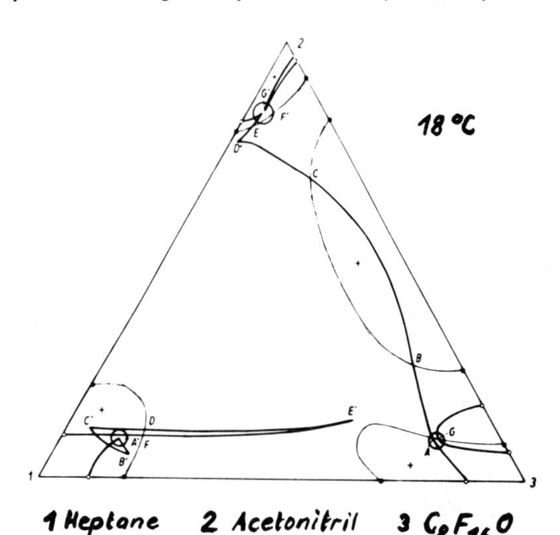

Figure 4 : Principal course of the branches of boundary calculated by parameters of Kikic and Alessi (o expected three phase equilibrium LLL)

- There are restrictions between the possible values of G^E and H^E following from the model equations, if the energy parameters are taken as temperature independent [9].

Figure 5 shows the influence of this effect on the calculated miscibility gap in the system heptane - aniline (NRTL equation). The curve 1 was calculated with temperature dependent parameters

$$\Delta g_{ij} = \Delta g_{ij}^C + \Delta g_{ij}^T\ (T - 273.15\ K)$$

resulting from p-x data (95 °C) and mixing enthalpies. The curve 2 is obtained, if the value of Δg_{ij} resulting from this relation at 60 °C is taken as temperature independent mean value. The critical miscibility temperature change more than 10 K.

- The transition from the binary subsystems to the more-component mixtures causes a variation of the influence of the parameters in the different terms of the equations, which can lead to a distorted description of the latter systems. If in the NRTL equation for one binary subsystem $\alpha_{ij} = 0$ is valid, the single terms τ_{ij} and τ_{ji} are undetermined in the binary system. But they act separately in the equation of the ternary system.

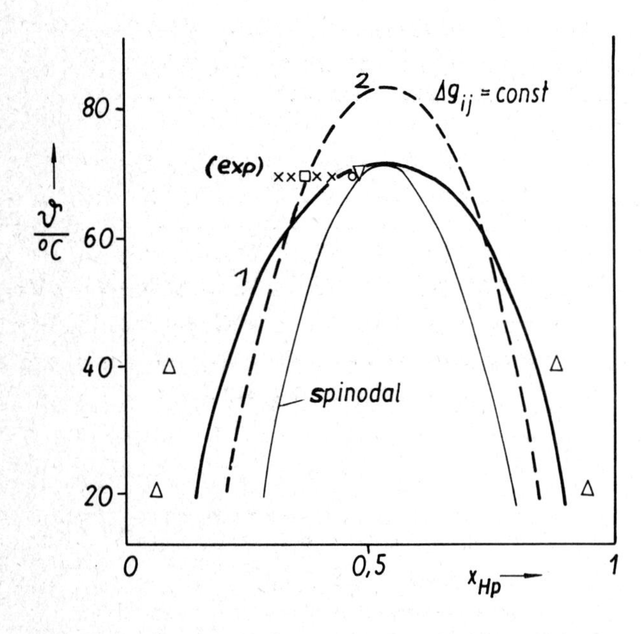

Figure 5: Calculated miscibility gap in the system heptane - aniline

Those problems can be illustrated with the limiting activity coefficients of hydrocarbons $\ln f_{\infty i}$ in the mixed solvent phenol (A) - aniline (B) (data taken from [10]). Although the location and the value of the extrema of G^E in the binary solvent system calculated from p-x data (VLE) shift only very little in dependence on the chosen α (NRTL equation), the differences

$$\Delta \ln f_{\infty i} = \ln f_{\infty i}^{(AB)} - (x_A \ln f_{\infty i}^{(A)} + x_B \ln f_{\infty i}^{(B)})$$

change even their signs (table 1).

Table 1: Limiting activity coefficients of hydrocarbons in a solvent mixture in dependence on the chosen α (NRTL equation)

α	$\ln f_{\infty i}$		
	hexane	cyclohexane	benzene
exp	0.122	0.087	0.058
0.4	- 0.079	-0.072	- 0.053
0.3	- 0.049	- 0.060	- 0.122
0.2	- 0.042	- 0.049	- 0.093
0	0.172	0.221	0.367
-0.1	0.322	0.383	0.540
-0.3	0.472	0.524	0.613
-0.47 x)	0.657	0.707	0.755

x) Optimal fitting of p-x data

Reversely the selection of reliable parameters is possible, if such data are determined experimentally.

4. PARAMETER ESTIMATION AND TESTS

The following experimental data are used for the estimation of the model parameters
1st binary LLE and VLE data, in some cases also limiting activity coefficients
2nd ternary tie lines
3rd binary LLE and VLE data for initial parameters and ternary data for readjusting.

The parameters obtained only from binary data give in some cases a satisfactory description of ternary and higher tie lines [11], but mostly the results are poor [12].

On the other hand the second method leads to a very good representation of the experimental data especially if the squares of the differences of mole fractions (SQX) instead of those of the activities (SQA) are used as an objective function in the regression [12]. However often it is not possible to describe also the VLE and other thermodynamic data by using the resulting parameters.

The third method is expensive but leads to a common description of either LLE or VLE data.

The parameter examination has to include
- the class of the quantitative representation (standard deviations)
- the evidence of a qualitative suitable representation (right location of the plaits)

- the comprehensive thermodynamic consistency
  - . with data of the subsystems
  - . with limiting quantities
  - . with the mixing enthalpy (in the case of temperature variation).

The application of these criteria to the model parameter estimation in the systems hydrocarbon - NMC - EG leads to the results given in table 2.

With the first method the ternary miscibility gaps are too large, especially in the systems with aromatic hydrocarbons. Furthermore the UNIQUAC equation has the disadvantage that the systems of higher aliphatic hydrocarbons with NMC show by calculation with the adjusted parameters a heterogeneous region while in fact they are homogeneous.

The second principle shows good quantitative results and could be taken for a data reduction and storing. But there is no thermodynamic significance as is shown by comparing the G^E and the $\ln f_{\infty HC}$ in the corresponding binary subsystems (table 2).

The best compromise is yielded by the third method. These parameters should be used for further purposes.

Table 2: Comparison of thermodynamic data calculated by different parameter sets for the ternary LLE (40 °C)

	VLE	Gaschrom.	UNIQUAC (LLE; SQX) $\overline{\Delta x}$ = 0.87 mol p.c.		NRTL (VLE + LLE; SQA) $\overline{\Delta x}$ = 2.92 mol p.c.	
System	G^E/RT (extrem.)	$\ln f_{\infty HC}$	G^E/RT (extr.)	$\ln f_{\infty HC}$	G^E/RT (extr.)	$\ln f_{\infty HC}$
H - NMC	0.500	1.773	0.464	1.479	0.500	1.880
H - EG		5.852	1.278	7.745	1.122	6.694
C - NMC	0.402	1.569	0.369	0.836	0.402	1.299
C - EG		5.493	1.441	5.617	1.168	5.858
N - NMC	0.494	2.111			0.494	2.299
N - EG		6.642			1.418	7.026
D - NMC	0.532	2.137	0.535	2.097	0.532	2.475
D - EG		7.125	1.725	8.447	1.216	7.375
B - NMC	0.039	-0.117	-0.262	-1.510	0.039	0.050
B - EG		3.517	0.991	3.829	0.789	4.015
T - NMC	0.057	0.049	-0.498	-0.688	0.057	0.107
T - EG		4.074	1.091	4.613	0.829	4.395
EB-NMC	0.114	0.231	-0.166	-0.713	0.114	0.286
EB- EG		4.634	1.170	5.758	0.862	4.715
NMC-EG	0.088		-0.278		0.088	

H hexane; C cyclohexane; N nonane; D decane; B benzene; T toluene; EB ethylbenzene; NMC N-methyl-ε-caprolactam; EG ethylene glycol

REFERENCES

[1] Lempe, D.; Graßmann, M.; Carl, C.; Georgi, P.: Calculation of Liquid -Liquid Equilibria in Ternary and Higher Systems Hydrocarbon(s) - N-Methyl-eps-caprolactam - Glycol. 7th Int. Congress Chem. Engng. Chem. Equ. Design and Automat. CHISA 81, Prague 1981 (Paper F 2.25)

[2] Kortüm, G.; Buchholz-Meisenheimer, H.: Die Theorie der Destillation und Extraktion von Flüssigkeiten. Springer-Verlag, Berlin, Göttingen, Heidelberg 1952

[3] Novák, J.P.; Suška, J.; Matouš,J.: CCCC 39 (1974),1943

[4] Mattelin, A.C.; Verhoeye, L.A.J.: Chem. Engng. Sci. 30 (1975), 193

[5] Lempe, D.; Feix, G.; Bittrich, H.-J.: Chem. Techn. 31 (1979),204

[6] Monfort, J.P.; Rojas, R.M.D.L.: Fluid Phase Equilibria 2 (1978), 181

[7] Kikic, I.; Alessi, P.: Annali chim. 64 (1974),363

[8] Novák, J.P.; Matouš, J.; Šobr, J.; Pick, J.: CCCC 46 (1981), 3003

[9] Voňka. P.; Novák, J.P.; Suška,J.; Pick, J.: Chem. Engng. Commun. 2 (1975), 51

[10] Vernier, P.: Thèse Doct. sci. appl. Fac. sci. Univ. Paris 1967

[11] Heinrich, J.; Dojčansky, J.: CCCC 40 (1975), 2221

[12] Sørensen, J.M.; Magnussen, T.; Rasmussen, P.; Fredenslund, A.: Fluid Phase Equilibria 3 (1979), 47

ACKNOWLEDGMENT

The authors are grateful to the VEB Leuna-Werke "Walter Ulbricht" for the financial support of and the interest in this work, and to Prof. Dr. K. Quitzsch, Karl Marx University of Leipzig, for carrying out the parameter estimation with the program of Sørensen.

Data for Science and Technology, P.S. Glaeser (ed.)
North-Holland Publishing Company

CORRELATION OF BINARY LIQUID-LIQUID EQUILIBRIUM AND/OR SOLUBILITY DATA

Roman Stryjek and Marek Łuszczyk

Institute of Physical Chemistry, Polish Academy of Sciences
01-224 Warsaw, Poland

A new equation for the calculation of liquid-liquid equilibrium data of binary systems directly from solubility data is proposed. The equation was tested for the data reduction of more than 50 binary hydrocarbon-polar compound systems covering a wide temperature range including the critical solution temperature. The procedure was applied as a subroutine of programs for the correlation of liquid-liquid equilibrium data through equations for the g^E function. The use of NRTL, UNIQUAC, LEMF and LCG equations for the LLE data correlation especially over a wide temperature range and in the vicinity of the critical solution temperature is discussed.

1. INTRODUCTION

Data on limited miscibility are reported in the literature at least for a few thousand binary systems; it is unfortunate that most of these are not equilibrium but solubility measurements. The statistics based on the data for over 600 systems, mostly of non-electrolyte mixtures, as reported by Sørensen and Arlt [1], are given in Table 1.

Table 1

Statistics of binary systems based on data in [1].

water-hydrocarbon	99	15.1%
water-polar (non-water)	264	40.3%
polar (non-water)-hydrocarbon	197	30.1%
polar (non-water)-polar (non-water)	95	14.5%

The number of systems and their percentages are reported in the first and second columns respectively. They were aggregated into four somewhat arbitrary groups, and water is individualized from polar compounds.

A close inspection of the data reported therein for binary hydrocarbon-polar (non-water) compound systems showed that almost all of them represent solubility measurements. Desired binary liquid-liquid equilibrium (LLE) data which can test theories of non-electrolyte solutions, especially equations for g^E function, and which are needed in designing of processes in which the liquid phase is separated, are obtained from the solubility data by graphical interpolation. Although this method is subjective it is widely used. Replacing of this method by empirical polynomial equations does not improve the procedure, especially if the data are scattered. Moreover, for most of the systems the temperature range was narrow and the number of experimental values limited (Table 2). It is thus evident that a good method for the calculation of LLE data from solubility ones is required.

Table 2

Statistics (%) of solubility data for hydrocarbon-polar (non-water) systems as a function of the temperature range (ΔT) and number of data points (N).

	$\Delta T\leqslant 10$	$10<\Delta T\leqslant 30$	$30<\Delta T\leqslant 60$	$\Delta T>60$
$1\leqslant N\leqslant 5$	19.40	12.93	4.48	-
$5<N\leqslant 10$	1.00	7.96	11.94	1.00
$10<N\leqslant 20$	-	11.94	11.94	6.92
$20<N\leqslant 30$	-	1.49	3.98	1.99
$N>30$	-	0.50	1.49	1.00

The second problem which appeared in the solubility and/or LLE data handling is their correlation with vapour-liquid equilibrium (VLE), activity coefficients at infinite dilution (γ^∞) and h^E data for binary and multicomponent systems. For these purposes equations for the g^E function, especially those based on the local composition model, are applied predominantly. The analysis of data reported in literature showed that the equations for the g^E function were tested mainly for the correlation of isothermal data.

The purpose of the study was to work out a semi-empirical method of calculation of reliable LLE data for binary systems from solubility data and to comment on the behaviour of the equations for the g^E function in LLE correlation over a wide temperature range as well as in the proximity of the critical solution temperature.

2. EQUATION FOR BINODAL AND DIAMETER CURVES

The representation of solubility data of binary systems as a function of temperature was an object of study in the past and at least two approaches should be mentioned:

Malesińska [2] found that mutual solubilities of

binary systems formed with nitromethane and members of homologous series of n-alkanes followed a common and symmetric curve if the solubilities were expressed as a function of critical solution parameters. One of us [3] applied that idea to various polar compound-hexadecane systems over a wide (up to 120 K) temperature range, and recently Wesołowska et al. [4] applied this method for the presentation of solubility data for a series of systems over a 250 K range. A close inspection of this method, however, showed its rather qualitative than quantitative character. Cox and Herington [5], on the basis of universal behaviour of critical phenomena, showed that the binodal coexistence curve can be expressed, but in the proximity of the critical solution point only, as a linear function of critical solution parameters and a coexistence curve critical exponent, β, where β=1/3. The development of the theory of critical phenomena in the last decade or so, especially Wegner's results [6], enabled an important improvement to be made by Ley-Koo and Green [7]. They proposed two expansions for fluids within an extended temperature range, namely:

$$x_i' - x_i'' = B_o\epsilon_i^{\beta} + B_1\epsilon_i^{(\beta+\Delta)} + B_2\epsilon_i^{(\beta+2\Delta)} + \ldots \quad (1)$$

for the difference in order parameters of the binodal curve, where ϵ = 1-T/T^C, x' and x" denote composition of the liquid phases in equilibrium and

$$x_i' + x_i'' = 2(A_o + A_1\epsilon_i + A_2\epsilon_i^{(1-\alpha)} + \ldots) \quad (2)$$

for the diameter of binodal curve. Mole fraction, volume fraction, density etc. can be applied in equations 1 and 2 as the order parameter, and amongst them the mole fraction is applied most often. A specific heat critical exponent: α=0.1 according to Levelt-Sengers [8], a coexistence curve exponent: β=0.325, and a correction exponent: Δ=0.5, the last two according to Greer [9], all as universal constants, were used. It is obvious that the composition of two liquid phases coexisting in the equilibrium have to be known to solve these equations.

One of us (M.Ł.) found that from the equation in the form:

$$x_i = A_o + A_1\epsilon + A_2\epsilon^{(1-\alpha)} + f\frac{B_o}{2}\epsilon^{\beta} + f\frac{B_1}{2}\epsilon^{(\beta+\Delta)} + f\frac{B_2}{2}\epsilon^{(\beta+2\Delta)} + \ldots \quad (3)$$

where: f=1 for $(x_i - A_0)>0$ and f=-1 for $(x_i - A_0)<0$, obtained through a simple summation of equations (1) and (2), one can get all parameters describing both (i.e. coexistence and diameter) curves directly from solubility data. Additionally, only the value of T^C estimated to about 0.1 K is needed, the value of x^C can be introduced if it is known, thereby reducing the number of adjusted parameters as x^C is put as A_0. Equation (3) was tested for the data of Nagarajan et al. [10] for heptane-acetic anhydride systems chosen as the most closely examined, as the 76 solubility data were determined with high accuracy within a 66 K range. The residuals are presented in Figure 1, and the parameters evaluated in the original paper and found from equations (1) and (2) and ours from equation (3), respectively, are presented in Table 2 for comparison. It is evident that excellent agreement was obtained; the A_0 value was put as x^C in the fitting procedure. It should be pointed out that the authors [10] had to get the equilibrium data by graphical interpolation; that difficulty was overcome by us through the method described.

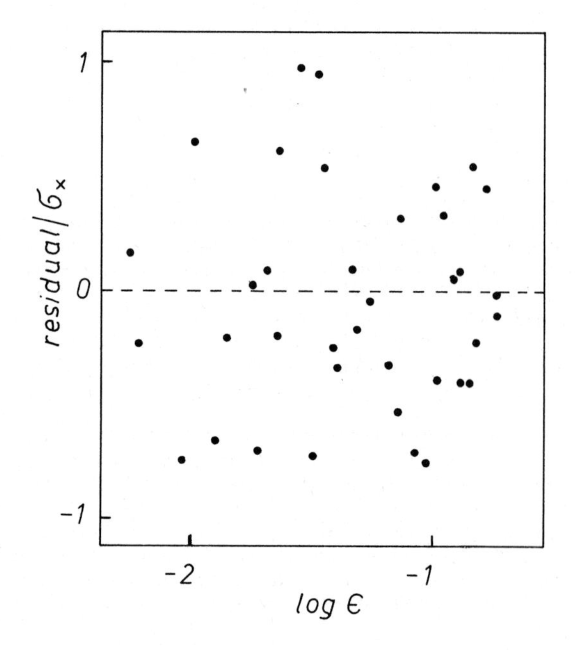

Figure 1: Residuals of Equation (3) for the n-heptane-acetic anyhydride system; solubility data from [10].

Table 3

Parameters of Equations (1) and (2) from [10] and Equation (3) - this work, for n-heptane-acetic anhydride system.

	A_o / σA_o	A_1 / σA_1	A_2 / σA_2	B_o / σB_o	B_1 / σB_1
Eq. (1)	-	-	-	1.89 / 0.01	-1.05 / 0.03
Eq. (2)	0.486 / 0.005	1.00 / 0.07	-0.91 / 0.06	-	-
Eq. (3)	0.486 / 0.001	0.95 / 0.09	-0.87 / 0.07	1.90 / 0.01	-1.07 / 0.01

The proposed method was later tested for the correlation of more than 50 binary systems for which at least 10 solubility data points representing both branches of the binodal curve were available within a temperature range permitting the estimation of the T^c value within an accuracy of about 0.1 K. For the chosen systems the T^c values could be estimated rather easily but the estimation of x^c values led to doubtful results; thus, in most cases, the procedure was applied in which A_0 was adjusted to the x^c value. In all the cases a good fit of the experimental data to equation (3) was found. The results in which β coefficients were adjusted showed that its values oscillated around 0.325, and the quality of the fit slightly depended on its value; it suggests that β=0.325 value can be applied as a universal constant for this kind of system in agreement with the theoretically postulated value. It significantly simplified the procedure as only the linear parameters can be adjusted. The number of parameters B depended on the temperature range ΔT of the correlated data. It has been found that for $\Delta T \leqslant 30$, $30 < \Delta T \leqslant 60$, and $\Delta T > 60$ K, B , B_1 and B_2, and B_1, B_2 and B_3, were needed respectively to assure a good representation of the binodal curve. The number of A_j parameters was independent of the temperature range of correlated data; for all systems two A_j parameters were sufficient to represent the diameter curve through equation (2) if x^c was known. The results for the β-methoxypropionitrile-hexadecane system, for which only 11 solubility data within about a 120 K range were available [3] illustrate

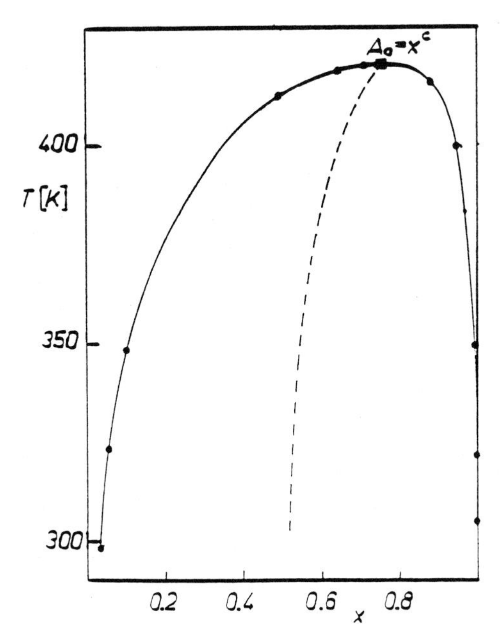

Figure 2: Solubility of the β-methoxypropionitrile-hexadecane system; $\bullet$ - experimental [3], full and dashed lines represent coexistence and diameter curves from Equation (3). Its parameters: A_1 = 7.063, A_2 = -6.997, B = 1.386, B_1 = 1.613, B_2 = -2.804 and standard deviations σ = 0.011 and σT = 4.62 for T^c = 419.75 K and x^c = 0.7668.

the utility of equation (3) in a case where a small number of experimental points exist within a wide temperature range (see Figure 2). In the presented case, three B_j parameters were fitted as $\Delta T > 60$ K.

3. CORRELATION OF BINARY LLE DATA

The correlation of binary LLE data by means of equations for the g^E function has been an object of intensive study since 1968 when Renon and Prausnitz [11] published their NRTL equation which could fulfill thermodynamic condition for liquid phase separation and then could be applied for correlation of LLE data. In the meantime more than ten equations, mostly based on the local composition model, were proposed in the literature. The problem of LLE correlation was comprehensively reviewed in a series of papers [12] and other interesting ones, considering various aspects of LLE, were also delivered at the 8th International CODATA Conference [13-15]. Our remarks are narrowed then to the problem so far rather neglected in the literature, namely, the problem of consistent correlation of binary LLE data over a wide temperature range and in a broad vicinity of UCST. Not only NRTL and UNIQUAC [16] equations, but also the LEMF [17] equation, being a modification of NRTL, and the LCG [18] equation were tested. It should be noted that these equations differ in the number of adjustable temperature-dependent parameters as follows: NRTL has three, UNIQUAC, LEMF and LCG have two, and moreover, in the latter case, one adjustable temperature-independent parameter was added. Our results based on LLE data for a series of binary polar compound-hexadecane systems [19] available within a wide (over 100 K) temperature range and activity coefficients at infinite dilution gave a chance to test these equations. An examination of results showed that all studied equations behaved qualitatively similarly but quantitative differences were not negligible. Although definitive advantages of any of them could not be claimed, some practical conclusions were reached: if a big enough set of data is correlated, the best equation should be fitted by a trial and error procedure; if LLE data and only a little additional data on VLE, γ^∞ and h^E are available, then the NRTL equation should be applied, if only LLE data are available, then the UNIQUAC equation should be recommended as giving us statistically the most reliable results. It should be noted that even within so wide a temperature range, adjustable parameters showed a linear dependence on temperature if the proximity of the critical solution temperature was excluded from the analysis.

The analyses in a broader range around the critical solution point were carried out using a set of data [20] on VLE, LLE, h^E and γ^∞ for binary C_5-hydrocarbon-β-methoxypropinitrile systems. Our results showed that neither the NRTL nor UNIQUAC equations (these two were examined in detail) were able to describe consistently properties in a broad range around UCST, and the quality of correlation depended strongly on the distance of

T and x vectors from this point.

4. CONCLUSION

The new equation and the corresponding program for the calculation of LLE data directly from solubility data were tested for data reduction covering both a wide temperature range and the proximity of the critical solution temperature. It is suggested that these can be applied as a subroutine of any program for the correlation of phase equilibria through equations for the g^E function, especially if only a small number of data points is available.

The work was performed within Research Project 03.10.1.4.1.

REFERENCES

[1] Sørensen, J.M. and Arlt, W., Liquid-liquid equilibrium data collection, V.1., (DECHEMA, Frankfurt/Main, 1979).

[2] Malesińska, B., Studies on the mutual solubility curve in binary liquid systems. I. Binary systems of nitromethane and representatives of n-paraffin series, Bull. Ac. Pol.: Chim. 8 (1960) 53-59.

[3] Rogalski, M. and Stryjek, R., Mutual solubility of binary n-hexadecane and polar compound systems, ibid., 28 (1980) 139-147.

[4] Wesołowska, M., Koliński, A. and Semeniuk, B, Liquid-liquid equilibrium data in some ethanediol-methylbenzene-phenol systems, (Presented at 8th International CODATA Conference, Jachranka, Poland, 4-7 Oct. 1982).

[5] Cox, J.D. and Herington, E.F.G., The Coexistence curve in liquid-liquid binary systems, Trans. Faraday Soc. 52 (1956) 926-930.

[6] Wegner, F.J., Corrections to scaling laws, Phys. Rev. B 5 (1972) 4529-4536.

[7] Ley-Koo, M. and Green, M.S., Revised and extended scaling for coexisting densities of SF_6, Phys. Rev. A 16 (1977) 2483-2487.

[8] Levelt Sengers, J.M. and Sengers, J.V., Universality of critical behavior in gases, Phys. Rev. A 12 (1975) 2622-2627.

[9] Greer, S.C., Liquid-liquid critical phenomena, Acc. Chem. Res. 11 (1978) 427-432.

[10] Nagarajan, N., Kumar, A., Gopal, E.S.R. and Greer, S.C., Liquid-liquid critical phenomena, the coexistence curve of n-heptane-acetic anhydride, J. Phys. Chem., 84 (1980) 2883-2887.

[11] Renon, H. and Prausnitz, J.M., Local compositions in thermodynamic excess functions for liquid mixtures, AIChE J. 14 (1968) 135-144.

[12] Sørensen, J.M., Magnussen, T., Rasmussen, P. and Fredenslund, A., Liquid-Liquid Equilibrium Data: their retrieval, correlation and prediction. Part 1: Retrieval, fluid phase equilibria 2 (1979) 297-309. Part II: Correlation, ibid., 3 (1979) 47-82. Part III: Prediction, ibid., 4 (1980) 151-163.

[13] Schuberth, H., The use of characteristic data for predicting phase equilibria behaviour, (Presented at 8th International CODATA Conference, Jachranka, Poland, 4-7 Oct. 1982).

[14] Bittrich, H.J. and Feix, G., Correlation of data from liquid-liquid equilibria, (Presented ibid.).

[15] Lempe, D., Grassmann, M., Kruger, H. and Carl, C., Estimation of critically evaluated model parameters for the calculation of multicomponent liquid-liquid equilibria, (Presented ibid.).

[16] Abrams, D.S. and Prausnitz, J.M., Statistical Thermodynamics of Liquid Mixtures: a new expression for the excess Gibbs energy of partly or completely miscible systems, AIChE J. 21 (1975) 116-128.

[17] Marina, J.M. and Tassios, D.P., Effective local compositions in phase equilibrium correlations, Ind. Eng. Chem., Process Des. Develop. 12 (1973) 67-71.

[18] Vera, J.H., Sayegh, S.G. and Ratcliff, G.A., A quasi lattice-local composition model for the excess free energy of liquid mixtures, Fluid Phase Equilibria 1 (1977) 113-135.

[19] Stryjek, R., Łuszczyk, M. and Fedorko-Antosik, M., Correlation of binary liquid-liquid equilibria, Bull. Ac. Pol.: Chim., 29 (1981) 203 - 211.

[20] Łuszczyk, M. and Stryjek, R., Correlations of phase equilibria and heats of mixing of the binary C_5-hydrocarbon-methoxypropionitrile systems, (7th CHISA'81 Congress, Praha, Aug. 31-Sept. 4, 1981, paper D2. 34).

Data for Science and Technology, P.S. Glaeser (ed.)
North-Holland Publishing Company

MODEL TESTING OF N/Ch/-I* PHASE EQUILIBRIUM

Zenon Ryszard Szczepanik, Piotr Miller

The Technology and Quality of Production Department
School of Planning and Statistics
Warsaw, Poland

The thermodynamic model of N/Ch/-I phase equilibrium for binary systems: liquid crystalline substance (LC) - nonmesomorphic component (L) was formulated. The model was based on Porter's equations applied separately for liquid phase and liquid crystalline phase. The N/Ch/-I transition curves in the systems: p-azoxyanisole (LC) - abietic acid, its esters and other chosen substances with different chemical constitution (L) were empirically determined. Porter's constants values obtained for investigated systems were correlated with the specific entropy of the substances L. The correlated values were found to agree well with approximation function. The obtained results allow prediction of the transition curves N/Ch/-I.

1. INTRODUCTION

Binary phase systems containing a component with a thermotropic nematic phase and a non-liquid crystalline component were studied. Phase equilibrium in these systems has been reported [1,2, 3]. The structure of liquid crystalline phase in the analyzed systems is dependent on the chemical constitution of the non-liquid crystalline component. When other substances are added to nematic substances, nematic phases N or cholesteric phases can appear. A cholesteric structure appears when the substance added has optically active molecules. The nematic and cholesteric structures constitute, according to the accepted classification rule [4], a phase of the same type: nematic type.

The range of occurrence of the mesomorphic phase in the mentioned systems seems to be of a particular interest on cognitive and practical grounds. This range results mainly from the position of the curves of equilibrium between nematic type phases N/Ch/ and isotropic liquid I. In spite of many works [5,6] the prediction and programming of the position of the mentioned curves and hence also the prediction of the range of occurrence of the mesophase in the analyzed systems still constitutes an open problem not solved entirely. For this reason the investigations comprising experimental and theoretical parts were performed to form the basis for determination and prediction of the position of the N/Ch/-I curves of equilibrium in the analyzed systems.

2. PROGRAM OF INVESTIGATIONS

The investigations performed have included:
- formulation of the thermodynamic model of N/Ch/-1 phase equilibrium in the analyzed systems;
- adaptation of the thermodynamic model for chosen phase systems; liquid crystalline p-azoxyanisole (PAA) - nonmesomorphic component, performed on the basis of the author's own empirical work;
- formulation of the conception of the quantitative model which would permit prediction of the position of the curves of N/Ch/-I equilibrium in the systems with PAA.

Among the investigated nonmesomorphic substances optically active abietic acid esters were used, which, in the systems with PAA, formed the cholesteric structures not yet investigated. The method of the synthesis of these esters was elaborated elsewhere [7,8].

The basis of prediction and determination of the analyzed curves of N/Ch/-I equilibrium, resulting from the applied investigation program and resulting calculation and data processing system, can be presented in the form of a processing program according to the achievements of theory of the quantification of the quality of processes [9] as shown in Figure 1.

Phenomenological thermodynamics in its applications has rather limited possibilities of prediction of phase equilibrium on the basis of the phenomenological parameters which characterize pure components of the system. In order to solve this problem it makes use of definite formal methods including application of the activity system. In order to determine the position of the curves of N/Ch/-I equilibrium it is necessary to know the activity coefficients of the components in the phases in the equilibrium state. It is possible then to make use of the standard thermodynamic formulas and of the thermodynamic model formulated on the basis of these formulas (process II). The thermodynamic model used in the investigations was based on Porter's equations applied separately for liquid phase and liquid crystalline phase. The application of this model is possible when Porter's constants values for liquid phase and for liquid crystalline phase are known.

---

* N/Ch/ - nematic phase type (N-nematic phase or Ch-cholesteric phase); I - isotropic liquid state.

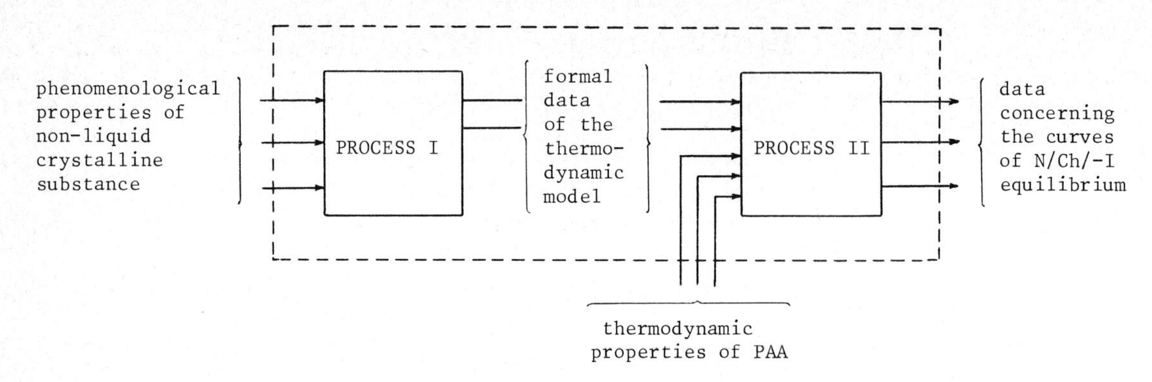

Figure 1: Logical system of data processing in the model of N/Ch/-I phase equilibrium

Process I: process of data processing on the basis of the quantitative model which determines formal parameters of the thermodynamic model.
Process II: process of data processing through the thermodynamic model of the phase system.

The identification of the parameters which constitute an activity system consists often in their empirical determination in the specific systems. This is the basis for some limited extrapolations made within the framework of the homologous series. The parameters of the activity system can also be determined by modelling the molecular structure of the analyzed systems. The theoretical models based on the statistical-mechanical theories of mixtures verify the real state to a limited degree only. An application of the mechanical-statistical methods is connected with many inconveniences resulting from limited knowledge of the values of the molecular parameters concerning their shape, intermolecular forces, etc. It results from the above that application of the activity system constitutes a formalism resulting from the fact that the laws of matter, of which the structure level is characteristic of the considerations taken within the framework of phenomenological thermodynamics, are unknown. A knowledge of the mentioned laws could permit connection of the definite phenomenological parameters of the components of the system in equilibrium states and hence prediction of the position of the curves of equilibrium on the basis of the phenomenological parameters of the pure components of the system. Taking into consideration a conception of dependence of the parameters which constitute the activity system, on the specific entropy of the used nonmesomorphic substances a concept was formulated. This conception was the consequence of the theoretical considerations which were verified by the experimental tests. An investigation of the systems in which PAA is a liquid crystalline component has enabled us to calculate, on the basis of the obtained empirical values, the Porter's constants which were next correlated with the specific entropy of the nonmesomorphic substances. The correlated values were found to agree well with the approximation function which constitutes a quantitative model for data processing (process I).

The detailed results of investigations presenting an argumentation for the thermodynamic model, the calculations and the method of use of the obtained mathematical formulas will be reported in other works. The presented conception induces the continuation of the investigations concerning verification and generalization of the obtained results.

3. REFERENCES

[1] Dave, J.S., Vasanth, K.L., Influence of molecular structure on liquid crystalline properties and phase transitions in mixed liquid crystals, Mol. Cryst. Liq. Cryst. 2 (1966) 125-133.

[2] Wóycicki, W., Stecki, J., Thermodynamics of solutions of liquid crystals.I. Enthalpy of mixing and solid-liquid equilibrium in mixtures of MBBA and some benzenes and naphthenes, Bull. Acad. Sci. Polon., Ser. Sci. chim. 22 (1974) 241-244.

[3] Zhdanov, S.I., Zhidkie kristally, Izd. Chimia, Moscow (1979).

[4] Demus, D., Richter, L., Textures of liquid crystals, Verlag Chemie, Weinheim, New York (1978).

[5] Peterson, H.T., Martire, D.E., Thermodynamics of solutions with liquid crystal solvents. VIII. Solute induced nematic-isotropic transitions, Mol. Cryst. Liq. Cryst. 25 (1974).

[6] Martire, D.E., Oweimreen, G.A., Agren, G.I., Ryan, S.G. and Peterson, H.T., The effect of quasispherical solutes on the nematic transition in liquid crystals, J. Chem. Phys. 64 (1976) 1456-1463.

[7] Miller, P., Phase systems liquid-liquid crystalline phase-solid state of the abietic acid and its derivatives with chosen liquid crystalline substances, doctoral thesis, Chemical Industry Institute, Warsaw (1981)

[8] Szczepanik, Z.R., Miller, P., Process of obtaining abietic acid esters with a programmed quality, Papers from III International Symposium on cooperation areas in commodity science and technology in view of human needs, Part II, Cracow (Sept. 1981).

[9] Szczepanik, Z.R., Cause and effect system in quantifying and control of complex technological manufacturing and exploitation processes, Zeszyty Naukowe SGPiS 93 (1973) 13-66.

Data for Science and Technology, P.S. Glaeser (ed.)
North-Holland Publishing Company

THERMODYNAMIC ANALYSIS AND OPTIMIZATION OF TITANIUM CARBIDE SYNTHESIS

S.A. Badrak, I.M. Barishevskaya, V.L. Briskin, T.Ya. Kosolapova
V.F. Perchik, E.V. Prilutsky, A.I. Rosenfeld, V.P. Turov

Institute for Problems of Materials Science, Kiev, U.S.S.R.

The results of thermodynamic calculations of chemical equilibria in the Ti-O-C system simulating titanium carbide synthesis involving carbothermal reduction of titanium dioxide are introduced. The possibility of titanium oxycarbide formation by the ideal solid solution model is studied. The dependence of the oxygen content in titanium carbide on temperature and pressure is calculated. This paper concludes that the two-stage method of titanium carbide production is optimal.

1. INTRODUCTION

The intense interest in titanium carbide powder production, connected with the severe and ever growing shortage of tungsten for the needs of the hard metals industry, gives rise to a considerable number of studies on methods of its production and that of powder. However, at the present time there is neither complete clarity on the question of the mechanism of titanium carbide formation during its reduction from oxide nor sufficiently effective technology, making the solution of the task of the mass production of titanium carbide powder of high quality possible. On those grounds the questions of titanium carbide production are of great theoretical and practical interest.

The purpose of our investigation is to make thermodynamic analysis and substantiate the possibilities of efficient production of titanium carbide powder, Tic, of high quality during carbothermal reduction. The system Ti-O-C was chosen as an object of investigation since the preliminary calculations showed that the influence of a hydrocarbon gas phase on titanium carbide formation is insignificant.

The thermodynamic analysis of chemical equilibrium in the TiO_2+3C system at different temperature and pressure values was made with the help of the ASTRA data bank. While making this experiment the equilibrium concentrations of gaseous and condensed products of reaction under given conditions were obtained.

The elements Ti, O, C, and their compounds [O_2, O_3, CO, CO_2, C_2O, C_2, C_3, C_3O_2, Ti_2, TiO, TiO_2 (gaseous), TiO, TiO_2, Ti_2O3, Ti_3O_5, Ti_4O_7, TiC (condensed)] were considered to be components of the system investigated.

2. BACKGROUND

The first industrial method for titanium carbide powder production developed as early as 1913 has been only slightly improved until now as the main process for production of titanium carbide powder, TiC. It consists in heating a titanium dioxide stoichiometric mixture in a graphite tube furnace in a pure hydrogen flow in the temperature range between 2000 and 2400 K.

One should refer to the merits of this method the relative simplicity of both the technology and the equipment used, and the cheapness of source materials. Its faults are connected mainly with the necessity of prolonged grinding and the classification of the powders produced. Moreover the purely "technological" drawback of this method is the high sensitivity of the quality of the powder produced to the deviation of the parameters of the process from the optimal ones.

It is possible that an unsintered or a slightly sintered final product may be formed during carbothermal reduction reaction under a reduced pressure condition. This makes it possible to produce high quality powder with a high degree of dispersion. The investigations which were carried out by G.A. Meerson and his colleagues, showed that one can practically completely remove oxygen from the TiO_2+3C system at the temperature of 1600 K, if one maintains the pressure within 10^{-5} - 10^{-6} MPa. However the vacuum method of titanium carbide production, which became known about half a century ago, has not yet become widely used as a consequence of its exceptionally low productivity.

3. INVESTIGATION OF THE CARBOTHERMAL REDUCTION OF TITANIUM DIOXIDE

The calculated values of equilibrium partial pressure of monatomic and molecular oxygen at different pressures in the analysed system are of greatest interest from the point of view of the investigation of the mechanism of the reaction. This interest is concerned with the possibility of comparing the values obtained with the common values of partial pressure of monatomic and nuclear oxygen over titanium oxides at different temperatures.

It appears from the present calculations that the equilibrium partial pressure of monatomic and molecular oxygen in the TiO_2+3C system over the entire investigated temperature range are several orders lower than the partial pressure of oxygen over titanium dioxide in the same temperature range. This gives a basis to the claim that the interaction between titanium dioxide and carbon proceeds according to the dissociative mechanism. In this case the role of carbon involves only binding of the "extra" oxygen and shifting of the dissociative reaction of oxides to the right. It is most probable that the transformation of titanium dioxide into titanium carbide proceeds stepwise

$$TiO_2 \rightarrow Ti_4O_7 \rightarrow Ti_3O_5 \rightarrow Ti_2O_3 \rightarrow TiO \rightarrow TiC_xO_y \rightarrow TiC$$

It is natural that every pair of oxides has a certain value of equilibrium partial pressure of oxygen (monatomic and molecular) and these very values characterize this or that depth of reduction according to the value of equilibrium concentration of oxygen. The calculated data on the partial pressure of oxygen over various titanium oxides are given in Figure 1.

The equilibrium concentration of oxygen in the TiO_2+3C system under different pressures intersects the graph (plot) as a function of partial oxygen pressure in the region which is close to the composition of oxide Ti_3O_5 at the temperature of 1300 K. On the basis of this one can state that under these conditions (T=1300 K, P= 0.1 MPa) interaction between titanium dioxide and carbon can develop only as far as forming oxide Ti_3O_5. For further development of the interaction process in this system there are two factors - the pressure and temperature in the system. Figure 1 shows that in this system reducing the pressure to 10^{-4} MPa leads to development of the interaction approximately to the composition $TiC_{0.5}O_{0.5}$. Under a pressure of 10^{-4} MPa in the TiO_2+3C system equilibrium the oxygen pressure is two orders lower than its partial pressure any oxygen-containing titanium compound. At this the reduction will take place until titanium carbide is formed. Alternatively increasing the temperature to 1600 K leads to such a situation when the equilibrium pressure (already at 0.1 MPa) is about half an order lower than the partial oxygen pressure over any oxygen-containing titanium compound.

One can observe an analogous situation while examining dependencies of partial and equilibrium pressures of molecular oxygen. For the sake of convenience of observing and owing to the fact that the prevailing concentration of monatomic oxygen is 3-4 orders higher, Figure 1 gives the

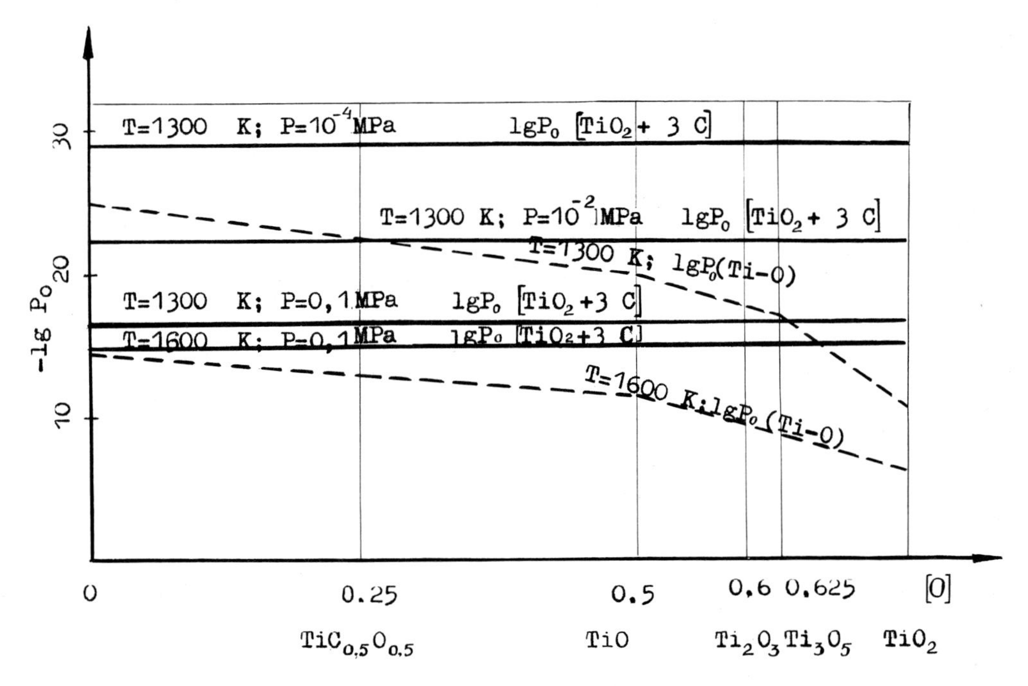

Figure 1: Equilibrium pressures of oxygen in the Ti-O and TiO_2+3C systems

data only for the oxygen atoms. The results of the calculations confirm the dissociative mechanism of interaction between titanium dioxide and carbon.

4. ACCOUNT OF TITANIUM OXYCARBIDE FORMATION

It is known that the formation of titanium oxycarbide is an inevitable stage of the production of titanium carbide from titanium dioxide. Furthermore the thermodynamic analysis of the given process cannot be considered to be correct without taking titanium oxycarbide into account. Unfortunately the thermodynamic data of titanium oxycarbide do not exist. We accepted the thermodynamic model of titanium oxycarbide as an ideal solid solution of titanium monoxide in titanium carbide. The available data of titanium oxycarbide stability and degree of order of its structure testify to the prevailing role of mixing entropy during oxycarbide formation, which confirms the correctness of using an ideal solid solution model. When this approach is adopted, the criterion of the final product quality is its oxygen content, which must not exceed the prescribed value.

Within the limits of the model accepted, chemical equilibrium in the Ti-O-C system was calculated with stoichiometric blend composition of $TiO_2 + 3C$ and also with excess and limited carbon over the temperature range between 1200 and 2200 K and the pressure $10^{-1} \div 10^{-4}$ MPa. Figure 2 shows the calculated dependence of the oxygen content of titanium oxycarbide at different temperatures and pressures. As the figure shows, under the pressure of 0.1 MPa the oxygen content in oxycarbide is considerably reduced while heating from 1200 K to 1800 K. At higher temperatures the equilibrium oxygen concentration in the product varies insignificantly. Moreover, as the calculations showed, even the presence of a considerable amount of of excess carbon in the blend does not change the character of dependence of equilibrium oxygen concentration in oxycarbide upon the temperature.

5. CONCLUSIONS

The calculated values show that carrying out the process at temperatures above 1800 K with the aim of refining the product from oxygen is effective only to a slight degree. Besides that, a high temperature process is not desirable from the technological point of view because in real conditions it leads to a high degree of sintering of the final product and to the necessity of its subsequent grinding.

At the same time, as it is seen from curves 2 and 3, for low pressures the equilibrium oxygen concentration in oxycarbide reaches a satisfactory value at temperatures not exceeding 1800 K.

The given results prove that although titanium carbide production under the pressure of 0.1 MPa is characterized by high efficiency, it is un-

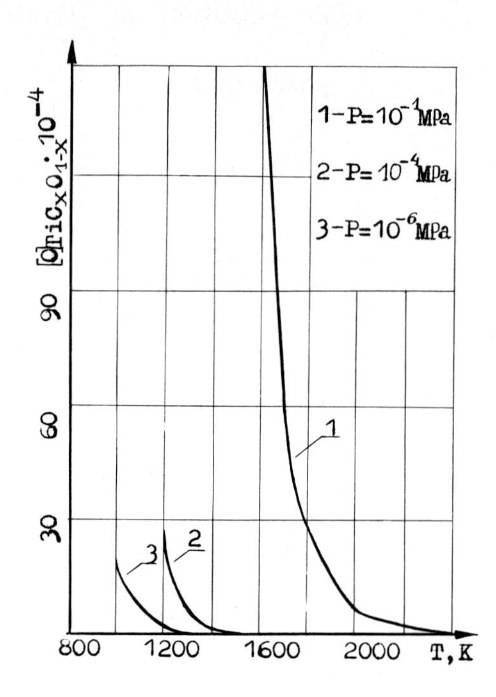

Figure 2: Calculated equilibrium concentration of oxygen in oxicarbide

satisfactory from the point of view of the quality of the product obtained. Moreover, usage of the vacuum technology, enabling production of high quality products, is marked by insufficient efficiency. In this connection the application of a two-stage method of titanium carbide production is considered expedient. The first stage consists in heating the stoichiometric blend to the temperature 1800 K under normal pressure. After this - during the second stage the refinement of the received semifinished item - this form is made by means of its exposure to the same temperature and a reduced pressure.

The proposed method makes it possible to combine the high efficiency of the first stage of the process with the requirement of the relevant purity of the product.

Data for Science and Technology, P.S. Glaeser (ed.)
North-Holland Publishing Company

SYSTEM OF EXPERIMENTALLY BASED EQUATIONS FOR THE CALCULATION OF THERMODYNAMIC PROPERTIES OF FLUIDS

V.V. Sytchev, A.A. Vasserman, A.D. Kozlov, G.A. Spiridonov and V.A. Tsymarniy

Soviet National CODATA Committee
Academy of Sciences of the U.S.S.R.

A system of experimentally based equations for a number of technically important fluids has been developed permitting the calculation of thermodynamic properties in the single phase region and phase equilibrium lines (melting and boiling lines). The equations were used for compilation of thermodynamic tables for nitrogen, air, oxygen, ethylene, methane, and ethane.

The system of equations is as follows:

$$Z = 1 + \sum_{i=1}^{r} \sum_{j=0}^{S_i} b_{ij}\omega^i/\tau^j \quad (1)$$

$$C_p^0 = \sum_{j=0}^{m} \alpha_j \theta^j + \sum_{j=1}^{n} \beta_j \theta^{-j} \quad (2)$$

$$\pi_s = \sum_{j=0}^{m} a_{Sj} \tau_s^{-j} \quad (3)$$

$$\pi_\lambda = \sum_{j=0}^{m} a_{\lambda j} \tau_\lambda^{j} \quad (4)$$

Where: (1) is the thermal equation of state; (2), (3), (4), are approximations for: isobaric heat capacity in ideal gas state, saturated vapour pressures and the pressure along the solidification line respectively; $Z = P/\rho RT$ - compressibility; $\omega = \rho/\rho_{cr}$ and $\tau = T/T_{cr}$ - reduced density and temperature; $\pi_s = P_s/P_{cr}$ and $\tau_s = T_s/T_{cr}$ - reduced pressure and temperature along the saturation line, $\pi_\lambda = P_\lambda/P_{cr}$ and $\tau_\lambda = T_\lambda/T_{cr}$ - reduced pressure and temperature along the solidification line; $\theta = T/100$; R = specific gas constant. The empirical coefficients b_{ij}, a_{Sj} and $a_{\lambda j}$ were determined by statistical processing of experimental data. The coefficients $\{\alpha_j\}$ and $\{\beta_j\}$ were determined by the least squares method from the tabulated data, obtained by spectroscopic analysis.

For developing the thermal equation of state (1) the method of equivalent equations was used. This method was elaborated earlier and can be found in [1,2]. Essentially it consists in averaging the equation of state based on the system of equivalent equations. The system of equivalent equations allows tracing of the influence of disturbing factors on the dispersion of the thermodynamic properties to be determined and thus to evaluate the tolerance for any thermodynamic function at fixed values for pressure and temperature.

While building the equation of state for a diversified group of technically important substances the problem of modelling the system of equivalent equations has been studied in great detail. Various algorithms of generating equations of the system have been considered (algorithm for varying number of coefficients, weighting function algorithm, algorithm depending on the type and number of the experimental data, etc.). The method proved especially efficient when compiling thermodynamic tables on the basis of compressibility measurements, since it provides a definite answer if the available compressibility data are sufficient to obtain the calculated values of thermodynamic quantities with the required accuracy, or whether additional information on caloric and acoustical properties should be included in the processing procedure.

The expression for free energy which is a thermodynamic potential in volume - temperature (density - temperature) variables with regard to the equation of state (1) is as follows:

$$F = F_0(T,P_{rt}) + RT\ln(\omega/\omega_0) + RT\sum_{i=1}^{r}\sum_{j=0}^{S_i}\frac{1}{i}b_{ij}\omega^i/\tau^j \quad (5)$$

Where $F_0(T,P_{rt})$ is Gibbs energy in ideal gas state, depending on the temperature of a standard isobar at P_{st} = 1.01325 bar; $\omega_0 = P_{st}/\rho_{cr}RT$. It can be easily demonstrated that any thermodynamic function can be presented as a combination of Gibbs energy F and its temperature and volumetric partial derivatives of the following type: $\partial F/\partial T$, $\partial F/\partial v$, $\partial^2F/\partial T^2$, $\partial^2F/\partial v^2$, $\partial^2F/\partial v\partial T$. In other words:

$$\chi = \chi(F, V, T, \partial F/\partial T, \partial F/\partial v, \partial^2F/\partial T^2, \partial^2F/\partial v^2, \partial^2F/\partial v\partial T) \quad (6)$$

$$\chi \in \{P, Z, h, S, C_v, C_p, W, k, \alpha, \beta, \gamma ...\} \quad (7)$$

The symbols in brackets in (7) are commonly known.

The relationships for enthalpy and the entropy in the ideal gas state with the stated approximations can be given as:
for enthalpy

$$h_0 = \int_{T_0}^{T} C_p^0 dT + h_{00} + h_0^0 \quad (8)$$

where h_{00} is enthalpy at temperature T_0 (in our case T = 100 K); h_0^0 is sublimation heat at T = O K.
for entropy

$$S_0 = \int_{T_0}^{T} C_p^0/TdT + S_{00} + S_0^0 \quad (9)$$

where S_{00} is entropy at temperature T_0; S_0^0 is a constant reading (in our case $S_0^0 = 0$).

The system of equations is available for nitrogen, air, oxygen, methane, ethylene, ethane and helium-4. At present the work on the system of equations for a number of other technically important substances is in progress. It is noteworthy that the thermodynamic equation of state in all cases except air permits the analysis of properties along the boiling and condensation lines based on Maxwell's rule. In this respect the equation for saturated vapour pressure (3) is necessary as the initial estimate in the iteration process.

Below a brief characteristic of the system of equations and thermodynamic tables developed for a number of technically important substances is given.

Nitrogen [3] . Experimental PVT data published in scientific literature before 1975 have been used while building the thermodynamic equation of state. The initial data array contained 2134 experimental data in the single phase region and 101 experimental data in phase equilibrium lines. Processing mistakes were discovered in some data arrays forming the principal data array. From then on corresponding data were assigned zero weight. A system consisting of 54 statistically equivalent equations was obtained. Standard deviations of experimental values of density from calculated ones computed for different equations of state are stable enough to testify to the statistical equivalence of equations of a system. The final equation of state obtained by averaging the system from 54 equivalent equations describes a data array of 1996 most reliable experimental values of density with the root-mean-square deviation $\sigma_\rho = 0.17\%$. The averaged equation of state satisfies Maxwell's rule with the root-mean-square deviation for the saturation pressure $\sigma_\beta = 0.11\%$. For the approximation (2) Hilsenrath's tabulated data (3) were used. Based on the resultant system of the experimentally based equations, tables of thermodynamic functions in a single phase region at $65 < T/K < 1500$, $P \leqslant 100$ MPa, as well as along the melting, boiling, and condensation lines were made. The list of tabulated properties includes density, compressibility, enthalpy, entropy, isochoric heat capacity, isobaric heat capacity, sound velocity and a number of other properties. Detailed information on the system of initial equations and thermodynamic tables of nitrogen is available in Prof. Sytchev et al.'s monograph [3].

Air [4]. When building the thermal equation of state, experimental PVT data published before 1977 were used. The initial data array contained 1089 experimental points in the single phase regions and 80 experimental points in the phase equilibrium lines. Wrong points were discovered in the course of preliminary analysis and processing. Later they were assigned zero weight. A system of 53 statistically equivalent equations was obtained. The final equations of state obtained by averaging the system of 53 equations give an array of 1169 most reliable experimental values of density with the root-mean-square deviation $\sigma_\rho = 0.11\%$. The calculations were performed for air containing 78.11% N_2, 20.29% O_2, and 0.93% Ar in volume. The tabulated values of isobaric heat capacity in the ideal gas state were computed for air of a particular composition using Hilsenrath's data [4] for separate components. Based on the obtained system of experimentally based equations the tables of thermodynamic functions in the single phase region at $70 < T/K < 1500$ and with pressure up to 100 MPa as well as along the boiling and condensation lines were compiled. The list of tabulated properties is similar to that for nitrogen. Detailed information on the system of equations and thermodynamic tables of air can be found in a monograph by Prof. V.V. Sytchev et al. [4].

Oxygen [7]. Besides experimental PVT data the values of the second and third virial coefficients, isochoric heat capacity, information of the derivatives $\partial P/\partial T$ and $\partial P/\partial V$ were used to get the thermal equation of state. The initial data array contained 2311 experimental data of density, 25 data for the second and third virial coefficients respectively, 148 data of isochoric heat capacity. 11 sets of equations were compiled. Corresponding sets differ in values of weighting function and sets of data which we have taken into account. Within a set the difference of the equations is conditioned by the number of empirical constants. All in all the system of equivalent equations contained 159 equations of state. The final equation averaged along the entire equivalent system gives an array of 1842 experimental data for density with the root-mean-square deviation $\sigma_\rho = 0.10\%$, an array of 137 experimental data for isochoric heat capacity with the root-mean-square deviation of $\sigma_{C_V} = 2\%$. Tabulated data obtained by Gurvich et al. [7] were used in building the approximation (2). Based on the resultant system of equations tables of thermodynamic properties of oxygen in the single phase region at $55 < T/K < 1500$ and with pressure of up to 100 MPa as well as along the melting, boiling and condensation lines were computed. The list of tabulated properties is similar to that for nitrogen. Detailed information on the systems of equations and thermodynamic tables of oxygen is available in a monograph by V.V. Sytchev et al. [7] published in 1981.

Methane [5]. 2303 experimental data for density, 247 data for isochoric heat capacity and 58 data on the second and third virial coefficients were used to get the thermal equation of state. As a result of computations a system consisting of 64 statistically equivalent equations was obtained. The averaged equation of state gives an array of 2206 experimental data for density with a root-mean-square deviation σ_ρ = 0.14%, an array of 347 experimental values of isochoric heat capacity with a root-mean-square deviation σ_{C_V} = 3%. The data obtained by McDuwell and Kruse [5] were used to get the approximation for $C_p^0/R = f(T)$. Based on the resultant system of experimentally based equations tables of thermodynamic properties of methane for the temperature range of 100 up to 1000 K with pressure of up to 100 MPa including melting, boiling, and condensation lines were compiled. The list of tabulated properties is similar to that for nitrogen. Detailed information on the system of equations and thermodynamic tables of methane is available in a monograph by Prof. V.V. Sytchev et al. [5] published in 1979.

Ethylene [6]. Experimental PVT data published before 1978 were used to get the thermal equation of state. Data on the second virial coefficients and the derivatives $\partial P/\partial V$ and $\partial P/\partial T$ were additionally taken into account. The initial experimental data array consisted of 2608 data of density, 26 data on the second virial coefficients, 54 data on the derivative $\partial P/\partial V$ and 517 data of the derivative $\partial P/\partial T$. A system of 91 statistically equivalent equations was compiled. The averaged equation of state gives an array of 2262 experimental data of density with root-mean-square deviation σ_ρ = 0.12%. Computed data of L.V. Gurvich et al. [6] were used to obtain the approximation. Based on the obtained system of experimentally based equations tables for thermodynamic properties of ethylene were compiled for the temperature range from 110 up to 600 K with the pressure of up to 300 MPa including the melting, boiling, and condensation lines. Detailed information on the system of equations and thermodynamic tables of ethylene is available in a monograph by Prof. V.V. Sytchev et al. [6] published in 1981.

At present similar calculations for ethane and helium-4 have been completed. The calculation results will soon be published as monographs by the Publishing House of Standards.

REFERENCES

[1] Spiridonov, G.A., Kozlov, A.D., Sytchev, V.V., Compilation and evaluation of thermophysical data on properties of fluids by computer numerical experiment. The Proceedings of the Fifth Biennial International CODATA Conference. Pergamon Press, Oxford and New York, 1977.

[2] Sytchev, V.V. , Spiridonov, G.A., Determination of thermodynamic functions of gases with experimental PVT data using computer mathematic modelling. Proceedings of the U.S.S.R. Academy of Sciences, vol. 211, N 4, pp. 808-811.

[3] Sytchev, V.V., et al. Thermodyanmic properties of nitrogen. M., Publishing House of Standards, 1977, p. 352.

[4] Sytchev, V.V., et al. Thermodynamic properties of Air, Publishing House of Standards, 1978, p. 276.

[5] Sytchev, V.V., et al. Thermodynamic properties of methane, M., Publishing House of Standards, 1979, p. 351.

[6] Sytchev, V.V., et al. Thermodynamic properties of ethylene, Publishing House of Standards, 1981, p. 280.

[7] Sytchev, V.V., et al. Thermodynamic properties of oxygen.,M., Publishing House of Standards, 1981, p. 304.

Data for Science and Technology, P.S. Glaeser (ed.)
North-Holland Publishing Company

THE ACCESSIBILITY OF NUTRITIONAL DATA – NECESSITY AND REALIZATION

Harald Haendler

Documentation Centre of Hohenheim University
Stuttgart, Federal Republic of Germany

The increasing world population and the naturally limited food production demand an economic utilization of the natural resources. This requires reliable nutritional data. Numerical data of this kind are scattered around the world. To make use of them, they must be collected, evaluated, collated and generated in suitable form. Activities of feed data documentation and the efforts undertaken by the International Network of Feed Information Centres are an example how this can be realized. Such activities should be extended to the whole field of nutrition and endorsed by responsible organizations and governments to prevent mankind from hunger and malnutrition.

1. INTRODUCTION

Some well known and simple facts may be mentioned at the beginning in order to introduce the topic:

(1) The first and really vital requirement human beings have before they can do anything - for example to work, to think, to love, to be creative in any way - is to eat, i.e. to provide their bodies with the nutrients necessary for life and performance.

(2) The world population is increasing in a logarithmic progression. Hence, the demand for food increases correspondingly.

(3) The part of the earth surface which can be used for plant growing is limited. Thus, plant material, the original and natural resource for food, cannot be extended arbitrarily.

The consequences of these facts are quite clear: there will not be enough food for humans in the future. The time when food production cannot cover the demand any longer may be a matter of speculation. But it is clear enough that this point is not very far from today. And it is obvious that a considerable proportion of the world population is already suffering from hunger and malnutrition.

The question of how mankind could be prevented from this fate has worried thinking people for centuries. Nearly 200 years ago Thomas R. Malthus emphasized the need of birth control. On the other hand proposals have been made on how to overwhelm the natural limitation of food production. But experience shows: that neither of these attempts can be realized to the necessary extent.

This situation can be regarded as the greatest challenge to mankind since its beginning. Therefore it is not only a subject of scientific research but also a cause of political conflicts. This obliges us to study very thoroughly the possibilities of solving the problem.

Here the question arises, are birth control and reckless increase of food production - both political matters in themselves - the only ways out of the dilemma? Perhaps the political tinder which lies in these matters has overshadowed the simple question, whether all possibilities have been exhausted to utilize the existing natural resources for food. A successful economical utilization of natural resources presupposes a very consequent utilization of all existing information on nutrition and related fields.

2. NUTRITIONAL DATA

The economic utilization of natural resources for food production has many aspects. These are on the one hand the different links of the nutritional chain: soil, plant, animal, food industry, man. On the other hand there are questions of nutrient content, availability and digestibility of nutrients, the requirements of nutrients in different physiological states. But also the avoidance of negative influences of food on health is an important aspect.

All these aspects have in common that reliable information is needed to optimize the nutritional conditions and the utilization of limited nutritional resources. Nutrition research has been developed to a very high level. It already passed the point in which the quantitative results extend in number and importance beyond the qualitative ones. In other words, the period, where the essential nutrients, for example special amino acids, minerals, vitamins, were discovered or recognized as essential is concluded. Recently the content of these and other substances in feeds and foods was analyzed, and the specific requirements under specific conditions of the different nutrients are known today. Thus, a lot of numerical data have been produced.

The complexity of nutritional aspects and the large number - some hundreds - of different essential nutrients provoked a division of labour among a multiplicity of research institutions and scientists in all parts of the world. Furthermore numerous laboratories are carrying out chemical analyses with foods or feeds for limited tasks for the purpose of control.

Thus, there is a large number of data, but they are scattered all over the world, sometimes hardly available because they are not published but only stored in institution files. The access to this reservoir of data is very difficult. Therefore it is necessary to acquire all relevant documents - published as well as unpublished ones - to scan them thoroughly, to extract, collect, and collate the data systematically, and process them appropriately.

3. ANIMAL NUTRITION

Before turning to questions of data documentation, a special field of nutrition should be treated separately.

The original nutritional resources, i.e. the organic matter of plants grown on the soil as far as available, are utilizable by humans only to a small proportion. More than half the available land - i.e. agrarian land in the broadest sense, depending on geological and climatic conditions - can only be used by animals especially by ruminants which convert bulky forage to qualified foods. This is the case for permanent grasslands, ranges, mountains, savannah, tundra and others. The existence of men in these areas depends on animals. An impressive example of this situation are the nomads who follow their livestock - cattle, sheep, reindeer - and this is the basis of their life. But also settled farmers in regions of permanent grasslands depend on their herds.

Furthermore, it should be borne in mind that even from crop plants only a small proportion is edible by humans. More than half the plant parts - wastes and by-products - can be converted to human food only by animals.

Finally, for the nutrition of man, i.e. for providing him with all essential nutrients, a certain proportion of the food must be of animal origin. Thus, under the aspect of the economic utilization of natural resources for human nutrition, the nutrition of animals plays a very important role. Furthermore, nutrition research can be done quite better with animals, and - of course - an optimal and economical nutrition can better be realized with animals than with humans.

As a consequence of these facts much more data about animal nutrition than about human nutrition have been produced, and in the past more efforts have been undertaken for collecting and collating feed data than food data.

4. FEED DATA DOCUMENTATION

After what was said before it is not surprising that systematic collection and collation of nutritional data were first realized within the field of animal nutrition. Following certain attempts in former times a consequent activity of feed data recording was started in 1949 in Germany. It may be noted that at this time we just had overcome a period of hunger and malnutrition, and people recognized the necessity to make efforts for improving the situation.

The main idea at that time was to collect the broadly scattered isolated data in archives - today we call it databank - and to generate them in such a manner that reliable new data for general purposes can be provided.

As often can be seen in pioneer situations the efforts, although made with very restricted means and simple methods but with great engagement, soon brought success: the first feed composition table - based on this data collection and officially accepted for the Federal Republic of Germany - was published in 1952, followed by several tables for special kinds of animals or special groups of nutrients like minerals, vitamins, amino acids, most of them in several editions. The systematic collection of data on feed analyses and physiological trials is continued nowadays at the Documentation Centre of the Hohenheim University near Stuttgart (FRG). The data store increases from year to year as to number and kind of data. This and the conditions of international cooperation demanded a thorough adaptation of methods of and tools for data processing. They were developed by taking into account experiences of several decades, new technical knowhow and recent scientific knowledge.

5. INTERNATIONAL COOPERATION

The use of carefully prepared feed composition tables in developed countries was undoubtedly an important progress in animal feeding, that means an economic progress of agriculture as well as of food production. The influence of such activities was well recognized by the Food and Agriculture Organization (FAO) of the UNO as they looked for solutions within the campaign against hunger in the world. FAO experts and consultants surveyed the activities in the field of feed data documentation. They found that only very few institutions have been active in this special field. Besides the named Hohenheim Centre, an institution of the Utah State University, Utah (USA), later called "International Feedstuffs Institute", may be mentioned here. But the FAO experts also learned that this kind of data collecting and processing is very troublesome and time consuming. Really succesful work can hardly be undertaken by single small institutions. This recognition demands sharing of work by international cooperation. Especially the utilization of collected and generated data for the benefit of the whole world cannot be realized without integrating all these efforts into a world-wide network.

A first meeting of representatives of certain institutions working in this field was organized by FAO in 1971. As a follow-up to this meeting an International Network of Feed Information Centres (INFIC) was formed.

During the first period the INFIC partners developed new methods and rules for cooperation and tried to adapt the individual methods to the new international ones. The immense efforts in this direction have been carried out in a more personal but very close exchange of opinions between the responsible persons. Only recently INFIC became more official, ratified a Constitution, fixed the structure and the steering body of the organization. Today INFIC comprises about twenty institutions as members.

The general aims of the Network have been defined in the Constitution as follows: "To contribute to more efficient animal production throughout the world by improving access to reliable information on the composition, nutritive value and practical use of feeds for animals".

In particular the Constitution ordains - among other items - that the network should promote the prerequisites for the collection, processing and dissemination of numerical data on chemical composition and nutritive value of feeds as well as for the exchange of those data between Member Centres. Membership is available to institutions or organizations which subscribe to the aims of INFIC and are actively engaged in the aspects of processing of information as separately defined for different types of members.

6. WORKS AND METHODS

Within the frame of this paper only a glance on the works and methods may be sufficient. To give a general idea it may be mentioned that the Hohenheim Databank on feeds comprises about 1 200 000 data units (each of them is a single and retrievable information), belonging to about 20 000 different types of feeds. The recorded different properties of these feeds concerning chemical composition and nutritive value are about 800.

These figures, which may be similar to those of other centres, may make understandable that special methods of recording, evaluating and processing of data are necessary to handle the material in an adequate manner. The participation of institutions of different parts of the world and from regions with different languages demands more: standardized methods and - as far as language is concerned - multilingual tools.

The preparation of such tools and methods was time consuming for the staff people of the pioneer centres. This is true because it was necessary to harmonize different ideas; and also it soon became obvious that only a very sophisticated system would meet all requirements of precise data handling.

As an example of what is meant here the development of the INFIC Feed Thesaurus may be mentioned. It is the tool for describing feeds independently of natural languages and idioms. Therefore the principle of conceptual analysis and denoting the conceptual constituents with standardized descriptors is applied. A facetted vocabulary is the main part of the Thesaurus, established partially in accordance with existing thesaurus guidelines |1|.

However, attention was paid to the fact that the function of a thesaurus for data documentation differs in some aspects from that of a thesaurus for literature documentation. This and other reflections have been treated more in detail within a separate publication |2|. Here just some notes should be given about the structure of the Thesaurus.

The conceptual analysis of feeds led consequently to six different kinds of characteristics, each of these forming a special facet:
(1) original material or origin
(2) parts of the material used as feed
(3) processes or treatments (the material has been subjected to)
(4) stage of maturity
(5) cutting or crop (for plants only)
(6) grade (quality).

Each of these facets offers descriptors for denoting the special characteristics. Within the Thesaurus descriptors for the same concept can be found in different languages. This was achieved by thorough semantic studies of the candidate terms for the vocabulary. Finally semantic equivalence has been obtained between corresponding descriptors.

The studies for and the work on the INFIC Feed Thesaurus consumed several man years and it must be admitted that it is not yet available as a printed volume but only in computer lists and other preliminary displays. But just the reflections on and the practised work for such a type of thesaurus for data documentation may be considered as an important step in developing retrieval-oriented databanks.

Using the INFIC Feed Thesaurus, about 20 000 feeds have been described and recorded within a publication of "International Feed Descriptions" |3|. This voluminous publication is an important handbook for identifying analyzed feeds.

Another basic tool is the Coding System for nutrients and feed values, also edited as a separate INFIC publication |4|. It is used for standardized recording of numerical feed data.

These tools and a set of adequate computer programs allow processing the data material in different ways. An important point thereby is, that the single unit can be retrieved or selected, in order to check it, compare it with others, use it for calculating averages, or generating new data. An example of generated data is the calculation of energy values on the basis of crude nutrient contents and digestibility, using special formulas.

For special purposes information material can be selected from the databank, generated, and compiled. This can be done for the publication of feed composition tables as well as for providing individual users with special data on request.

All these types of information are used to improve the feeding of animals and to better utilize available feeds. Thus, planning and management of animal production are influenced by using information of these kinds.

7. REGRESSION OR PROGRESSION IN THE FUTURE?

Doubtless the efforts carried out in the field of animal nutrition had a considerable effect. But it should not be ignored that efforts like these have not always the necessary endorsement of the responsible administrations and agencies. Activities as described here, are initiated by foreseeing scientists who mostly are convinced that such efforts are really necessary. Unfortunately they often have no official mandate for this task. This may be less important as long as the expenditure is kept in limits. But if data documentation should be practised effectively it soon becomes more expensive. The more the economic situation of public budgets is deteriorating, the more the administrations responsible for science and research hesitate to provide funds for those activities.

This is the point where it becomes a question of morality to use parts of the small budget to continue and extend efforts for the benefit of humans in other parts of the world. Since morality is just not an outstanding feature of public administrations, institutions active in this mission often suffer from shortage of funds. What single institutions have experienced in this aspect seems to apply still more to international networks. For a long time, without success, INFIC has been seeking financial means for a coordinating centre or secretariat. The system is functioning solely by voluntary efforts of idealistic persons working in member institutions. History shows that idealism can be a good force for starting something but for the continuation a concrete basis is necessary.

Bearing in mind the overall problem of a satisfying nutrition for all humans in this world - doubtless the basis of welfare, freedom and peace - it must be claimed that not only the existing activities can be consolidated but that a consequent utilization of all kinds of nutritional data including those for human nutrition, including those about the influence of nutrition on health and performance of humans, including those of nutrient requirements of men and animals under specific conditions, including those of preparing better diets, should take place.

All these tasks seem to form a large program, which causes high costs. However if not considered in isolation but compared with the benefit it can bring, the funds are worthy of being spent. It should be understandable that the financial efforts necessary for successful handling of nutritional data are a modest amount compared with the financial expenses for treating undernourished children in hospitals, for curing starving humans or nursing humans with irreversible damages caused by malnutrition. And it can also be said that - under an economic as well as a moral aspect - means for nutritional databanks are better invested than those for police actions against hungry people.

8. CONCLUSIONS

In order to ensure sufficient nutrition of mankind, better utilization of natural food resources is necessary. This demands the overall accessibility of all relevant and reliable nutritional data as produced by researchers or laboratory activities. Examples are given how such a data collection can be organized and used. More efforts are necessary in this direction. This can only be achieved, if governments, public administrations and agencies can be persuaded to provide the funds for these efforts. CODATA as an officially recognized international organization within the respectable ICSU family is appealed to to arouse worldwide public conscience so that activities essential for the survival of mankind can be secured.

REFERENCES

|1| Haendler, H. et al., INFIC Feed Thesaurus, a multilingual thesaurus for describing feeds for the databank of the International Network of Feed Information Centers, Publ.4 (Dokumentationsstelle der Universität Hohenheim, Stuttgart, FRG (in preparation))

|2| Haendler, H., The INFIC Feed Thesaurus. Proceedings, INFIC Workshop London 1981 (in press)

|3| Harris, L.E. et al., International feed descriptions, international feed names, and country feed names. INFIC Publ.5 (International Feedstuffs Institute, Utah State University, Logan, Utah, USA, 1980)

|4| Kearl, L.C. et al., International feed databank system, coding instructions and processing procedures. INFIC Publ.3 (International Feedstuffs Institute, Utah State University, Logan, Utah, USA, 1980)

Data for Science and Technology, P.S. Glaeser (ed.)
North-Holland Publishing Company

ADVANCEMENT OF A NATIONAL INFORMATION SYSTEM OF LABORATORY ORGANISMS (NISLO)

Hideaki Sugawara and Yoshio Tateno

Institute of Physical and Chemical Research
2-1 Hirosawa, Wako-shi, Saitama-ken 351, Japan

We have developed a National Information System of Laboratory Organisms (NISLO) which covers laboratory animals, microorganisms, plants, algae, animal tissue and cell cultures, and plant tissue and cell cultures. In this paper we outline the present status of the NISLO mainly focusing on laboratory animals, microorganisms and animal tissue and cell culture. The phase of the development has come to the point at which consideration has to be made concerning how the NISLO should be used by researchers in foreign countries as well as in Japan. The paper concludes that both international and domestic co-operation is very important to develop and disseminate a data bank like the NISLO.

1. INTRODUCTION

Life science is an interdisciplinary science which is built upon the foundation of biology, chemistry, physics, medicine, agriculture and engineering with a common idea that it elucidates the phenomena of life by testable explanations. This immediately implies that great amounts and various kinds of data are produced and used in the pursuit of life science. As is true for other sciences, any discussion may be futile or sometimes harmful to life science, if it is not based on experiments and thus on data.

Reliable data, once produced, are not changed unless they become obsolete, and can be used by anyone, provided that they are accessible. It is a common practice to use other's data from papers and books. This practice has undoubtedly promoted research in science, and life science is no exception. Particularly noteworthy is the citation of a book, Atlas of Protein Sequence and Structure compiled by M. O. Dayhoff. Since the publication of its first volume in 1969, this book has greatly contributed to advancement of many areas in life science. For instance, it may be quite natural to mention that evolutionary biology would not have advanced to the present phase, if Dayhoff's book had not appeared. Recent progress in life science has consequently changed the attitude of collecting sequence data from proteins to nucleic acids. Considerable amounts of nucleotide sequence data, several hundred thousand bases, have then been accumulated in the past several years. There is, however, a notable difference in collecting data between protein and nucleotide sequences. In the collection of nucleotide sequence data a computer plays such a powerful role that computer-aided sequence data bases have been established in the U.S.A. and in Europe.

To use a computer is not, of course, confined to management of the sequence data, but is applied to that of many other kinds. As a matter of fact, there has been a growing demand for such data that are not directly related with scientific outcomes yet quite helpful to researchers, like the accession information about strains of mice. Advantages of a computer-aided database over a data book are easily pointed out. First of all, it is much easier and faster to update the data in the former than in the latter. Second, arrangement of the data in the former can easily be modified if necessary, while that in the latter can hardly be. Third, accessing the data in the former is free from difficulties that are created when using the data in the latter. For example, comparison of two nucleotide sequences in the former can be carried out easily, but that in the latter should be quite cumbersome.

Recognizing a strong demand for a nationally unified system of information on laboratory organisms in Japan, the Institute of Physical and Chemical Research decided in 1976 to develop a National Information System of Laboratory Organisms (NISLO). Actually, the development of the NISLO has been implemented by the Life Science Research Information Section (LSRIS) in the Institute with the collaboration of six study groups which specialize in laboratory animals, microorganisms, plants, algae, animal tissue and cell culture and plant tissue and cell culture.

The LSRIS has employed various database techniques to establish the NISLO which is mainly concerned with accession information and characteristic data on laboratory organisms. The basic procedure in developing the NISLO is as follows: A study group first selects organisms for which data are to be processed. The selected organisms may be such a taxonomic group as genus, species, or a group used in a research project. The study group then chooses and standardizes data items to describe the accession information and characteristics of the organisms. Following this a coding system is devised to set up an input format for the collection of data.

The collected data are converted into a data file in the main computer of the Institute by a file-creating program which is in accordance with the coding system. Data in the file can be accessed by the use of information retrieval, report-generating and other application programs.

In this paper we report the development and the present status of the NISLO mainly for laboratory animals, microorganisms and animal tissue and cell culture. We shall also discuss feasible ways by which the NISLO is accessed internationally as well as domestically.

2.DEVELOPMENT AND PRESENT STATUS OF NISLO

2.1 Laboratory Animals

As reported in a previous paper (Tateno, Sugawara and Sakamoto, 1981) the LSRIS collected and edited the accession information of mice and rats which had been maintained in 96 laboratories of universities, national institutions and private institutions in Japan. The accession information contains about 500 mouse strains and 100 rat strains in the aggregate numbers. It was found that among the 96 laboratories the National Institute of Genetics, Institute of Medical Science of the University of Tokyo, Central Institute for Experimental Animals and Department of Agriculture of Nagoya University were representatives, each of which maintained at least 40 strains of mice and rats. The information has been filed in the computer in a format shown in Fig. 1. The information of any laboratory in the computer is thus easily accessible, and tells a researcher about which laboratory he should ask to obtain a strain of his interest.

The LSRIS has started collecting the characteristic data of those strains of mice and rats. The data include such items as (1) history of a strain, (2) typical genes to a strain, (3) preventive device against bacterial contamination; germfree, gnotobiote, specific-pathogen-free, or conventional, (4) breeding and maintenance conditions including litter size, weight, longevity, nutrition, temperature, humidity and lighting duration, (5) physiological and pathological traits, and (6) references to a strain. These data will be installed in the computer and serve a researcher in many ways, when he carries out an experiment using the animal.

One more activity the LSRIS is pursuing now in this field is to survey the number of animals used in research experiments, examinations, education, diagnosis and in vaccine production at laboratories all over the country in a period of a year. The survey started in 1957 and has been carried out once every five years. The main purpose of this survey is to grasp an ever-changing attitude to using animals including primates, mammals, amphibians, reptiles, birds, fish, insects and other invertebrates in the pursuit of life science, and to advise a proper organization when the establishment of a new laboratory animal or a new institution of animal preservation comes into consideration. The past four surveys were carried out by different working groups and the last two were supported by the Ministry of Education. However, none of the past results were installed in a computer,

		Name of strain	Synonym	Maintenance	Availability
1	1m	CF#1		c	e
2	2m	C3Hf		a	e
3	3m	C57BL/6		a	e
4	4m	C3H/He		a	e
5	5m	DBA/2		a	e
6	6m	ddOM/Rw		a	e
7	7m	NC		a	e
—	— —	— — —		—	—

Figure 1. Computer file for accession information of mice and rats.

This is a part of the file for Institute of Physical and Chemical Research. First column shows serial numbers and second those for mice and rats separately where a letter m stands for a mouse strain and r for a rat strain. An alphabet of Maintenance means; a) inbreeding, b) conjenic, c) outbreeding, d) mutant and e) chromosomal aberrant. An alphabet of Availability implies; a) to sell, b) to give surplus strains, c) to give a seed strain, d) not available and e) to give on condition.

though they were published in journals. This had made updating the results so difficult that a survey had to be carried out independently each time.

The LSRIS has taken over the survey from the fifth time, since it was considered that the LSRIS was more suitable to implement such a survey than the previous working groups. It has been a firm belief in the LSRIS that the results of the survey should be stored in the computer to make editing and updating them very easy. The survey is being conducted on about 800 laboratories in our country by using an input format shown in Fig. 2.

A	Name of institution		
B	Address		
C	Name of recorder		
D	Telephone number		
E	Common name of an animal	F	Scientific name of the animal
G	Name of the strain	H	Number of the animal used
I	Comments		

Figure 2. Input format used for the survey.

It is needless to say that the function of a databank is two-fold; one is to collect and edit data, and the other is to disseminate them to researchers. One way to provide data is to form a computer network between related institutions. This has become quite feasible, since a recent development of a microcomputer has raised its performance remarkably. A plan conceived now in the LSRIS is to build a microcomputer network between the LSRIS and some of the representative institutions mentioned above. An outline of the network is presented in Fig. 3. In this way, the passage of data between those institutions becomes simple and fast, and many things can be realized besides just accessing data. For instance, such a network enables one to monitor a stream of an animal strain from one institution to another, and to find the best way to obtain a strain of one's interest by analyzing the data thus accumulated.

2.2 Microorganisms

In collaboration with a study group of microorganisms (Chairman, Prof. T. Mitsuoka of the University of Tokyo) the LSRIS has been collecting characteristic data on *Lactobacillus*, *Enterobacteriaceae* and coryneform bacteria. The characteristic data on *Lactobacillus* have been coded following the RKC coding system (Rogosa, Krichevsky and Colwell, 1971), and been stored in the computer using a computer program, MICRO-IS developed by Krichevsky and his colleagues at the National Institutes of Health, U.S.A. The data cover 148 strains of *Lactobacillus* for 157 characteristics. A part of the computer file of those strains is shown in Fig. 4. The data on the other two taxa of microorganisms will soon be installed in the computer in a format similar to that of *Lacto-*

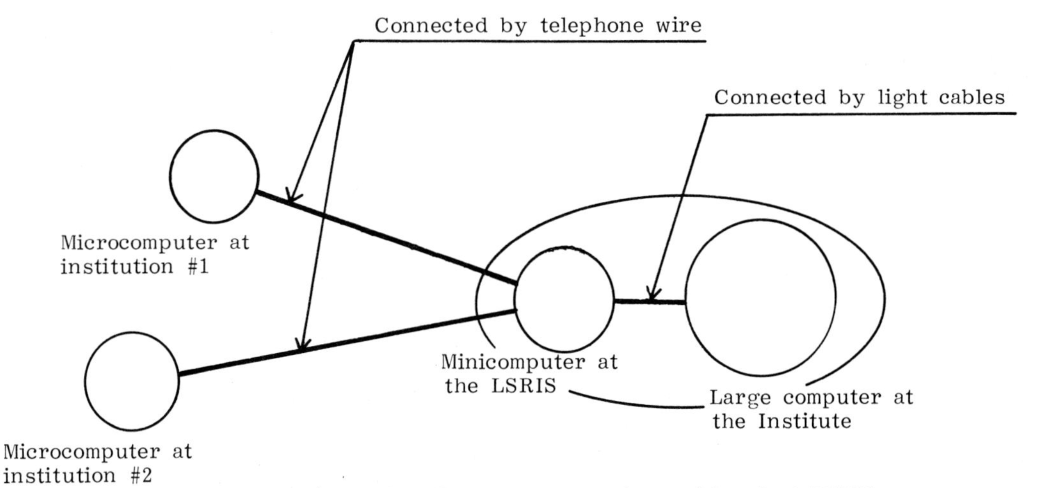

Figure 3. Computer network considered at LSRIS.

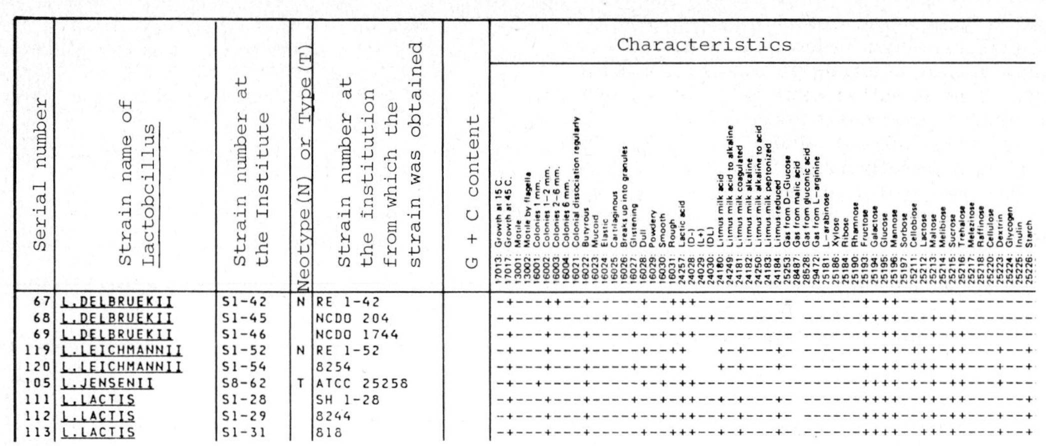

Serial number	Strain name of Lactobcillus	Strain number at the Institute	Neotype (N) or Type (T)	Strain number at the institution from which the strain was obtained	G + C content	Characteristics (17013: Growth at 15 C. / 17017: Growth at 45 C. / 13001: Motile / 13002: Motile by flagella / 16001: Colonies 1 mm. / 16002: Colonies 1–2 mm. / 16003: Colonies 2–6 mm. / 16004: Colonies 6 mm. / 16021: Colonial dissociation regularly / 16022: Butyrous / 16023: Mucoid / 16024: Elastic / 16025: Cartilaginous / 16026: Breaks up into granules / 16027: Glistening / 16028: Dull / 16029: Powdery / 16030: Smooth / 16031: Rough / 24257: Lactic acid / 24028: (D-) / 24029: (L+) / 24030: (DL) / 24180: Litmus milk acid / 24249: Litmus milk acid to alkaline / 24181: Litmus milk coagulated / 24182: Litmus milk alkaline / 24250: Litmus milk alkaline to acid / 24183: Litmus milk peptonized / 24184: Litmus reduced / 25253: Gas from D–Glucose / 28487: Gas from malic acid / 28528: Gas from gluconic acid / 29472: Gas from L–arginine / 25181: L–arabinose / 25186: Xylose / 25184: Ribose / 25190: Rhamnose / 25193: Fructose / 25194: Galactose / 25195: Glucose / 25196: Mannose / 25197: Sorbose / 25211: Cellobiose / 25212: Lactose / 25213: Maltose / 25214: Melibiose / 25215: Sucrose / 25216: Trehalose / 25217: Melezitose / 25218: Raffinose / 25220: Cellulose / 25223: Dextrin / 25224: Glycogen / 25225: Inulin / 25226: Starch)
67	L.DELBRUEKII	S1-42	N	RE 1-42		-+---++--+-----+--+++---------- ------+-++-----+--------
68	L.DELBRUEKII	S1-45		NCDO 204		-+---+-----+---+--++--+-------- ------++++---+-+--------
69	L.DELBRUEKII	S1-46		NCDO 1744		-+---+---+----+--+-++---------- ------+-++---+-++-------
119	L.LEICHMANNII	S1-52	N	RE 1-52		-+---+---+-----+--++ +-+---+- ------+-++-+++-++---+--+
120	L.LEICHMANNII	S1-54		8254		-+---+---+----+--+-+ +-+---+- ------+-++-+++-++---+--+
105	L.JENSENII	S8-62	T	ATCC 25258		-+--+----+-----+-+-++-----------------++++-+-+-++---+---
111	L.LACTIS	S1-28		SH 1-28		-+---+---+----+--+-++--+-+---+- ------++++--++-++------+
112	L.LACTIS	S1-29		8244		-+---+---+----+--+-++--+-+---+- ------++++--++-++---+--+
113	L.LACTIS	S1-31		818		-+---+---+-----+--+++--+-+---+- ------++++--++-+----+--+

Figure 4. A part of computer file for *Lactobacillus*.
Number in Characteristics is the RKC code.

bacillus. The characteristic data could be used for identification of an unknown strain and for acquiring a wider knowledge of a strain of one's interest.

The LSRIS is also in collaboration with the Japan Collection of Microorganisms (JCM), which was established in the Institute in 1980, to develop its data processing system. Since the JCM has been established as the central Japanese culture collection with the chief aim to collect, preserve and supply strains of microorganisms, it necessarily has a number of tasks which should better be processed by using a computer. Particularly, it is almost inevitable to employ computer-aided data processing for editing a catalogue of the strains, making a careful inventory of the strains and for general management of the JCM. This is because as the number of strains increases, it takes a greater amount of man-power and time to carry out a constant revision of the strain data and a daily checking of the inventory. Besides that, nowadays the computer output is good enough to be used directly for printing thanks to the advancement of the printing device. This greatly facilitates the publication of a catalogue. It should be noted that the characteristic data mentioned above are also quite helpful for preservation and supply of the strains in the JCM.

It is hoped that the first edition of the JCM catalogue will be published in April, 1983. The catalogue will include about 500 strains of bacteria and fungi. This number certainly increases as time goes on, and the final number will be 18,000. A tentative format of the catalogue is presented in Fig. 5.

Clostridium clostridiiforme (Burri and Ankersmit 1906) Kaneuchi, Watanabe, Terada, Benno and Mitsuoka 1976
(*Clostridium clostridiiforme* Approved list No.1 1980)

10002

--C. Kaneuchi SVI-68 (10) --ATCC 25537 -- L. Holdeman VPI 0316 (15) --M. P. Bryant T90 (28) From Calf rumen (18).
Type strain (55)
%GC 35 (11).
Medium 10, 37C, anaerobic.
(= ATCC 25537 = DSM 933 = CCM 6066)

Figure 5. Computer output of the format for JCM catalogue.
This is an example of *Clostridium*.

2.3 Animal Tissue and Cell Culture

Animal cell lines have recently become quite usable experimental materials in many areas of life science. This is rather expected, since research in life science progresses towards human welfare, and scientific results obtained based on the cell line are more directly related with this purpose than those obtained using prokaryotes. With increasing demand for the cell lines, more attention has been paid than ever before to establish a cell bank and a cell data bank in Japan. A cell bank here means to collect, preserve and supply the cell lines, while a cell data bank is to collect, maintain and supply accession information and characteristic data on the cell lines. It is desirable that the two banks work co-operatively.

For the first step to establishing a cell data bank, the LSRIS has developed an information system of the cell line. The program of this system is written in BASIC to be adopted in a microcomputer. There are two reasons for this; 1) performance of a microcomputer has greatly advanced and 2) microcomputers are now affordable to many researchers. The system is called the ANCLES (Animal Cell Line Enquiry System), and works interactively in installing, editing and retrieving the data on animal cell lines. The format of the computer file is shown in Fig. 6. It could be modified according to the grade of computer employed.

To explain the ANCLES, two examples shown on the display machine are presented in Fig. 7. Fig. 7a displays 'Job-selection', the first step to using this system. If a user wants to add data on a cell line to the computer file, then he keys in number 3 in the figure. This results in proceeding to the next step shown in Fig. 7b. This figure shows a serial record number (10, in this case) which is automatically assigned for a newly added cell line. The figure also shows that the newly added cell line is WI 38, and that the process paused at the place where the file identification is to be entered. If the file identification is entered, then the process goes to the next step where the name of the establisher is asked. In this way the process of data entry advances. If an asterisk is given at 'cell line', then this process ends, and shown is 'Job-selection' where the next execution is chosen.

If a user wants to retrieve the data, he gives a keyword of his interest. For example, if he feeds a keyword, T, for cell line, then he could obtain the information about such cell lines as 3T3 and 3T6. Editing of the data can also be executed easily. Now data on 2000 cell lines can be stored in a standard floppy disk in the format given in Fig. 6 due to advances in microcomputers.

The ANCLES can be applied to data management for fused cells like hybridomas which have been explosively used in many experiments in life science.

Item	Content	Record length
Cell line	Name of the line	20
File id.	File number	2
Establisher	Name of establisher of the line	20
Journal	Information about the journal in which the line is reported.	30
Volume		4
Page		4
Year		4
Curator	Information about the place where the line is preserved.	20
Institution		100
Address		
Telephone		18
Keyword(1) … Keyword(9)	Keywords	32 each

Figure 6. Computer file of ANCLES. Record length is presented in a unit of byte, length of a letter. Nine keywords whose length is at most 32 bytes each can be stored in this file.

3. CONCLUDING REMARKS

In the above we have mentioned the development of the NISLO focusing on laboratory animals, microorganisms and animal tissue and cell culture. It may be worthwhile here to refer briefly to the other three, plants, algae and plant tissue and cell culture. As to plants, working in close co-operation with the Institute of Agricultural and Biological Sciences at Okayama University, the LSRIS has developed an information system of barley, which deals with genetic information about this plant. The data on about 4000 strains of barley have been installed in the computer, and are now ready for genetical work like linkage study. This system can be applied to other plants such as wheat and rice. Another aspect in this field is to develop an information system by which trees native to Japan are identified. The system contains the information necessary to identify about 500 species. This is a result of co-operative work with Makino Herbarium at Tokyo Metropolitan University.

```
ANCLES-(Version 0.1)      TIME 00:00:30

       * job-selection *

list cell lines ------ 1

search & print ------- 2

add new cell lines --- 3

correct old data ----- 4

end of ANCLES -------- *

-----> select number ? ▮


9 record(s) in the file
```

a

```
ANCLES-(INPUT DESCRIPTION)

                              record no. 10
cell line (*)? WI 38

file id? ▮

establisher:

journal:

vol:          page:          year:

curator:

institution & address:


tel:
```

b

Figure 7. Computer displays of ANCLES.

a) This shows a user five jobs that can be executed in the ANCLES. When each job is finished, the system always returns to this display. The closed square indicates the present position of the cursor. The last line means that the data have been stored for nine cell lines. b) This is shown after giving number 3 in Fig. 7a. At the present working position, a sign ":" changes to a sign "?", and the file identification is in question in this case.

It has been considered that image data processing is needed to devise an information system for algae. Equipment necessary for the system have, however, not fully advanced to the phase in which they are affordable, The LSRIS seeks a feasibility to develop an image information system by the use of a video system. As to plant tissue and cell culture, a study group for this field (Chairman, Prof. M. Takeuchi of Saitama University) and the LSRIS have created a bibliographical data base. This system is called the IRIS (Information Retrieval Interactive System) and about 2000 papers on this subject are ready for retrieval at present. The number of papers, of course, increases annually. Readers interested in the IRIS may refer to Tateno, Nemoto and Sugawara (1982).

The NISLO contains not only published data but also the original information. It is, however, useless, if not disseminated widely. We have put our efforts mainly on developing the NISLO in the past several years. Now, we think about ways by which the NISLO could be used by researchers in overseas countries as well as in Japan. There may be the following three ways to realize the dissemination of a data bank like the NISLO.

(1) To record data and programs on magnetic tape or on floppy disk and send it to a concerned researcher.
(2) To form computer networks in a country, so that a researcher is able to access a data bank by the closest computer terminal. A computer network may include a large computer and microcomputers. The large computer stores all data and programs of a data bank and the data are accessed by using a microcomputer which works as a computer terminal. A researcher could contribute to enriching and correcting of the data in the large computer through the terminal.
(3) To establish an international computer network. The network may be formed between a large computer in a country and that in another. A researcher could access the data in a computer in another country through a terminal and a large computer in his country.

There is not much difficulty in employing the first way. EMBL (European Molecular Biology Laboratory) employs this method to provide researchers with the nucleotide sequence data. The LSRIS also provides the ANCLES on a floppy disk. Concerning the second way,there is an inter-university computer network in Japan. Most people in Japanese universities can access data and programs using the network. The tree identification system mentioned above has been installed in a computer of the network. The IRIS will also be installed in the computer. Besides that we are planning to form a computer network for building a data bank of laboratory animals, as mentioned earlier. Compared with these two ways, it is much harder to build an international computer network. It is not, however, impossible to form it, since now many commercially available databases are used internationally. Our final goal is to install the NISLO in an international computer network.

REFERENCES

Rogosa, M., Krichevsky, M. I. and Colwell, R. R., Methods for coding data on microbial strain for computers, Int. J. Sys. Bacteriol., 21, 1A-175A (1971).

Tateno, Y., Sugawara, H. and Sakamoto, N. Data contents and processing in a National Information System of Laboratory Organisms, In: Data for Science and Technology, 98-101, Glaeser, P. S. ed., Pergamon Press, Oxford, 1981.

Tateno, Y., Nemoto, Y. and Sugawara, H. Bibliographical data base for plant tissue and cell culture (IRIS), Proceedings of the Fifth International Conference of Plant Tissue and Cell Culture, (in press).

Data for Science and Technology, P.S. Glaeser (ed.)
North-Holland Publishing Company

INTERNATIONAL COLLABORATION IN DATA COMPILATION FOR ELEMENTARY PARTICLE PHYSICS

F.D. Gault

Particle Data Group, Department of Physics
University of Durham, DH1 3LE, U.K.

There is no international coordinating agency for data compilation, evaluation and dissemination in elementary particle physics. Historically, independent groups using different standards have covered data in their own areas of interest and not all data have been compiled. In the past five years groups in various countries and organizations have begun to exchange compiled data and to evolve standards. This paper looks at the mechanisms of such collaboration and considers the usefulness of international coordination.

1. INTRODUCTION

Elementary particle physics is 'big science'. It costs countries which participate large amounts of money and in this respect it resembles the nuclear and fusion research programmes. A significant difference is that nuclear and fusion research data are seen to have potential economic and strategic value and consequently they are compiled, evaluated and disseminated under the direction of international agencies (1). Compilation in particle physics proceeds on a more ad hoc basis and why this is so is considered in what follows.

The economic return to the society in elementary particle physics lies not in the data but in the knowledge gained from the high technology and computing techniques required to carry out the experiments and analyse the results. Students involved in the experiments are seen to 'receive the excellent experience of being exposed to a wide range of advanced technology and of working to realistic time scales and controlled budgetary planning in a competitive international environment' (2) and it is observed 'that the products of such training are attractive to employers in our more successful high technology industries'. (2)

What then happens to the data? The answer that they are published and preserved in learned journals is increasingly less credible as journals encourage the publication of graphical representations rather than tables of numerical data. Even where data are preserved on paper their accessibility is reduced because of the need to transfer them to a computer file before the data can be used for further research. Experimental groups might be expected to retain their own data in machine readable form but experimental collaborations are dynamic entities and by the time the data are emerging from the analysis which follows the taking of raw data in the experiment, the group is dissolving as members seek new collaborations. No one is responsible for preserving in perpetuity the data. The laboratory where data are taken might be expected to support an archive facility but few do. The arguments against such archives range from the philosophical, that only tapes of raw data are worth compiling, to the pragmatic, that in times of financial stringency resources are better used in producing new science than in preserving the old.

Data compilation is done. It divides into two categories, particle properties and reaction (or scattering) data. Particle properties include the best values for the various quantum numbers of the elementary particles, like mass, lifetime, spin, parity, etc., and their deduction requires the critical analysis of all of the experimental measurements reported in the literature. This work is the result of an international collaboration of individuals coordinated by the Particle Data Group (PDG) at the Lawrence Berkeley Laboratory. Reaction or scattering data include all of the measurements carried out as a result of scattering one particle off another such as cross sections, multiplicities, polarizations, etc. Not all of this information is compiled by any one group and with about 150 experiments completing a year and 1500 papers or preprints being published the volume of data is considerably greater than that required for particle properties. It is in this area that compilation and international collaboration are increasingly more important if the data are to be preserved.

2. THE PARTICLE DATA GROUP

For the past twenty-five years the Particle Data Group has developed and coordinated an international collaboration to compile, evaluate and report data on particle properties (3). The Review of Particle Properties (4) published every two years alternately in Physics Letters or Reviews of Modern Physics, and in condensed form as the Particle Properties Data Booklet, is regarded as the authoritative source of particle properties information. This work was begun by Rosenfeld and Barkas in 1957 to provide reliable particle properties information for the analysis of data from particle detectors and in this respect its motivation was similar to that in nuclear and fusion research where the best value is required for the continuation of the work. The PDG grew and attracted support from the National Science Foundation, the Department of Energy and the Office of Standard Reference Data, of the U.S. National Bureau of Standards. (The OSRD coordinates and supports groups in the U.S. engaged in the development of reliable physical and chemical data bases (5).)

In 1972 the PDG at LBL with G.C. Fox at Cal. Tech. decided to expand into compiling reaction data and related bibliographic information as well as continuing with particle properties and to use the techniques of computer based data management to control the work. Available manpower limited these objectives but the next nine years saw the publication of an indexed compilation of the high energy physics literature (6), an annual compilation of experimental proposals (7), and an index of data compilations with the data on microfiche (8). The group also produced the Berkeley Database Management System (9) to support the computer searchable data bases.

The international contribution to the various projects of the PDG at LBL has been considerable but it has mainly been in support of bibliographic compilation (literature and experimental proposals) and the Review of Particle Properties which is described by Rosenfeld (3). The compilation and marketing of reaction data was undertaken by a UK group at Durham (1) which had collaborated with the PDG at LBL from 1972 and became, since 1975, the PDG(UK). After the most recent financial cuts suffered by the PDG at LBL it would appear that only the Review of Particle Properties will survive there as an ongoing programme.

3. REACTION DATA COMPILATION GROUPS

After much activity in the late 60's and early 70's compilation of reaction data is now done by a small number of organizations and a few individuals with specialized interests. There are groups at the European Organization for Nuclear Research (CERN), in the Federal Republic of Germany, Japan, the UK and the USSR. They vary in the type of data they compile, the range of data within a particular type, whether they actually compile data from published sources or collect the compilations of others for local use, or, whether their compilations, however acquired, are machine readable or not. In what follows the work of the various organizations is outlined.

The CERN High Energy Reaction Analysis (HERA) group under the direction of D.R.O. Morrison has a seventeen year history of producing printed data compilations which concentrate on reaction cross sections. The group produces tabular and graphical reports and uses a small computer to store 28 000 data points and associated bibliographic material. Five people of whom three are physicists are associated with the work which is carried out at CERN and Pisa. The HERA group derives its support from existing CERN programmes and the voluntary involvement of its participants. It is not a separate CERN programme but it is ongoing and the compiled data are used both by high energy physicists and by astrophysicists.

In the Federal Republic of Germany the Fachinformationszentrum, Energie, Physik, Mathematik (FIZ) at Karlsruhe (10) provides a central organization for compiling data of all kinds. In particle physics it commissions compilations of particular reactions for publication in the Physics Data series produced by FIZ. Some compilations are also available on magnetic tape and the use of floppy discs for data dissemination is also being considered. Some data, although not for particle physics, are in computer searchable data bases and are available on EURONET-DIANE.

In addition to FIZ, particle physics compilation is supported in the FRG by the Landolt-Börnstein organization which publishes the data in books. The two organizations serve a research community within the FRG estimated at about 400 in 1979 (2). In both cases the compilers are individuals or groups working with the data either as experimentalists or phenomenologists as is the case with Professor G. Höhler's pion-nucleon compilation group at the Institut für Theoretische Kernphysik der Universität Karlsruhe.

In Japan at the National Laboratory, KEK, there is a compilation group which gathers and publishes (11) bibliographic material relevant to high energy physics and collaborates in this respect with the PDG at LBL. Reaction data are not compiled although existing compilations of other groups are made available. That data, including those produced by the KEK accelerator, are not compiled would appear to be a policy decision based on limited manpower.

In the UK particle physics reaction data compilation is supported by the UK Science and Engineering Research Council and is directed from the University of Durham by F.D. Gault. At Durham one physicist and one coding assistant funded by SERC compile data from about 300 published or preprinted papers a year and manage the group's computer searchable data bases which are held on the Rutherford Appleton Laboratory central computers. One RAL physicist, R.G. Roberts, works half time on compilation and other physicists in the UK contribute to the project, known since 1975 as the PDG(UK).

The PDG(UK) collaborates with the PDG at LBL to maintain a standard particle and observable vocabulary and does not publish books, preferring to make the data and associated bibliographic information directly available throughout the UK on the SERC computer network which reaches CERN and DESY in Hamburg. User guides are published (12) and the work is described in (1,13). The purpose of the PDG(UK) is to compile all elementary particle scattering data and to make the resulting compilations easily available to the physics community. With available manpower however this goal is impossible to realize without the collaboration of other compilation groups.

In the USSR at the Institute of High Energy Physics, Serpukhov, the Compilation, Analysis and Systematization group (COMPAS) led by V.V. Ezhela compiles all reaction data produced in the USSR and, under the CERN-USSR agreement, collaborates with the PDG(UK) and with the PDG at LBL (14). With ten physicists working in the COMPAS group only part of their time is spent on compilation, the rest goes on maintaining and using the group's interactive graphics system, ATLAS, and its library of theoretical models. COMPAS differs from the other groups described in that the group has a deliberate policy of working with the data and of building up physical models as well as engaging in compilation. It also enjoys a good relationship with Serpukhov experimentalists who provide the group with data and consult it on standards of data reporting. COMPAS is a separately supported group within the IHEP.

4. COLLABORATION IN REACTION DATA COMPILATION

Without central direction collaboration in particle physics reaction data compilation has evolved slowly. At the lowest level group members have met informally to exchange views in Berkeley, Durham, Geneva and Serpukhov but since 1980 the groups active in Europe have started to meet formally along with other interested parties to look for ways of collaborating. The last such meeting was held in the UK in April 1982 and involved representatives of seven organizations of which four, HERA, FIZ, PDG(UK) and COMPAS, were active in reaction data compilation. The meeting was reported in the CERN Courier (15).

At a different level the PDG(UK) and the COMPAS group, under the CERN-USSR agreement, have since 1980 evolved a collaborative exchange of data and software for the mutual benefit of both groups and their users. To achieve this three things were required: a common data base management system, a common data model, or schema, including a common vocabulary, and agreement on areas of compilation and exchange.

BDMS (9) is written in FORTRAN and is designed to be transportable. Freely available within the scientific community it runs on a variety of machines throughout the world and a version has recently been taken to China. Its use defines the exchange formats for the data once the data model is agreed (16). This model, along with the internationally agreed vocabulary maintained by the PDG at LBL for its work in particle properties and bibliographic compilation constitutes the Particle Physics Data Language (PPDL) which is used by the PDG(UK) and the COMPAS group to code published and preprinted data and to store the data in computer searchable data bases.

Once the data are stored in data bases the query language facility of BDMS is used by the PDG(UK) to give direct access to the data to physicists throughout the UK, at CERN and DESY using the SERC network. In the USSR it is the transportability of BDMS which makes it attractive as, in the absence of computer networks, the COMPAS group plans to transfer it, the data bases and related software to the Institute of Theoretical and Experimental Physics (ITEP) and to the Joint Institute of Nuclear Research (JINR) and eventually to other high energy physics centres with the intention of encouraging colleagues at these centres to code experimental data in PPDL.

COMPAS compiles and codes all particle physics data published or preprinted in the USSR and sends the data on magnetic tape to the PDG(UK) quarterly. These data are merged with PDG(UK) data bases and annually the complete set of data bases is sent to the USSR. This is a satisfactory arrangement as the increased overhead to the PDG(UK) in tape handling is more than compensated for by the coded data received and the data bases are more complete as a consequence. COMPAS has also converted the HERA compilations into a computer searchable data base coded in PPDL and managed by BDMS and made the result available to HERA and the PDG(UK).

5. THE FUTURE

The PDG(UK)-COMPAS collaboration which has evolved out of mutual self interest and not as a result of a central direction by an international agency is a model of how this work can be done. It suggests an obvious extension to the USA which resembles the USSR in its use of geographically separate computers in support of particle physics research as compared with the European model of central computers linked to research workers by networks. This approach has not been taken as yet because of the uncertain future of the PDG at LBL as far as its involvement in reaction data is concerned.

Direct contact between the experimentalists who supply the data and the people who compile the data is desirable as it provides a rapid transfer of data to the data bases and also an influence on how data are reported (1,5,13). COMPAS and HERA have direct access to experimentalists at Serpukhov and CERN while the PDG(UK) makes contact through UK experimentalists, attendance at international conferences and visits to national and international laboratories. At the last meeting of European compilers (15) it was suggested that compilation might best be centred at and coordinated from an international laboratory, like CERN, to facilitate this dialogue between experimentalists and compilers.

In the long term a common data base of all published measurements of elementary particle scattering easily available in computer searchable form is the goal. However this will require both more compilers and more coordination to prevent duplication of effort and to maintain the standards of the existing data bases.

6. CONCLUSION

Particle physics reaction data must be preserved in support of scholarship and no single existing compilation organization is able to compile all of the available data. This problem can be resolved through collaboration between these groups in a manner already begun. The long term goal must be a complete and common data base easily available throughout the world to the particle physics community.

ACKNOWLEDGEMENT

This work is supported in part by the Science and Engineering Research Council grant NG 1027.9 and a Royal Society travel grant is acknowledged.

REFERENCES

(1) Gault, F.D., Physics data bases and their use, Comput. Phys. Commun. 22 (1981) 125-132.

(2) Butterworth, I., Particle physics in the 1980s, Phys. Bull. 33 (1982) 293-295.

(3) Rosenfeld, A.H., The Particle Data Group: growth and operations - eighteen years of particle physics, Ann. Rev. Nuc. Sci. 25 (1975) 555-598.

(4) Particle Data Group, Review of Particle Properties, Phys. Lett. 111B (1982) i-xxi & 1-294.

(5) Lide, D.R., Critical data for critical needs, Science 212 (1981) 1343-1349.

(6) Horne, C.P., Yost, G.P., Rittenberg, A., Armstrong, F.E., Hutchinson, M.S., Richards, D.R., Trippe, T.G., Uchiyama, F., Chew, D.M., Coffeen, T.A., Crawford, R.L., Enstrom, J.E., Kelly, R.L., Kingston, J.-L., Lasinski, T.A., Lasinski, M.J., Losty, M.J., Pollard, D.L., Rosenfeld, A.H., White, V.A., Wohl, C.G., Barash-Schmidt, N., Fox, G.C., Stevens, P.R., Cooper, C.S., Gault, F.D., Read, B.J., Oyanagi, Y., Roberts, R.G., Addis, L. and Row, G.M., An indexed compilation of experimental high energy physics literature. Lawrence Berkeley Laboratory Report LBL-90, Berkeley. (September 1978).

(7) Wohl, C.G., Kelly, R.L., Armstrong, F.E., Horne, C.P., Hutchinson, M.S., Rittenberg, A., Trippe, T.G., Yost, G.P., Addis, L., Ward, C.E.W., Baggett, N., Goldschmidt-Clermont, Y., Joos, P., Gelfand, N., Oyanagi, Y., Grudtsin, S.N., Ryabov, Yu.G., Compilation of current high energy experiments. Lawrence Berkeley Laboratory Report LBL-91, Berkeley. (May 1981).

(8) Fox, G.C., Stevens, P.R., Read, B.J., Gault, F.D., Rittenberg, A., Hutchinson, M.S., Horne, C.P., Kelly, R.L., Armstrong, F.E., Richards, D.R., Trippe, T.G., Yost, G.P., Roberts, R.G. and Crawford, R.L., Compilation of high energy physics reaction data: Inventory of Particle Data Group holdings 1980, Lawrence Berkeley Laboratory Report LBL-92, Berkeley. (December 1980).

(9) Richards, D.R., BDMS user's manual. Lawrence Berkeley Laboratory Report LBL-4683 (revision), Berkeley. (1977).

(10) Behrens, H. and Ebel, G., Numerical data banks in physics: what has been done, what should be done, in Glaeser, P. (ed.), Data for Science & Technology (Pergamon Press, Oxford, 1981).

(11) KEK Particle Data Group, πN inelastic reaction index, 1978, National Laboratory for High Energy Physics Report. KEK-79-10, Oho-machi, Tskuba-gun, Ibaraki, Japan. (June 1979).

(12) Gault, F.D., Lotts, A.P., Read, B.J., Crawford, R.L. and Roberts, R.G., Guide to the Durham-Rutherford high energy physics databases (2nd ed.). Rutherford Laboratory Report RL-79-094, Chilton. (December 1979).

(13) Gault, F.D., The impact of the data base on elementary particle physics research in the UK, in Glaeser, P. (ed.), Data for Science & Technology (Pergamon Press, Oxford, 1981).

(14) Alekin, S.I., Grudtsin, S.N., Demidov, S.G., and Ezhela, V.V., Particle physics data system at IHEP, Institute of High Energy Physics Report, IHEP-81-048, Serpukhov (1981).

(15) Data compilation meeting, CERN Courier 22 (1982) 238.

(16) Stevens, P.R., Rittenberg, A., Gault, F.D., and Read, B.J., The use of database management systems in particle physics, in Dreyfus, B. (ed.), Proceedings of the Sixth International CODATA Conference (Pergamon Press, Oxford, 1979).

Data for Science and Technology, P.S. Glaeser (ed.)
North-Holland Publishing Company

DATA FILE FOR DISORDERED CRYSTAL STRUCTURES

Karl-Otto Backhaus, Hannelore Schrauber and Hildegard Grell

Central Institute of Physical Chemistry, Computing Centre
Academy of Sciences of the GDR, Berlin-Adlershof
German Democratic Republic

A growing number of substances has been investigated with crystals forming polytypes with OD character (ordered or disordered). In existing data files neither the OD character nor the relation between different polytypes are noted in detail. This is why the proposed OD data file aims at storing data referring to crystals of the mentioned sort, stressing in particular the features connected with their OD character, missing in the Cambridge Crystallographic Data Base (1) and the Inorganic Crystal Structure Base (2) as well as in the Powder Diffraction File (3).

1. INTRODUCTION

OD structures (OD abbrevation for Order-Disorder) have been defined by Dornberger-Schiff (4) in 1966. These structures are characterized as consisting of building units linked by partial coincidence operations where the arrangement of the building units is not uniquely determined. The notation of OD crystals includes structures consisting of rods, finite blocks or layers. The latter are of particular practical interest, because all known polytypic substances can be described as OD structures consisting of layers. This holds not only for ordered but also for disordered samples of a polytypic substance.

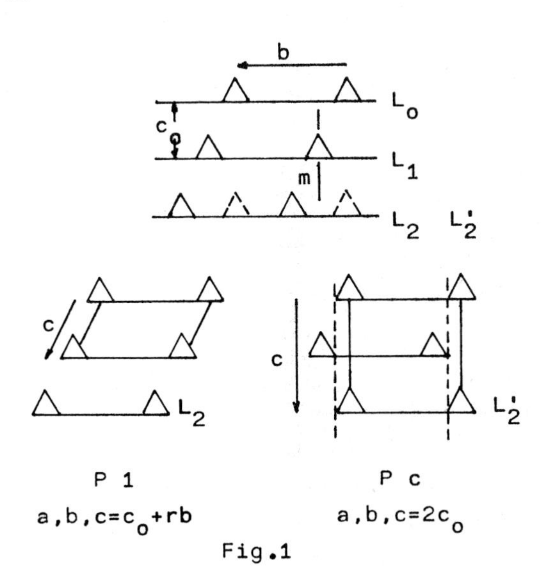

Fig.1

Fig. 1 shows a simple schematic example. Any two isosceles triangles linked by a horizontal line represent a layer with the layer group P 1m(1) periodic in two dimensions. Due to the partial mirror plane of L_1 there are two possible positions of the layer L_2 relative to L_1, each leading to a layer pair $(L_1;L_2)$ equivalent to $(L_0;L_1)$. There are two kinds of stackings as indicated in Fig. 1 leading to ordered structures with all n-tuples of layers (n=1,...,m) equivalent. Other structures containing both of the triples shown in Fig. 1 in a random way are possible members of this polytype family. In general they are disordered. Obviously, these possibilities are not included in the symmetry description by a space

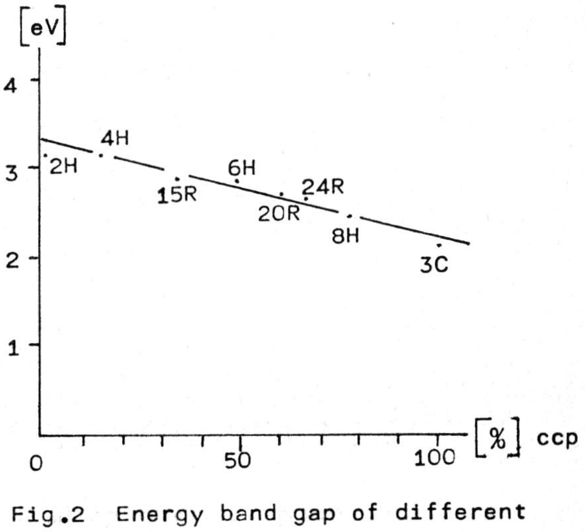

Fig.2 Energy band gap of different SiC polytypes, correlation coeff. 0.98

group. The coincidence operations of a disordered structure consisting of equivalent layers form a groupoid in the sense of Brandt (5). Structures consisting of the same layers forming the same layer pairs belong to a family of OD structures and are described by an OD groupoid family. Besides, the description of a polytype by a groupoid family needs a further indication of the particular stacking and that is done by a fully descriptive polytype symbol. (Dornberger-Schiff, Durovic & Zvyagin, 1982) (6).

2. SOME PROPERTIES OF OD STRUCTURES (POLYTYPES)

The relations between the stacking of layers within polytypes and physical or chemical properties are of high practical importance. Classical examples are the polytypes of SiC and ZnS. Fig. 2 shows the relation between the energy gap (eV) and the percentage of cubic close packed layers (ccp) in SiC (7). The relation is nearly linear and it is possible to predict the energy gap for any given polytype. Similar relations exist for ZnS for the optical double refraction (8) and the percentage of hexagonal close packed layers (Fig. 3). Other examples are CdI_2 with its dielectric constants and lepidolites with the relation between the content of LiO_2 and polytypism. It is well known that some technological properties of kaolinites depend on disorder. All these examples mentioned are OD crystals.

3. PROPOSED CONTENT OF THE OD DATA FILE

It is planned to develop the OD data file in two stages. As a first step, for a limited number of OD crystals (about 200) we are working out the best way to select and to store characteristic data of OD structures, to indicate their relationships and to retrieve them for further use. Experiences with limited data file will then be used to find out the final version how to proceed in storing and how to organize retrieving and connection of the data.

Problems connected with the characterization of crystallographic data of structures showing disorder, polytypism or twinning were discussed by H. Fichtner-Schmittler et al. (9). The given conclusions have been taken into account for the concept of the OD data file.

During the first step, three entries are provided per substance: BIBLIOGRAPHY, LATTICE and ATOM. Entries STACKING and METHODS are planned. The data entry BIBLIOGRAPHY is divided into two parts. The first corresponds closely to the bibliographic entry of the Cambridge Crystallographic Data Base: reference code, number of publication, accession date, substance name, substance synonym, chemical formula, name(s) and/or OD interpreter(s) and bibliographic data. The second part contains OD suspicious features, e.g. diffuse streaks (direction and which selection of intensities), polytypism, twinning, non-space group extinctions, pseudosymmetry, additional symmetry of a class of selected reflections. All these features are taken from data contained in structural papers. Furthermore, the entry contains the OD groupoid family with parameters r, s, (if any), isogonal point group of the family and the space group and isogonal point group of the superposition structure. Their determination is the task of the interpreter(s). The data entry LATTICE contains three kinds of lattice parameters: LAYER, UCELL and SCELL with data referring to a single layer, a unit cell of actually investigated structures or to the superposition structure, respectively.

The notation of the type of parameter is followed by NPUBL, the number of publication from which the parameters are taken. The next, NSYM denotes the number of the plane space group according to Weber (10), in the case of parameter type LAYER. In the case of parameter types UCELL and SCELL it denotes the respective number of the space group as it is given in ITX. NREF numbers the reference system. These figures are followed by the sequence of parameters ($A,B,C,\alpha,\beta,\gamma$). For LAYER, CO

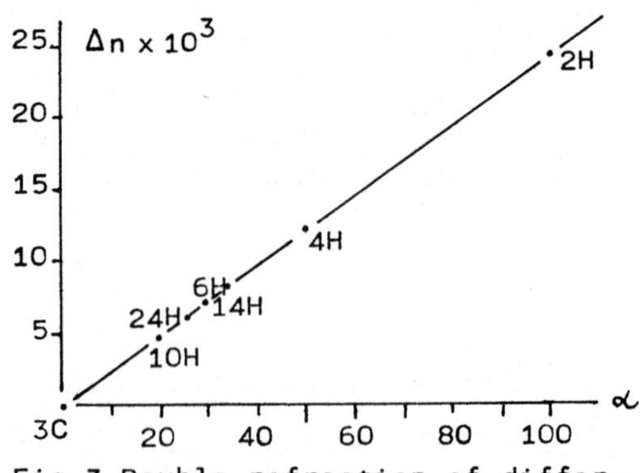

Fig.3 Double refraction of different ZnS polytypes (= percentage of h.c.p.)

is given at the site of C; α and β are assumed as $\alpha=\beta=90^{\circ}$, corresponding to the meaning of CO.

The data entry ATOM is followed by NPUBL, NREF, NLAY (the number of the layer to which the atom is assumed to belong), element symbol with number, atomic coordinates, isotropic temperature factors BISO or U11, occupation factor and standard deviations (the last three values are not given in the following example). The crystal orientation is chosen in agreement with the author(s).

In the data entry STACKING, there will be stored the stacking of layers in observed polytypes by means of polytypic symbols. The data entry METHODS will be referred to the methods of the intensity measurement and solution or refinement of the corresponding structure.

4. STRUCTURE OF THE DATA FILE

The logical structure of the data file is shown in Fig. 4. The physical storage realized in terms of single linked segments of predefined constant length is done with the aid of the program system MASD (11).

The data entries - each of them in a directly accessible part of the file - are (like relations) two-dimensional tables consisting of horizontal rows and vertical columns. No two rows in a table are identical. There are some combinations of columns - a key - whose values uniquely identify each row. The names, types and lengths of data elements in each column of the table are stored in the data description table.

The different data entries (tables) for a given structure are linked by the aid of the reference code. The reference code is stored in the first column of the bibliographic data entry and is also used as part of the table names. The table names are stored in the catalogue of the file and are used to retrieve the data entries.

Data input and output, selection of special rows and projection of special columns of selected tables are done by means of a framework of programs and are controlled by the data description tables.

The selected data can be used for further calculations. As an example, we may quote the possibility to calculate powder diagrams for polytypes of a given family, indicated by their polytype symbols (12).

All programs used for the first stage of the data file are written for the computer BESM6 and its operating system. For the second step, an implementation on ESER computer is planned with some changes in the structure and content of the data file.

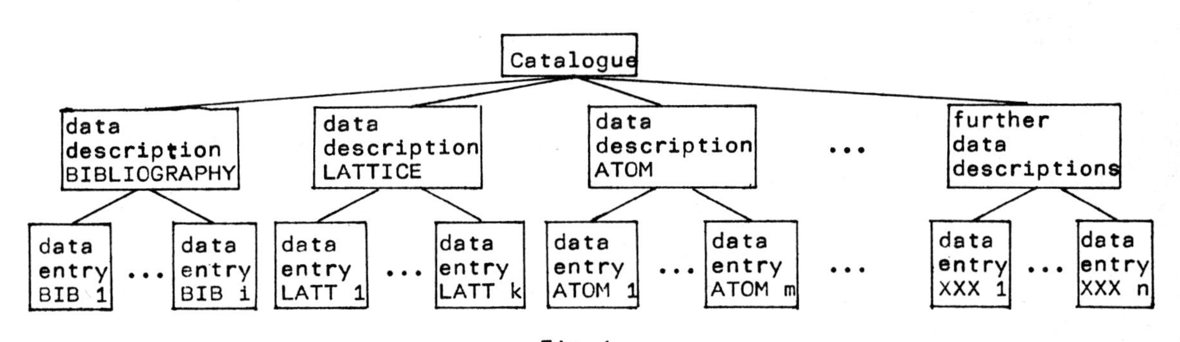

Fig.4

5. EXAMPLES FOR DATA ENTRIES

5.1. Entry BIBLIOGRAPHY for potassium silicate

```
POSI  1  800728  POTASSIUMSILICATE  XXX  K4 SI8 018
      SCHWEINSBERG H.,LIEBAU F.,107;30,2206,1974,ACTA CRYST.B
      DIFFUSE STREAKS PARALLEL C*,TWINNING;SPACE GROUP P-1
```

```
POSI  2  800728  POTASSIUMSILICATE  XXX  K4 SI8 O18
      DUROVIC S.,107;30,2214,1974,Acta CRYST.B OD INTERPRETATION
      DIFFUSE STREAKS PARALLEL C*,TWINNING;FAMILY SYMBOL P21/M1(1)!1:X=-0.2455;Y=0
      2 MDO STRUCTURES:P-1 BASIC VECTORS A,B,CO-0.2455A Z=1;P21/C11 BASIC VECTORS
      A,B,2CO Z=2;SUPERPOSITION STRUCTURE B2/M11 BASIC VECTORS A/2,B,2CO;
```

5.2. Entry LATTICE for potassium silicate

```
LAYER  2  15C 1 12.3200  4.9430  7.8033  90.00  90.00  90.00
UCELL  1    2  2 12.3200  4.9430  8.3690  90.80 111.19  89.69
UCELL  2   14  3 12.3200  4.9430 15.6066  90.00  90.00  90.00
SCELL  2   12  4  6.1600  4.9430 15.6066  90.00  90.00  90.00
```

5.3. Entry ATOM for potassium silicate

```
2  1  1   K   1   0.7500  -0.2637   0.4237
2  1  1   K   2   0.2500  -0.2410   0.4346
2  1  1  SI   1   0.0064  -0.2126   0.2339
2  1  1  SI   2   0.1245   0.2899   0.0920
2  1  1   O   1   0.0495   0.1001   0.2052
2  1  1   O   2   0.1041  -0.3988   0.1538
2  1  1   O   5   0.1001   0.2537  -0.1104
2  1  1   O   8  -0.0198  -0.2609   0.4244
```

REFERENCES

[1] Crystallographic Data Centre, University Chemical Laboratory, Cambridge (Manual Book).

[2] Bergerhoff, G., Siewers, R., Hundt, R. and Brown, I., Zs. Kristallographie 149 (1979) 184-185.

[3] Joint Committee on Powder Diffraction Standards, Swarthmore, Pennsylvania, Powder Diffraction File.

[4] Dornberger-Schiff, K., Lehrgang über OD-Strukturen (Berlin, 1966).

[5] Brandt, H., Math. Ann. 96 (1926) 360.

[6] Dornberger-Schiff, K., Durovic, S. and Zvyagin, B.B., (accepted for publication in "Crystal Research and Technology").

[7] Verma, A.R. and Krishna, P., Polymorphism and Polytypism in Crystals (New York, London, Sydney: 1966).

[8] Brafman, O., Steinberger, I.T., Physical Review 143 (1966) 501-505

[9] Fichtner-Schmittler, H., Dornberger-Schiff, K. and Fichtner, K., The Proceedings of the 6th International Codata Conference (1979) 155-157.

[10] Weber, L., Zs. Kristallographie 70 (1929) 309-327.

[11] Assmann, W., MASD-Memory Administration System for Direct Access, BESM6 Program Information 1977, Centre of Computing Technique, Academy of Sciences of the GDR.

[12] Weiss, Z. and Miklos, D., 8th Conference on Clay Mineralogy and Petrology, Teplice (1979) 105-110.

AUTHOR INDEX

SUBJECT INDEX

LIST OF PARTICIPANTS

ABBEL, Robert, Gesellschaft für Information und Dokumentation, Lyoner Strasse 44-48, D-6000 Frankfurt, F.R.G.

ALCOCK, Charles B., Department of Metallurgy & Materials Science, University of Toronto, Toronto, Ontario, Canada M52 1A4

ARTBAUER, O., Czechoslovak Institute of Metrology, Geologicka 1, 825 62 Bratislava, Czechoslovakia

ATLANI, Charley, ITODYS, Université de Paris VII, 1, rue Guy de la Brosse, 75005 Paris, France

AULEYTNER, J., Institute of Physics, Polish Academy of Sciences, al. Lotnikow 32/46, 02-668 Warsaw, Poland

BACKHAUS, Karl-Otto, Zentralinstitut für Physikalische Chemie der AdW der G.D.R., Rudower Chaussée 5, Berlin-Adlershof, G.D.R.

BADRAK, Sergey, A., Institute for Problems of Materials Science (SKTB IS), Academy of Sciences Ukranian SSR, ul. Krzhyzhanovskogo 3, Kiev 180, U.S.S.R.

BAILEY, Angel, Sea Grant Program of the University of Maryland, 1224 H.J. Patterson Hall, University of Maryland, College Park, Maryland 20742, U.S.A.

BAKER, F.W.G., ICSU, 51, Bd. de Montmorency, 75016 Paris, France

BALAREW, Christo, Inorganic Salts Research Laboratory, Department of Chemistry, Bulgarian Academy of Sciences, 1040 Sofia, Bulgaria

BAŃKOWSKI, J., Institute for Scientific, Technical and Economic Information, P.O. Box 123, 00-950 Warsaw, Poland

BAULCH, Donald L., Department of Physical Chemistry, University of Leeds, Leeds L52 9JT, U.K.

BEHRENS, H., Fachinformationszentrum 4, (Energie, Physik, Mathmatik), D-7514 Eggenstein-Leopoldshafen 2, F.R.G.

BERNDT, S., Chemical Synthesis Design Centre PROSYNCHEM, Gliwice, Poland

BERTAINA, P., Istituto di Scienze dell'Informazione, University of Torino, C. Massimo d'Azeglio 42, 10125 Torino, Italy

BIEŁEK , J., Department of Physics, Electrotechnical Faculty, Slovak Technical University, Gottw. nam. 19, 812 19 Bratislava, Czechoslovakia

BRAFMAN, Marc, CRES, Chemin du Canal, P.O. Box 22, 69360 St. Symphorien d'Ozon, France

BROWN, I. David, Institute for Materials Research, McMaster University, 1280 Main Street West, Hamilton, Ontario, Canada L8S 4M1

BUCCI, Giacomo, ISIS, Istituto di Informatica e Sistemistica, Facoltà di Ingegneria, University of Florence, Via di Santa Marta 3, 50139 Florence, Italy

BUSSARD, Alain, Service d'Immunologie Cellulaire, Institut Pasteur, 75015 Paris, France

BYLICKI, Andrzej, Institute of Physical Chemistry of the Polish Academy of Sciences, ul. Kasprzaka 44/52, P.O. Box 49, 01-224 Warsaw, Poland

CARAPEZZA, Marcello, University of Palermo, Istituto di Geochimica, Via Archirafi 36, 90123 Palermo, Italy

CARTER, G. Cynthia, Numerical Data Advisory Board, National Research Council, National Academy of Sciences, 2101 Constitution Avenue, N.W., Washington, DC 20418, U.S.A.

CHIKÁNY, Gabor, Technical University of Budapest, Department of Chemical Engineering, 1521 Budapest, Hungary

CHOLIŃSKI, J., Institute of Industrial Chemistry, Rydygiera 6, 01-793 Warsaw, Poland

CHYLIŃSKI, K., Institute of Physical Chemistry of the Polish Academy of Sciences, ul. Kasprzaka 44/52, P.O. Box 49, 01-224 Warsaw, Poland

CICHY, R., Central Board of Geology, 00-950 Warsaw, Poland

CLARK, Allen L., IIRD - International Institute for Resource Development, Belverdergrasse 2, A-1040 Vienna, Austria

COHEN, E. Richard, Rockwell International Science Center, P.O. Box 1085, Thousand Oaks, CA 91360, U.S.A.

COUNIOUX, Jean-Jacques, Université Calude-Bernard - Lyon I, Laboratoire de Physico-Chimie Minérale II, 43, Bd. du 11 Novembre 1918, 69622 Villeurbanne Cedex, France

CREASE, James, Institute of Oceanographic Sciences, Brook Road, Wormley, Godalming, Surrey, GU8 5UG, U.K.

DAVID, Antoinette, 55, avenue de la Motte-Picquet, 75015 Paris, France

DAVISON, Peter S., Scientific Documentation Centre Ltd., Halbeath House, Dunfermline, Fife, KY12 OTZ, U.K.

DEMBOWSKI, Z., Central Board of Geology, 00-950 Warsaw, Poland

DE SANTIS, Roberto, Istituto di Chimica Applicata et Industriale, University of Rome, Via Eudossiana 18, Rome, Italy

DI LEVA, A., Istituto di Scienze dell'Informazione, University of Torino, C. Massimo d'Aseglio 42, 10125 Torino, Italy

DOBRUCKI, Marek, Institute of Ferrous Metallurgy, Gliwice, Poland

DUBOIS, J.E., ITODYS, Université de Paris VII, 1, rue Guy de la Brosse, 75005 Paris, France

DUDICH, Endre, MAFI, Hungarian Geological Institute, Nepstadion ul. 14, Pf. 106, Budapest XIV, Hungary

DYNOWSKA, Elzbieta, Institute of Physics, Polish Academy of Sciences, al. Lotników 32/46, 02-668 Warsaw, Poland

EBEL, G., Fachinformationszentrum 4, (Energie, Physik, Mathematik), D-7514 Eggenstein-Leopoldshafen 2, F.R.G.

FLORCZAK, Tadeusz, Research and Development Centre for Machine Technology and Design "Tekoma", ul. Lucerny 108, 04-686 Warsaw, Poland

FOGL, J., Central Office for Scientific, Technical and Economical Information, Havelkova 24, 13000 Prague 3, Czechoslovakia

FREYDANK, Harri, VEB Leuna-Werke "Walter Ulbricht", Thälmannplatz, 4220 Leuna, G.D.R.

FUJIWARA, Yuzuru, Institute of Information Sciences and Electronics, University of Tsukuba, Sakura-Mura, Niihari-gun, Ibaraki 305, Japan

GAULT, Frederick D., Department of Physics, University of Durham, Durham, DH1 3LE, U.K.

GAWOR, Jan, Institute of Control and Systems Engineering, Technical University of Wrocław, 50-370 Wroclaw, Poland

GIBBINS, Patrick John, Pergamon Infoline Ltd., 12 Vandy Street, London EC2A 2DE, U.K.

GIERYCZ, P., Institute of Physical Chemistry of the Polish Academy of Sciences, ul. Kasprzaka 44/52, P.O. Box 49, 01-224 Warsaw, Poland

GIRONI, Fausto, Istituto di Chimica Applicata, Facoltà di Ingegneria, University of Rome, Via Eudossiana 18, 00184 Rome, Italy

GLAESER, Phyllis, CODATA Secretariat, 51, Bd. de Montmorency, 75016 Paris, France

GORAL, M., Department of Chemistry, Pasteura 1, Warsaw University, 02-093 Warsaw, Poland

GURVICH, L.V., Institute for High Temperature, Korovinskoye Shosse, Moscow I-412, U.S.S.R. 127412

HAENDLER, Harald, Documentation Centre of Hohenheim University, Postfach 70 05 62, D-7000 Stuttgart 70, F.R.G.

HAUCK, B., Institut d'Astronomie, Université de Lausanne, 1290 Chavannes-des-Bois, Switzerland

HIPPE, Zdzislaw, I. Lukasiewicz Technical University, 35-959 Rzeszów, Poland

HUANG, Zhengzhi, Division of Earth Sciences, Unesco, 7, Place de Fontenoy, 75700 Paris

HUTCHISON, W.W., Earth Sciences Sector, Energy, Mines and Resources Canada, 580 Booth Street, Ottawa, Canada K1A OE4

IVANOV, Kiril P., Pavlov Institute of Physiology, U.S.S.R. Academy of Sciences, Nab Makarova 6, 199164 Leningrad, U.S.S.R.

IWATA, Shuichi, Department of Nuclear Engineering, Faculty of Engineering, University of Tokyo, 7-3-1 Hongo, Bunkyo-ku, Tokyo, Japan 113

JANICKI, Andrzej, Institute of Mathematical Machines, Warsaw, Poland

JELJASZEWICZ, J., National Institute of Hygiene, 24 Chocimska Street, Warsaw, Poland

JÜNTGEN, Harald, Bergbau-Forschung GmbH, Franz-Fischer-Weg 61, 4300 Essen-Kray, F.R.G.

KACZMAREK, A., Technical University of Rzeszow, 35-959 Rzeszow, Poland

KASPRZYK, J., Chemical Synthesis Design Center PROSYNCHEM, Gliwice, Poland

KEHIAIAN, V. Henry, ITODYS, Université de Paris VII, 1, rue Guy de la Brosse, 75005 Paris

KEMÉNY, Sandor, Department of Chemical Engineering, Technical University, 1521 Budapest, Hungary

KERSCHACKEL, Günther, Institut für Maschinelle Dokumentation - Rechenzentrum Graz, Steyrergasse 25a/I, A-8010 Graz, Austria

KHODAKOVSKY, I.L. Vernadsky Institut of Geochemistry and Analytical Chemistry, U.S.S.R. Academy of Sciences, ul. Kosigina B-19, 117334 Moscow, U.S.S.R.

KIERZKOWSKI, Z., Computer Centre, Poznań Technical University, 60-965 Poznań, Poland

KIZAWA, Makoto, University of Library and Information Science, Yatabe-machi, Ibaraki-ken, 305 Japan

KLON PALCZEWSKA, M., Institute of Industrial Chemistry, Rydygiera 6, 01-793 Warsaw, Poland

KOLENDOWSKI, J., Computation Center CYFRONET, 4 Reymonta Street, 30-059 Krakow, Poland

KOLIŃSKI, J., Department of Chemistry, Pasteura 1, Warsaw University, 02-093 Warsaw, Poland

KOTANI, Masao, 36-4 Sanno, 3-chome, Ota-ku , Tokyo 143, Japan

KOTCHANOVA, W., VINITI, Baltijskaja ul. 19, Moscow, U.S.S.R.

KOTOWSKI, W., Institute of Organic Synthesis "BLACHOWNIA", 47-232 Kędzierzyn-Koźle, Poland

KOWALCZYK, A., Institute of Organic Synthesis "BLACHOWNIA", 47-232 Kędzierzyn-Koźle, Poland

KOWALSKI, A., Institute of Geodesy and Cartography, 00-950 Warsaw, Poland

KOŻDOŃ, A.F., Committee on Standardization Measures and Quality, 00-139 Warsaw, Poland

KOZLOV, Alexander D., Gosstandart, VNIIMS, Ezdakov Per. 1, Moscow, U.S.S.R.

KRĘGLEWSKI, Alexander, Thermodynamics Research Center, Texas A & M University, College Station, Texas 77843, U.S.A.

KROP, E.J., Chemical Synthesis Design Centre PROSYNCHEM, Gliwice, Poland

KRZYZANKOWSKA, A., Chemical Synthesis Design Centre PROSYNCHEM, Gliwice, Poland

LAMBERT, I., Commissariat à l'Energie Atomique, CEN Saclay, 911191 Gif sur Yvette Cedex, France

LELAKOWSKA, J., Institute of Industrial Chemistry, 01-793 Warsaw, Poland

LEMPE, Dicter, Technical University "Carl Schorlemmer" Leuna-Merseburg, Geusaer Strasse, 4200 Merseburg, G.D.R.

LE NEINDRE, Bernard, CNRS-LIMHP, Université Paris-Nord, 1, av. J.B. Clément, 93430 Villetaneuse, France

ŁUKASZEWICZ, K., Institute of Low Temperature and Structural Investigations, 50-950 Wrocław, Poland

ŁUSZCZYK, M., Institute of Physical Chemistry of the Polish Academy of Sciences, ul. Kasprzaka 44/52, P.O. Box 49, 01-224 Warsaw, Poland

MĄCZYŃSKA, Z., Institute of Physical Chemistry of the Polish Academy of Sciences, ul. Kasprzaka 44/52, P.O. Box 49, 01-224 Warsaw, Poland

MĄCZYŃSKI, A., Institute of Physical Chemistry of the Polish Academy of Sciences, ul. Kasprzaka 44/52, P.O. Box 49, 01-224 Warsaw, Poland

MAIO, Dario, Istituto di Elettronica, University of Bologna, Viale Risorgimento 2, 40136 Bologna, Italy

MALANOWSKI, S., Institute of Physical Chemistry of the Polish Academy of Sciences, ul. Kasprzaka 44/52, P.O. Box 49, 01-224 Warsaw, Poland

MAŁCZYŃSKI, J., Research and Development Centre for Standard Reference Materials, 00-139 Warsaw, Poland

MAŁEK, H., Computing Centre, Nicholas Copernicus University, 97-100 Toruń, Poland

MARRELLI, L., Istituto di Chimica Applicata e Indstriale, University of Rome, Via Eudossiana 18, Rome, Italy

MASHIKO, Yo-ichiro, Japan Society for CODATA, Daiichi Kanamori Building, 1-5-31 Yushima, Bunkyo-ku, Tokyo 113, Japan

MATULEWICZ, Andrzej, "Chemoautomatyka", ul. Rydygiera 8, 01-793 Warsaw, Poland

MISIASZEK, Leopold, Institute of Control and Systems Engineering, Technical University of Wrocław, 50-370 Wroclaw, Poland

MOSER, Z., Institute of Metallurgy, Polish Academy of Sciences, Reymonta 25, 30-059 Kraków, Poland

NATER, K., Netherlands Energy Research Foundation, 1755 Z.G. Petten, Netherlands

NEAU, Evelyne, Laboratoire de Chimie Physique, Faculté des Sciences de Luminy, Route Léon Lachamp, 13288 Marseille Cedex 9, France

NĚMEC, Václav, Geoindustria, 17004 Prague, Czechoslovakia

NERI, G., Istituto di Automatica, University of Bologna, Viale Risorgimento 2, 40136 Bologna, Italy

NEVEROVA, G., Soviet National CODATA Committee, U.S.S.R. Academy of Sciences Leninski Prospekt 14, 117901 Moscow, U.S.S.R.

NEZBEDA, I., Institute of Chemical Fundamentals, Prague 16000, Czechoslovakia

NISHIWAKI, Niichi, Department of Geology and Mineralogy, Faculty of Science, Kyoto University, Kitashirakawa, Sakyo Ward, Kyoto 606, Japan

OBUKOWICZ, J., Pharmaceutical Insitute, 01-793 Warsaw, Poland

ORACZ, Pawel, Department of Chemistry, Pasteura 1, Warsaw University, 02-093 Warsaw, Poland

ÖSTBERG, Gustaf, Lunds Tekniska Hogskola, Institutionem for Konstructionsmaterial, Box 725, 220 07 Lund, Sweden

PICCHIOTTINO, Roland, ITODYS, Université de Paris VII, 1, rue Guy de la Brosse, 75005 Paris

PLEBAŃSKI, T., Research and Development Centre for Standard Reference Materials, 00-139 Warsaw, Poland

PROST, Chantal, Institut de Programmation, Université de Paris VI, 75230 Paris Cedex 05, France

RAJECKI, Maciej K., Computer Science Department, Institute of Geology, Warsaw, Poland

RAKOWSKI, J., Chemical Synthesis Design Centre PROSYNCHEM, Gliwice, Poland

RAMBIDI, Nikolai G., U.S.S.R. Research Centre for Surface and Vacuum Investigation, Ezdakov Pereulok 1, Moscow 117334, U.S.S.R.

RAO, P.V. Krishna, Department of Chemistry, Andhra University, Visakhapatnam 530 003, India

ROGALSKI, M., Institute of Physical Chemistry of the Polish Academy of Sciences, ul. Kasprzaka 44/52, P.O. Box 49, 01-224 Warsaw, Poland

ROHLOFF, Hans, Betriebsforschungsinstitut, VDEh - Institut für Angewandte Forschung GmbH, Düsseldorf, F.R.G.

ROSE, John, Unesco, 7, Place de Fontenoy, 75700 Paris, France

ROSSELIN, Jacques, MIDIST, 280, Bd. Saint Germain, 75997 Paris, Frànce

RYBIŃSKI, H., Institute for Scientific, Technical and Economic Information, 188 Neipodległości Street, Warsaw, Poland

ŠACH, J., Scientific Information Centre, Polish Academy of Sciences, 00-330 Warsaw, Poland

SATA, Histada, Institute of Physcial and Chemical Research, 2-1 Hirosawa, Wak-shi, Saitama-ken 351, Japan

SCHIRMER, Wolfgang, Zentralinstitut für Physikalische Chemie, Rudower Chaussée 5, 1199 Berlin-Adlershof, G.D.R.

SCHMIDT, Josef Johannes, Nuclear Data Section, International Atomic Energy Agency, P.O. Box 100, Wagramerstr 5, A-1400 Vienna, Austria

SCHÖNBERG, Manfred, Hoechst A.G., Postfach 800320, D-6230 Frankfurt/Main 80, F.R.G.

SCHRAUBER, Hannelore, AdW der D.D.R., Zentrum für Rechentechnik, Rudower Chaussée 5, 1199 Berlin-Adlershof, G.D.R.

SCHUBERTH, Horst, Martin Luther Universität Halle-Wittenberg, Schlobberg 2, 4020 Halle, G.D.R.

SHULMAN, Michael, Freie Universität Berlin, Institut für Geologie, Malteserstrasse 74/100, D-1000 Berlin 46, F.R.G.

SIFNER, O., Institute of Thermodynamics, Puskinovo n. 9, 160 00 Prague, Czechoslovakia

SIMONSEN, Jørgen, Water Quality Institute (URI), Agern Alle AA, DK-2970 Hørsholm, Denmark

SINDING-LARSEN, Richard, Department of Geology, Norwegian Institute of Technology, Høgskoleringen 6, N-7034 Trondheim - Nth, Norway

SKALIŃSKI, T., Institute of Physics, Polish Academy of Sciences, al. Lotnikow 32/46, 02-668 Warsaw, Poland

SKRZECZ, A., Institute of Physical Chemistry of the Polish Academy of Sciences, ul. Kasprzaka 44/52, P.O. Box 49, 01-224 Warsaw, Poland

STRYJEK, R., Institue of Physical Chemistry of the Polish Academy of Sciences, ul. Kasprzaka 44/52, P.O. Box 49, 01-224 Warsaw, Poland

SÜHNEL, Klaus, Karl Marx University, Department of Chemistry, Linné Strasse 2, 7010 Liepzig, G.D.R.

SUFFCZYŃSKI, M., Institute of Physics, Polish Academy of Sciences, al. Lotnikow 32/46, 02-668 Warsaw, Poland

SYTCHEV, V.V., Soviet National CODATA Committee, Academy of Sciences of the U.S.S.R., 14, Leninsky Prospekt, 117901 Moscow B-71, U.S.S.R.

SZAFRAŃSKI, A., Institute of Industrial Chemistry, Rydygiera 6, 01-793 Warsaw, Poland

SZCZEPAŃSKI, Richard, Department of Chemical Engineering and Chemical Technology, Imperial College, Prince Consort Road, London SW7 2BY, U.K.

SZIRTES, L., Central Institute for the Development of Mining, P.O. Box 83, 1525 Budapest, Hungary

SZULCZEWSKI, M., Department of Earth Sciences, Polish Academy of Sciences, 00-950 Warsaw, Poland

TATENO, Yoshio, Institute of Physical and Chemical Research, 2-1 Hirosawa, Wako-shi, Saitama-ken, Japan

TENU, Richard, Laboratoire de Physico-Chemie Minérale II, Université Claude Bernard - Lyon I, 43, Bd. du 11 Novembre 1918, 69622 Villeurbanne

TRĄBCZYŃSKI, W., Research and Development Centre for Standard Reference Materials, 00-139 Warsaw, Poland

TRABELSI, Malika, Ecole Normale Supérieure, 43, rue de la Liberté, Le Bardo, Tunis, Tunisia

TRESZCZANOWICZ, A., Institute of Physical Chemistry of the Polish Academy of Sciences, ul. Kasprzaka 44/52, P.O. Box 49, 01-224 Warsaw, Poland

TRESZCZANOWICZ, T., Institute of Physical Chemistry of the Polish Academy of Sciences, ul. Kasprzaka 44/52, P.O. Box 49, 01-224 Warsaw, Poland

TRUSOV, Boris G., Moscow Higher Technical School, 2-Baumanskaja 5, 107005 Moscow, U.S.S.R.

URBAŃCZYK, K., Research Centre for the Exploitation of Chemical Raw Materials, Kraców, Poland

VISHNYAKOV, Y., Department of Computer Equipment, U.S.S.R. Academy of Sciences, Leninsky per. 14, 117901 Moscow, U.S.S.R.

VOGELY, William A., Pennsylvania State University, 220 Walker Building, University Park, Pa. 16802, U.S.A.

VOROB'EV, A.F., Moscow Chemical Technological Institue, Miusskaja squ 9, Moscow, U.S.S.R.

WATERMAN, Norman A., Michael Neale & Associates Ltd., 43, Downing Street, Farnham, Surrey GU9 7PH, U.K.

WATSON, D.G., University Chemical Laboratory, Lensfield Road, Cambridge CB2 1EW, U.K.

WEISE, Alfons, Akademieder Wissenschaften der DDR, AdW der DDR, Zentralinsitut für Organische Chemie, Rudower Chaussée 5, 1199 Berlin, G.D.R.

WERNER, Z., Geological Institute, Noakowskiego St. 8/15, 00-660 Warsaw, Poland

WESLEY-TANASKOVIC, Ines, Medical Academy Belgrade, Nyegoseva 41, 11000 Belgrade, Yougoslavia

WESTBROOK, J.H., Materials Information Services, FNB Building, 120 Erie Boulevard, General Electric, Schenectady, New York 12305, U.S.A.

WESTRUM Edgar, F., Jr., Department of Chemistry, University of Michigan, Ann Arbor, Michigan 48109, U.S.A.

WIECZORKOWSKI, K., Computing Centre, Nicholas Copernicus University, 97-100 Toruń, Poland

WIŚNIEWSKA, J., Research and Development Centre for Standard Reference Materials, 00-139 Warsaw, Poland

WIŚNIEWSKI, J., Computing Centre, Nicholas Copernicus University, 97-100 Torun, Poland

WOKROJ, A., "Chemoautomatyka", Rydygiera 8, 01-793 Warsaw, Poland

WRÓBLEWSKI, K., Chemical Synthesis Design Centre PROSYNCHEM, Gliwice, Poland

WYRZYKOWSKA-STANKIEWICZ, Institute of Industrial Chemistry, 01-793 Warsaw, Poland

YAMAMOTO, Ryozaburo, Laboratory for Climatic Change Research, 17-1 Ohomie-cho, Kitakazan, Yamashina-ku, Kyoto 607, Japan

YAMAZAKI, Masato, Electrotechnical Laboratory, 1-1-4 Umezono, Sakura-mura, Ibaraki, 305 Japan

ZAWISZA, A., Institute of Physical Chemistry of the Polish Academy of Sciences, ul. Kasprzaka 44/52, P.O. Box 49, 01-224 Warsaw, Poland

ZIĘBORAK, K., Institute of Industrial Chemistry, 01-793 Warsaw, Poland

ŻYTOWIECKI, J.M., Department of Physics, Chemistry and Mathematics, 00-959 Warsaw, Poland